3/5/92

Prentice Hall Advanced Reference Series

Physical and Life Sciences

The Scientist's Handbook
for Writing Papers
and Dissertations

Antoinette Miele Wilkinson

Cornell University

PRENTICE HALL, Englewood Cliffs, New Jersey 07632

Library of Congress Cataloging-in-Publication Data

Wilkinson, Antoinette M.
   The Scientist's handbook for writing papers and dissertations /
Antoinette M. Wilkinson.
     p. cm.—(Prentice Hall advanced reference series. Physical
and life sciences)
   Includes bibliographical references.
   ISBN 0-13-969411-0
   1. Technical writing. I. Title. II. Series.
T11.W48 1991
808'.0666—dc20
                                   90–7196
                                    CIP

Editorial/production supervision
  and interior design: MARY P. ROTTINO
Cover design: GEORGE CORNELL
Manufacturing buyers: KELLY BEHR & SUSAN BRUNKE

Prentice Hall Advanced Reference Series

© 1991 by Prentice-Hall, Inc.
A Division of Simon & Schuster
Englewood Cliffs, New Jersey 07632

The publisher offers discounts on this book when ordered
in bulk quantities. For more information, write:

Special Sales/College Marketing
College Technical and Reference Division
Prentice Hall
Englewood Cliffs, New Jersey 07632

Printed in the United States of America

10 9 8 7 6 5 4 3 2 1

ISBN 0-13-969411-0

PRENTICE-HALL INTERNATIONAL (UK) LIMITED, London
PRENTICE-HALL OF AUSTRALIA PTY. LIMITED, Sydney
PRENTICE-HALL CANADA INC., Toronto
PRENTICE-HALL HISPANOAMERICANA, S.A., Mexico
PRENTICE-HALL OF INDIA PRIVATE LIMITED, New Delhi
PRENTICE-HALL OF JAPAN, INC., Tokyo
SIMON & SCHUSTER ASIA PTE. LTD., Singapore
EDITORA PRENTICE-HALL DO BRASIL, LTDA., Rio de Janeiro

To
R. E. Wilkinson
and to the memory of
P. G. Miele and G. S. Miele

Contents

7 RESULTS AND DISCUSSION   316

8 ABSTRACT AND OTHER PARTS OF PAPER   347

Preface

This book is designed for anyone who plans to write a research paper or dissertation in the social, biological, or physical sciences. It can serve as a text for the novice or a reference for the seasoned research scientist. Papers in the natural sciences are taken as the model, and papers in the social sciences and dissertations are discussed in relation to them.

The approach is an attempt to look to the rhetorical setting in which scientific research is embedded as the source for scientific writing. The setting determines the function and content of scientific papers, and establishes a universe of discourse, which sets up imperatives for the language forms and structures adapted to writing the paper.

The chief objective is to help users understand the setting so that they can make language serve their scientific objectives. The book is not intended to reform scientists or their writing, or to present them with universal principles of writing; it has little to say about style, or rules. It presents objectives for the paper and its parts, describes common pitfalls, and suggests language forms and structures for achieving the objectives and avoiding the pitfalls.

All sections of the paper are treated, as well as related writing, editorial, and publication processes. Some parts are treated in greater detail than is customary, like the abstract and tables and illustrations, and some subjects are included that are not customarily addressed, like scientific terminology and equations.

The book depends on examples to explain and illustrate ends and means, usages and misusages. These have been chosen primarily from the scientific literature to avoid the contrived and artificial character of invented examples. They have been taken randomly from as wide a variety of journals and disciplines as possible. However, there has been some bias toward short examples, examples that are easy to set into type, and those from journals not requiring a high fee for permission. The last has reduced examples from social science journals.

For faulty writing, faulty examples are used rather than model ones, as being more instructive. However, the commentary might be different if the full passage were considered or if we had full knowledge of the subject matter or the usages of the particular discipline.

Most examples have been used with little change. They may have been shortened or simplified; the point being made may have been heightened; sometimes the example has been *made* faulty. Ellipsis dots, have been used to give readers a notion of the original rather than to quote precisely. The examples have not been revised to improve them or make them consistent with usages discussed in the book. They were chosen to illustrate a particular usage or misusage, and so can be used only as models for such usage.

The use of actual examples has influenced the citation of sources. To have cited the source for every example and given detailed citations for all of them would have given the book a spurious scholarly character. To have modified originals drastically to avoid citation would have vitiated efforts to keep the examples authentic. Besides it seemed unlikely that readers would wish to consult the original sources.

For these reasons, the citation of sources is greatly reduced. They are cited according to length: for one or two (sometimes three) sentences, the source is usually not cited; for longer examples, the source is cited only enough to retrieve it: the abbreviated name of the journal, the volume, year, and page number. I must therefore ask the indulgence of the authors of examples for the minimal citation and for selecting the one faulty example out of their whole paper. I hope that the pedagogic usefulness of such a lapse will make up for any concern about having allowed it to escape their notice. The omission of authors' names in citations should help screen authors from remark.

Although this book is intended to be comprehensive, it cannot be definitive. Nevertheless it is hoped that the book will address most of the concerns of readers and that an understanding of the approach developed in this book will help them address problems in scientific writing not discussed here.

ACKNOWLEDGMENTS

I thank the authors who knowingly or unknowingly contributed examples. Their contributions have given something of a laboratory character to this discussion of scientific writing. I especially appreciate the scientific attitude of authors who granted permission for examples, even though they might be used as faulty examples. I owe much to my students for teaching me how to help them, and I thank them. I would like to express my appreciation to the librarians in the Cornell University libraries, especially those at the Albert R. Mann Library for their assistance and their patience with my chronically overdue books. I appreciate the care and attention of the numerous typists who deciphered my scrawl and frequent tortuous reorganizations, particularly N. Riesbeck, R. Campbell, W. Oliver, and N. Jacobs.

I am grateful to the administrative officers of the College of Agriculture and Life Sciences at Cornell Univesity for released time and funding, particularly to K. E. Wing, Associate Dean, and H. L. Wardeberg, formerly Associate Director of Instruction, for their support, both tangible and intangible. I am especially grateful to H. L. Wardeberg for understanding the nature of the task I had undertaken. I would also like to thank M. P. Rottino and K. J. Tennity for their editorial assistance in guiding the book through the press, and K. W. Torgeson for the preparation of the index.

Several colleagues gave me invaluable help, and I would like to express my gratitude for the improvements that they made possible. W. J. Dress served as consultant for Latin and Greek. M. A. Martin gave Chapter 2 an extremely careful reading and helped me to avoid the errors of an amateur in linguistics. J. Millman and R. B. Thompson reviewed sections on the social sciences and made thoughtful and incisive comments and excellent suggestions. They must not be blamed, however, for my not always heeding their good advice.

I am deeply grateful to M. P. Rouse, who could be depended on to substitute for me effectively when she assisted me. She ferreted out sources and examples with just a hint of what I was looking for or only a scrap of information to go on. May she someday have as dedicated and intelligent an assistant.

To R. E. Wilkinson, I owe more than I can express here, the freeing of my free time to work on the book being only a tittle of his assistance and support.

A. M. Wilkinson
Ithaca, NY

1

Scientific Research

This is a book about writing scientific research papers and dissertations in the social, the biological, and physical sciences. Scientific papers are discussed here as the primary form of communication for new scientific information, from which other forms of scientific communication are derived. Research papers in the natural sciences are taken as the model, and papers in the social sciences and dissertations are discussed in relation to them.

Experimental papers in the natural sciences are described and discussed in detail, and papers in the social sciences and dissertations in the sciences are described in relation to them. Other types of writing in science or about science are briefly described in the next section, but they are not otherwise treated in this book.

1.1 WRITING IN SCIENCE AND WRITING ABOUT SCIENCE

Writing related to science may be writing *in* science or writing *about* science. The most important writing *in* science is the writing of scientific papers. These are written by research scientists to report their research to their peers. Research scientists also write research papers for oral presentation at conferences and symposia. They may also write books about current research for their peers. These may be reviews of literature or state-of-the-art books, or they may be general syntheses that elaborate upon a theory or a model. For journals, they may write short notes on current scientific concepts or issues, personal views, or speculative ideas, as well as book reviews, editorials, letters to the editors, and so on. In all such writing they are writing *in* science or "writing science."

Research scientists also write reference texts for colleagues and professionals and advanced texts for graduate students. The writing in such books may be more general than the writing in a scientific paper. They may also write elementary college textbooks. Such textbooks introduce a science or some branch of it, and are designed to give students a general knowledge of the subject. The writing is therefore influenced by pedagogical objectives. It requires simplication, but there is little attempt to eliminate scientific terminology. One of the main functions of an elementary text is to introduce the principal concepts, principles, and entities in the science, and these are so closely associated with the scientific terms for them that elementary textbooks are partly an introduction to the terminology and nomenclature of the science. Some attempt is also made to interest the reader, but this objective is subordinate to the scientific objectives.

Research scientists may write for science magazines, such as *Scientific American,* intended for general readers having a strong background or interest in science. In such magazines, some attempt is made to interest the reader, at least with an interesting or general introduction. The writing in such articles is therefore simplified, with as little loss of essential accuracy as possible, but it is usually not simplified enough to make all articles fully accessible even to scientists. For example, the reader may be enticed by a gentle introduction but may soon flounder when he or she encounters difficult or unfamiliar concepts and terminology.

Scientists do write for general audiences, that is, *about* science, though much less frequently than for a scientific audience. Lewis Thomas, for instance, writes on scientific subjects in a manner that is accessible to nonscientists. This kind of writing about science, at its best, comes as close to literature as writing about science can, but it is not a kind of writing suited to reporting research. Similarly, Hoyle, Huxley, and even Asimov have been recommended as "great scientific stylists" (O'Connor, 1979, p. 50). Although such authors certainly set the standard in their type or "genre" of writing about science, it cannot be the standard for scientific papers. The writing of such authors can only come *after* the writing of scientific papers. Such writing is writing *about* science.

Most writing on science for the general reader is writing *about* science. Such science writing is found in books about science for the general reader, natural science magazines such as *Natural History,* science magazines for a mass audience, general magazines, and newspapers.

Most articles written for such mass media publications, however, are written by professional writers, who are usually not scientists and who often have little or no training in science. The audience consists of general readers who have little if any science background, and often a limited interest in science. Science articles for such an audience require the greatest simplification, both in degree of detail and complexity. This rhetorical setting, the limited scientific knowledge of the writer and readers, and the overriding requirement of interesting the reader often result in sacrificing content to form, so that such an article, though interesting, or even sensational, may not be very informative or very accurate. This inaccuracy and sensationalism is so antithetical to the values of scientists, that they are often unwilling to grant writers from such publications interviews and may in this way contribute indirectly to the inaccuracy of the articles written about science.

Writing about science for newspapers and magazines that have a mass audience is designated as *science writing*. *Scientific writing* has been broadly defined here as the writing of scientists addressed to peer scientists, but it is used in the narrow sense in this book to refer to the writing of research scientists reporting their research to peer scientists in the same discipline. The narrower definition is chosen because the scientific paper is the primary form of this kind of writing.

1.2 SCIENTIFIC RESEARCH AND THE SCIENTIFIC METHOD

The writing of a scientific paper or dissertation is not simply a writing task like any other: it differs in having its roots in scientific research. Writing is an integral part of research. It seems useful, therefore, to begin by examining research to determine how it contributes to forming the canons of scientific writing.

Scientific research is the actualization of scientific thinking, and scientific papers are the end point of that actualization. Research scientists are engaged in developing a structure of knowledge by systematically observing, manipulating, and measuring natural phenomena and then explaining and interpreting the results of such observations, measurements, and manipulations. Scientific research is a continuing and unending search for scientific knowledge and understanding. To direct their search, scientists use the scientific method.

1.2.1 Scientific Method

The scientific method entails observation, hypothesis, and experimentation. There is discussion about the actual method of science, that is, how hypothesis, observation, and experimentation form the structure of science. It has been proposed that scientists begin with hypotheses and then test their hypotheses by making observations or performing experiments; they then classify, interpret, or explain the results of the experiment. Or they begin with observations, then formulate hypotheses, and then test these. Or they may follow paradigms in the particular discipline. These various methods may be idealizations imposed upon a more complex and more variable process.

Despite differences about the exact method of science, there is little question about the kinds of activities characterizing it. Few would question that scientists formulate hypotheses and engage in observation, if not experimentation. Observation both antecedent to or consequent from experimentation is fundamental to the scientific method. Science is, after all, basically empirical; without its empirical character, it would be philosophy. Formulating hypotheses is fundamental to "making sense" of observations, and given hypotheses, experimental or other empirical research to test them is inescapable.

For the purpose of this discussion, it is assumed that the research scientist begins by identifying a problem or asking a question. Scientists develop a *tentative* explanation, and on the basis of their explanation, they can formulate an hypothesis that if certain conditions hold, then one would expect particular consequences to follow. They next develop the methods for manipulating experimental conditions and controlling them so that the results will be direct consequences of the conditions and manipulations. The actual results obtained are then compared to predicted results, and the hypothesis may be confirmed. This confirmation may, however, raise further questions or problems, and so the results renew the process of inquiry. The scientific method is, therefore, more a method of inquiry than a method of obtaining final, fixed, or definite answers. Nonscientists are so impressed with science as discovery that they may overlook the central role that hypotheses play in science and, as a consequence, may be unaware of the tentativeness of scientific findings and discoveries.

Observation.   Science begins with observation and ends with observation. Whatever cannot be observed, directly or indirectly—and preferably measured— falls outside of science. Experimentation may follow upon observation, and it results in further observations, that is, the findings or results of the experimentation. Research is sometimes said to consist of systematic observation.

1.2.2 Hypotheses and Scientific Inquiry

Hypothesis and induction.   Hypotheses are the driving force for scientific research. Without hypotheses, science would be little more than a directory of orderly but arbitrarily arranged observations. In the broad general sense, scientific hypotheses are *tentative,* explanatory, interpretative generalizations about natural phenomena. They arise out of past or present observation, experimentation, and scientific thinking. Their tentativeness requires that they be confirmed or verified and thus exposes them to testing, because they can only be supported by further observation and experimentation; they cannot be supported by logic alone. As generalizations, hypotheses set up expectations for subsequent observations and operations. They join given conditions to predicted consequences and are inherently conditional and predictive. In the strict sense, therefore, hypotheses are predictive, if-then statements, and when they are tested, they are most accurately stated as conditional or predictive statements. Indeed the ability of an hypothesis to predict outcomes successfully is a measure of its usefulness. It is useful to distinguish between a *general* or *"working" hypothesis,* in the broad sense and a specific and *predictive hypothesis* in the narrow sense. In this book we will be discussing primarily the predictive hypothesis. Hypotheses are critical to the development of science, because they can serve as bridges between the known and the unknown, and between past observations and future expected observations. Hypotheses are derived by inductive reasoning. Although scientists use both deductive and inductive reasoning in their research, inductive reasoning characterizes and is fundamental to the scientific method.

Inductive reasoning allows research scientists to use any observations, interpretations, generalizations, or theory to reason from the specific and the known to the general and unknown. Induction is essentially an act of creation or a synthesis, but this very creative or synthesizing function is the source of its weakness logically. Whereas in deduction, the conclusion is inherent in the original generalization or premise, and an inevitable logical consequence of the reasoning, in induction, the generalization is not the *inevitable* logical consequence of the reasoning. Inductive logic is therefore not regarded as pure logic. The research scientist can demonstrate whether the hypothesis is supported or holds "true"—that is by developing a predictive hypothesis. But even with the conditions stipulated, induction does not ensure closure. If the hypothesis is not confirmed by experiment, it may still hold "true"; for example, the design or performance of the experiment may not have adequately tested the hypothesis. The inductive logic of science therefore does not have the inevitability of pure logic; however, neither is it confined within its own borders, as is deduction. Inductive reasoning has demonstrated itself to be a powerful tool in the systematic development of new scientific knowledge.

Because inductive reasoning is not pure logic, scientific research must be object-oriented and follow the rules of evidence at all stages: in the genesis of the problem, the development of hypotheses, the design of the research, and the conclusions drawn. Without such an object-oriented, logical structure, research scientists could not achieve a shared body of theory and data that describes and interprets natural phenomena. This logical structuring of research makes it a public rather than a private enterprise and requires the establishment of rigorous standards for scientific evidence.

Hypothesis and experiment.   Experimentation, the operations on natural entities and processes to discover their structure, functioning, or relationships, is used to test hypotheses. When research scientists perform an experiment, they base the experiment on one or more hypotheses, implicit or explicit. The hypothesis, stated predictively in a form for testing, may include a *general* hypothesis in a conditional form (*Ex 1.1a,b*), or it may be restricted to the predictive hypothesis, as in *Ex. 1.1c*:

Ex. 1.1

    a. If cholesterol intake is directly related to the formation of plaques in the arterial walls, then if rats are fed diets with increasing concentrations of cholesterol, the arterial walls will show corresponding increases in the development of plaques.

b. If cholesterol intake is directly related to the formation of plaques in the arterial walls, then feeding rats diets with increasing concentrations of cholesterol will show corresponding increases in the development of plaques in the arterial walls.

c. If rats are fed diets with increasing concentrations of cholesterol, the arterial walls will show corresponding increases in the development of plaques.

Usually the introduction to the paper discusses the general hypothesis, and only the specific hypothesis derived from it is stated in predictive form, as in *Ex. 1.1c.*

If the result of the experiment is that the development of plaques in the arterial walls does vary directly with the amount ingested, then the hypothesis is said to be confirmed, supported, substantiated, or verified. It is inaccurate and misleading, scientifically, to say that the hypothesis is true or proven. Actual proof can be derived in mathematical operations, because a mathematical system is a closed system, within which logical proof can operate. Empirical science is an open system, and the research scientist is at the mercy of Nature. Confirmation of the hypothesis can be accepted only as long as the hypothesis continues to be confirmed. Hypotheses are, therefore, forever open to disconfirmation. This infinite vulnerability of scientific hypotheses is fundamental to science. The lack of understanding about the vulnerability of hypotheses and the impossibility of definitive proof in empirical research leads to misunderstandings about the nature of scientific knowledge. Science is often mistakenly taken to be the discovery of "truth" the "real" truth, or "facts," and many a neophyte scientist has been led into science by this mirage of certitude. But science is constitutionally provisional and uncertain; any generalization, conclusion, and even finding is open at all times to modification or reversal by new observations, experiments, or evidence; therefore no generalization, conclusion, or finding is final or immutable. Scientific writing must reflect this vulnerability of scientific research.

1.2.3 Nature of Research

Scientific research must be based on the scientific method and address problems and questions related to observable natural phenomena. Research is first of all innovative, in the problem or question addressed, the methods of investigation, the results, their interpretation, or the conclusion. Because of this, accuracy is essential and makes measurement an essential requirement; indeed some experiments are important primarily for the accuracy or precision of their measurements. Although the repetition of experiments is considered important in supporting and strengthening hypotheses, most journal editors would not accept a paper based on research that simply repeated an experiment, because research is expected to be innovative. Rather than lose an opportunity to inch forward on the problem by simply replicating it, research scientists may modify the design of the experiment.

New knowledge makes accuracy an essential and indispensable requirement, but accuracy is not absolute. The research scientist must determine the appropriate or optimal level of accuracy and maintain it at consistent orders of magnitude. The need for accuracy results in an emphasis on quantitative over qualitative research. The application of mathematics to research has made it possible to model natural systems and so allow a shortcut to experimentation.

1.2.4 Research in the Social Sciences

Empirical research in the social sciences differs from that in the natural sciences because the subject of study is the social behavior of human beings, individually or in social groups. Social scientists study human beings—intelligent, social, interacting animals with language capacities that allow them to think symbolically and to express themselves in verbal symbols. Moreover, scientists study their behavior—the behavior of beings with feelings, thoughts, imaginings, reactions, and memories that influence their behavior, and with linguistic capacities that allow them to report on these psychological and emotional mental states and their behavior. These attributes make human beings extremely variable, mutable, and complex, and therefore unstable subjects for scientific study. This makes it more difficult to control the conditions of a study in the social sciences and so makes it more difficult to design and perform research and to analyze and interpret the results.

Social scientists must therefore make greater efforts to randomize the participants and standardize the conditions in their research than scientists in the natural sciences. Even with the best of experimental designs, the results are less reliable, because they cannot depend on the degree of stability in nature that natural scientists can. Once a natural scientist has discovered a property of a chemical compound, subsequent investigators can count on the compound having that property when designing research. There are no groups of human beings, however, that show a comparable uniformity of behavior (hence the great interest in identical twins). Moreover, the group studied—or similar groups—may change as a result of the study or may respond differently at a later time because of subsequent events and interactions, or may view their responses, behavior, attitudes, constructions, and institutions in a different light. Such subjects, therefore, make the results less stable and certain than in the natural sciences.

Social scientists also lack universal, standardized, objective, and generally accepted measures. Although social scientists can use body temperature, skin differentials, brain waves, or other physical measures, for most social science research, such measurements are not appropriate. Standardized *objective*, if not physical, measures are needed. Yet the variable and mutable character of human beings makes it very difficult to develop instruments that can reliably measure the phenomenon they are intended to measure. Research scientists in the social sciences devote much effort to developing accurate measures and standardizing them. Even with reliable measures, the intractable character of human subjects makes it necessary to use statistical analytical procedures to validate the methods. Research methodology, therefore, is of such central concern, that it is a subject of study in its own right.

The methodological difficulties influence the relative dominance of the different types of studies in the social sciences. Experimental studies, for example, are not as common as they are in the natural sciences. Descriptive studies, especially case studies, survey studies, and analytical studies are much more common.

1.2.5 Dissemination of Research

Since research is innovative, it is imperative that it be disseminated as widely as possible. It has little meaning except as part of the structure of scientific knowledge, and it is not always possible to know how significant or useful a research study is likely to be—either scientifically or practically, until it is made available to peers.

Also research is a costly activity. Research scientists are highly trained in very specialized work, and research, because it deals with the new and sometimes unique, is not an efficient process. Moreover, it is rarely a pencil-and-paper activity: most research is highly dependent on advanced technology, and even pencil-and-paper theoreticians have abandoned the simplest of tools for the most complex of computers. If research is not disseminated widely, (1) research dependent on it is likely to be delayed, (2) the same type of research is likely to be repeated, and (3) other research scientists may fall into pitfalls that they might have avoided.

So far it has been implicitly assumed that scientific research will be published. A book on the writing of scientific papers clearly attests to the importance of such research papers. Writing a scientific paper is one of the major steps in scientific research and a critical step in the advancement of science.

Scientifically, research does not really exist, unless it is disseminated. Moreover, it should be disseminated widely. A scientific paper presented orally, even at a large conference, cannot reach scientists not present at the conference, contemporary or future. Until research is published, therefore, it cannot contribute to scientific knowledge. For the widest dissemination of research, publication is essential.

1.3 RHETORICAL SETTING

The dissemination of their research places research scientists in a rhetorical setting that includes (1) the purpose of the research, (2) other scientists as peers and readers, (3) the research scientist as scientist and writer, and (4) the communication, that is, the scientific paper that binds the other three into a completed communication. Because of the nature of scientific research, the communication is the dominant element in scientific writing.

1.3.1 Scientific Paper

The scientific paper is the written report of original research that a research scientist presents to the scientific community, more specifically, to scientists in the discipline. The paper is published in a scientific journal, usually a journal of one of the professional societies in the research scientist's discipline. The published paper is the original record of the research, and it becomes the primary source of reference for the research; it is the only authentic source of the research for both scientists and nonscientists. This central position of scientific papers makes them a crucial element in the structure of scientific knowledge and establishes the criterion for the accuracy of any writing about science. Because of the new knowledge presented, content determines form. This centrality of content makes transmission of the message the primary objective in scientific writing. Unless a paper is well written, it cannot transmit the research accurately.

Most empirical scientific papers have a conventional structure, which is considered to reflect the process of scientific inquiry. It provides an established framework for writers and readers, and most empirical and experimental studies are well adapted to it. A scientific paper conventionally has four sections: introduction, methods, results, and discussion. The last two may be combined, and some of the major subsections may emerge as separate sections, for example, the theoretical framework and the materials and methods. Other subsections vary with the paper.

1.3.2 Purpose and Objectives

The primary purpose of a scientific paper is to disseminate the results of scientific research to peer scientists, both in the present and in the future. Research scientists understandably enjoy receiving recognition and seeing their paper published in a scientific journal, but the publication may become important for other than scientific reasons. The desire for numerous papers may lead to publishing a research study as two papers instead of one, or worse, to publishing it in more than one journal in the guise of different contributions. Moreover, the dubious prestige of a long paper may tempt the writer to publish research prematurely.

1.3.3 Readers and Writers as Scientists

Readers and writers of scientific papers have similar scientific training, and scientific and professional experience. The similarities most pertinent for scientific writing are those related to the research. Many of the similarities derive from the writers' use of the scientific method, their attitude toward research, and their values as scientists. This includes commitment to rationality, with a strong dose of skepticism.

The importance of observation requires that research scientists have a constantly attending eye. For example, it is not enough to see colonies of bacteria on an agar plate where the untrained eye sees glistening disks. The scientist must be able to notice the clear ring around some colonies. Nor is it enough to notice the clear area. Had the discoverer of penicillin not wondered why there was a clear area and hypothesized that the bacteria might release a substance that inhibited bacterial growth, the development of antibiotics might have been delayed or taken a different turn. The consequences of such unexpected observations account for the importance of unflagging observation in scientists. They can never stop attending to what they observe, because they never know when Nature will present them with a major question or hint at an answer to a puzzling one.

Observations lead the scientist to formulate hypotheses to place the observations in a meaningful context. In formulating hypotheses, scientists must exercise their creative imaginations. They cannot depend on observations alone or even on logical inference from the evidence, because there is no inherent logic in the development of a hypothesis; it is often an intuitive leap, the scientist's "hunch." Creativity and imagination are also required for creating theoretical models and developing research designs.

The development of hypotheses leads to experimentation, which makes demands on scientists' imaginative faculties for designing experiments and equipment and on their mechanical and manual skills for executing them. Research scientists must be able to envision an experimental design that addresses the crux of the hypothesis. They must develop any new procedures in minute detail and, if necessary, invent and design equipment for performing the procedures.

The similarities between readers and writers of scientific journals results in a shared frame of reference. This allows the writer to take the framework as a given in writing the paper, but tends to exclude readers who do not share it.

1.3.4 Readers

The readers of a scientific paper consist of peers in the writer's immediate discipline and closely related disciplines, as *Ex 1.3* indicates. Once research scientists have chosen a journal for the publication of their paper, the readers and scanners of that journal become their primary audience. Moreover the audience extends indefinitely into the future, because the present research forms part of the framework that future scientists studying related problems must take into account.

Research scientists read scientific papers for various reasons: (1) to learn about research related to their particular research interests, (2) to keep abreast of research in the discipline in general, (3) to keep current with research related to their teaching interests, and (4) to keep informed about the scientific literature in related disciplines. Readers' interest, therefore, centers on the content of the paper, that is, the research. This centrality of the content has much to do with the exclusionary character of the scientific paper for readers who are not scientists or who are scientists but not in the discipline represented by the paper.

Readers occupied with their own research, teaching, and associated responsibilities, have only a limited amount of time for reading papers. Most readers cannot hope to read all the papers they would like; therefore, they read papers differentially. Papers closely related to their research or teaching they may read carefully; others they read more casually or skim. For most papers, therefore, they will read only the title and possibly the abstract. This has obvious implications for writing abstracts and titles; it also emphasizes the importance of conciseness and clarity in writing scientific papers.

1.3.5 Writer

In the rhetorical setting of the research, the writer is important primarily as a facilitator. The writer's task is reporting the research with the purpose of the paper and the needs of the reader in mind.

It is particularly incumbent on the writers of scientific papers to reach their readers, because their papers are the primary source of the research. Given the shared background and interests, however, this may lead them to write to themselves. They may focus so much on how best to explain their research—to their *own* satisfaction—that they may forget that they must explain it to their readers. Even though readers are peers, however, they are not as familiar with the research as the writer.

1.3.6 Consequences of Rhetorical Setting for Scientific Writing

The rhetorical setting of scientific research has direct consequences for scientific writing. It calls for a type of discourse, scientific discourse, in which the writing objectives, values, and style are different from those arising out of other kinds of rhetorical settings. Here the term "scientific discourse" has a more restrictive and highly specific meaning than that assigned to it in other uses of the term (Kinneavy, 1971).

Scientific research affects writing (1) by the generation of new scientific knowledge, (2) by its object-oriented character, (3) by the relationship of scientists as readers and writers, and (4) by the need for dissemination of research. The first necessitates the development of new scientific terms and accurate reporting; the second requires that the writing be focused on the objects of research and be logically structured. The third establishes a shared framework for writer and reader, and the last makes clarity the overriding requirement in scientific writing. Together, they make accuracy fundamental to scientific writing and are related to the use of the passive voice, the past tense, and direct description. Object-oriented research is discussed here; the others are discussed in Chapters 2 and 3.

Object-oriented research.   Because research is focused on entities, events, and processes that are studied as *objects* viewed as outside the research scientist, science is said to be objective. Scientific writing cannot, therefore, be the vehicle for the writer's personal relation to the objects studied or personal objectives in the research. The writing does not even focus on personal opinions about the research, because readers are interested in inferences that can be supported by scientific evidence. This orientation to the subjects of research as objects, together with the rules of evidence followed in research, is the basis for the objectivity valued and required in scientific research. Therefore, as used here, the term "objective" for scientific research and writing refers to this object-oriented approach; it is not used to refer to some superhuman detachment in scientists. Science does not deal with absolutes, not even absolute objectivity. That such Olympian detachment is not possible does not, however, preclude scientists' attempting to study natural phenomena as objects outside themselves, in as detached a manner as is possible.

1.3.7 Dissertation

The dissertation differs from a scientific paper in the elements of the rhetorical setting. It is the final outcome of the student's graduate program. Scientifically, its primary purpose, like that of the scientific paper, is to report scientific research. Academically and professionally, its primary purpose is to demonstrate the student's ability to conduct independent research, and so to qualify the candidate as a research scientist and to gain him or her entrance into the profession. Consequently, the research is conducted under the direction of members of the graduate faculty who are experienced research scientists. The faculty members act as monitors and eventually as guarantors of the student's research expertise. Therefore, despite variations in standards for acceptance of the dissertation, it is reviewed very closely. Because of its pedagogic function, the research for the dissertation is expected to be comprehensive.

The function of the dissertation makes for a more local and smaller readership than for the scientific paper. The audience for a dissertation consists of the student's faculty committee, current and subsequent graduate students in the laboratory, and research scientists that might be interested in the research.

The graduate student, as a candidate for a degree, is a novice in the discipline, who must demonstrate to the faculty that he or she is capable of conducting independent research, can report the research accurately and clearly, and follow the conventions for reporting research in the discipline. The dissertation is longer and more comprehensive than a scientific paper because of (1) the greater and sometimes minute detail; (2) the more extensive discussions and explanations; (3) the more numerous tables and illustrations; (4) the inclusion of supplementary material, such as preliminary or pilot studies, detailed procedures, or a related study that was not pursued; and (5) the storage of excess, and sometimes voluminous information in appendices.

Dissertations are now rarely published in their entirety. Usually the abstract of the dissertation is published with other abstracts of dissertations. For most doctoral dissertations, however, one or more scientific papers may be extracted from the dissertation for publication in scientific journals.

Sometimes dissertations consist of several scientific papers, some of which may have been published before the dissertation is completed and the degree awarded. In this book, dissertations are taken to mean a comprehensive or definitive study of a scientific problem or question, which is investigated and reported as a unit.

1.4 CRITICISMS OF SCIENTIFIC WRITING

1.4.1 Nature of Criticism

Criticism of the writing of scientists is common. The deficiencies are considered so evident that little substantive evidence is advanced to support the criticism. The criticism of scientific writing often becomes a criticism of scientists. Critics regularly censure scientists for being pompous and pretentious, and wanting to impress, because they use jargon and the passive voice, and for writing that is opaque and verbose.

Scientists are also criticized for having little facility with the language, for disregarding the genius of the language, and for lacking imagination, taste, grace, and so on, for example:

Ex. 1.2

> Most of the students who have signed up for your course will have done so because they are afraid of and dislike writing: they have no instinctive feeling for the power and beauty of words and no gift for putting them together tellingly. They like science because it is full of experimental action and definite, verifiable facts; writing, on the contrary, seems to be a less exciting activity requiring intuition, for which they have small use, and taste, for which they have no use at all [Woodward, 1980, p. 3]

It is not germane here to question whether there is evidence for such deficiencies or to discuss the stereotypic view of science and scientists in this excerpt—except for intuition.

It may be that scientists have little use for intuition, yet they regularly count on their intuition—which they more often refer to as ''hunches''—to develop hypotheses and make connections between what is known and what is not known. Scientists also use their intuition in their theory and model building to develop creative, imaginative, or innovative solutions to scientific problems or answers to scientific questions. That this creativity and imaginativeness is not reflected in their writing is a consequence of the rhetorical setting. That is, they write about the end products of their intuition and only briefly take their peer readers along through the creative process leading to the research. The paper, the final product, is devoted to describing the research, not the creative process.

Considering stereotypes (*Ex. 1.2*), it is not surprising that critics of scientific writing often assume a prescriptive stance. Their prescriptions to scientists imply that language is static and that writing has ''correct'' forms. Such a static orientation toward language leads to a belief in general, universal principles of writing expressed as ''writing is writing''; that is, if one can write well in one form, one can write well in other forms. Yet novelists, dramatists, and poets often spend a lifetime mastering a single form, and even good writers produce doggerel as amateur poets. Writing differs as its purpose, subject matter, readers, and writer differ; consequently, scientific writing is different from other kinds of writing.

1.4.2 Style

Scientific writing is criticized for being deficient in style, and much of this criticism is so laced with platitudes, stereotypes, and generalizations that it precludes reasonable address. Because of the rhetorical setting of research, style is a consequence, not a goal, of scientific writing. It is a result of matching the form (1) to the content of the communication, (2) to the objectives of the writer, and (3) to the requirements of the readers, that is, the scientific community. Form thus follows function; it is not the yardstick by which the writing is measured. Critics not recognizing this dominance of content and function over form, expect the style of scientific papers to meet criteria for a quasi-literary style in which form is a goal. But scientific writing is not a form of literary writing.

Style is often used to refer to a person's form of writing, that is, his or her *personal* style. Strictly used, this term refers to the form of writing that professional writers develop by directing serious, intentional efforts toward controlling the form of their writing to mold it to their personal objectives and needs. Such a personal style is part of a writer's achievement as a writer, and being an end in itself, it is considered inviolate. The term is used more loosely for any person's writing style, that is, for the incidental accumulation of personal idiosyncratic writing habits, mannerisms, predilections, infelicities, and faults. By a combination of loose use of the term and confused egalitarianism, any person's "style" of writing is sometimes considered inviolate. In scientific papers, the characteristic rhetorical setting and conventional form allow little latitude for developing a personal style or for readers to respond to personal style. A personal style is irrelevant, except as it facilitates or impedes achieving scientific objectives.

Style also refers to a set of forms, adopted as conventions by a publisher, journal editor, or author to achieve consistency and a certain standard of usage. This set of forms becomes the *style* or *style format* of the journal, which may be incorporated into a style sheet, or manual. When *style* is used hereafter in this book, it is used in this sense.

1.4.3 Interest

Scientific writing is regularly criticized for being dull, boring, heavy, that is, not interesting. This criticism can be ascribed partly to the incomprehensible terminology, partly to the specialized, complex, and unfamiliar subject matter, but largely to critics' mistaking the purpose of scientific writing and its audience.

The incomprehensibility of scientific terminology is admittedly a major impediment in making a scientific paper accessible to nonscientist readers. However, scientific entities, concepts, theories, and models are often specialized and complex and would be difficult to understand, even if the forbidding terminology were simplified. Many papers are too mathematical to be accessible to readers, unless they have a strong background in science or mathematics. Besides, esoteric theories of economics or mathematical models of fluid flow might still be of little interest to readers concerned about unemployment or pollution.

More important, the purpose of a scientific paper is to record and transmit new scientific information and concepts to peer scientists, and so stylistic techniques to interest readers are subordinate to those that conduce to clarity. The following statement of purpose

Ex 1.3

STATEMENT OF PURPOSE

The *Annals of Physics* is intended to provide a medium for original work of broad significance in which the author will be able to give attention to clarity and intelligibility, so that his paper may be read by the widest possible audience. It is, of course, impossible to require that papers in the *Annals* be readily understandable by all. However, a reasonable criterion might be that sufficient detail and background should be given so that a reader who had worked in the field within the last few years would be able to follow the arguments and understand their significance without undue hardship. In this way we hope the *Annals* can encourage those who have interests in several fields of physics and who like to think of physics as a unified structure of high aesthetic appeal. [Ann. Phys. 168, 1986, i]

makes clear that the audience for the journal is physicists, but not even *all* physicists, let alone scientists in other disciplines or general readers.

Yet critics expect scientists to be writing to reach and interest nonscientist readers. They imply that only the impenetrable and uninteresting writing keeps readers from reading scientific papers. But it is doubtful that most scientific papers can be simplified enough for the general reader and still be accurate and informative enough for research scientists or, that even serious, educated, nonscientist readers would be an audience for such papers.

Scientists read papers because the subject matter is of interest. They constitute a captive audience, but a self-chosen captive audience. They are already interested in the writer's paper because of sharing the writer's interests. Consequently scientists regularly read seemingly dull papers and reports with interest, just as book collectors avidly read bibliographic book catalogs and stock brokers read lists of the stock quotations. When scientists speak of an ''interesting'' paper, they are referring to its content, for example, an imaginative solution to a difficult scientific problem, not to its style. Consequently in writing a scientific paper clearly, logically, and precisely, research scientists are doing as writers what they do as scientists. They do not have to make the research or the writing interesting to any reader.

1.4.4 Responsibility to General Readers

At the root of much of the criticism of scientific writing is the expectation that scientists have a responsibility to communicate with the general reader, an entirely defensible expectation. After all, scientists are the primary source for new scientific knowledge, and so they can be held responsible for its transmission. However, if each research scientist were expected to report his or her research interestingly, in everyday language, it would be a disadvantage to readers as well as scientists. Not all scientists have the skill to write for a general readership, and the general reader is not interested in the work of every scientist. Besides, if research is written in everyday language, it is not rigorous enough for other scientists. One can certainly expect, however, that when scientists *do* write for the general reader, they will at least use plain, simple, everyday language—but that is more a matter of courtesy than of style.

That scientists have failed to communicate with the public is apparent from misconceptions about science and scientists and from the criticisms of scientific writing. Even the educated reader-critic does not understand the rhetorical setting of the scientific paper and so mistakes its purpose and audience. This failure has allowed science to become one more superstition among others. For the general reader, science is discovery, "facts," "truth," even magic—like science fiction— rather than searching or problem solving.

It is easy to blame the schools for the general ignorance about science or the media for not educating the public, but ultimately, scientists are responsible for the dissemination of scientific knowledge. However, writing scientific papers so that the general educated reader can understand them is not the solution to this problem.

2

Scientific Terminology

Words must be found or developed for the new entities and concepts revealed by scientific research. One of the inevitable consequences of research, therefore, is the addition of new scientific terms. If they cannot be found among existing words, new words must be developed or invented. The new terms become part of the scientific vocabulary or terminology that scientists in a discipline share.

A scientific terminology develops by the specialization of a vocabulary and successive additions and use of new scientific terms for new entities, concepts, and so on. This special terminology is the "jargon" of science, or of a discipline.

2.1 SCIENTIFIC TERMINOLOGY AND JARGON

2.1.1 Jargon in Scientific Writing

Most scientific terminology is comprehensible only to those in the discipline and related disciplines. Critics view the long, unfamiliar, complex scientific terms as jargon, that is, as pretentious words used to mystify, obfuscate, exclude, and impress readers rather than to communicate with them. Such criticism indicates confusion about the denotative and connotative meanings of jargon, and about the nature of language.

In criticizing jargon in science, critics often start with a passage from the social sciences as their example and then satirize it by explaining what the pretentious, high-sounding terms are really saying in "plain, simple language." More intrepid critics may use a passage with highly specialized vocabulary from a natural science. They begin by berating the writer for the incomprehensibility of the writing; then they simplify it to make it comprehensible. In simplifying the passage, they usually do not make clear for whom they are writing. More important, they mistake their simplification for synonymy, whereas it is usually a summary and translation.

2.1.2 Jargon and the Nature of Language

A language is a system of symbolic forms used by members of a national, ethnical, or cultural group to communicate with one another about their shared culture. When a subgroup within the larger group is engaged in specialized activities (e.g., scientists), the general, shared language is usually not adequate for the "tools" of their trade. The members of the subgroup then modify the general language to adapt it to their special needs. If the larger group is not familiar with the activities of the subgroup, the modified language of the subgroup will not spread to the larger group. When the modifications are extensive, the subgroup will ultimately develop a specialized language that will remain the special language—mostly a specialized vocabulary—of the subgroup.

In communicating about their special activities, the members of the subgroup have at their disposal both the general language and their specialized language, but the members of the main group have only the general language. Therefore, if members of the main group or subgroup wish to communicate with one other about the special activity, they are limited to the general language, which they share. If they wish to share more about the subgroup's activities than is possible with the general language, either the members of the subgroup must translate their special language into the general language or the members of the main group must learn the language of the subgroup. Members of the main group are therefore dependent on members of the subgroup to translate for them. The main group can expect, and even insist, that members of the subgroup, in communicating with the main group about their special activities, use the general language that they all share. However, they cannot prescribe that members of the subgroup shall restrict themselves to the general language when communicating among *themselves*. As long as members of the subgroup find the special language more effective than the general language for communicating among themselves, they are not likely to abandon using it.

In communicating their special knowledge to the main group, therefore, members of the subgroup must use words from the general vocabulary that approximate the meaning of the words of their special vocabulary; that is, they must translate their message into the general language. If members do not find words that are close enough in meaning to those of the special language, they must adapt their message to the words that are available in the general language. They may have to explain and elaborate the message to find words to express it. They may have to adapt it by eliminating complexities, subtleties, or details that are too distant from general activities and ideas to have words in the general vocabulary to denote them. The resulting translation is, therefore, simplified, but not synonymous. Moreover, the simplification may be a variable reduction of the original message, depending on the words and concepts available in the general language. This makes the simplification an inaccurate reduction of the message, and the inaccuracy may be great enough to distort the message. Members of the main group therefore often receive a message that is variably reduced in detail and complexity and that may be inaccurate or even distorted. More important, the members of the main group cannot determine the nature or the degree of inaccuracy or distortion.

The miscommunication is greatly compounded when a member of the general group translates the specialized knowledge. Then to the reduction and distortion resulting from the inadequacies of the general language is added the distortion due to a translator with limited scientific knowledge. That is, any deficiencies in the translator are magnified in the translation. Consequently the less knowledgeable the general audience, and the more complex the concepts, the greater the simplification required (and the inaccuracy resulting). And the less knowledgeable the translator, the greater the inaccuracy and likelihood of distortion. This is the kernel of the problem that scientists have with journalists and nonscientist writers of science. For very complex concepts or for concepts that require a great deal of background scientific knowledge, the simplification results in oversimplification or distortion.

2.1.3 Criticism of Jargon

The special language of a subgroup is its jargon. Much of the criticism of jargon in scientific writing is due to critics confusing the denotative and connotative meanings of the word. In its denotative sense, as the vocabulary or "characteristic idiom of specialists or workers in a particular activity or area of knowledge" (*Webster's Third New International Dictionary*), jargon is acceptable in scientific writing; in its connotative sense, as pretentious verbiage, it is not.

Scientific or technical terms are blanketed as "jargon" by critics whether they are specialized or pretentious words. The reasoning seems to be that specialized words are jargon, and jargon is bad. Yet such critics would hardly fault Melville for the sailing jargon in *Moby Dick,* which only a sailing enthusiast can understand.

The aversion to the term "jargon" proliferates the confusion about it. Consider the authors who advised, "Ruthlessly cut out all jargon . . . and substitute words and phrases that say something definite" (Menzel et al., 1961; p. 24); yet scientific terms are highly specific and very definite. Critics confuse "engineerese" and "geolegese," derogatory terms for two particular scientific terminologies (which their practitioners cannot avoid using), with "gobbledegook" and "bafflegat," derogatory terms for circumlocutory, pretentious writing. Writers of "scientific" jargon are accused of inventing *intransitive* verbs when the language has many explicit verbs! Jargon often serves as a whipping boy for any words that critics find undesirable or unacceptable, or that they do not approve or sanction.

Some critics believe that words should not be used unless sanctioned by the dictionary—as though dictionary makers can keep up with the constant stream of new scientific terms. Others believe that new words must not be in the "wrong" direction and must not "debase" the language. For some critics, any long, complex word, especially when made up of foreign roots, is pretentious jargon.

Jargon has even been criticized for its "stringent limitations"; for example, "It is clearly nothing more than vulgar superstition to suppose, as most professionals do, that their particular jargon is infinitely superior to ordinary language or to other professional jargons" (Bross et al., 1972). But, indeed, for the purposes of the *particular* research scientists, the jargon of the discipline *is* infinitely superior to ordinary language, as any specialist can attest; it arose out of the inadequacy of ordinary language for their particular activities. That scientific terms are limiting because of their *analytic* character is unquestionable, but that is a limitation of the scientific method, which cannot be overcome by replacing scientific terminology with everyday language.

Scientific jargon is commonly criticized as the antithesis of plain simple language, yet it is not clear what critics consider plain simple language to be. Many scientific words borrowed from the general vocabulary are short familiar words (*Table 2.6*), though they have a specialized scientific meaning.

Actually the specialized character of scientific terms has the advantage of isolating them and maintaining their accuracy and precision of meaning. If scientific terms were used in the general vocabulary, they would tend to be used more flexibly, and so would become less accurate and precise. The difficulty that general readers have in understanding scientific writing is due as much to the difficulty of understanding the entities, concepts, ideas, theories, and models denoted by the scientific terms as to the unfamiliar terms.

In part, the reason for the greater pressure on scientists to abandon their professional jargon than on sailors or musicians is, of course, that the jargon of science may have practical or life-and-death consequences incident upon not understanding the terminology, for example, in medicine, nutrition, and space science. Nonscientists are entirely reasonable in expecting scientists, physicians, and other specialists dealing with human welfare to write so that the general reader can understand them. But this is really raising the issue of how scientists should speak to nonscientists. Nonscientists can scarcely insist that scientists use the general language for their peers.

Actually research scientists have no choice about using jargon, that is, their scientific terminology, in their scientific papers. It is not possible to address two such different audiences as general readers and scientists in one's discipline in the same language. One cannot address one type of reader without losing the other. Papers should therefore be written for peers in the discipline broadly conceived. Writers can then use the terms that they share with their peers without defining, translating, or simplifying them. If writers wish to write a paper so that it can be read by anyone, they must abandon most of their scientific terminology and clearly define and explain any scientific terms that they do use.

2.1.4 Function of Scientific Terms

Scientific terms are largely the result of new knowledge and new precisions. When biologists wanted a word for *antigen* or *antibody*, or later for *deoxyribonucleic acid (DNA)* or *interferon*, there were, of course, no words in the language for these entities or concepts. Scientists had to invent them. "Deoxyribonucleic acid" may seem unnecessarily long and pretentious; however, it does provide information about the chemical structure of the molecule, whereas a simple everyday term would not. Scientific terminology, or jargon, also provides a high degree of accuracy and precision, a requirement without which scientific inquiry would not be possible.

Scientific terminology, or jargon, makes for great economy. It may take several words to a paragraph, or more to express some scientific term or the concept in plain, simple language. For example, erythrocytes are *red blood cells,* leukocytes are *(phagocytic) white blood cells,* and plasma is the *fluid part of the blood.* Moreover, translating the sentence, "Lymphocytes give rise to plasma cells that produce antibodies against antigens" into simpler language gives:

Ex. 2.1

> Lymphocytes, a particular kind of white blood cells (called phagocytes), give rise to plasma cells. These are special cells in the blood that produce antibodies, which are proteins (usually gamma globulins) in the blood plasma. Antibodies are produced in response to antigens, usually also proteins, that are foreign to the body of the organism.

In this simplification, *lymphocytes, plasma cells, antibodies,* and *antigens* are grossly reduced in meaning. Moreover, the plain simple language has not explained *phagocytes* (except as a particular kind of white blood cell), or *gamma globulin* (except as a kind of protein), or *plasma.* And one might question whether *protein, cell,* or *organism,* though familiar, are really well understood by the general reader. One may, of course, omit all the scientific terms:

Ex. 2.2

> A particular type of white cell in the blood gives rise to a special type of cell, which produces a kind of protein. This kind of protein is produced in response to a foreign substance in the body. This substance is usually a protein, which is foreign to the body of the plant or animal that reacts to it.

Even in this very simplified version, one is assuming that the reader understands the terms "cell" and "protein." It is so reduced, however, that a biologist would find it difficult to translate this "simplification" back to the original sentence.

The sentence used as the basis for *Ex. 2.1* and *2.2* was paraphrased from an elementary biology textbook (Keeton, 1972). It is in itself a simplification from the scientific literature. How accurate the simplification (translation) in the textbook is, only scientists versed in immunology can say. The example illustrates the loss of accuracy, precision, specificity, and information with simplification and demonstrates how effective scientific terminology is for the scientist's purpose. The economy of "jargon" is not a trivial advantage. In a storm at sea few would insist that sailors speak to one another so that passengers could understand them.

2.2 DEVELOPMENT OF NEW SCIENTIFIC TERMS

2.2.1 Frequency of Coinage

New words are such a regular part of scientific research and so generally accepted that coining them is not recognized as a scientific event. Furthermore, such coinages have become an increasingly common activity among scientists, because scientific research is a more widespread activity and because of the increasing complexity of modern research. The emphasis on quantitative studies, with the associated increase in statistical analyses, has required the formation of new terms to denote measures, variables, quantities, and analyses. These may be temporary descriptive terms formed for their convenience in the particular study. However, they may persist, and their formation then becomes important. The increasing complexity has affected all aspects of scientific research and is reflected not only in a great expansion of scientific terminology, but also in the complexity of new scientific terms, which are now often descriptive compound terms rather than single words.

2.2.2 Nature of Scientific Terms

Scientific terms are inherently analytical; they denote concepts or entities that research scientists have abstracted from nature. In their research, scientists conceptually isolate and identify natural phenomena that can be observed and measured, variables that can be manipulated experimentally or analyzed statistically, or concepts that can be used in interpreting findings or structuring the framework of a field of research or a discipline. To be treated scientifically, these factors must first be sharply and clearly delimited conceptually. The terms to denote them must also be sharply and clearly delimited, so that scientists have sharply defined terms with which to think clearly about their research and to communicate their findings and ideas accurately.

A scientific term does not encompass the full complexity of the entity or concept. It is not intended as a holistic representation of the entity, but a carving out of an area of meaning that can be treated or manipulated scientifically. The delimitation is inherently restrictive, however. In science, scientists cannot treat or study the whole; the whole is synthesized from its parts.

2.2.3 Accuracy of Scientific Terms

The centrality of new knowledge in scientific research mandates accuracy and precision of meaning in terms used for reporting research. A scientific term is a word or phrase that scientists have agreed to use to designate a particular referent and *only* that referent; moreover they have agreed implicitly to use no other term for that referent. The term therefore has one meaning and only one meaning. *Ideally,* therefore, there are no synonyms in scientific terminology. As a result, where in writing outside of science, the writer looks for the exact word among synonyms, the research scientist must focus on the entity and use the scientific term that accurately denotes it—repeatedly, if necessary. Because of this, scientific terms come close to the linguistic utopia—that a word have only one meaning. Some utopians would like ideally to curb the language of its perverse tendency to steadily vary, mutate, and evolve. To achieve such immutability, users of a language would have to agree to use a word with only one meaning and to adhere to this restricted meaning consistently. Among scientists, that agreement has been reached—in principle—for scientific terms, although some scientific terms do have synonyms, and some have more than one meaning, and thus act to erode the accuracy and precision of scientific terminology.

2.2.4 Research, Terminology, and Nomenclature

Changes in terminology and in the meaning of scientific terms result from scientific research itself and from scientists being embedded in the scientific thinking of the time. It is not possible to make a definitive delimitation of an entity or concept in the early stages of research. The original term, as defined, may not continue to be an adequate denotation of the entity or concept. As more is learned about an entity, the delimitation of the term denoting it may be modified, and with further development of that branch of research, the term may actually become so imprecise that it may be abandoned. But it is difficult to displace a scientific term after it becomes established. For example, engineers may use different terms from physicists for the same entity, or they may delimit the entities differently, and medical researchers may use a different term from medical practitioners.

This multiplication of scientific terms and meanings eventually makes it necessary to make order in the terminology of a discipline. Attempts are made to order and group entities systematically and to regularize the terminology correspondingly. This usually results in the development of a system or a taxonomy, ideally a natural one, for classifying the various entities delimited in the discipline. In this process, terms that are synonyms may be redefined, displaced, or withdrawn, if they are no longer appropriate or acceptable. The resulting more or less systematic terminology becomes the nomenclature of the discipline. In some disciplines, in fact, the ordering and naming of entities is so important that taxonomy is a separate branch of research.

An accurate and systematic terminology is considered so important that major disciplines have established international bodies for developing a systematic nomenclature and standardizing terminology to eliminate synonyms and terms that cause confusion. Such bodies establish rules for nomenclature and for the formation of terms in the discipline, and so regulate the form that new terms will take. Bodies such as the International Zoological Congress, the National Conference on Nomenclature of Diseases, the American Society for Testing and Materials, the Joint Commission on Biochemical Nomenclature, and so on, review problems of nomenclature periodically and publish the results of their deliberations, as in *Enzyme Nomenclature* and *Standard Nomenclature of Diseases and Operations.* Their decisions thus establish the official terminology or standard nomenclature for the discipline.

2.2.5 Developing a Scientific Term

Synonyms are undesirable not only because of the imprecision or confusion that they cause, but also because of the proliferation of terms. Advances in scientific research, therefore, make great demands on research scientists to adapt the language to new scientific knowledge rather than proliferate terms. Research scientists needing a new term must therefore carefully first consider existing terms. However, the likelihood of finding a word in the language that denotes the new entity is small, so that the research scientist must often develop a new term.

In forming a new scientific term, the research scientist must try to form it so that it sounds idiomatic and fits comfortably into the language. Both the meaning of the words, stems, or affixes in the new term and their form and function should be in accord with customary and common meanings and usages. The objective is to avoid stretching the language so much that it is unidiomatic or disturbing to readers. The process of delimiting a concept, and then managing the resources of the language to develop a term that will accurately denote the new entity or concept and that will, at the same time, be consonant with the structure and idiom of the language, requires a niceness of judgment that is both scientific and linguistic. The elegant solution to the problem is a term that clearly delimits the entity but does not deviate widely from the idiom of the language.

2.3 STRETCHING THE LANGUAGE

In seeking words for new entities and concepts, research scientists naturally turn first to existing words to find one that encompasses the meaning of the entity. They may choose one either from the general vocabulary or from a special vocabulary. For example, social scientists often borrow terms from the natural sciences and adapt them for use in social relations, as in *niche, symbiosis,* and *synergism.* The obvious advantage of this choice is that they can count on readers having some familiarity with the word. It also avoids the multiplication of scientific terms and the difficulties of coining a new term. It has the disadvantage of using a word that already has one or more meanings, and so multiplying meanings.

It is difficult, however, to find existing words that can denote the new entities and concepts, without stretching them in form or limiting them in meaning to fit the new meanings. However, the language can only be stretched within limits, which though indefinite are quite real. If writers go too far beyond the limits of customary manipulations, the stretching will result in unidiomatic or infelicitous forms. It is, after all, the conservatism of the language—its relatively fixed conventional structure—that makes the language effective in communication.

Furthermore, although writers outside of science can draw on the full gamut of spoken and written language for examples of forms and patterns to use as models to make the language fit their thought, research scientists are more restricted. The requirements of accuracy and precision make many models inappropriate for reporting scientific research, for example, metaphors and other figures of speech, ellipses, and humor (see Chapter 3). In developing new scientific terms, therefore, research scientists must maneuver within scientific constraints and linguistic constraints, with which they are less familiar and within which it is difficult to maneuver because of the varying flexibility of the constraints.

2.3.1 Old Words, New Meanings

In choosing an existing word for a new scientific entity, it is usually necessary to redefine it, since the existing word is not likely to represent the new scientific referent accurately or precisely enough.

Meanings related.   Usually the objective is to try to find a word with a meaning approaching the scientific meaning intended. Words are embedded in a spectrum of meanings and connotations that form part of the full meaning of the word. These may intrude on the scientific meaning and so influence the meaning of the scientific term to a greater or less degree.

The word chosen must then be defined to make the *scientific* meaning clear and to differentiate it from the general meaning. This differentiation of scientific meaning from the general meaning is especially important when the two meanings are close enough to be confused, as they frequently are in the social sciences. Because the general language is flexible and changing, the meanings of words are usually broader than is needed in scientific terminology; therefore, the redefinition of words for scientific use is often a restriction of the general meaning:

Ex. 2.3

    a. *Restricted:* orbit, ball, resistance, field, work, energy, power, fusion, gain

    b. *Expanded:* fruit, ovary, spike, family, salt, acid, alcohol

Less frequently, the redefinition is an expansion of the meaning.

The general vocabulary often has no everyday word for a new scientific term. The following example illustrates the limitations of existing words for denoting new entities, the use of a temporary term which does not denote a persisting, naturally distinct entity or phenomenon, and how the lack of an existing word can lead to a compound term.

In a study of the dispersion of toxic substances in a lake, the research scientist chose *substrate* for the bottommost three meters of water. But "substrate" is too specialized to refer to this area. The term is commonly used to mean the physical, and often nutritional, base for microorganisms. This meaning creates a dissonance when the word is used to designate merely the lowest layer of water in the lake. Therefore, "bottom," "base," or "layer" would probably have been closer to the meaning intended than "substrate." However, these nouns, even though defined, are not narrow enough to denote the definite three-meter layer of water above the floor of the lake. A compound term can supply this deficiency, as in "bottom layer" or "base layer." The term "base layer" might even prove useful in subsequent studies, for designating bottom layers of different heights, as in 1-m base layer, or high-turbidity base layer, or low-temperature base layer. The new term *base layer* is therefore not only more specific than "base," "bottom," or "layer," but it has latitude enough to serve as a class term that can be variously modified, should new categories arise.

Meanings not related.   Research scientists may sometimes, though rarely, choose an existing word without relation to its meaning, especially if it is difficult to assign a meaning to the new entity. Words thus chosen are being chosen like mathematical symbols, as though they had no meaning or previous history of use.

This method of choosing an existing word for a new scientific meaning is most likely to be successful if the new meaning bears no relation at all to the old meaning. For example, the following words,

Ex. 2.4

u	up	t	top (truth)	c	charm
d	down	b	bottom (beauty)	s	strange
					(strangeness)

describe fundamental or quasi-fundamental particles, most of them types of "quarks." In a discipline that is strongly mathematical, and in which the entities, as in subatomic particles, are difficult to observe, visualize, and characterize or describe physically, the concrete denotation of the everyday words is not likely to clash with the unimaginable character of the particles. In a discipline in which the entities are at least imaginable, however, if not actually perceptible, this use of words as meaningless, unrestricted symbols would be too discordant to allow the words to function successfully for scientific use. The "meaning" of these terms in nuclear physics has little relation to the meaning of the words in the general vocabulary. Furthermore, not only is their original meaning disregarded, but their grammatical form is also ignored. To denote particles, one would expect nouns, not adjectives (*strange*), or prepositions (*up, down*). To complete the topsy-turvy choice, the two nouns of position are synonymous with two nouns designating qualities.

Few research scientists elect this arbitrary method of forming a new term. Such arbitrary choices for scientific terms are likely to be successful as long as they remain relatively isolated examples of such arbitrariness. If scientific terms were regularly chosen arbitrarily, scientific writing might *look* more readable, but would be even more confusing than the present complex scientific terminology.

2.3.2 Old Words, New Forms and Functions

The writer can modify the form of an existing word and so change the meaning or change the function, as in noun to verb, adjective to adverb. These changes are often made by adding affixes to words. In such changes the prefix usually expresses a change in meaning; the suffix may express a change in function as well as meaning. Some of these changes may be elective; that is, writers may use them freely to form temporary compounds, but most prefixes and suffixes are used only in permanent formations.

When several affixes convey similar meanings, it is necessary in using them to be alert to their differences. For example, several prefixes denote negation, that is, *in-, non-, un-*. The prefix *non-* can be most freely attached to words as a temporary form. The prefix *un-* cannot be as freely used, and *in-* is not used in temporary formations. Moreover, these prefixes are not interchangeable; the prefix *-in* is more consonant with Latin stems, as in *inelastic, insolubility; un-,* the Anglo-Saxon negative, is more appropriate with English words, as in *uncrop, uneven, unbend*. There are many exceptions, however, especially where the stem is not markedly Latin. Often both forms exist as synonyms, but in some pairs, each may be reserved for a slightly different meaning. Also the prefix *in-* for the negative can be confused with the prefix *in-,* which means "in" or serves as an intensive; therefore, new words with the prefix *in-* may not be readily interpretable by readers.

Changes in function may include changes from nouns to verbs, and vice versa, from concrete nouns to abstract nouns, and from adjectives to nouns, and vice versa (*Table 2.1*). Some of the changes are elective, and one can form them at will for many words. For example, adjectives are readily changed to adverbs by the suffix *-ly*, and nouns can be readily changed to adjectives by the suffixes *-al, -ic,* or *-ical,* as from *neuron* to *neuronal*. Other changes cannot be as freely made.

Changes from noun to adjective are the most common and most readily made. Suffixes that can be used for this adjective-to-noun change include

Ex. 2.5

    a. *-cy:* interstitialcy, malignancy, frequency   c. *-ism:* radicalism

    b. *-ity:* conductivity, transversity, expansivity   d. *-ness:* fineness, laboriousness

The verb-to-noun change is common and has long been with us in the following suffixes:

Ex. 2.6

    a. *-er:* transformer, interrupter, transducer, impeller

    b. *-or:* insulator, integrator, reactor, commutator

    c. *-ion:* torsion, emersion, ionization, sedimentation, oxidation, hybridization

    d. *-ment:* encirclement, development, engorgement

    e. *-ance (-ence):* conductance, emergence, transmittance

    f. *-ate:* leachate, condensate, distillate

    g. *-age:* drainage, lavage, tillage, breakage

    h. *-less:* windless, hairless

TABLE 2.1   Nouns, adjectives, and verbs showing changes in function associated with changes in form

a. *Noun to Adjective*		**c.** *Noun to Verb*		**f.** *Adjective to Adverb*	
neuron	neuronal	crystal	crystallize	caudal	caudad
geology	geological	colony	colonize	steep	steeply
entropy	entropic	oxide	oxidize		
phobia	phobic	fraction	fractionate	**g.** *Verb to Noun*	
cybernetics	cybernetic	ligation	ligate		
embryo	embryonic	electrophoresis	electrophorese	bombard	bombardment
knowledge	knowledgeable			transmit	transmission
seed	seedless	**d.** *Adjective to Noun*		react	reactance
evolution	evolutionary			propagate	propagation
labor	laborious	dwarf	dwarfism	exude	exudate
clock	clockwise	redundant	redundancy	cleave	cleavage
		paternal	paternalism		
b. *Noun to Noun*		abrasive	abrasiveness	**h.** *Verb to Adjective*	
pigment	pigmentation	**e.** *Adjective to Verb*		need	needless
capital	capitalism			contract	contractile
physics	physicist	fluorescent	fluoresce		
appendix	appendicitis	permeable	permeate		
		stable	stabilize		

The *-ate* suffix has the disadvantage of changing verbs to nouns, as in *exudate, eluate,* as well as the reverse, as in *fractionate;* therefore, the function of new words formed with this suffix may not be clear to readers. It is also used as a suffix with various Latin and Greek stems to form adjectives, as in *dentate, uncinate, biflagellate,* as well as nouns, as in *salicylate, linolenate,* and verbs, as in *hydrate, precipitate.* Consequently the same word may have various functions, as in *chelate,* which can function as a verb, adjective, and noun.

The noun-to-verb changes must be most circumspectly made. Coinages like *exude* to *exudate* and *elute* to *eluate* are more readily accepted than those made with words of Anglo-Saxon origin. One should also avoid using this suffix in a back formation, that is, the formation of a word from a longer term by removal of an affix to form a word that is a synonym with an existing word, as in *orientate* for *orient* from *orientation* and *fixate* (as in staining) for *fix* from *fixation.* The suffix *-ize,* is to be avoided because it is so indiscriminately used that it does not allow a precise denotation of meaning. From its central meaning of "cause to be," as in *crystallize,* it diverges as far as "to place into" in *"hospitalize."*

The verb-to-adjective change is usually effected with a suffix, *-able, -ant, -ent, -ive.* (*Table D.17,18* at end of the book). The suffix *-able,* as in *collectable,* can be freely used for this conversion. Noun-to-noun changes are most commonly used to differentiate an abstract from a concrete noun, as in *biochemist, biochemistry.* The change may be made from concrete to abstract noun, *alcohol* to *alcoholism,* or vice versa, in *pathology* to *pathologist.* Redundant forms such as *competency* for *competence* and *saliency* for *salience* are to be avoided.

Formal scientific (Latin) names may be changed to common or vernacular names for more informal use. Although many species (as in *gorse, praying mantis, ptarmigan*) and groups (as in *oaks, rodents, reptiles, daffodils*) have a common name, many do not. When a scientific group does not have a common name, one can derive a common or vernacular name from the Latin name. For genera, the change requires only decapitalization of the scientific name, as in "necturus" "drosophila," and "vaccinium." For groups more elevated in the hierarchy, the decapitalization is accompanied by the elimination of the ending that designates the level of the hierarchy, as in *agarics* from *Agaricaceae, chordates* from *Chordata, crucifers* from *Cruciferae,* and *gastropods* from *Gastropoda*. The resulting common names may then be used as stems to form adjectives, as in

Ex. 2.7

Scientific name	Common name	Adjective
Brucella	brucella	brucellar
Streptococcus	streptococcus	streptococcal, -ic
Gymnospermae	gymnosperms	gymnospermous
Helminthes	helminths	helminthoid
Amphibia	amphibia	amphibian

A change in the function of a word need not be accompanied by a change in form. Nouns are often used as adjectives without change in form in two-word terms in the general language, as in *apple butter* and *cottage industry*. In scientific writing, this change is so common, that several successive nouns may be used to form compound terms. In informal usage, adjectives may be separated from the nouns they modify and used alone as nouns for the compound term (*Ex. 2.8a–c*):

Ex. 2.8

    a. a *capillary* tube: a capillary

    b. a *canine* animal, a *canine* tooth: a canine (dog)

    c. *antihelminthic* drugs: antihelminthics

    d. *diabetic:* person having diabetic condition or diabetes

    e. *arthritic:* person having arthritic symptoms or arthritis

Adjectives describing a person having a diseased condition or certain symptoms may be used alone informally to refer just to persons (*Ex. 2.8 d,e*), but this is not acceptable usage in scientific papers. Other nouns formed from adjectives include *schizophrenics, syphilitics,* and *alcoholics*. Some adjectives that describe age are regularly used as nouns, as in *juvenile* and *adolescent; convalescent* is regularly used for a person recovering from an illness. However, to use this adjectival form freely, as in "The Superorganic in American Cultural Geography," where *superorganic* refers to the superorganic mode of explanation, is stretching the language beyond effective bounds. Using nouns as verbs, as in *to doctor,* and vice versa, is less common and is likely to result in unidiomatic expressions.

2.3.3 Stretching the Language Versus Coining New Terms

Stretching the language to give special meanings to words or modifying the meaning of old words for new entities cannot supply the continuing and expanding need for new terms. It is not an accurate method, because an existing word can rarely be as sharply delimited in its meaning as a new word coined specifically for the new entity. The meaning of an existing word may be too broad for the new entity, so that a modifying term may be needed; this is the first step in the development of a compound term. The research scientist can undoubtedly *describe* the new entity accurately. The description can then be reduced to its essential elements to form a term to denote the entity. In the past, such a descriptive label was translated into Greek and Latin stems and affixes to coin a new scientific term. This method of coining a new scientific term is still used, as in *pheromone, nucleotide,* and *ribosome.* However, both the less general knowledge of Latin and Greek and the more frequent use of temporary descriptive terms for variables, concepts, and so on, contribute to the retention of the descriptive phrase. Commonly, the descriptive phrase is condensed into a compound term, usually a base noun and its modifiers.

2.4 COMPOUND TERMS

2.4.1 Definitions and Development of Compound Terms

In restricting the meaning of an existing word to develop a new scientific term, the research scientist may use one or more words to modify it, thus forming a compound term. A *compound term* is here used to mean a word consisting of two or more *separate* words, that is, standing alone or hyphenated, and *not* joined, as in *back focus,* not *backcross.* In the following discussion, the set of words used to designate an entity or concept is the compound term, sometimes referred to as the *compound* for short. The *base term* is the noun or the noun plus modifier(s) that constitutes the central concept of the term, that is, the word or term that the other words modify. When the base term is a noun, the term *base noun* may be used. For example, in the term "nervous system," *system* is the base term (or noun); in "autonomic nervous system," *nervous system,* is "the base term, but in "deep palmar arch," *arch* is the base term (or noun).

The base term is determined by usage, not by structure; therefore, it is not always possible for one outside the discipline to identify it; indeed, research scientists in a discipline may not always agree on the base term of a compound term in their discipline. There is also no definitive work on what constitutes the base term in scientific terms. Since the compound terms included in this book are taken from a wide variety of disciplines, it was not always possible to determine the base term, and it is not always designated.* The intent has been not to make scientific state-

*For terms of three or more words, scientific dictionaries were consulted to determine whether the last two or three words of a compound were listed as a term. The shortest compound so listed was

ments about the structure of the compound terms but to use them as examples to illustrate how words are formed into compound terms, and how the formulation is functional or malfunctional in transmitting scientific meaning.

Compound terms range from two to six words, which are usually nouns and adjectives, but which may include other parts of speech. The noun-adjective terms are designated as *abridged compound terms* and those with other parts of speech, which are really phrases retained as scientific terms, as *unabridged compound terms*. These words are discussed at two levels of analysis: (1) the "part of speech" in isolation and (2) the function of a word in the term or phrase. When the compound term *spectrum analyzer* is examined as to parts of speech, each word is a noun; when examined as to the function of the words, *analyzer* functions as a noun and *spectrum* functions as an adjective modifying *analyzer*.

Compound terms are not peculiar to scientific terminology; they are found in the general language:

Ex. 2.9

    a. life insurance, dirt farmer, rush hour, summer school

    b. dead air space, horse and buggy, round robin letter

    c. lily of the valley, freedom of the press

    d. public works and ways system

Two-word compounds illustrate the advantages of a compound term. The modifier increases the specificity and therefore the accuracy and precision of denotation:

Ex. 2.10

    bed table: table beside the bed

    life insurance = insurance on a life against death

    rush hour = hour during which most workers travel to and from work

    dirt farmer = farmer who actually works the land

They also provide a context, as in *rush hour* and *dirt farmer*. Most important, they provide brevity by condensing phrases and clauses to a short term (*Ex. 2.10*). Scientific terminology includes many more two-word terms than the general language and a great many more terms with three or more words.

Process of development.   Compound terms are often the final stage in a process that begins with a concept or some data that can be variously stated and can be condensed to a clause or sentence. This may be reduced to a temporary phrase or catch-all or shorthand phrase that is used during the research. At the time of writing, however, such phrases must be formalized into a precise term. When such descriptive terms are too long, as in the following italicized phrases:

considered the base term. Other determinations have been based on personal knowledge and on the obvious function of words in the compound.

Ex. 2.11

> No significant difference was detected between the untreated control and the *treatment group receiving T⁴ for the first three weeks.* However, for the *treatment group administered T⁴ for the full four weeks* the mean thymus weights were significantly greater.

they must be shortened if they are to be referred to repeatedly. In formalizing them, writers condense the descriptive words and phrases into a compound term, for example, full-term and short-term group, or simply groups G1, G2 or groups A, B.

To keep the term as short as possible, the descriptive phrase, which is usually an unabridged compound term, is reduced to a term usually consisting of a noun and modifying adjectives and nouns, an abridged compound term. Prepositions are omitted, thus requiring the rearrangement of nouns. For example, *cycle of sedimentation* becomes *sedimentation cycle.* The compound term used in the paper now becomes the scientific name for the entity.

Such compound terms may have started out as ad hoc descriptive designations. The intent may have been merely to *label* or *describe* the variable for convenience. Such descriptive terms, however, may become established as regular scientific terms, and they may then generate other compound terms, as in:

Ex. 2.12

parafollicular cells	*parafollicular cell* inactivation
parafollicular chief *cells*	*parafollicular cell* secretory granules
parafollicular cell studies	bat *parafollicular cell* studies
parafollicular cell secretion	bat thyroid *parafollicular cells*

The descriptive modifiers are powerful tools for accurately and precisely delimiting scientific terms. The descriptive character of many compound terms may therefore make them more meaningful than a single word coined specifically for the entity and probably accounts in part for the dominance of compound terms in current scientific teminology. However, compound terms may be difficult to decode *accurately* and so may not be really effective for transmitting scientific information.

The discussion that follows first addresses compound terms in which the individual words have a noun or adjectival form, that is, abridged compound terms; then unabridged compound terms are discussed. The latter demonstrate how words of different parts of speech, and hyphens, serve as markers that clarify the structure of a term and so help readers to decode its meaning.

2.4.2 Abridged Compound Terms

Two-word compound terms.   Two-word terms are a common construction and, therefore, well understood. They appear most commonly as the adjective-noun combination, as in "social security," but not uncommonly as the noun-noun combination, in which the first noun has an attributive or adjectival function, as in "book club."

Two-word terms are extremely common in scientific terminology (1) because their idiomatic character makes them easy to coin, (2) because if the base noun is an existing word, adding the modifier provides accuracy and avoids the invention of an entirely new word, but (3) primarily because the adjective or attributive noun increases the information content and the precision of the base noun (*Table 2A.1*). The strong contribution of the modifying first word to the meaning of the compound term is readily seen in words in which the base noun is a specific, concrete noun, as in "mass spectrometer" or "periodic table," but it is even more apparent in words in which the base noun is a general or abstract term:

Ex. 2.13

steady state	barodinic disturbance	tunnel effect
membrane potential	operant conditioning	community types

In such words, the general meaning of the base noun, which alone would not be specific enough to be meaningful, is immediately made definite and specific by the modifying adjective or noun.

Two-word terms are usually easy to decode. The adjectives may be transformed into complements of the nouns, as in "strains (that) are mesophilic" for *mesophilic strains,* and the modifying noun can be transformed into the object of a preposition, as in "impulse of the nerve" for *nerve impulse.*

Three-word compound terms.   Three-word terms are more difficult for the reader to decode than two-word terms because (1) they are less common in everyday language and so the pattern of decoding is less conventional and familiar and (2) readers must decide whether the base term consists of one or two words (*Table 2.2*). Terms with words having adjectival modifiers are more easily decoded, since the reader need only determine the noun modified. The adjective-adjective-noun compound (*Table 2A.2a*) is the easiest to decode, because the base term is often the adjective-noun term and the first adjective modifies the two-word base term (*Table 2.2a*):

Ex. 2.14

peripheral *nervous system*	equivalent *partial pressure*	closed *ecological system*
colloidal *osmotic pressure*	gross *national product*	absolute *stereoscopic parallax*

However, the base term may consist only of the noun, and the adjectives may modify it independently, as in "ternary incremental *representation*" and "structural functional *analysis*." When the compound has only one adjective, and this is between two nouns (*Table 2.2c, 2A.2b*), the adjective can usually be taken as modifying the base noun:

Ex. 2.15

messenger ribonucleic *acid*	absorption optical *system*	vacuum fluorescent *lamp*
prolactin regulatory *hormone*	target molecular *weight*	gas thermostatic *switch*

TABLE 2.2   Three-word compound terms showing relationships of adjectives (participles) and nouns

a. *Adj. + Adj. + Noun*	**c.** *Noun + Adj. + Noun*
apogean tidal currents	unit magnetic pole
latent social identity	vacuum condensing point
b. *Adj. + Noun + Noun*	**d.** *Noun + Noun + Noun*
zero population growth	receiver noise threshold
perpetual frost climate	rhesus blood group

However, the adjective may modify the first noun, as perhaps in *"suppressor sensitive mutant"* and *"isotope derivative method"* (*Table 2A.2b*). Even when a participle takes the place of the adjective, it cannot be determined from the form whether the participle modifies the final noun and forms a two-word base term,

Ex. 2.16

a. pressure *melting temperature*   viewer *mounting bracket*   median *effective dose*
b. *ribosome binding* technique   *penicillin treated* disks   *liquid holding* recovery

or whether it is joined to the initial noun in modifying the final noun. The adjective-noun-noun compound (*Tables 2.2b, 2A.2c*) introduces further uncertainty. It is not clear whether the noun in the middle is modified by the adjective, the two together modifying the base noun, as in *Example 2.17a,* or whether it functions as an adjective and modifies the last noun, forming a two-word base term, as in *Example 2.17b*:

Ex. 2.17

a. environmental impact *statement*   b. zero *population growth*
   connective tissue *proliferation*   artificial *radio aurora*
   occupational prestige *scale*   total *lung capacity*
   variable thickness *microbridge*   alkaline *storage battery*

When a compound term consists of three nouns (*Table 2.2d, 2.3*), the reader cannot use the form of the words to distinguish whether nouns are used as nouns or attributively. Such terms may have a one-word or two-word base term (*Table 2.3*), so that their decoding depends on the context or prior knowledge. A reader decoding such noun structures solely on the basis of structure would have various logical alternatives in trying to identify meaningful pairs of words depending on the word or words chosen as the base term:

Ex. 2.18

a. *soil moisture regime*
   regime: regime (with, for) soil moisture, regime with moisture of soil
   moisture regime: moisture regime (for, of, with) soil

b. *computer voice recognition*
recognition: recognition of computer voice, recognition (by, of) computer by voice
voice recognition: voice recognition (by, of, with) computer

Compound terms of four or more words.   Decoding a term from its structure becomes increasingly difficult in compounds of four or more words. Such terms allow even more variability in the relationship of the words constituting the term, so that it is almost impossible to determine the relationships of many words in such compounds unless one has some knowledge of the subject (*Tables 2.4, 2A.3*). They may consist of a single base noun (*Example 2.19a*) or a two-word base term variously modified (*Example 2.19b*):

Ex. 2.19

a. mean ion activity *coefficient*   b. closed source *field diaphragm*
   low molecular weight *proteins*   quantum mechanical *tunneling process*
   multiple benign superficial *epithelioma*   silicone bath *heat source*

Identifying the base term reduces the decoding of the term to an analysis of the remaining two or three words.

Terms with two-word base terms (*Example 2.19b*) are easier to decode because they reduce the decoding problem to the remaining two words, which can modify the base term individually or conjointly. *Tables 2.4* and *2A.3* illustrate the difficulties of decoding the relationships of the words in four-word terms because of the differing configurations of nouns and adjectives.

The problem is especially acute when the base noun is preceded by three nouns, and sometimes even by three adjectives. In "multiple benign superficial epithelioma," the three adjectives seem clearly serial; in "linear variable differential transformer," it is not possible to determine their relationship. In the N-N-N-N group (*Table 2.4*), if one had no prior knowledge of the subject to help in recognizing a unit, such as "data base," one might have various alternatives in decoding the terms. In terms with both adjectives and nouns modifying the base noun, the reader

TABLE 2.3   Three-noun compound terms with (a) one-word base term and (b) two-word base term

a. *One-Word Base Term*	b. *Two-Word Base Term*
[trade wind] cumulus	rhesus [blood group]
[mass transfer] coefficient	vacuum [circuit breaker]
[glucose tolerance] test	sibling [sex ratio]
[radar signal] spectrograph	plasma [electrolyte balance]
[culture epoch] theory	voltage [reflection coefficient]
[zone position] indicator	omnibus [cycle test]
[surface membrane] immunoglobulins	aircraft [weather reconnaissance]
[catabolite activator] protein	emission [electron microscope]
[sickle cell] anemia	chill [wind factor]

Note: In this and other tables, brackets enclose unit terms.

TABLE 2.4   Four-word compound terms in various configurations showing probable relationship among nouns and adjectives (or participles)

a. $A + A + A + N$	**d.** $A + N + N + N$
multiple benign superficial epithelioma	volumetric mass transfer coefficient
linear variable differential transformer	potential transformer phase angle
bipolar magnetic driving unit	differential blood cell count
b. $N + N + N + N$	windowless gas flow counter
data base storage structure	**e.** $N + A + N + N$
enzyme multiplicity feedback inhibition	instrument landing system localizer
c. $A + A + N + N$	surface acoustic wave device
low molecular weight proteins	**f.** $N + N + A + N$
versatile automatic test equipment	time reference scanning beam
digital intercontinental conversion equipment	**g.** $N + A + A + N$
reduced mean residue rotation	quantum mechanical tunneling process

is often helped by familiar two-word units, such as "acoustic wave," "data base," or a participle, such as "penicillin *treated*." In "low molecular weight proteins," (*Table 2.4c*) the first two adjectives are not serial; rather "molecular weight" is a unit, "low" modifies it, and all three words together modify *proteins* as a unit.

It should be clear from these examples (1) how the difficulty of decoding words increases with the number of words making up the compound, (2) how important structural cues are for long compounds, and more important, (3) how much the meaning of a long compound term depends on knowledge of the subject. Ironically, therefore, even in using long descriptive terms for new entities and concepts, research scientists are requiring that readers already know something about the new entities and concepts named.

Five-word compounds are even more difficult to decode (*Table 2.5*). Six-word compounds consisting only of nouns and adjectives, such as *trapped plasma avalanche transit time diode* (TRAPATT diode) are very uncommon. Both their length and the difficulty of decoding them discourage their use, and in fact, they are likely to be abbreviated, as is this one. For such long terms with only nouns and adjectives, a recasting to include a prepositional phrase should be considered. For example, in *albino rat liver oxidase activity,* the *albino rat* can be separated into a prepositional phrase, making the meaning of the resulting term, *liver oxidase activity of albino rat,* much clearer.

TABLE 2.5 Five- and six-word terms in various configurations showing probable relationships among nouns and adjectives (or participles)

a. $A + N + N + N + N$

diagonal screen cathode ray tube

normalized time pulse logic signal

b. $A + A + A + A + N$

multidither coherent adaptive optical techniques

inferior lateral brachial cutaneous nerve

c. $N + N + N + N + N$

sodium water pressure release system

d. $N + A + N + N + N$

instrument landing system reference point

e. $A + N + N + N + N + N$

trapped plasma avalanche transit time diode

2.4.3 Unabridged Compound Terms

With prepositions.   The preposition is the most common of the other parts of speech used in compound terms. It can be found in any compound term of more than two words, and its most common position is between two nouns, as in "angle *of* elevation":

Ex. 2.20

line of collimation	center of percussion	sociology of language
range of motion	calms of Cancer	theory of equations
structure of culture	rites of passage	origin of life
time over target		

One can readily abridge such prepositional phrases to two-word terms, as in

Ex. 2.21

period of vibration	threshold of audibility	zone of saturation
vibration period	audibility threshold	saturation zone

Note that the scientific terms in Example 2.21 are the existing scientific terms *with* prepositions. The two-word terms show their form if they were to be abridged.

In four-word compound terms (*Table 2A.4*), the preposition joins the initial word or unit with a one- or two-word base term:

Ex. 2.22

conflict model *of society*	order *of phase transition*
angle *of descent*	zone *of maximum precipitation*
law of *large numbers*	center *of pressure* coefficient

However, the relationship may be more complex. In "center *of pressure* coefficient," the prepositional phrase , *of pressure,* with the word it modifies, *center,* functions as a unit modifying the base noun *coefficient.* In compound terms of more than four words, it is not common to find terms that are formed only with prepositions, adjectives, and nouns, as in:

Ex. 2.23

thermodynamic potential *at* constant volume	actual cubic feet *per* minute
center *of* momentum coordinate system	equivalence law *of* ordered sampling
multiple nuclei pattern *of* city land use	

Usually such long terms include other parts of speech.

With other parts of speech.   The article is often part of the prepositional phrase:

Ex. 2.24

resolution *of the* identity	perpetual motion machine *of the* third kind
dilution value *of a* buffer	finite intersection property *of a* family *of* sets

The compound term may also include a conjunction, which makes clear the relationship between the two conjoined elements, as in:

Ex. 2.25

basic sediment *and* water	adiabatic lower-energy injection *and* capture experiment
vertical takeoff *and* landing	

or an adverb:

Ex. 2.26

almost periodic function	*perfectly* inelastic collision	*logically* connected topological sequence

The contribution of the adverb to clarity can be seen by substituting the adjectives "perfect" and "logical" for the adverbs.

2.4.4 Compounds with Hyphens

Types of hyphenation.   The use of hyphens in compound terms is of marked assistance in decoding compound terms because they make the relation of the hyphenated words explicit. The joined words can then be taken as acting as a unit in their relationship to the other words.

In the general language, two or more words used together as one word may be hyphenated to show their connection, and they are listed in dictionaries as hyphenated words: *by-pass, warm-bloodedness, square-planar, space-time, mid-term.* In time these may become agglutinated into one word (e.g., *bypass, midterm*). The tendency has been to join two-word terms into one word without their passing through a hyphenated phase, such as *view finder* to *viewfinder*. Indeed the hyphenated terms used here are not consistently hyphenated in various dictionaries.*

Writers can also use the hyphen *electively,* or *facultatively,* just as any other mark of punctuation, to make the relationship of words clear. Such hyphens facilitate the reader's understanding of the connection between the words in compound terms. Note that the hyphen is not electively or facultatively used in two-word terms; it is used electively mostly to join two or more words that together modify a word or phrase. The hyphenated words may consist of an adjective and a noun (*thin-layer* chromatography), two adjectives (*blue-green* algae), or two nouns (*land-use* map). However, the hyphenated modifying term may be more complex, and so may be an extended phrase, as in "*beyond-the-horizon* communication" (*Table D.23*). The elective use of the hyphen is a powerful tool in making clear the relation of words in a compound scientific term.

The most common use of the elective hyphen is in the adjective (or participle) and noun combination (*Table 2A.5*). The adjective or participle usually precedes the noun (*Table 2A.5a),* as in "*steady-state* current" and "*activated-sludge* effluent," but it may follow the noun, as in "*time-invariant* system," "*gate-controlled* rectifier" (*Table 2A.5b*). The hyphenated modifying term may consist of two adjectives,

Ex. 2.27

rural-urban fringe	blue-green algae	manic-depressive psychosis
personal-social learning	basic-nonbasic ratio	structural-functional analysis
aerobic-anaerobic lagoon	solar-terminal phenomena	

but compounds with two nouns hyphenated are much more common (*Table 2A.5c*):

Ex. 2.28

  a. cell-surface ionization, time-pulse distributor, vacuum-tube rectifier
  b. age-sex pyramid, stress-strain concept, sand-shale ratio

The first noun usually modifies the second, but the two nouns may be coordinate, as in *Ex. 2.28b.*

In a compound of four or more words, the hyphenated words may be embedded in the compound. Such internal hyphenation divides the term into two- to three-word units, thus greatly simplifying the decoding of the term. Internal hyphenation also makes it easier to identify the base term and relate the modifying words to it. A common and easily decoded compound is the internal hyphenated unit preceded by an adjective (*Table 2A.6e*):

---

*No attempt was made to survey the various scientific dictionaries or collate the spellings from the different dictionaries. The examples are presented as illustrations of the hyphenation of different types of compound terms; they are not offered as standardized spellings.

Ex. 2.29

unidirectional log-periodic antenna	coherent moving-target indicator
multiple sister-group rule	Asian wet-rice cultivation
latent ion-binding capacity	unidirectional phase-amplitude modulation

The hyphen may however join the initial two or three (*Table 2A.6*) words:

Ex. 2.30

open-loop control system	sky-wave transmission delay
cell-free protein synthesis	rough-surfaced endoplasmic reticulum
forced-choice rating scale	sea-air temperature diffusion correction

In these various hyphenated 4-5-word terms, the hyphenated units may be adjective-noun, noun-adjective, or noun-noun units (*Table 2A.6*). When three initial words are hyphenated (*Table 2A.6d*), the hyphenated words may be coordinate or independent as in *Ex. 2.31a*, but usually they are interdependent, as in *Ex. 2.31b*:

Ex. 2.31

a. carbon-nitrogen-phosphorus ratio	b. tuned-radio-frequency receiver
centigram-gram-second system	narrow-band-pass filter

Compounds of five words or more may include an initial or internal hyphenated unit:

Ex. 2.32

anti-egg-white injury factor	liquid-globular protein mosaic model
medium-long-chain fatty acid thiokinase	emergency position-indicating radio beam

or two hyphenated units, as in,

Ex. 2.33

five-level start-stop operation	permanent-magnet moving-coil instrument
tape-float liquid-level gauge	tuned-grid tuned-anode oscillator
light-gating cathode-ray tube	light-activated silicon-controlled rectifier

Finally, hyphens may be used to join nouns and adjectives with other parts of speech (*Table D.23*):

Ex. 2.34

beyond-the-horizon communications	gene-for-gene concept
time-of-flight mass spectrometer	portal-to-portal pay
second-time-around echo	pillar-and-breast system
wolf-in-sheep's clothing strategy	all-or-none law
mean-square end-to-end distance	blacker-than-black level
glove-and-stocking anaesthesia	

It should be even clearer how helpful hyphens are, when they are inserted in unhyphenated forms to clarify the relation of parts, as follows:

Ex. 2.35

binary-coded octal system	field-desorption mass spectroscopy
carbon-pile pressure transducer	potential-transformer phase angle
vane-motor rotary actuator	instrument-landing-system reference point

Here the addition of one or two hyphens has immediately clarified the relationship between the words for the reader.

Hyphenating compound terms.   It is surprising that hyphens are not more commonly used, especially in compounds of four or more words. This is due partly to an understandable aversion to the obtrusive appearance of repeated hyphens and to the understandable uncertainty about which words to hyphenate. Inconsistencies found in hyphenation are evidence of this. For example, in the following example,

Ex. 2.36

  a. light-activated silicon controlled rectifier

  b. first-estimate—second-estimate method

  c. first estimate-second estimate method

  d. first-estimate second-estimate method

  e. first-estimate-second estimate method

the noun "silicon" and participle "controlled" are in the same relation as "light-activated," yet the term is used without the second hyphen. Such a term raises the problem of hyphenation in terms with two paired terms modifying the base noun, as in *Ex. 2.36b–e.*

Of the alternatives, *Ex. 2.36c,e* offend logic; both imply that *estimate method* is the base term and so misrepresent the meaning intended. The form in *Ex. 2.36b* shows the relations most clearly, but requires the writer to separate the paired hyphenated elements by a longer hyphen or dash. The third form (*Ex. 2.36d*) seems analogous to the usage in some other compounds,

Ex. 2.37

a. tuned-grid tuned-anode oscillator	b. viscous-drag gas-density meter
three-phase three-wire system	tape-float liquid-level gauge

and seems an accepted substitute for the form in *Ex. 2.36b.*

Sometimes hyphens are omitted because a two- or three-word term is so familiar that the writer hesitates to hyphenate it when combining it with other words. For example, in "total white blood cell count" "white blood cell" is left unhyphenated, with the expectation that the reader will recognize the well-known term. Consequently it is not considered necessary to hyphenate it to "total white-blood-cell count." The relationship is not often so obvious, and hyphenating similar less familiar terms would help decode the term for the readers. Correctly used, therefore, hyphens can contribute significantly toward clarifying scientific terms.

The lesson to be learned about compound terms is that brevity is purchased at the high cost of intelligibility and that even such small additions as one or two hyphens or a two- or three-letter preposition can increase intelligibility in geometrical proportion.

2.4.5 Eponyms

An eponym is a special type of compound term. It consists of one or more words regularly associated with a proper name, usually a person's name, as in *Doppler* shift (*Table 2A.7*). Eponyms are the easiest type of compound term to form. The scientific term, whether a single word or compound term, is usually already at hand, so that one needs only to add the proper name to it. Eponyms are therefore analogous to a two-word formation, a common pattern of word formation in the language, which may help to account for their ready use. The eponym may take different forms (*Table 2A.7*).

As a new formation, the eponym is an escape from the accurate denotation or description of the new entity or concept. The proper name is the least informative of the words forming the eponym and lengthens the word without adding scientific information. The name makes a historic rather than a semantic contribution to the term. However, an eponym can be useful when so little is known about the new entity or concept that it is difficult to describe or characterize it or to distinguish it from similar or related ones; the eponym names it without committing the research scientist to an inaccurate descriptive epithet. It is also useful to designate mathematical manipulations because they are difficult to characterize in a descriptive term. The eponym may also be simply a convenient designation to distinguish a new type of a known entity, such as one anemia from another. Eponyms are very commonly used for the scientific (Latin) names of organisms, such as *Polinices Lewisii, Stemonitis Webberi.* They are especially useful in a genus with a large number of species or with species showing great similarities. In such a genus, it may be difficult to assign a descriptive epithet to a species without duplicating an existing epithet.

The eponym may be referential, historical, or honorific; it may designate the discoverer, the person studied, or the person honored. In "Kidd blood group system," the proper name refers to the mother of the infant patient. The commonest use of eponyms is as an honorific designation—for the originator or discoverer, or for a distinguished investigator or leader in the discipline. This use contributes to the disfavor with which they are sometimes viewed.

Efforts are therefore sometimes made, such as in medicine, to reduce their number, and an official body may be set up to reform them. When eponyms have noneponymous synonyms, the official body can try to select the most appropriate of the synonyms. If they have no synonyms, a choice must be made between eliminating a widely used eponym and adopting an unfamiliar noneponymous term. Replacing the eponymous term with a more descriptive term may not result in its adoption, despite its official status, and so may only add another synonym to the literature.

2.4.6 Abbreviated Terms

Definitions, advantages and disadvantages.   It is not a large step from a compound term to an abbreviated term. When a key term consists of more than three words, its advantage as a descriptive term may be counterbalanced by the inconvenience of writing a long term repeatedly through the text. If a writer had repeatedly to write words like

Ex. 2.38

    a. enzyme-linked immunoabsorbent assay (ELISA)
    b. external rotation abduction stress test (EAST)
    c. hydrogen ion concentration (pH)
    d. light amplification by stimulated emission of radiation (laser)

one can scarcely quarrel with the wish to resort to an abbreviation, especially when the terms chosen are as felicitous as "laser" or "ELISA," or as convenient as "pH" (power of hydrogen). The convenience of the abbreviated form gives it currency, so that it becomes the standard form, as with DNA. In disciplines in which long, compound terms are common, a nomenclatural body may establish the standard abbreviated forms for the discipline.

    The two main types of abbreviations that have become common in scientific terminology are acronyms, such as ELISA, and initialisms, such as DNA. Both are capitalized to mark them as abbreviated forms. Acronyms are formed by combining the initial letters of the words in the compound term together with such succeeding letters as will form a term that *sounds* like a word in the language, such as *radar, scuba.* In fact, they fit into the sound system of the language so well that they may lose their capitals and slip into the language, becoming assimilated and losing all recognizable evidence of their multiword origin. Then they are treated like any other word and may even be used as a model to form another word.

    Initialisms are usually formed by combining the initial letters of the main words in the compound:

Ex. 2.39

BWO backward wave oscillator	LCAO linear combination of atomic orbitals
CNS central nervous system	MICR magnetic ink character recognition
REM rapid eye movement	RBE relative biological effectiveness

The second or third letter of some of the words may also be included, but this is not a common practice, and such letters are often not capitalized (*Ex. 2.41*); however, there is no attempt to make the letters form a pronounceable word, and they are verbalized letter by letter, as in D - N - A, and so they rarely enter the language as words. In common use, the term "acronym" is used for both acronyms and initialisms. Traditional, standard abbreviations differ from acronyms and initialisms in that (1) the abbreviation of a word may not be reduced to a single letter and may recall the word such as *vol.* for volume; (2) the abbreviation of each word is followed by a period; and (3) the word is read or spoken as though written in full, such as "post office" for "P.O."

The reduction of a compound term to an abbreviated form eliminates the advantage of the descriptive term, because the resulting abbreviated form becomes as meaningless as the compound term was meaningful. Therefore, abbreviated forms are impossible to decode, hence the importance of writing out abbreviated terms at the first use. In acronyms, the fact that the term sounds like a word in the language obscures its origin even more, and the reader is often not aware that the meaning is related to the letters. Abbreviated forms differ from most words, which retain vestiges of their earlier form that provide some clues to their origins, and often their meanings. They carry little trace of their age or experience on their face.

Use of abbreviated terms.   Abbreviated forms in frequent and general use are a convenience to both readers and writers. Those that are not familiar tend to exclude even scientists in related disciplines from the readership, yet their convenience has led research scientists to form them even for temporary use. Such temporary abbreviated terms must be defined.

For a dissertation, the writer has more latitude in using temporary abbreviated terms than for a research paper. The dissertation is likely to be long, so that the abbreviated forms will be repeated frequently enough for readers to become familiar with them. Also only a relatively small circle of specialized readers read it, so that the writer is not imposing a temporary abbreviated form on many readers, as in a journal.

For the publication of research, the writer must consider the wide readership and the editor's policy on abbreviated terms. Most editors discourage abbreviated forms in principle. In most journals, standard, widely used abbreviated terms are permitted and are used without definition. In specialized journals, with a narrower audience, the specialized abbreviated forms common in the discipline are also permitted without definition; journals often specify acceptable abbreviations. Temporary abbreviated forms are regularly discouraged and should be avoided. If used, they should be defined. They may be justified if the compound term is a key term and must appear frequently throughout the paper, or if the compound is so long that even if it is not used frequently, its length would be intrusive and distracting to readers. Such long compound terms that do not appear repeatedly can often be named indirectly by a more general term, such as *the polymer, the assay, the independent variable,* and so on, to avoid repeating the full name.

The objective in electing to use an abbreviated form should always be the convenience of the reader. However, long and complex names, such as "ultra-high molecular wieght polyethylene" used repeatedly in a paper might make even the ungainly UHMPE acceptable to an editor, who must balance the cost of repeatedly setting the compound term in type against his or her objections to the abbreviated form. Single-letter initialisms must also be avoided, as in P for progesterone or T for tocopherol. They are too inconspicuous and they are apt to be mistaken for mathematical notation or be confused with symbols for chemical elements, which besides having priority, are readily recognized because of their long-standing and international use, and so preempt most letters of the alphabet.

For abbreviated terms that are not standard or widely recognized, it is preferable to write out the full compound term at its first use. The abbreviated term may be included in parentheses or, less commonly, set in apposition to the compound term, following a comma:

Ex. 2.40

rotating ring-disk electrode (RRDE)	RRDE (rotating ring-disk electrode)
rotating ring-disk electrode, RRDE,	RRDE, the rotating ring-disk electrode

If one wishes to make certain that the abbreviated term is understood, then at its first use,the full compound term is included in parentheses or in apposition following a comma (*Ex. 2.40*). Where there are many such terms, they can be defined in a footnote or appendix and then used in the text without definition.

Although initialisms are derived from the initial letters of a compound term, similar abbreviated forms may not be derived entirely from the initial letters. Some seeming initialisms are derived from a single term, as in EKG (electrocardiogram), and EEG (electroencephalogram), in which the initial letter of the main parts of the term are extracted. Certain abbreviated forms are only partly capitalized, as in

Ex. 2.41

dUTP deoxyuridine triphosphate	mtDNA mitrochondrial deoxyriobnucleic acid
Rnase ribonuclease	dHpuA deoxyheptulosonic acid
mRNA messenger ribonucleic acid	

In forms with "H," such as NADH (nicotinamide adenine dinucleotide reduced), the "H" is used for hydrogen to signify reduction, instead of "R" for *reduced* in the full term.

2.4.7 New Words From Other Languages

Words borrowed or derived from other languages have some advantage in being somewhat segregated from words in the general vocabulary. This helps to keep the scientific terms from having their meaning eroded by common use and so contributes to maintaining their accuracy.

Direct borrowing into English is far less common than the reverse borrowing. When a word is borrowed from a modern language, it is not usually due to a research scientist seeking for a word in the language; rather, the word is already in use in the discipline in the foreign language and is introduced into English because it seems to denote the meaning of the new entity or concept better than a translation of the term:

Ex. 2.42

| zeitgeist | gestalt | aiguille | breccia | sarcoma |
| gemeinschaft | veldt | kaolin | mesa | tsunami |

When such words are first introduced into English, they retain their original form, such as their capitalization and their diacritical marks, but they are italicized to mark their external origin. Eventually they become Englished and lose these marks of their alien origin.

Some scientific terms are of course Latin or Greek words from ancient times; many older medical terms, especially anatomical ones, are Latin terms. Many more are terms subsequently derived from classical languages, when Latin was the language of scholars and Latin and Greek part of the training of the educated elite. The use of classical stems and affixes continued into modern times. Therefore until the scientific training of research scientists outstripped their training in Latin, the customary method of coining new scientific terms was to form them by combining stems and affixes from Latin and Greek. Scientific terms are still being coined from Latin and Greek stems and affixes, for example,

Ex. 2.43

| microspectrophotometer | endopolyploidy | fractal | pheromone |
| hypercholesterolemia | phytohemaglutinin | ribosome | cryotron |

or the names of most new chemical compounds. However, whereas it was once the rule to form new scientific terms from Latin and Greek, it is now more exceptional. The development of such terms is analogous to the development of a compound term. In both cases, the research scientist begins with a descriptive phrase that is reduced to make it more economical. In the coinage from the classical languages, the meaningful elements of the phrase are translated into words constituted from stems and affixes of the classical languages, or each noun or adjective is translated into a stem or affix to form one word. The word is often a long word because of the compound character of the word, and may even be a sesquipedalian word, as in *acrocephalosyndactalia*.

Because research scientists are familiar with most Latin and Greek stems and affixes, they can decode and translate such new terms easily and derive the essential meaning. Such terms can, of course, be decoded much more readily than can abbreviated terms, since the stems and affixes have meaning, whereas the letters of an initialism or acronym are meaningless. The word formed may consist of one or more stems. As in two-word compound terms, they may be composed of an adjective and a noun, (*erythrocyte*), two nouns, (*photosynthesis*), or two adjectives (*cardiovascular*).

More frequently affixes are joined to the stem(s). These may be prefixes or suffixes (*Ex. 2.5, 2.6*). Many prefixes are prepositional or quantitative:

Ex. 2.44

Prepositional: ante-, hypo-, para-, pre-, sub-, inter-
Numerical: quadri-, octo-, deci-, milli-, mega-, quinque-

Quantitative: micro-, multi-, pan-, poly-, semi-, macro-
Functional: -ic, -al, -ize, -ate, -ant, -able, -or, -cy
Substantive: -cyte, -lysis, -meter, -ology, -plasia

Suffixes are often functional or substantive (*Ex. 2.44*). Many scientific terms consist of both stem and affixes:

Ex. 2.45

microtome	polydactyl	pandemic	macromolecule	permeable
reactor	ionic	diluent	hypermetropia	ternate

2.5 PROBLEMS IN DEVELOPMENT OF NEW TERMS

Developing a scientific term requires a careful examination and judgment of alternatives. Scientific terms are not merely words that name; they are in the nature of measures and so must delimit a segment of nature meaningfully. Official bodies for nomenclature can help the research scientist to make both scientific decisions and decisions about scientific terms by establishing guidelines for types of words common in the discipline. Research scientists can call on the expertise of colleagues to avoid misconstruction of new scientific terms. For words in English, a colleague in linguistics, specializing in the English language or textual analysis, can help in both stretching the language, so that one stays within the idiom of the language, and in combining descriptive phrases into compound terms that fit idiomatically into the language. For words to be coined from classical languages, colleagues in the classics can lead one through the pitfalls of the various declensions and connecting vowels. These interdisciplinary exchanges are likely not only to result in more felicitous scientific terms, but also to lead to mutual understanding, interest, and respect.

2.5.1 Unidiomatic Forms

New terms are formed by analogy with existing words and forms, but "analogy" is mistaken to mean "logically." The new formations may then be unidiomatic or infelicitous and tend to elicit criticism, which reinforces the criticism of jargon and tends to blanket scientific terminology in general. If unidiomatic terms are then used as models to develop new terms, the unidiomatic forms proliferate and make it increasingly difficult for research scientists to differentiate them from idiomatic models.

Stretching the language, by the very fact that one extends the language to its limits, may press on meaning or form and make a term unidiomatic or infelicitous. If a research scientist borrows a term from the general vocabulary, its different meanings and its familiarity may mislead the writer into using the scientific term in some of its different general meanings. Scientific terms borrowed from the general vocabulary have a more highly specific or specialized meaning or even a different meaning from that of the everyday word (*Table 2.6*):

Ex. 2.46

competence (embryo)	section (anatomy)	spike (flowers)
orbit (atoms)	valve (blood vessels)	stigma (ovary)
doping (surface)	wave (electromagnetic energy)	process (bone)
fatigue (metals)	mantle (brain, earth, molluscs)	vessels (plants)

Moreover, general terms may be borrowed and adapted for use in different disciplines in research, engineering, and applied disciplines or in different industries, with a different meaning in each:

TABLE 2.6   Familiar words adopted from the general vocabulary into various disciplines, illustrating their more specialized meanings as scientific terms

---

a. *Social Sciences*

descent: (1) allocation at birth to a group of relatives, (2) genealogical criteria for membership of unilineally bounded groups, (3) actual or socially defined genealogical connection with ancestor or ancestors, (4) actual genetic relationship to an ancestor.

instinct: complex, species-specific, invariant, innate disposition.

rope: descent group resulting from alternating descent.

sign (b): an object, condition, action, or sound that indicates something more complicated because it is intrinisically associated with its referent. A symbol, as such, would be completely divorced from its referent. For example, a fever is a sign of the flu (but not a symbol of the flu) because it is an intrinsic part of the flu (Zadrosny, 1959).

signal: (1) sign indicating action to be performed, (2) medium carrying encoded message, (3) pattern of stimuli eliciting response.

symbol: (1) a thing that stands for something else, (2) a thing that stands by intention and convention for something else, (3) best possible expression of a relatively unknown fact.

b. *Various Disciplines*

base: Chemical species capable of accepting or receiving proton [CHEM]. Primary substance in solution in crude oil [CHEM ENGR]. Region between emitter and collector of transistor [ELECTR] . . .

bed: Ion-exchange resin contained in column [CHEM]. Layer of mortar on which masonry unit is set [CIV ENGR]. Smallest division of stratified rock series, marked by well-defined divisional plane from neighbors [GEOL]. Bottom of channel for passage of water [HYDRO].

key: Projecting part that prevents movement of parts at construction joint [CIV ENG]. Datum that uniquely identifies a data record [COMP SCI]. Device used to secure or tighten [ENG]. Hand-operated switch for transmitting code signals [ELEC]. Cay, especially one of islets off Florida [GEOL]. Arrangement of distinguishing features of taxonomic group as guide for identifying members of group [SYST].

pit: Cavity in secondary wall of plant cell; stone of drupaceous fruit [BOT]. Quarry, mine, or excavation area, worked by open-cut method [MIN].

set: Radio or television receiver [COMMUNIC]. Placement of storage device in a prescribed state [ELECTR]. Group of conformable strata separated from other sedimentary units by surfaces of erosion or nondeposition [GEOL]. A collection of objects, for which, given any thing, it can be determined whether it is in the collection [MATH]. Direction toward which ocean current flows [OCEANOGR].

Source: *a:* Adapted from Reading (1977), except as indicated; *b:* Adapted from Parker (1984).

Ex. 2.47

   tree: botany, electricity, mathematics, networks
   key: engineering, computer science, electricity, geology, systematics
   mole: chemistry, engineering, medicine, vertebrate, zoology

The definitions of a few such words in different disciplines are given in *Table 2.6b.*
   In using an existing word for scientific use, a research scientist may so expand or limit the meaning that it may diverge from its usual meaning. In *Ex. 2.48a,*

Ex. 2.48

   a. Electrophoresis of DNA samples was carried out using .7% agarose minigels stained with ethidium bromide and *visualized* with ultraviolet light.
   b. The measure was *piloted* by the investigator and two observers who were hired to code. . . .

visualize means "to make visible," but usually "visualize" means "to form a mental image," a quite different meaning. However, using the term with the new meaning avoids the verbal phrase, and this has helped the usage to become established in some disciplines. Shifting the function of words is a common pitfall in stretching the language for scientific terms. In *Ex. 2.48b,* the verb *pilot* represents a shortening of *to pilot test,* derived from a *pilot study.* This two-word noun is frequently converted to the verb, *to pilot-test. To pilot* appears to be more idiomatic in form, but it is less accurate and more confusing because the existing verb "to pilot" (e.g., an airplane or boat) has a distinctly different meaning.
   In general, when a term is infelicitous because of the liberties the writer has taken in stretching the language, the criteria for electing the term should be whether no idiomatic term or usage is available and whether its precision and conciseness excuse its infelicity. On this basis, the verb *electrophorese* from *electrophoresis* is permissible, even though it is not felicitous, because it substitutes for a cumbersome phrase. However, this does not apply to "to sequence" from "the sequence."
   New words formed by the modification of words, either by the addition or subtraction of parts may be unidiomatic or infelicitous. In the following sentences,

Ex. 2.49

   a. The concentration of the o-phthaldialdehyde in the *derivatizing* reagent was increased because of the high levels of Cys in some of the samples.
   b. After *discretizing* the lifting surface into panel elements, equation (8) is specified for each collocation point resulting in the following location.
   c. Variation of the *complexation* energy per nucleotide pair with ligand bending density.

derivatize (*Ex. 2.49a*) has presumably been coined from the noun or adjective *derivative.* But the reader is not likely to understand why *derivatize* was preferred to *derive.* If "derivatizing reagent" is intended to mean "the reagent that produces derivatives, " then the originator of the term was stretching the language beyond its limits. Readers cannot decode it—except by knowing in advance what the writer *meant* and so ignoring what the writer *said.* Similarly in *discretizing* (*Ex. 2.49b*), if the lifting surface is being *divided* into discrete panel elements, then "after the lifting surface is divided into discrete panel elements'' is much clearer and requires only the addition of "is divided," a very small addition for a marked increase in clarity. In *Ex. 2.49c,* the term "complexation" is difficult to decode. Similarly it is doubtful that such infelicitous coinages as "distanciation" and "biologize" can be justified.

2.5.2 Compound Terms and Nominalization

Compound terms are often infelicitous and regularly derogated for their nounal character and their consequent incomprehensibility. Their weakness, from a scientific standpoint, is that often they cannot be decoded with accuracy, except by those who are already familiar with their meaning. The greatest difficulty confronting the writer in forming compound terms is making them (1) short so that they will not be reduced to initialisms and (2) easily decodable. In fact, if they are short, they will be easier to decode.

Nominalization.   The practice of using nouns as modifiers has been referred to as nominalization, noun stacking, nounspeak, nounsense, and other derogatory terms. A compound term of three or more nouns is almost impossible to decode without prior knowledge of the term or the subject. It is very difficult in such terms to distinguish between a noun functioning as a noun and one functioning as an adjective.

The problem in making attributive nouns clear is apparent even in two-word terms. In the following terms,

Ex. 2.50

red house	old house	beautiful house	clean house
small house	low house	palatial house	sturdy house

all the adjectives clearly describe the noun, *house;* that is, the house is red, low, and so on. However, in words, in which the noun *house* is modified by a *noun* acting as an adjective, the modifying noun does not describe the house:

Ex. 2.51

farm house = house on a farm	spring house = house with a spring
garden house = house in a garden	dog house = house for a dog
tree house = house in a tree	toy house = house as a toy
stone house = house (made) of stone	ice house = house for (storing) ice
	wheel house = house with a (steering) wheel

One cannot say the house *is* farm, tree, wheel, and so on. The modifying nouns may be associated with prepositions, which are different but implicit, so that the characterization of the house varies with the unstated preposition.

Where one has a choice therefore, an adjective is to be preferred to an attributive noun because it makes the relation clearer than the noun. This greater uncertainty about decoding the function of a noun has led some to inveigh against the use of nouns as adjectives. However, adjectives are not always interchangeable with the corresponding noun used as an adjective. More important, many of the nouns used are scientific terms that lack an adjectival form. To coin temporary adjectival forms would exchange the unidiomatic character of the nounal term for the unidiomatic character of the coined adjectives, such as *emissional electronic microscope.* In general, however, the intelligibility can be maximized, if research scientists (1) replace attributive nouns by adjectives when this can be done effectively, (2) use prepositions or other parts of speech to avoid overreduction of terms, and (3) use hyphens to clarify relationships. The reason for discussing compound terms is not to abolish them, but to encourage more expanded forms and a more careful selection of words making up the compounds, so that they can be more readily decoded by peers working in related disciplines.

Infelicitous syntactic use.   A compound term may be infelicitous because of its accidental development. For example, shortened, unidiomatic labels tend to replace the conventional descriptive terms. In a study on resistance to stress, the variables studied were the "number of branch roots" and "weight of the branch roots," but these descriptive terms were telescoped to "branch roots no." and "branch roots wt." in the column headings of the tables. Then the telescoped forms were used in the text instead of the more appropriate expanded form; that is, "Growth conditions affected branch root weights," instead of "Growth conditions affected the weight of branch roots" (see also *Ex. 2.52b*).

Compound terms are often difficult to use in grammatical transformations. Though the compound term is a well-formed compound, it may be stretched to unidiomatic use:

Ex. 2.52

> a. The ethanol concentration in the water and solvent phase, and the water content of the solvent phase, were determined *gas chromatographically.*
>
> b. *Forest land percent* and *cropland percent* decreased with depth of soils with frangipan

Although the adverb *chromatographically* can modify the verb *determined,* the noun *gas* cannot function idiomatically in relation to it or the adverb. The construction cannot be justified, because *by gas chromatography* is more idiomatic and even slightly shorter.

When abbreviated forms are used, the words they represent may be forgotten, and may lead to redundancy, as in

Ex. 2.53

> An ELIS*A assay* specific for IgC was set up on day 9 of culture.

where the final letter in ELISA represents *assay.*

2.6 SCIENTIFIC TERMINOLOGY IN THE SOCIAL SCIENCES

2.6.1 Problems in Development and Use

Like other scientists, social scientists must differentiate a new entity or concept sharply from similar or related ones. This scientific conceptualization impinges on the writing because the terms ultimately chosen will reflect any lack of accuracy and precision in the delimitation of the entity and concepts. As has been indicated (Chapter 1), it is especially difficult to delimit concepts about human groups or institutions. Consequently it is difficult to develop precise new names for such indefinite or fluid entities and to ensure that they will be used consistently.

Also, social scientists continue to derive their terminology from the everyday language. They can "make over" everyday words into scientific terms, because they can more easily adopt or adapt words from the general vocabulary to their scientific purposes than scientists in the natural sciences. What everyday words could one use for *adenosine* or *photodiode?* Most scientific terms in the social sciences, therefore, are terms derived by giving words in the everyday language more precise meanings and stretching the language (*Table 2.6a*).

The social scientist therefore pursues the inherently difficult task of finding words in the general vocabulary that precisely cover the meaning intended. Yet such words cannot be easily detached from their general use; they come enveloped in their multiple meanings, which hover closely around ready to intrude on the delimited scientific meaning intended. Social scientists must therefore take a less than adequate word and define it so that it sharply delimits the area of meaning intended. Even if they are able to define a word precisely for scientific use, the related meanings are often so close at hand that they can mislead the writer into imprecisions in writing and the reader into misinterpretation of the term.

The ease with which synonyms can be used in the social sciences and the variability in the subjects of research add to the difficulty of using scientific terms accurately and consistently. Scientific terms adopted from the general language are likely to have synonyms, which invite themselves in for a more precise shade of meaning. The writer thus forgoes consistency to achieve greater precision of expression. However, using synonyms is likely to dilute, broaden, or becloud the meaning of the scientific term, as originally defined, so that ultimately the striving for precision of expression erodes the precision of the scientific term.

Also, social scientists have at their disposal the older discursive tradition (see Chapter 3), in which writers can use all the resources of the language to discuss their subject in all its subtleties. This allows them not only to use the multiple meanings of words, but also to draw on synonyms for their slightly different meanings. This older tradition is reinforced by writing instruction, in which writers are urged to use the full spectrum of words and forms, and are explicitly encouraged to use synonyms for variation. Although this tradition makes possible a holistic consideration of a topic, it is antithetical to the analytic accuracy and precision required for reporting empirical research, where the universe of discourse is much more restricted. In scientific research, social scientists are examining human beings, their behavior, and concepts about them in terms abstracted and detached from the actual social setting.

Using synonyms for variation is inimical to scientific accuracy. In scientific research, the scientist delimits the essential meaning of a term and then adheres to its restricted meaning as defined; the terms act like measures. This use of scientific terms is especially important in the social sciences, where actual physical measurement is usually not possible. Research scientists cannot substitute synonyms for a term, because readers must then infer whether the writer is using the synonym rhetorically for variation or scientifically for the meaning embedded in the synonym. They must infer the intended meaning, yet cannot be certain that the meaning inferred is the meaning intended. This uncertainty erodes the accuracy and precision of meaning essential for reporting scientific research. Yet the very difficulty of expressing one's meaning often leads to using a synonym, because it more precisely expresses one's meaning than the scientific term. This may actually lead to a proliferation of terms because social scientists delimiting a concept in a slightly different way from earlier investigators may use a quasi-synonymous term rather than redefine or qualify a term already in use. This makes consistent terminology particularly imperative in the social sciences.

The tension between consistency to provide specificity and accuracy and variation to provide precision of expression, and the dependence of clarity on the accurate, precise, consistent use of terms is illustrated in the following example. Assume that a term A has two related meanings a and b and that the synonym, term B, has two related meanings, c and d, i.e. *Aab, Bcd,* and that all four meanings are related, with b and c close synonyms in the range from a to d. Also assume that term A, for a particular scientific use, has been restricted to meaning a, i.e. *Aa*. Now at some point in the paper, the key term *Aa* does not quite express the meaning intended; what is intended is meaning b. Although meaning b is included in the meanings of term A in general use, i.e. *Ab*, the meaning of b intended approaches meaning c; therefore the writer uses the term, A with meaning c, i.e. *Ac*. Later in the paper the term A may again refer to the original meaning, i.e., *Aa*, or it may include both meaning a and c, i.e. *Aac*. With further use by the writer or by colleagues following the writer's usage, the term A may be used to mean *Aa, Ab, Ac, Aab, Aac, Abc, Ad,* and so on. The original delimitation of meaning is thus eventually blurred, and it may no longer be clear *exactly* what a writer means in using the term A. It cannot therefore be accurately used in the strict denotation of the referent or as a basis for measurement. The writer may of course use the synonym B also. Here the desire for precision may lead to the introduction of meanings *Bc, Bd, Bbc, Bac,* and so on. This can now result in using different terms for the same meaning, e.g., *Ac* and *Bc*. This erosion of the precision of the scientific terms hinders the establishment of a stable framework for research. Thus the multiple meanings of the term and the overlapping of meanings in synonyms, which makes them so effective in expressing nuances of meaning in writing outside of science, makes the scientific term less precise and therefore less scientific.

The preceding illustration of the process is based on two simple words. For a sense of the magnitude of the problem, consider that words in everyday language often have several synonyms and that each often has several meanings. Consider also that subsequent users of a term may not understand the exact scientific meaning intended, and so may misuse it. Or they may understand it but use it less rigorously, or they may disagree with previous usage and use the term differently intentionally. Therefore, social scientists have to make a special effort to be accurate, precise, consistent, and clear in using scientific terms.

Besides communicating with readers, social scientists must communicate with the subjects of their research when they conduct the research. When communications are the means of gathering data for the research, as in interview schedules, questionnaires, analyses of texts, and so on, language becomes a direct instrument of measurement and must therefore be used precisely and consistently. The writer must frame the statements, directions, questions, and responses, so that the variables can be measured reliably and accurately. The meaning of words that the social scientist uses in conducting the research must be the same for participants and investigator. Otherwise the participants will be responding to a different variable from that which the investigator is measuring. Even *at best,* this is barely possible; therefore, the writer must make every effort to achieve as close a correspondence as possible.

2.6.2 Stretching the Language

In the social sciences, because a word from the general language often comes with a spectrum of meanings, the writer may be led to using it with several of its related meanings. For example the term "office setting" was used for three different meanings in one paper: (1) for the physical walls or partitions and enclosed space; (2) for the employees' experience of crowding, concentration, and privacy; and (3) for the physical character of openness, density (employees per square foot), workspace within 25 ft. of the employee, and light conditions. Using one term to include several distinct meanings makes it difficult to treat the variable rigorously. Using synonyms for the same variable has a similar effect:

Ex. 2.54

   a. Upgrading, *downgrading,* and conditional effects are the three positions in the debate. The upgrading proponents view skill as substantive complexity and measure skill trends using aggregate studies. The *deskilling* theorists view skill as autonomy and rely mainly on case studies.
   b. work place, office, work area, workspace,
   c. program, organization, system

Similarly *environment,* and other terms borrowed from the natural sciences, such as *valence, niche, ecology,* are often used loosely by analogy, without being adequately distinguished.

Because language is extremely malleable, coiners of terms in the social science may stretch it beyond permissible usages:

Ex. 2.55

a. The daily caloric intake of foods of the *sample infants* is reported in Table 3.

b. Occupants of positions having responsibility in *executive resourcing* influence the problem development.

c. *Novelty of demand:* degree to which role permits exercise of prior knowledge, practical skills, and established habits.

In *Ex. 2.55b,* the term *resourcing,* refers to selecting executives (as a resource); *selecting executives* or *executive resources* would have been more idiomatic and more effective. Similarly, readers would expect *novelty of demand* (*Ex. 2.55c*) to be a measure of "novelty," that might be measured as high, moderate, or low. Actually the term refers to the *role* of a person, which permits the person to use many, some, few of his or her skills, in responding to the demands on the role (person). Whenever readers see the term, they must extrapolate and say to themselves "it's not novelty; it's responding attributes" or some such clarification. This is not an instance of confusing close synonyms, which at least have a similar base meaning; this is saying "*demand* is *response,*" which is neither accurate nor logical. Defining such a term is not enough. Language is too conventional to allow such wide deviation from accepted usage.

Appendix 2A

Structure of Compound Terms

The following tables list examples of different configurations of words in compound scientific terms. Brackets and braces show the relationships of words within the compound terms. The relationships were derived primarily from listings in scientific dictionaries. However, the dictionaries are not always in agreement on the treatment of a particular term and its parts, and research scientists do not always agree on the relationships of the words in a compound term in their discipline. No comprehensive study was attempted of the structure of the compounds listed, and no attempt was made to collate the usage in different dictionaries. They are presented here primarily as a sampling of the types of construction of compound scientific terms.

TABLE 2A.1 Two-word compound terms

a. *Adjective (Participle) + Noun*

dimensional analysis
futile cycle
horizontal evolution
impounding reservoir
incomplete antibody
interpersonal relations
melting point
parabolic dune
ritual pollution
scheduling algorithm
social causation
stellar scintillation
temperate phage
topographic anatomy
upper quantide
variable nebula
warm front
white damp
metabolic pathway
morganatic marriage
symbolic language
yielding arches
yoked basin

b. *Noun + Noun*

vasomotor center
windmill anemometer
estuary sediments
urea cycle
kinship system
population pyramid
rift valley
sand wave
tundra climate
doomsday model
spectrum analyzer
zenith coordinates
membrane potential
block diagram
path analysis
personality integration
dependency ratio
vacuum filtration

TABLE 2A.2 Three-word compound terms in various configurations

a. *Adjective + Adjective + Noun*

median [lethal dose]
bacterial [soft rot]
primary [mental ability]
negative [social control]
acute monocytic leukemia
stationary [stochastic process]
transcontinental [ballistic missile]
rural [social organization]
concrete [social class]
layered [metabolic pathway]
solstitial [tidal current]
distal [convoluted tubule]

recommended [dietary allowance]
urban [social planning]
horizontal [social distance]

b. *Noun + Adjective (Participle) + Noun*

unit [normal curve]
[segment long] spacing
[suppressor sensitive] mutant
soil [diagnostic factors]
[gas controlled] atmosphere
[thyroid activating] hormone
[wire-wound] rheostat
[codon recognizing] site

c. *Adjective (Participle) + Noun + Noun*

circular [particle accelerator]
[coastal flow] index
[pressurized water] reactor
scanning [electron microscopy]
[free energy] change
[partial fraction] coefficients
partitioned [data set]
[small group] research
[social distance] scale
[thematic aperception] test
urban [heat island]
digital [carrier system]

TABLE 2A.3 Four-word compound terms with nouns and adjectives in various configurations

a. *A + A + A + N*

hierarchical {distributed [processing system]}
total immediate [ancestral longevity]

b. *N + N + N + N*

{[complement fixation] inhibition} test
[field desorption] [mass spectroscopy]
{[data base] management} system
[rank order] [correlation coefficient]
{[signet ring] cell} carcinoma

c. *A + A + N + N*

adenoid {[squamous cell] carcinoma}
hybrid {[algebraic manipulation] language}
instantaneous {automatic [gain control]}
psychophysiologic [nervous system] reaction

d. *A + N + N + N*

[magnetic ink] [character recognition]
{ebony [body color]} phenotype
interstitial {[cell wall] water}

TABLE 2A.3   (Continued)

potential transformer [phase angle]
ternary [pulse code modulation]
[hot chamber] [die casting]
[whole leaf] [diffusion model]
[white pine] [blister rust]

e. $N + A + N + N$

[penicillin treated] gauze disks
surface [acoustic wave] device
serum free [bilirubin level]
community {[mental health] center}

f. $N + N + A + N$

[check list] [rating scale]
time reference [scanning beam]

g. $A + N + A + N$

[video-data] [digital processing]

TABLE 2A.4   Compound terms of four to five words with preposition and article
in various configurations

1. *With Preposition*

a. *One-word object*

apparent movement of faults
second law of thermodynamics
membron theory of cancer
bulk modulus of elasticity

thermodynamic function of state
iron law of oligarchy
anomic division of labor
[right to work] laws

b. *Two-word object*

interaction of heme groups
belt of soil water
index of hydrogen deficiency
coefficient of sliding friction
reduction of tidal current

zone of optimal proportion
principle of cultural possibilities
theory of antecedent conflicts
calculus of finite differences

c. *Two words + preposition + two words*

axiate hypothesis of urban growth
equivalence law of ordered sampling

thermodynamic potential at constant volume
polygenetic theory of race origin

d. *Other*

great man theory of history
center of mass coordinate system

emission beam angle between half-power points
multiple nuclei pattern of city land use

(continued)

TABLE 2A.4   (Continued)

2. *With Preposition and Article*

a. *Noun + preposition + article + noun*

resolution of a vector	rank of an observation
state of the sea	order of a polynomial
sense of a place	rules of the game

b. *Two words + preposition + article + noun*

arc secant of a number	tangent plane to a surface
duty classificatin of a relay	dilution value of a buffer

c. *Noun + preposition + article + two nouns*

partition of a positive integer	[action at a distance] theory
period of a variable star	[offense against the sine] condition
order of a differential equation	[half of the sites] phenomenon

TABLE 2A.5   Three-word compound terms with hyphenated units modifying base noun

a. *Adjective (Participle)-Noun*	b. *Noun-Adjective (Participle)*	c. *Noun-Noun*
hanging-drop preparation	time-dependent drift	approach-avoidance conflict
paired-comparison scale	set-theoretic model	antigen-template theory
split-ballot technique	affectivity-affective neutrality	church-sect typology
vibrating-reed tachometer	folk-urban typology	land-use map
great-circle distance	logic-tight compartment	salt-spray climax
low-energy physics	melanocyte-stimulating hormone	solubility-product constant
random-number generator	duck-billed dinosaur	valence-bond method
renal-cell carcinoma	tradition-directed society	viscosity-temperature chart
steady-state model	ego-involved motive	vorticity-transport hypothesis
thin-layer chromatography	vapor-dominated hydrothermal reservoir	waveform-amplitude distortion
variable-reluctance pickup		work-flow chart
wet-bulb temperature		winter-talus ridge
yellow-dog contract		zero-sum game

TABLE 2A.6 Four-word compound terms with initial and internal hyphenated two-word and three-word units

a. *Initial: Adjective (Participle)-Noun*

cold-cathode ionization gage
enhanced-purity enzyme protein
high-energy phosphate donor
solid-propellant rocket engine
standing-wave loss factor
tilting-type boxcar unloader
variable-density sound track

b. *Initial: Noun-Noun*

slot-mask picture tube
air-earth conduction current
cathode-ray tuning indicator
gas-liquid partition chromatography
range-height indicator display
sea-air temperature difference
valence-bond resonance method

c. *Initial: Noun (Adjective)-Adjective (Participle)*

dipole-induced dipole interaction
line-controlled blocking oscillator
protein-bound iodine test
self-rating social class
tick-borne thyphus fever
solar-thermal unit theory
staggered-intermittent fillet welding

d. *Initial Three-Word Unit*

standing-wave-ratio meter
rapid-eye-movement sleep
metal-nitride-oxide semiconductor
contagious-disease-buffer hypothesis
variable-focal-length lens
quasi-square-wave static inverter

e. *Internal*

cold lime-soda process
fetal fat-cell lipoma
mechanized dew-point meter
multistator watt-hour meter
vertical field-strength diagram
matrilateral cross-cousin marriage
symmetrical band-pass filter
translation error-ambiguity theory

TABLE 2A.7 Types of eponyms

a. *With One-Word Term*

Abel theorem	Venn diagram	Gram stain
Bessel function	Euler equation	Richter scale

b. *With Two-Word Term*

Gibbs free energy	Hanus iodine number	Longsworth scanning method
Debye relaxation time	Newcastle disease virus	Pauli exclusion principle

c. *With Three-Word Term or More*

Auberger blood group system	Zener diode voltage regulator
Wentworth quick-return motion	Meyer atomic volume curve
Wilcoxon paired comparison test scale	Brooks standard-cell comparator potentiometer

d. *Adjectival Form*

Besselian elements	Toricellian vacuum	Lombrosian theory
Riemannian geometry	Pavlovian conditioning	Hermitian conjugate operation

(continued)

TABLE 2A.7   (Continued)

e. *As Possessive*

Faraday's constant	Abbe's sine condition	Piaget's motor stage
Tyson's gland	Tollens' aldehyde test	Young's two-slit interference
	Girard's reagent theory	Hopkin's host-selection princple

f. *Internal Eponym*

optical Doppler effect	collisionless Boltzmann equation
universal transverse Mercator grid	inverted Marcus Gunn syndrome

g. *With Preposition, Conjunction, or Other Parts of Speech*

Wentworth quick-return motion	canal of Hovius	vestibular membrane of Reissner
Hermitian conjugate of a matrix	Dow process for magnesium	Doppler velocity of position

h. *Multiple Eponym*

DeBye-Jauncey scattering	North-Hatt scale of occupational prestige
Frazier-Spiller operation	Roska-De Toni-Caffey-Smith disease
Cailletet and Mathias law	Lineweaver-Burk plot of biological rate equation

3

Scientific Writing

The rhetorical setting of research has consequences for scientific writing as well as scientific terminology. It establishes a universe of discourse which mandates accuracy and clarity, and an object-oriented point of view.

3.1 ACCURACY

Efforts to achieve accuracy in research are ineffectual unless they are communicated to readers as accurately and closely as the language permits. Accuracy is therefore central to scientific writing. For accuracy, the first requirement in reporting research is that the data reported be absolutely accurate. The accuracy of numerical values is especially important in reporting measurements or values derived from statistical data or analyses. Data cited from earlier research must be verified to ensure that no error is made in transcription. All pertinent data should be presented. An omission may be as important to readers as the data included.

Falsification is a cardinal sin in scientific research. When a research scientist intentionally fabricates or falsifies data, the falsification is considered so heinous an offense that it causes serious disturbance in the discipline and in science at large. It raises questions about scientists, the conducting of research, the training of scientists, and about the scientific enterprise in general. It may undermine the confidence of nonscientists in science and scientists, and raise questions about their reliability, credibility, and even about monitoring. Research scientists are more likely to be led into *unintentional* misrepresentation, because of bias. They may not adequately report exceptions or variations that threaten their favored interpretation, or they may present their research in a biased manner. While less flagrant than intentional falsification, such biased reporting muddies the waters and impedes the advancement of science. Therefore, although accuracy in reporting measurable entities can be achieved simply by reporting quantities and units of measurement correctly, in all else, accuracy depends on the use of words, and in science, words must function with the accuracy and precision of measurements. Because of this dependence on language, accuracy is inextricably linked with clarity and is not possible without clarity.

3.2 CLARITY

3.2.1 Accuracy, Clarity, and Validity

Clarity is the central requirement in writing a scientific paper. It is not merely a desideratum of style. It is important for fundamental scientific reasons, because of the innovative character of scientific research and the need to validate it. Clarity is unqualifiedly necessary for validity, and is thus the canon for scientific writing. Clarity is also essential because a scientific paper is the primary source for the research. The research is not fully available to other scientists unless the scientific paper is written clearly.

It is important to recognize that the accuracy is antecedent to clarity, that clarity has only to do with accuracy in transmission. Even with clear denotation and clear reporting, it may be difficult to validate the research. However, if the research is not clearly presented, it is *ipso facto* impossible to validate it. For example in *Table 3.1c* and *d,* it is not possible to address the observations reported because they include the inaccuracy in transmission as well as any original inaccuracy. More important, there is no certain way for readers to distinguish between the two sources of inaccuracy. Clarity cannot correct for inaccuracies or confusion in the original observations or concepts. It is the importance of accuracy of transmission that makes clarity central to scientific writing. If a research scientist's report is accurate, then it will result in an accurate transmission of the observation to the reader, whether the observation was accurate (*Table 3.1a*) or inaccurate (*Table 3.1b*). If the transmission is inaccurate, or unclear, then the original accurate observation will be inaccurate when received, regardless of the accuracy of the observation (*Table 3.1c,d*).

Consequently, to achieve the necessary accuracy in transmitting information to readers, the writer must aim for maximum clarity. This means first that words must clearly denote the meaning of the entity that they represent. Then they must be so ordered into sentences and paragraphs that the language (1) matches the writer's intended meaning precisely and (2) can be decoded by the intended reader to match closely the writer's meaning.

TABLE 3.1 Effect of accuracy* of observation and transmission† of observations on report of observation and reception by peers

Observation	Transmission	Report	Reception by peers
a. Accurate	Accurate	Accurate	Accurate
A	T	AT	A_T
b. Inaccurate	Accurate	Accurate	Inaccurate
a	T	aT	a_T
c. Accurate	Inaccurate	Inaccurate	Inaccurate
A	t	At	A_t
d. Inaccurate	Inaccurate	Inaccurate	Inaccurate
a	t	at	a_t

*A = accurate, a = inaccurate relative to observation.
†T = accurate (clear), t = inaccurate (unclear) relative to transmission.

3.2.2 Precision of Language: Words and Meaning

Clarity in scientific writing encompasses the meaningfulness of words, attained by the precision of language at the word and phrase level, and the meaningfulness of relationships, attained by a logical conceptual structure at the sentence and paragraph level and beyond.

Precision of language is an absolute requirement for scientific writing. It is made possible by the scientific terminology that has been developed in each discipline. However in scientific writing, even everyday words, which are not as strictly delimited in meaning as scientific terms, should be used precisely.

Denotation versus connotation.   Scientific terms constitute the first step in the objective delimitation of natural phenomena for scientific measurement and operations, and for scientific communication. Ideally, a scientific term will have only one meaning. This is decisive in limiting scientific writing to denotative rather than connotative or figurative language. The need for accurate, precise language, therefore, precludes uses of the language that are often effective in general writing, because they *are* imprecise or indefinite. Such usages may be imprecise because of (1) their elaboration in figures of speech—metaphors, similes, analogies, euphemisms, cliches, and other forms of connotative language, or (2) their indefiniteness, vagueness, abstraction, or generality. The figure of speech may magnify, diminish, emphasize, heighten, or color the idea expressed, or it may cast a particular light on it or give it a particular tone in a way that the displaced denotative word or term cannot—and in scientific writing, should not.

The connotative use of language is also not consonant with the international character of scientific writing and impedes the dissemination of research. Connotative and figurative language requires that readers know the language well enough to recognize the play on words. For research scientists for whom English is not a first language, such language, like colloquial and informal usages, is confusing or incomprehensible.

Using words anthropomorphically, that is, to ascribe human attributes to non-human entities, also departs from denotative language (*Table 3.2*). The use of anthropomorphic language is partly due to making an object of scientific import the agent of human action. For example, in scientific papers, statements such as the following may be made:

Ex. 3.1

This *study, research, report, paper, investigation:*

reports	determines	disputes	plans to
demonstrates	examines	intends	agrees with
displays	establishes	compares	proposes to
shows	attempts	questions	takes issue with
indicates	displays	claims	hypothesizes
outlines	relates	investigates	asked two questions
describes	found	argues	objects

where the verbs make the statements increasingly anthropomorphic from left to right. The anthropomorphic language can be avoided by providing an agent or focusing on the substance of the research as in *Ex. 3.2b:*

Ex. 3.2

  a. Critical attribution *research has* largely *ignored* the different functions that attribution serves and *has* usually *been content* to take them at face value as expressions of casual belief.

  b. The different *functions* that attribution serves *have* largely *been ignored* in critical attribution research, and social scientists *have been content* to take . . .

The most serious problem with anthropomorphic statements arises when they are teleological, that is, giving nonhuman entities intention toward a goal or objective, as in *Table 3.2* no. *4–6.*

Pretentious writing is also not consonant with denotative language. Scientific writing is formal writing, because it is the serious, permanent documenting of scientific research. It therefore has a formal structure, and scientists use formal words suited to this formal structure. However, some formal words may be pretentious or even pompous:

Ex. 3.3

aforementioned	delineate	endeavor	possess	upon
amongst	depict	exhibit	prior to	utilize
anticipate	display	hereafter	reveal	whence
circa	elevated	herein	sacrifice	wherein
commence	employ	inaugurate	therein	whereof
concerning	encounter	initiate	thereof	

It is no less pretentious, however, to use language that is too informal. When writers use language to draw attention to themselves or their writing, whether casual or punctilious, they are no longer focusing on the research, but about themselves.

Humor is a departure from the unity of structure, form, and tone of the dis-

TABLE 3.2   Sentences in which words are used anthropomorphically

1. The current long-term persistence *models can have no trouble* reproducing the Hurst coefficients.
2. The *research concluded* that there was a strong correlation between the two variables.
3. The *farms* in the sample *packed* their silage differently.
4. The important idea to emerge is that complex *physical systems* with infinitely many available degrees of freedom *choose* to execute a dynamical motion.
5. The *enzyme needs* a pH 4 to break the bond.
6. Figure 3 shows that in line with experimental indicators (18) *putrescine prefers* binding at the T sequence.
7. The extensive theoretical *studies examine* this anomaly, but only a few *studies have tried* to solve the problem empirically.

course in a scientific paper. A scientific paper has an implacably declarative and documentary character, which precludes humor. To believe that tossing in a bit of humor lightens the formality or "heaviness" of scientific writing is to misunderstand the rhetorical setting, the universe of scientific discourse, and the nature of humor. A scientist in the discipline interested in the research does not find it dull and may even find the humor intrusive.

Humor is also topical and cultural. Readers who are not insiders may not recognize the humor or be confused by it. These include readers outside the discipline, nonnative speakers of the language, and future readers. Ultimately, only the historian of science may know that humor was intended.

Sex-biased language.   The use of masculine forms (*he, man, mankind*) as neuter forms is inaccurate and therefore not appropriate for scientific writing. One of the commonest problems in avoiding sex-biased language is the use of personal pronouns with gender (*he, she, him, her, his, him*). Since the plural pronouns (*they, them, their, theirs*) lack gender, then whenever a sentence can be structured so that the pronoun can be plural, bias can be avoided. When it is necessary to use the singular pronouns, both pronouns are used, for example, *he or she, his or her, himself or herself.* When such phrases are used repeatedly, they may be abbreviated. The form *he-she* or *his-her* is to be preferred to the form with a slash.

Two important sex-biased nouns are often used in the biological sciences: *man (Man)* and *mankind (Mankind).* Both are synonyms for *human beings,* and "mankind" is a synonym for the little-used *humankind.* Both make the part stand for the whole; they also make the name associated with the distinctive anatomy of one sex stand for the equally distinctive and different anatomy of the other sex. They are therefore less accurate than the "human" forms. It is true, of course, that *human beings* and *humankind* do not carry the connotations of *man (Man)* and *mankind,* but that is due to their having been long used in their particular niches, so that they have preempted that space for their particular associations of meanings. With similar usage and enough time, the unbiased terms can come to have the connotations of *man* and *mankind.*

Dehumanizing terms are also inaccurate, as well as diminishing. Referring to persons as *cases* or *subjects* instead of clients, participants, or respondents is depersonalizing.

3.2.3 Clarity and Structure

Words are not meaningful unless they are meaningfully ordered; therefore, clarity is dependent on structure as well as words. The two main structural prerequisites for clarity are coherence and conciseness. Conciseness is also required to save space in journals and to save readers' time and effort.

Coherence.   Coherence is the ordering of words into sentences, sentences into paragraphs, and so on, so that they develop a closely reasoned, logical, line of thought, both within and between units. Coherence is fundamental to clarity and makes the greatest demands on writers to think, write, and read clearly. Writers must formulate their ideas and then find the words and syntactic structures to express them so that they represent the conceptual structure of their thinking. This translation is very difficult, because there is no direct correspondence between thoughts and their formal expression in words.

The conventional sections of the paper establish the overall structure and so lay the ground for a coherent presentation. Once the line of development has been established for a section, the order of the paragraphs follows, and each paragraph contributes to the line of development. Therefore, much of the writing is focused on structuring paragraphs and the sentences in them.

The paragraph must have unity if it is to be coherent. A collection of sentences on different topics is not a paragraph; neither is a collection of sentences on one topic—the related sentences must be coherently ordered. That is, the sentences within the paragraph must be so ordered and structured that there is a line of thought that begins with the opening sentence and terminates with the last sentence of a paragraph. The paragraph may be variously developed. It has no prescribed shape except that which is prescribed by the structure of the message, but it must have an orderly structure.

In establishing order in a paragraph, the research scientist cannot depend on the topic sentence, the key structuring device recommended for much general writing. A topic sentence is a general statement that states the topic or subject of the paragraph and is developed by succeeding sentences. Instead of topic sentences, research scientists require a kind of organizing structure that allows the specific, detailed, and complex relationships found in scientific research to be developed into a logical line of thought. This can be accomplished to some extent, *externally,* by connectives, that is, transitional words or phrases. Scientists sometimes express the connections implicitly by structuring successive sentences so that their content and meaning expresses the relationship. Such implicit connections leave readers adrift or require them to supply connections from the context and structure alone. This is not conducive to clarity, since the reader's connection may only approximate the connection that the writer intended.

Carbohydrate loading on the High Performance Diet was developed in the United States based on studies by a team of Swedish physiologists. These studies show that the average concentration of glycogen stores is 1.75 g/100 ml with a normal diet. If this diet is then changed for 3 days to one of high fat and high protein, then the glycogen level drops to .6 g/100 ml. If the diet is modified again to include large amounts of carbohydrates for 3 days, then the glycogen stores will increase to 3.5 g/100 ml. If this carbohydrate phase is accompanied by strenuous exercise, then the glycogen level will rise to 4.7 g/100 ml. This is almost a three-fold increase in glycogen stores compared to a normal diet.

Figure 3.1.   Paragraph illustrating coherence achieved by hook-and-eye linkage.
Word (in oval) in one sentence linked to word (in rectangle) in preceding sentence.

Internal connections are more effective and important constructions for achieving coherence in scientific writing. A kind of hook-and-eye construction, in which one sentence is explicitly connected to the preceding sentence, clearly establishes the relation between the two sentences. This is accomplished by relating one sentence, often the beginning of it, to the preceding sentence by repeating a word or phrase from the preceding sentence, or by referring to a word directly or indirectly.

The example in *Fig. 3.1* illustrates this kind of connection in an almost exaggerated form. Near the end of each sentence is a word (the eye), which is in the central line of development of the sentence. Near the beginning of the next sentence the same or a similar word (the hook) links the rest of the sentence to the preceding one and thus continues to develop the line of thought. The hook and eye need not be the same word (see *Ex. 11.8*), but they must be clearly and closely enough related that the link between them can be readily recognized.

Coherence at the sentence level, referred to as cohesion, requires the internal ordering of a sentence syntactically and meaningfully so that the parts clearly make the statement intended and the sentence forms part of the logical progression of the paragraph.

Coherence is also effected by consistency. A consistent point of view avoids distracting the reader from the main approach in the presentation. Consistency of form makes clear the coordinate character of the elements in a series or list, or the parallelism of concepts. Consistency in the titles of tables and figures, in footnotes, and in bibliographic references provides a formal order, which allows the various parts to emerge as ordered units, and thus contributes to clarity.

Conciseness.   Writing is concise when everything that needs to be said is stated in whatever detail is needed in as few words as possible—that is, just the right words and the right number of words, in the right order—no more, no less. This economical matching of words to message contributes to clarity, because it eliminates verbiage, excessive details, and repetition.

Forms of Verbiage.   Verbiage may consist of excess words or of words with little meaning. It constitutes noise in writing and stands in the way of clarity and conciseness. The wordiness may be due to common empty, filler words or phrases:

Ex. 3.4

| due to the fact that | the test in question |
| in a considerable number of cases | at this point in time |

(see also *Appendix A*). Such phrases are readily recognized and easily eliminated without loss. Many are vacuous prepositional phrases; others are introductory phrases that the writer seems to need to wind up before getting into the substance of the sentence. They may be used as transitions, but they are feeble or pseudotransitions that may be superfluous, as in the phrase, "it is of interest to note" -- if it is not, saying so will not make it so.

Repetition is a form of verbiage. It may be contiguous; for example, it is not uncommon to state an idea in one sentence and then restate it in the next sentence in different words, as though it were a different idea. Such tautology is more likely in writing about concepts than concrete observations and measures. Or the repeated material may be distant. In a paper that is not well organized, the same idea or information may be repeated at different points in the paper. When more than one idea is repeated, the line of development is likely to become confused. In fact, repetition is a good clue that the organization may be faulty, a clue that the writer can use in revising and editing the paper.

Verbiage may consist of excessive detail, which also interferes with clarity, for example, long series of words or phrases or long lists of numerical quantities or values. Writers must show readers a pattern in the details or series; otherwise, the paragraph becomes a confusing conglomeration of details.

Conciseness versus Brevity.   The need for conciseness must not override the need for clarity. It is not a call for brevity. The objective is not simply to save words or space; the objective is to make *optimal* use of words, that is, to use every word that is needed for expressing an idea, no more, no less.

This confusion of brevity for conciseness leads to writing elliptically, implicitly, or in summary. Ellipsis is the omission of words. An attempt at brevity may lead to misconstructions such as shifts and dangling or misplaced modifiers. Stating ideas implicitly omits connections and leaves the reader to make connections and interpretations that have not been expressed. Writing in summary is minimal writing; it is writing what one essentially means, but it omits some of the information, interpretation, and connections that readers need to arrive at the essential meaning. Making brevity a dominant objective may lead to a telegraphic style that distorts normal idiomatic language patterns and so interferes with clarity.

Writers who aim for brevity may follow personal guidelines. The cultivation of short simple sentences is often a cult of brevity and simplicity, and it results in an abandonment of the writer's responsibility—to determine the necessary relationships and to state them explicitly and clearly for readers. Some writers even omit articles, but the automatic omission of so regular a particle in the language is unidiomatic. Since it is impossible to omit all articles, readers are left to interpret the writer's intent whenever an article is omitted or included.

The choosing of short forms in preference to long forms, for example, the preference of the *-ic* form of the adjective to the *-ical* form, illustrates the difficulty of following "logical" guidelines for language. The two forms are not merely long and short forms; they often differentiate between meanings (historic, historical, economic, economical), between usages (optic, optical), and between functions (statistic, *n*; statistical, *adj*). In sum, if writing is prolix, the verbiage obscures the meaning; if it is general or indefinite, the vagueness makes the meaning unclear; and if it is terse, the omissions leave gaps in meaning.

3.2.4 Types of Development

Deductive versus inductive.   Scientific research is essentially inductive, moving from the known to hypotheses about the unknown, and from specific findings to generalizations. This method is particularly well adapted to the development of parts of the paper that are short, simple, or straightforward and that conform to the conventional structure of the paper. For example, hypotheses are effectively developed by following the inductive method. When the development is a long and complex discussion requiring elaborating on the conceptualization of a problem, and drawing on different ideas and material, starting with the general concept or statement may make it easier for the reader to follow the argument, and so makes for clarity. This deductive method is adapted to theoretical and complex papers or sections, for example, a discussion section with a complicated explanation and interpretation.

In the inductive method of development, readers do not know the destination until they arrive at it; therefore, they cannot recognize or verify a wrong turn along the way. They must follow the writer very attentively to avoid becoming lost. The writer must therefore be absolutely clear and logical in the development to avoid confusing or misleading readers.

In the deductive method, readers are told the destination at the start and so have it as a point of reference along the way. As the development unfolds, they mentally nod their heads in agreement, unless the path seems to diverge from the logical route to the destination. They can then pause to determine whether they have missed a turn or whether the writer has digressed from the logical route. Because the deductive development thus deviates from the form that would reflect the research, and starts with the end result, the writer must be wary of giving the development the authenticity or validity of definitiveness, generality, or universality.

Sequential.   One may follow a sequential development, which is an easy type of development because it corresponds to the sequential, linear character of writing. Such a development is adapted to temporal events, spatial series, or two-dimensional entities. Three-dimensional entities, such as equipment, do not lend themselves to the linear, sequential character of sentences but can be reduced to a two-dimensional path diagram. The development may then proceed linearly from base to top, for objects oriented relative to a base, as for geographical periods; from top to bottom, as for organizational charts; or from side to side, as for block diagrams.

Comparative.   Comparative relationships are readily adapted to the linear character of writing. One is either comparing a linear series of entities, A to B to C . . . , or comparing two or more entities relative to a series of attributes, for example, A_1, A_2, A_3 to B_1, B_2, B_3. This serial comparison may be vertical, with the attributes of one entity, A_1, A_2, A_3, compared to the attributes of the second entity, B_1, B_2, B_3. It may be horizontal, with entities A and B compared for one attribute before the next attribute is considered. This allows the writer to shift the order from B to A, if B merits emphasis.

Analytical.   Analytical conceptualizations may include a variety of relationships. The development of a concept may include chronological elements, an explanation, cause-and-effect relationships, and comparisons. The conceptual complexity of such multifaceted relationships is analogous to the structural complexity of physical two- or three-dimensional objects, and so one can transform the multifaceted conceptualization into a path diagram that lends itself to the linear form of writing.

3.3 PASSIVE VOICE

The passive voice is commonly used in scientific writing, and scientists are regularly criticized for using it. Since even critics sometimes confuse the passive voice with the linking verb "to be," it seems useful to differentiate the passive from the active voice.

3.3.1 Passive versus Active Voice

In the active voice, the subject acts, and the verb carries the action, often to an object:

Ex. 3.5

    a. A control protocol *executed* the transactions.

    b. High levels of the drug *may inhibit* synthesis of proteins in the brain.

    c. The male birds *surrounded* the female birds.

    d. The curve *rises* sharply after each administration of the drug.

In the passive voice, the subject is acted upon, and the action of the verb falls on the subject. The focus of the sentence is on what happens to the subject, not what the subject does:

Ex. 3.6

a. The transactions *were executed* by a control protocol.

b. Synthesis of proteins in the brain *may be inhibited* by high levels of the drug.

c. The female birds *were surrounded* by the male birds.

In the passive voice, the verb is formed with the auxiliary "to be" (is, are; was, were; be) together with the past participle. The verb in *Ex. 3.5d* is in the active voice, but it is an intransitive verb that does not take an object and cannot be transformed to the passive voice.

The passive voice is variously criticized. Three criticisms directed to scientists are addressed here: (1) that they should not use the passive voice because it is weak, (2) that they use the passive voice to give a specious objectivity to their research, and (3) that in not providing an agent, they are avoiding being accountable for their research.

The passive voice is, of course, weaker and less direct than the active voice, as can be seen from *Ex. 3.5a–c*. This weakness is considered a defect in current criteria of "good style," in which an active, vivid, lively, "verby" style is espoused. The active voice is such an important stylistic device in present academic writing, that the passive voice is proscribed, as though it were an aberrant form in the language. The passive does have a function in the language, which is not easily performed by other syntactic structures and which is particularly useful for reporting scientific research.

The passive voice is the linguistic construction available to writers when the object is acted upon or when the object is the topic of interest. Sentences in the passive voice may be general statements which cannot be converted to the active voice, for example: "Aluminum is readily oxidized." The passive voice may be used more specifically:

Ex. 3.7

1. The nitrogen laser, based on the traveling design, *was composed* of a folded Blumlein acting as the storage capacity, a resonator, and a low-inductance spark gap.

2. The formation of tubercles *was stimulated* by the removal of the apical meristem.

Both sentences in *Ex. 3.7* can be revised to the active voice:

Ex. 3.8

1. A folded Blumlein acting as the storage capacitor, a resonator, and a low-inductance spark gap *composed* the nitrogen laser, based on the traveling design.

2. The removal of the apical meristem *stimulated* the formation of tubercles.

but this shifts the topic from *nitrogen laser* to its parts, and from *tubercle formation* to the apical meristem. Although in isolated sentences this shift in topic may be considered negligible in relation to the improvement resulting from a more direct, active construction, the shift is not negligible in a discourse focused on the nitrogen laser or tubercle formation.

As indicated in *Ex. 3.7,* some sentences in the passive voice cannot be written in the active voice:

Ex. 3.9

    a. Pycnidia *were produced* copiously on the bark lesions and were morphologically similar to those produced on eucalyptus under conditions of natural infection.

    b. The instar *was* most commonly *found* on the soil surface.

    c. Proteolysis of azocasein *was* completely *abrogated* when the cells *were separated* from the substrate.

In such sentences, the agent required for the subject of the sentence in the active voice is unknown, irrelevant, or nonexistent.

The following sentences illustrate how the passive voice allows the writer to focus on the subject of interest:

Ex. 3.10

    a. Green* (1985) *has used* interferon in a new treatment of cancer to control the division of cells.

    b. A new treatment of cancer *has been used* [by Green] in which interferon is used to control the division of cells.

    c. Interferon *has been used* [by Green] to control the division of cells in the treatment of cancer.

    d. The division of cells *has been controlled* by using interferon in the treatment of cancer (Green, 1985).

Here the sentence in the active voice (*Ex. 3.10a*), focuses on Green, not on cancer research. The remaining three sentences focus on different aspects of the research on interferon in treating cancer. These sentences could not be substituted for one another at a particular point in the development of the paragraph. The examples illustrate how the passive voice makes for a precision of focus that cannot be achieved by the use of the active voice. They also illustrate why the passive voice is particularly suited to reporting scientific research, and is in fact unavoidable (*Table 3.3*). However, in some sentences, the active voice is preferable in the context of the text (*Table 3.4*). Such sentences should be revised so that the verbs are in the active voice.

Other attributes of scientific writing tend to promote the use of the passive voice. Scientific writing is largely descriptive, reportorial, and conceptual. The materials, methods, and results of the research are described and reference is made to those of earlier research. Such description does not make for action verbs or vivid nouns or adjectives. And even the most exciting concepts have more to do with relations and states than actions. Also research focuses on *objects,* not agents, and on new entities that constitute additional *objects* to be named.

Green, DeBretor, DeBrion, DeCanet, DeLacor. Fictitious names used in fictitious reference citations.

TABLE 3.3   Sentences in which passive voice is appropriate, functional, or required for rigorous scientific writing

1. Only sound protoplasts with intact membranes *were included* in the counting.
2. Villages representing only 2 of Alaska's 12 native regions *were selected* for the study for reason of controls, comparisons, and economy.
3. Numerical solutions *are obtained* by means of the sequential gradient-restoration algorithm for optimal control problems.
4. Equation (9) *may be used* to describe the relationship between the measured and suspended volume.
5. When bromoergocryptina *was administered* after parturition, a marked increase in milk yield *was observed* in rats and rabbits.
6. Evidence of validity *must be considered* relative to specific purposes for which simulations *are to be used.*

3.3.2 Specious Objectivity

The focus on the object rather than the agent is the basis for the objective reporting required for scientific research. The term *objective* is used here denotatively, but it is used generally with a positive association for unbiased or impartial. To this latter use of the word for scientists, critics take exception. They believe that scientists are imposing on their readers, and on nonscientists, by presenting themselves as objective, when they unquestionably do have a personal investment in their research. Nevertheless, however subjective scientists may be about their research, their writing must focus on the research, and must satisfy canons of meticulous observation and measurement and of rigorous logic in drawing inferences. Moreover, the research must be replicable or confirmed by others, a requirement that places strong restrictions on subjectivity. The writing of research scientists therefore focuses on the *objects* of research rather than on themselves as agents, and the presentation is thus objective in the sense of being *object-oriented.*

TABLE 3.4   Sentences in which passive voice is not required for scientific or rhetorical purposes

1. They showed the effects of the hormone on the *Drosophila* Kc cell line, an established line of embyronic cells. Five stages of cell growth in culture *were described.* [They *showed* . . . and *described.* . . .]
2. The stomata in the leaves were the principal openings through which water vapor *was diffused* [diffused].
3. The pathogen is most commonly parasitic on clovers, both *Trifolium* and *Melilotus* species, but alfalfa can also *be* severely *damaged.* [but can also severely *damage* alfalfa]
4. It *was inferred* by Green* (1986) [inferred] that the low rate of synthesis of these products was a consequence of the low temperature.
5. The size had to be small enough that the implant *could be* easily *adjusted* to by the tissue. [that the tissue *could* easily *adjust* to the implant]
6. For illustration purposes, Fig. 1 shows part of the spectrum examined *by us.* (Delete *by us.*)

*Fictitious author

Some critics make the more sophisticated argument that, scientifically, an observation cannot be made in isolation, without considering the position of the observer. They extrapolate this to mean that scientists cannot make objective observations and therefore should include themselves in the observation in reporting it; that is, they should use the first person. To reify this theoretical position in a rigid practice would reduce the word "objective" to a philosophical abstraction that could not be attributed to *anyone.*

If research scientists' objectivity is spurious, then their subjectivity should bias their results and their interpretations and make them undependable; yet a gigantic space program is built on such undependable, subjective research, to say nothing of the material artifacts of modern civilization. And if research scientists, who operate within the controlled confines of the scientific method are subjective and biased, how unbiased and objective are the judgments of those who operate under no such restrictions? This is not to say that research scientists are without bias in reporting their research, but when they are, subsequent research or the skepticism or objectivity of their peers is likely to bring their bias to light.

3.3.3 Agent and First Person

Critics who fault scientists for using the passive voice argue that scientists use the passive voice to make scientific research appear more objective than it is and to avoid responsibility or accountability for their research. The critics maintain that an agent—the researcher—can be provided and therefore that scientists should write in the first person and in the active voice.

Actually, there may be no person as agent. In most reporting of scientific research (*Ex. 3.4*), the agent may be of secondary or no importance or relevance. The interest of most readers of scientific papers is focused on the research, not the agent.

The use of the first person creates problems both substantive and stylistic. Scientists are often seen as having the "facts" or knowing the "truth," terms that have little meaning in a scientific context. Scientists are therefore accorded much more authority than they claim or can claim. Nonscientists challenging such authority see the use of the first person as restricting that authority and making scientists accountable. But scientists are less apt to think of their empirical research as being "truth" than nonscientists. They know from their own and their peers' professional experience, and from the history of science, that original observations or measurements or interpretations and explanations have had to be modified, and that factors unknown at the time had influenced their earlier findings and interpretations. Moreover, the subject of discourse in scientific papers is not the research scientists; therefore the "I" is not the real subject of discourse.

Moreover, just as the frequent use of the passive voice tends to lead writers into using it when it is neither needed nor appropriate, so the prescription of the first person leads to its use even when it is inappropriate or not needed to avoid the passive (*Table 3.4*). In the following sentences:

Ex. 3.11

   a. We combined extensive records with clinical, serological, and pathological examinations, and we were able to document a high incidence of autoimmune diseases in the animals studied.

   b. Combining extensive records with clinical, serological, and pathological examinations made it possible to document a high incidence of autoimmune diseases in the animals studied.

   c. Extensive records combined with clinical, serological, and pathological examinations showed a high incidence of autoimmune diseases in the animals studied.

sentence *3.11a,* in the first person would be more appropriate for an announcement in a lecture than a written paper. A paper reporting the methods, findings, and interpretations of the scientist in the first person is a paper about the writer, not about the research. In the procedure described in *Table 3.5a,* every sentence is in the third person and the passive voice. When this excerpt is revised to put it in the active voice and provided with an agent (*Table 3.5b*), the subject of every sentence is "I," instead of *cells, coverslips, chamber, micropipette, pipette, pressure,* and so on. Although this revision may have made the excerpt more direct, active, and stronger, and the scientist more accountable, the subject has shifted from the various objects, which are the foci of interest to readers, to the author, who is of little interest relative to the procedure described. In fact, at the beginning of each sentence, the reader meets with a barrier, the "I" and its verb, before getting to the topic of interest.

TABLE 3.5   Excerpts from methods section written in (a) passive voice and revised to (b) active voice, with subject of sentences in italics

a. *Passive voice*

Cells were placed in a thin chamber ($10 \times 20 \times 1.5$ mm) consisting of two glass coverslips separated by a U-shaped brass spacer and held together with vacuum grease. The *coverslips* were siliconized to reduce adhesion of the cells to the glass. The *chamber* was placed on the stage of an inverted (Nikon-M) microscope. A *glass micropipette* formed by breaking off the tip of a glass microneedle and filled with PBS was inserted into the chamber through the open side of the U. The *pipette* was connected to a water-filled reservoir, the height of which could be adjusted with a micrometer. The *pressure* inside the pipette was controlled by adjusting the height of the reservoir. *Zero pressure* (± 25 dyn/cm^2) was determined by stopping the motion of cells or particles in the fluid at the pipette tip. [Biophys. J., 51, 1987, 364]

b. *Active voice*

I placed cells in a thin chamber . . . consisting of two glass coverslips separated by a U-shaped brass spacer and held together with vacuum grease. *I* siliconized the coverslips to reduce adhesion of the cells to the glass. *I* placed the chamber on the stage of an inverted . . . microscope. *I* formed a glass micropipette by breaking off the tip of a glass microneedle and filling it with PBS, and *I* inserted it into the chamber through the open side of the U. *I* connected the pipette to a water-filled reservoir, the height of which *I* could adjust with a micrometer. *I* controlled the pressure inside the pipette by adjusting the height of the reservoirs. *I* determined zero pressure . . . by stopping the motion of cells or particles in the fluid at the pipette tip.

In any case, accountability, in the sense of who performed the research, is not strictly relevant to scientific research. The research is expected to stand by itself, capable of being repeated, developed, expanded, retested, et cetera, no matter who performed it originally. In this sense, scientists may have a nicer perception than their critics of how avoiding the first person helps them to affirm the importance of their research over their importance as individuals or scientists. The use of "I" prevents them from attaining their central objective of focusing on their research.

Finally, the strong efforts made to eliminate racist and sex-biased language represent a recognition of how strongly and subtlely language does affect thinking and values. It can be expected that language that focuses on the object rather than on the person is more likely to promote objectivity than language that focuses on the writer.

The first person singular is appropriate when the personal element is strong, for example, when taking a position in a controversy. But this tends to weaken the writer's credibility. The writer usually wants to make clear that *anyone* considering the same evidence would take the same position. Using the third person helps to express the logical impersonal character and generality of an author's position, whereas using the first person makes it seem more like personal opinion. When it is necessary to cite oneself, it is almost a convention to refer to "the author," or the "writer," but one can use "I." The "present research (work, study)" can substitute for these. References to the writer's publications may be made by number or date, as in any reference: "Green and DeBretor showed. . . . ."

3.4 TENSE AND SCIENTIFIC RIGOR

A scientific paper is primarily a report of observations and experimental manipulations performed in the past, and any literature referred to is also in the past. However, the paper usually also includes discussion, which may include present commentary. The tense of verbs is an important means of differentiating between reporting and commentary. The writer must avoid two pitfalls: unexpected shifts in tense and using the present tense for particular observations and relationships. This gives them an aura of authenticity and authority that is not consistent with the rigor of scientific inquiry.

In the introduction and discussion sections, the present tense is used for statements of purpose, importance, generally accepted knowledge, or conclusions (*Appendix 3A*):

Ex. 3.12

   a. The purpose of this paper is to examine the less explored nonsyntactic portion of the mind-machine analogy in language processing.

   b. It is important to classify employment in terms of compatibility with child care.

   c. Like hydrocarbons, oxides of nitrogen contribute to the formation of photochemical smog.

   d. The source is, therefore, either transient or variable by at least a factor of 10.

3.4.1 Reporting Results in the Present Tense

It is the use of the present tense in reporting results or drawing conclusions that raises questions about scientific rigor. A scientific paper is a report of research which is in the past at the time the writer writes the paper and in the past at the time the reader reads the paper; therefore the appropriate tense for reporting the findings is the past tense. The observations and findings in empirical research can never be considered final, because they are forever subject to modification as a consequence of subsequent research. One cannot therefore generalize and report the findings from an experiment in the present tense, as though they were universal or general truths. Consequently, avoiding the present tense is not simply a grammatical decision about tense, but a recognition of the severe limitations of scientific research, that is, its narrow sphere and its limited generalizability.

Research scientists sometimes find it difficult to report findings in the past tense, when they are still "true," as presumably they are, until further research shows them to be otherwise. However, it is this proviso that necessitates the past tense. For example, the past tense in the statement in *Ex. 3.13a* makes it a rigorously accurate scientific statement of a result:

Ex. 3.13

   a. Dohler and Wuttke (1974) found [find] a sharp increase in prolactin at day 21 (about 40 ng/ml on day 20 and 80 ng/ml on day 21).

   b. There is a sharp increase in prolactin at day 21 (. . .).

   c. Prolactin increases sharply after treatment with. . . .

When the past tense is replaced by the present tense, "find", the statement becomes inaccurate, since Dohler and Wuttke are not continuing the experiment, and it is not certain that they would obtain the same result if they did. The present tense in *Ex. 3.13b* changes the statement of a research finding into a statement of generally accepted knowledge. The statement in *Ex. 3.13c* would be a very broad generalization, without adequate scientific support.

Writers may try to express the past event of a finding in the past but its continued "truth" in the present by using the present tense *Ex. 3.14a*:

Ex. 3.14

   a. It *was found* that the concentration of luteinizing hormone in the plasma *increases* with the injection of higher doses of progesterone.

   b. The increase in the dose of progesterone injected *results* in an increase in luteinizing hormone.

However, these results might not be obtained in future research. Reporting the results entirely in the past would not only be accurate at the time of writing, but would also ensure that the statement would remain accurate in the future, even with different results. The statement in *Ex. 3.14c* generalizes and so is not accurate.

3.4.2 Past versus Present Tense in Other Sections

In the introduction, it is easy to confuse the paper (present) with research performed (past). When the research scientist writes the *paper* it is in his-her present. Consequently the present tense is used in statements about the paper (*Ex. 4.10*), but the past tense is used in statements related to the research (*Ex. 4.11*). Similarly, the writer should not confuse what was done in the research with what is being said in the paper and should not write "a mean sex ratio of 1:1 female per male *will be used*."

Since references to the literature introduce the research of others, which precedes the research being reported, the past tense is clearly the tense of choice. The present tense is sometimes used for the literature in a general discussion,

Ex. 3.15

"Green states that feedback facilitates meaningful learning through. . . . However, he points out that feedback is less important than. . . . "

but is not appropriate for reporting the literature.

Some research scientists believe that using the present tense to refer to the literature indicates confidence in the research and is a courtesy due to the authors. This is moving out of the frame of scientific inquiry to the frame of collegial solidarity, which is not at issue in reporting research.

A scientific paper does not in itself confer authority. Scientific papers report particular experiments, and the results obtained are the results pro tem. They become part of the body of scientific research and can be used as evidence and for application, but they cannot be taken as given. To report earlier research in the past tense is therefore rigorous scientific practice. It allows the research to assert its authority on the basis of its rigor, validity, the evidence, and ultimately, subsequent empirical support, whereas reporting it in the present tense confers authority without substantiation.

When the literature must serve to set up the theoretical framework for the research being reported, it may become an introductory discussion on the subject, and may be sometimes treated in essay form. But the universe of discourse in an essay is different from that in a report on scientific research. An essay is essentially general and allows a great deal of freedom, whereas a scientific paper is specific and much more restricted in content, form, development, and point of view. In an essay one might make the statement in *Ex. 3.16a*:

Ex. 3.16

a. Drugs that are effective against human leukemia cells are also effective against L1210 cells.
b. Drugs that had been found to be effective against human leukemia cells were effective against L1210 cells.
c. The drugs that were effective against human leukemia are effective against L1210 cells.

The statement is a generalization about drugs in general. The statement in *Ex. 3.16b* is an accurate report of the literature, and the generalization in *Ex. 3.16c* more closely approaches this accuracy than *Ex. 3.16a*.

Discussions are too often largely in the present tense, presumably because the findings, both of the present and past research, are accepted as fact. In such a treatment the writer has stepped out of the role of a scientific inquirer and reporter into that of an authority. The discussion then becomes a discussion of a position, rather than a consideration of evidence. If the discussion is in the present tense (see *Appendix 7A*), it may assume the form of an essay, which is then focusing on the *subject* rather than the research.

Conclusions raise the same question as results about the present versus the past tense:

Ex. 3.17

    a. It is concluded that altitude *did* [does] not *affect* the color.

    b. Thus even controlling for skills, unions *provided* new jobs and promotion ladders to advance mobility of men's work.

    c. The results of the present study *seem* to provide some evidence about the reliabilities and validities that *can* be achieved by standardized case simulations.

    d. The close correspondence between the chemical uptake by plants and the RWD *indicates* [indicated] that the rate of root growth *was* more important than the specific absorption rate.

The statement about the conclusion may be past or present, but the tense of the substantive statement of the conclusion determines the rigor of the statement. The statement in *Ex. 3.17a* maintains the conclusion as a past inference of the past results; the present tense would make the conclusion a general statement about the results, which is scientifically less accurate. Specific conclusions are stated in the past tense (*Ex. 3.17b*). Conclusions about the research may be in the present tense (*Ex. 3.17c*), but those that generalize the finding are best kept in the past tense (*Ex. 3.17d*). The statements in the past tense are more rigorous scientifically than are those in the present tense.

The present lends itself to general discussion. It is the conventional tense for mathematical and theoretical models, and for the discussion of relationships that are not determined by the time of operations:

Ex. 3.18

Reactivity Control in Polymerized Vesicles

There are four extreme sites of reactant localization in polymeric vesicles. Hydrophobic molecules can be distributed among the hydrocarbon bilayers of the vesicles. Alternatively, they can be anchored by a long chain terminating in a polar headgroup. Polar molecules, particularly those that are electrostatically repelled from the inner surface of the vesicles, may move about relatively freely in the vesicle-entrapped water pools or they may be associated with or bound to the inner and outer surfaces of vesicles. Polar molecules can also be anchored to the vesicle surface by a long hydrocarbon tail. A large variety of reactivity control can be realized in polymeric vesicles. Conceivably, the position of a reacting substrate will be different from that of the transition state and that from the product formed in the reaction. Such spacial relocation of molecules as they progress along their reaction coordinates can be exploited in catalyses and product separation. A type of functionally polymerized vesicle-reactant interaction can be visualized in which the reactant, an organic ester, for example, would enter the vesicle. Hydrolysis would then occur in the matrix of polymeric vesicles and the products would subsequently be expelled into the bulk solution. With use of nonpermeable reagents, reactions can be limited to sites located at outer vesicle surfaces. Alternatively, finely tuned processes can be realized by allowing reactions to occur consecutively (at controllable rates) in the separate halves of the bilayer in the vesicles. This type of flexibility is only feasible in polymerized vesicles. [Accts. Chem. Res. 17, 1984, 8]

The present tense is of course regularly used in discussions and developments that include equations (*Chapter 6, Appendix 8C*).

3.5 DESCRIPTION IN SCIENTIFIC WRITING

Because scientific research consists of new operations and observations, description is an important part of scientific writing, and scientific papers regularly include description. Descriptive writing in scientific papers differs from general descriptive writing because of its subject matter and purpose. It differs in three ways: (1) the focus is conceptual, cognitive, or abstract, rather than sensory; (2) it is analytical and selective; and (3) it is direct; that is, it is predicative, rather than attributive.

3.5.1 Conceptual Character

In most general descriptive writing, the sensory characteristics of the objects described are the important and may even be the central characteristics of the description. In a scientific paper, however, the sensory characteristics are described only as they might be pertinent to the research. Readers do not need to be able to see, touch, hear, or feel what is being described; they need only be able to develop a concept of it, largely a visual concept. But even when sensory attributes are described, the description is written for cognitive, not for affective or expressive purposes.

This difference is seen in the ordering of the description. A linear, sequential, walk-through, visual description, which is a common type of description in general writing, is not suited to scientific description:

Ex. 3.19

> As one enters the growth chamber one is met by a dazzle of lights and enveloped in a humid atmosphere pervaded with the pungent aroma of the foliage. To the right are rows of tall bean vines; around to the left is a forest of tomato foliage. Passing through between the benches to the next chamber, brings a quarter acre of flats of seedlings into view. . . .

Such a walk-through description imitates the movement of the eye through space. It is too unsystematic for scientific research. It does not give the reader an understanding of the whole or of the separate parts, their relative dominance, their connections, nor does it focus on a part of particular interest.

When this sequential method is used in the description of an instrument, it becomes a description by construction, in which one part is described and then a second, adjacent part added, and then another, without regard to the relationships of the parts.

3.5.2 Analytic Character

Scientific descriptions are analytical in that the writer usually describes only those parts that are new or pertinent to the research. For example, when an apparatus is described, the writer may not describe the whole apparatus or may describe it only in outline, or only refer to it and then describe the new part(s) in detail. In a general description, this would put the part described in limbo. In scientific writing since writer and reader have a shared background, the writer can take this as a given in writing descriptions. For example, a research scientist can simply write "The experimental animals were male weanling, 28-day-old, Fisher-334 rats," and research scientists in the discipline working with rats require no further description. If they visualize the rats, they are more likely to "see" the rats as having certain scientific and research attributes than as little, white, soft, furry, baby boy rats. The analytic character of descriptions in a scientific paper is also reflected in the graphics that regularly accompany descriptions and in the type of graphics. When graphics are used, they too present only the details of interest.

3.5.3 Predicative Character

Finally scientific description is direct and predicative; research scientists cannot describe by indirection. The writer of a story can write the sentence in *Ex. 3.20a:*

Ex. 3.20

   a. Her deep blue eyes mirrored the blue eyes of a long line of Norsemen.

   b. The female offspring had blue eyes, a genetic character found in both parents in four generations. All of these progenitors had been born and raised in Norway.

   c. The 107 Navajo women who participated in the study were primarily self-selected according to whether they volunteered at a given site.

   d. The participants in the study were 107 Navajo women, who were primarily self-selected. They constituted all the women who volunteered to participate in the study when the investigator visited their villages.

A scientist writing about the inheritance of eye color would have to state explicitly that she had blue eyes, as in *Ex 3.20b.* In a story, such a statement would be unthinkable (except possibly to satirize scientists).

    This difference is not merely stylistic. It is important scientifically because direct description is more precise than indirect description. If in describing the Navajo participants in a study, the research scientist writes the sentence in *Ex. 3.20c:* The "107 Navajo women" may be part of a larger cohort of Navajo women, some of whom did not participate. They may be the Navajo women who participated in the study as distinct from some other group of women, for example, 107 Hopi or non-Indian women. Actually the 107 Navajo women were simply 107 Navajo women who agreed to participate in the study when the research scientist traveled to various villages to recruit participants. A direct description such as that in *Ex 3.20d* makes this clear.

Indirect description of time (*Ex. 3.21a,c*) is also not accurate enough for scientific description. Within a sequential series of steps in a procedure, statements such as those in *Ex. 3.22b, d* are more accurate:

Ex. 3.21

    a. An incision was made beginning at the base of the sternum . . . *after* the anaesthetic was injected into the cannula in the jugular vein.

    b. The anaesthetic was injected into the cannula in the jugular vein; then an incision. . . .

    c. In a 96-well plate, 100 μl of the cell solution, adjusted to 2×10^6 cell/ml was added to the well, *which already contained* 100 μl of diluted ConA, PHN, PWM, LSP, and TSST-1.

    d. To the wells in a 96-well plate, 100 μl of diluted ConA . . . were added. Then 100 μl of the cell solution, adjusted. . . .

In descriptive studies, observations may be very extensive and result in long, detailed descriptions. Some scientific descriptions are highly stylized and follow a conventional format:

Ex. 3.22

CARDINALIS CARDINALIS CARDINALIS (Linnæus).

CARDINAL GROSBEAK.

Adult male.—Lores, anterior portion of forehead, anterior part of malar region, chin, and throat, black, forming a conspicuous *capistrum,* entirely surrounding the bill; rest of head vermilion red, duller on pileum (including crest), brighter on auricular region and cheeks; under parts pure vermilion red, becoming slightly paler posteriorly, the flanks slightly tinged with grayish; hindneck, back, scapulars, rump, and upper tail-coverts dull vermilion red, the feathers margined terminally with olive-grayish (wearing away in midsummer); wings and tail dull red, still duller on greater coverts and secondaries, the tertials usually, and sometimes the reetrices, more or less edged with olive-grayish; bill bright orange-red or red-orange in life, fading to orange or yellowish in dried skins; iris, deep brown; legs and feet, horn-color. . . . [Ridgway, 1901]

The conventional position of noun and adjectives is reversed; the descriptive adjectives *follow* nouns, instead of preceding them. Other differences between scientific writing and a comparable type of general writing, that is, academic writing, are summarized in *Table 3.6.*

3.6 SCIENTIFIC WRITING IN THE SOCIAL SCIENCES

One of the major consequences of the rhetorical setting of social science research is that it makes social scientists more dependent on words than natural scientists. Whereas much research in the natural sciences is quantitative, in the social sciences

TABLE 3.6   Characteristics of academic and scientific writing

Characteristic	Academic writing	Scientific writing
Purpose	Expression, exposition	Communication
Generality	Often general, rarely highly specific and detailed	Highly specific, concrete, detailed; infrequently general; abstract for theory
Writer vs. subject	Personal, subjective, or objective	Impersonal, objective, i.e. object-oriented
Audience	Everyman, selected but unspecialized, author	Writer's scientific peers
Rhetorical setting	Writer most important; purpose, subject, readers less important	Writer least important; subject, readers, and purpose more important
Form	Intrinsic, chosen by author, molded in process of composition	Extrinsic; determined by convention, material, structure of discipline
Realism of content	Reflective, sometimes realistic, or imaginary, often imaginative	Observational, factual, reportorial, not imaginary
Form vs. content	Shaped for literary, aesthetic, or other objectives	Constrained by scientific content and purpose, closely matched to content
Interest of readers	Designed to interest	Inherent in content; readers self-selected by interest in subject
Accuracy, clarity	Not central requirements	Central requirements
Language	Expressive, connotative, vivid, metaphorical	Accurate, precise, denotative, concise
Variation, synonyms	Desirable for expressiveness, variety, interest	Avoided for conflict with accuracy, precision, and clarity
Jargon (scientific terminology)	Unacceptable, except for literary or aesthetic purposes	Essential for precision of meaning among peers
Passive voice	Proscribed because weak, not direct	Required to focus on object as topic of discourse
Coherence	Effected by topic sentences and transitional elements	Effected by internal hook-and-eye connections and transitional elements
Process of writing	Largely composing, writing, rewriting; finally revision and editing	Largely writing, revision, and editing, relatively little composing
Source of material	Writer's knowledge and experience	Discrete body of scientific data and concepts in present and past research
Graphics	Exceptional, supplementary, complementary, or embellishing	Required for empirical demonstration; integral part of writing
Format	Writing integrated, headings not common	Headings essential, often numerous

case histories, descriptive, analytical, and other types of qualitative studies are also common. Even when their research is more quantitative, social scientists must use language as a tool for their scientific measurements. Variables are frequently measured by words or by a numerical scale that represents verbal measures. Data to be analyzed are derived from questionnaires, interviews, or observation—all forms of verbal measures. The description of participants and sites and theoretical models are largely verbal. The consequences of this dependence on words is that the standardization of terminology and nomenclature, which is extremely important in the natural sciences becomes crucially important in the social sciences, where it is most difficult to achieve.

Social science papers also tend to include more discussion throughout, and it is difficult to report results accurately with so much discussion. This makes organizational structure more critical for clarity at the same time that clarity is difficult to achieve. With so much discussion, it becomes easy to move into abstraction both in terminology and in conceptualizing. Writing a paper in the social sciences therefore requires very close attention to structure and meaning, and social scientists have to make a special effort to achieve coherence and conciseness.

The strong dependence on words has its roots in the earlier scholarly tradition. Before the development of the scientific method, nature was the purview of the natural philosophers, scholars who were interested in natural phenomena. Similarly, the study of social groups was the domain of scholars interested in social phenomena—social philosophers, often political philosophers. Their writing was general, reflective, learned, philosophical, and accessible to the educated elite generally:

Ex. 3.23

Neighborliness is not restricted to social equals. Voluntary labor (*Bittarbeit*), which has great practical importance, is not only given to the needy but also to the economic powers-that-be, especially at harvest time, when the big landowner needs it most. In return, the helpers expect that he protect their common interests against other powers, and also that he grant surplus land free of charge or for the usual labor assistance—the *precarium* was land for the asking. The helpers trust that he will give them food during a famine and show charity in other ways, which he indeed does since he too is time and again dependent on them. In time this purely customary labor may become the basis of manorial services and thus give rise to patrimonial domination if the lord's power and the indispensability of his protection increase, and if he succeeds in turning custom into a right.

Even though the neighborhood is the typical locus of brotherhood, neighbors do not necessarily maintain "brotherly" relations. On the contrary: Wherever popularly prescribed behavior is vitiated by personal enmity and conflicting interests, hostility tends to be extreme and lasting, exactly because the opponents are aware of their breach of common ethics and seek to justify themselves, and also because the personal relations had been particularly close and frequent.

The neighborhood may amount to an amorphous social action, with fluctuating participation, hence be "open" and intermittent. Firm boundaries tend to arise only when a closed association emerges, and this occurs as a rule when the neighborhood becomes an economic group proper or an economically regulatory group. This may happen for economic reasons, in the typical fashion familiar to us; for example, when pastures and forests become scarce, their use may be regulated in a "co-operative" (*genossenschaftlich*) manner, that means, monopolistically. However, the neighborhood is not necessarily an economic, or a regulatory, group, and where it is, it is so in greatly varying degrees. [Weber, vol. 1, 1978, 362]

Scholars studying natural phenomena early adopted the scientific method, which gradually displaced the earlier general scholarly studies. Scholars studying social phenomena did not adopt empirical methods until later. Moreover, these did not displace studies in the earlier scholarly tradition. In the social sciences, therefore, the two traditions have continued side by side, so that studies in the social sciences vary from scholarly essays to rigorous experiments.

Social scientists continue to draw on the older tradition. They continue to use words from the general language for scientific terms. Furthermore, they may draw on the older tradition in the interpretation of their research and blend ideas derived from the scholarly tradition with theoretical concepts derived from empirical studies. This may lead to a blending of essay-style and reportorial writing. The dual tradition therefore makes it more difficult for social scientists to report on their research, and it subjects them to criticism on three fronts: Peers in the older tradition dismiss the empiricists as reductionists. Natural scientists may criticize them because their research is not as controlled and rigorous as they expect. Nonscientists criticize them for using pretentious language for commonsense notions.

3.7 MULTIPLE STUDIES OR EXPERIMENTS

Scientific papers may report preliminary or preparatory experiments or several studies or experiments that are too closely related to be published separately. Such papers are confusing to follow unless the various experiments are clearly designated and ordered. The studies can be characterized by name, for example, the plasma study, the high-protein, high-fat experiment, and so on. Or they can be designated by numbers or letters, for example, Experiment 1, 2, 3, or Experiment A, B, C. The ordering of the different parts depends on their relation. When they are independent, they are best treated in parallel; when they are interdependent, they are best treated in series.

3.7.1 Independent Studies

In independent studies, each study is treated separately and completely: the introduction, methods, results, and discussion are presented for Experiment A; then the introduction, methods, results, and discussion for Experiment B and similarly for Experiment C, as illustrated in Model I in *Table 3.7.* A general introduction precedes the series and a general discussion follows it.

For example, in a study of a new natural preserve in Africa to determine its present character as a base line for maintaining the preserve and possibly introducing endangered species, three separate, independent studies may be performed: one of the geology and soils, one of the fauna, and one of the flora. The three studies may be conducted in any order or simultaneously; no one study depends on the other. The paper begins with a general introduction. Then the three studies are presented, separately. They may be presented in any order, since they are independent.

TABLE 3.7   Models of scientific papers reporting several experiments, illustrating differences in ordering of parts

I. Independent	II. Interdependent	III. Independent-Related
Introduction	*Introduction*	*Introduction*
Experiment A	*Experiment A*	*Methods*
Introduction	Methods	Common to A, B, C
Methods	Results	Experiment A
Results	Discussion A	Experiment B
Discussion A	+ Introd. to B	Experiment C
Experiment B	*Experiment B*	*Results*
Introduction	Methods	Experiment A
Methods	Results	Experiment B
Results	Discussion B	Experiment C
Discussion B	+ Introd. to C	
Experiment C	*Experiment C*	*Discussion*
Introduction	Methods	Experiment A
Methods	Results	Experiment B
Results	Discussion C	Experiment C
Discussion C		
General discussion	*General discussion*	*General discussion*

However, they may be interrelated even if not interdependent; then the order reflects the interrelation. For example, the plants are likely to be closely related to the types of soil, and the animals to the kinds of plants. If so, the soil study may be presented first, then the study of the flora, and last, the study of the fauna. The general discussion then addresses the interrelation of the three studies and their relation to the overall problem of the preservation, introduction, and reintroduction of species, and so on.

3.7.2 Interdependent Studies

In an interdependent series of studies, each succeeding study depends on the preceding one (Model II, *Table 3.7*). The organization is essentially the same as in Model I, in that each study is completely described, but its order is fixed, and the discussion of each experiment has a double role: (1) to discuss the previous experiment(s) and (2) to provide the background for the next experiment. The general discussion is not an equal treatment of the various experiments, as in Model I, because the discussion after each experiment has been cumulative.

Sometimes an unplanned procedure or analysis is performed as a result of the findings. If it is very brief, it may be included in the results; otherwise it is described in the methods section, prefaced by a statement that the results made the procedure or analysis desirable. In the social sciences, research in which a program is tested or in which each analysis is followed by an analysis suggested by the results of previous analysis follows the interdependent pattern. For example, when the data are analyzed statistically, the result may suggest a different type of analysis, more rigorous or refined analysis, or a different configuration of variables. The paper then proceeds as follows:

Ex. 3.24

analysis 1 $\longrightarrow$ results 1 $\longrightarrow$ discussion of results and analysis 1, type of analysis needed for (2) $\longrightarrow$

analysis 2 $\longrightarrow$ results 2 $\longrightarrow$ discussion of results and analysis 1, 2, type of analysis needed for (3) $\longrightarrow$

analysis 3 $\longrightarrow$ results 3 $\longrightarrow$ . . .

3.7.3 Independent-Related Studies

A series of experiments may be independent experiments but too closely related to be treated as the separate experiments. In this model (III, *Table 3.7),* the methods for the experiments are described separately in the methods section, the results are presented separately in the results section, but there is no separate discussion of the different experiments until after the last experiment. Such discussion is followed by a general discussion. In this type of research, procedures may be varied, while the subject of research remains the same. For example, in a study to determine the effect of temperature, humidity, light cycle, and water stress on growth in plants, the common methods of treating the plants in preparation for the different experiments are described. Then the methods for the temperature, humidity, and water stress experiments are described separately in the methods section. The results for each experiment are presented separately in the results section, and they may be discussed separately in the discussion section. Then the general discussion addresses the interrelationships among the various experiments.

3.7.4 Preliminary Studies

A preliminary study or experiment sometimes precedes the main experiment. Planned, formal preliminary studies are reported; incidental preliminary studies are described if rigorously performed, and if their relation to the research warrants their being reported. Preliminary studies may be preparatory studies necessary to performing the main experiment. They may be exploratory experiments to help make the main experiment more rigorous or to suggest modifications:

Ex. 3.25

> The river stage was raised, because the preliminary experiment indicated that detention time, as determined by dividing a given inflow rate by pond volume, was too low for maximum pond efficiency.

Preliminary studies are sometimes performed as a trial of the method, as in some pilot studies, which are often small-scale trials of the experiment.

The position of a preliminary study depends upon its objective and results. When the preliminary study is important for indicating the direction of the research, it may be made a subsection of the introduction or a separate section between the introduction and methods section. The discussion of this preliminary study focuses on the meaning or importance of the results for the subsequent main study. When the preliminary experiment is decisive for the methods finally adopted in the main study, it may be preceded by a discussion of the research design, which leads to the necessity for a pretrial, or it may be part of the discussion of the choice of method. The discussion of the preliminary experiment evaluates it to determine the method to be followed for the main experiment. When the results of the preliminary study are related to the results of the main study or raise questions for discussion, it may then be discussed briefly, in the discussion section.

3.7.5 Theoretical Model

A theoretical model is often developed as a basis for empirical research. Most commonly the model is developed and then applied. The paper then consists essentially of two parts: one that describes the development of the model and the other that reports on its application (Model A, *Table 3.8*). The introduction addresses the problem and ends with an explanation of the kind of model needed or proposed. In the next section, the theoretical model, which may be quantitative or qualitative, is then developed. This section may then end with a discussion of the possibile applications or limitations of the model.

The next section reports on the particular application of the model. The discussion of the applied study may be a part of the section on the application of the model (*Table 3.8*) or part of the discussion section that follows, which is a general discussion section. Here the results of the application of the model relative to the predictions of the model, that is, its validity, promise, modifications needed, and so on, are discussed.

In some studies, data are collected as a basis for developing a theoretical model. The paper may then have a section reporting on the collection of data (Model B, *Table 3.8*). This section precedes and is directed toward the development of the theoretical model. For example, in a study of browse, data might be collected to relate browse to rainfall and other conditions. These data together with data of other studies may be used to develop a mathematical model for predicting browse under a variety of conditions. The applicability of the model and its limitations are then discussed. Research can then be designed to apply and test the model. The results are then discussed relative to the applicability and predictive power of the model and to possible modifications of it. In this type of study, the development of the model is not simply a preliminary to the empirical research, but part of it.

In disciplines or problems in which experimentation is difficult but prediction desirable, data may already be available for use in developing a theoretical model

TABLE 3.8   Models of scientific papers with theoretical model illustrating different positions of model

A. *Theoretical* Beginning with model	B. *Empirical* Preceded by collection of data	C. *Applied* Based on accumulated data
Introduction	*Introduction*	*Introduction*
Background Problem Criteria for model	Background Problem Need for data for model	Background Problem Types of data available
Model	*Collection of Data*	*Model*
Assumptions, data Development of model Discussion: applicability	(Introduction) Methods Results Discussion relative to model	Collected data Implications of data Development of model Discussion, limitations
Application	*Model*	*Application*
Methods *Results* *Discussion: results*	Discussion if not already discussed Development of model Discussion: applicability, limitations	Methods Results
Discussion	*Applications*	*Discussion*
Results if not discussed earlier Model Application General discussion and modifications needed	Methods Results	Predictive power Model and application Modifications needed
	Discussion	
	Predictive power of model Model and application Modifications needed	

(Model C, *Table 3.8*). In hydrological studies, extensive data have been collected regularly over many years on water table, water-bearing strata, rainfall, river levels, ice flows, silting, and so on. It is not possible to experiment with such unpredictable and enormous natural phenomena; yet it is important to be able to predict underground water reservoirs, the subterranean paths of pollution, possible collapse of dams, and so on. Research draws on the extensive data and on theory to develop a predictive model. The model can then be tested both against past records and future events and conditions. In this type of treatment, the model is developed after assembling and analyzing the available data. It can then be applied to other existing data and to subsequent observations. Its predictive power and application are then discussed.

3.8 STARTING TO WRITE

3.8.1 Process of Writing

Writers think that writing is simply recording one's thoughts, whereas writing is part of the process of formulating and clarifying them. One's thoughts are multifarious and many-sided, and they are embedded in a network of other multifaceted thoughts. In writing, the writer must extricate the thought, together with all its external and subtle internal relations, from this complex network of thoughts and try to express it in words. In doing this, the writer must face the basic difficulty of writing—trying to fix immaterial thoughts in material words, which themselves are only symbols of actual referents.

In preparation for writing, the writer should establish conditions conducive to writing. The ideal conditions are whatever conditions the writer finds congenial. Guidelines for making optimum use of such ideal conditions have not been better stated than by Barzun and Graff (1977, pp. 324–328). Just as the conditions of writing should be adapted to a writer's preferences, so should the method of writing. No one method of writing is adapted to everyone, and no one method will guarantee a well-written paper. Some writers prefer some kind of structure—notes, an outline, or summary; other writers find that these fetter them and prefer to write spontaneously. Whatever one's preferences, in the end, the writing must be tightly structured. Moreover there is no magical way to transform research into a written paper. Writing requires effort and attention.

3.8.2 Methods of Writing

Unstructured and structured.   The least structured form of writing is to write rapidly and spontaneously, without a definite plan or outline, composing as one goes, without stopping to organize ideas logically, develop paragraphs coherently, or structure sentences concisely. This kind of writing is most useful when one gains insight into a process or concept or when the material begins to come together and take shape. Writers can eliminate the barrier of the paper altogether, by dictating. This method calls at least for the amount of structure of an informal discussion with a colleague. The presentation should be structured at least as to sections or subsections. The writing will of course be rather informal, loosely structured, clumsily phrased, wordy, and repetitious. However, focusing on an invisible listener will help to maintain a consistent tone and point of view.

One may write by expanding a smaller piece of writing that is in summary or skeletal form. This establishes the central structure of the paper, which can then be elaborated, while keeping the elements in proper relation and proportion to each other and to the whole. This can start as no more than a series of summaries of the major parts of the paper, which can then be expanded into a minipaper. In expanding such a summary, one may expand the whole summary at each step, filling in with all the necessary details as one follows the summary. Or one may expand it by accretion, expanding now here, now there, as one collects material or develops one's ideas. Proposals can also be expanded into a paper, if they describe specific studies rather than research projects. A proposal has a well-developed introduction, review of literature, and description of the methods and materials. These are not *substitutes* for the sections of the paper, but can serve as rough drafts for them. The results, discussion, and conclusions must be added, and then the whole can be further expanded and developed into a rough draft.

Outline.   The most structured and time-honored method for structuring a piece of writing is to prepare an outline. This usually structures the whole paper, and is customarily prepared before the paper, but the outlines for the various sections may be prepared just before each section is written.

An outline serves writers as a guide to the content and the order of the paper. Preparing it helps writers to collect their material, organize their thinking, and specify the relationship of the parts. Then when they are writing, they can devote their full attention to the writing without having to stop to organize. The outline also serves as a record, which is useful simply to ensure that nothing is omitted that should be included in the paper. It also helps writers to note the data or graphics that will be needed for each section. A well-prepared outline thus reduces the amount of major reorganization and rewriting.

Some writers use an outline as a finished structure and adhere to it closely. This requires that the outline be complete and developed in full detail. It is a difficult type of outline to develop successfully, especially for a whole paper, but it can be used effectively for a section or subsection, and it makes writing easier and expeditious. Others use the outline only as a guideline and modify it as they write. This type of outline is easiest to prepare, but carries the greatest risk of a confusion in development and omission of information or conceptual points. Others view the outline as a working outline. They develop a skeletal structure and use it as a basis for fitting in additions, reorderings, and other changes before each part is written. This is useful in a long paper, for which all the details are not available at the time of writing or for which it is not possible to see how all the details are related. This method is therefore useful in writing concurrently with research.

The outline is not the answer to a writer's prayer, however. Writers who are resistant to structure may not find them useful. Also, an outline can only ensure unity, that is, that each part will be devoted to a particular topic, and a linear ordering of the parts. An outline lists elements and shows their order and hierarchical position, but it does not make clear *how* the elements are related. It cannot ensure the coherence that it is designed to achieve, because it cannot reflect the internal relationships that are not fully conceptualized and structured until the writer actually writes. This gap between outline and written text is part of the difficulty of converting the multidimensional and multirelational structure of one's thoughts to the sequential, linear structure of the outline, which is structurally maladapted to reflecting such structure.

The elements in the outline and their hierarchy may be designated by numerals and letters or by a decimal system. The decimal system is well adapted to theoretical, mathematical, or closely reasoned papers; it should not be used merely as a means of dividing the paper superficially into sections. Presumably the final draft of the outline reflects the proportional treatment of the parts. Actually the space given to a section in the outline need not reflect the emphasis intended. It is easy to overexpand sections for which one has much information, those that one understands well, and those that are easy to develop. And some types of information may take up proportionately more space in the outline than in the paper. A list or a series of elements may take up more space in the outline than a concept, yet in the paper, the explanation of the concept may require as much space as the list or more.

The outline may be developed from a path diagram, or a path diagram may substitute for the outline. This method is particularly useful for a section or paper with complex interrelationships. First, the elements of the paper are laid out spatially in clusters, for example, hypotheses, methods, collection, theoretical structure, findings, et cetera. The clusters are laid out so that arrows can be drawn to show relationships both between clusters and within them. The clusters, the elements in the clusters, and the arrows will probably have to be restructured several times before the relationships displayed match one's conceptual structure of the paper. Separate path diagrams may be developed for different sections. If the path diagram is subtly structured, it may be more effective than the outline as a basis for writing. If it is possible to lay out the path diagram in a tree diagram, the translation to the writing will be even easier.

3.8.3 Writing Concurrently with Research

One method of making writing less onerous is to make it part of the research process and write throughout the course of the research.* Writing concurrently with the research has several advantages. It makes writing an integral part of the research, as it should be. The reporting is likely to be easier, because the writer usually has equipment, material, data, sources at hand and can write from them directly. This method also helps research scientists to clarify their ideas during the research. This clarification feeds back on the research while it is in progress. Such feedback is especially important for graduate students, because it reveals gaps, raises questions, or suggests changes in the research before the research is completed. With this method, by the end of the research, the paper is largely written, at least as a preliminary draft, with only the results and discussion sections left to write. This method also allows time to elapse between drafts so that the writer can return to the paper with a fresh eye. It thus can reduce the number of major revisions, and it certainly improves the writing.

Writers should start to write as soon as they have something to write. The best

---

*This method is similar to one termed "writing in increments" (Michaelson, 1974). Writing concurrently with research differs in that the paper is not necessarily written in an orderly sequence from beginning to end.

place to start is *not* the introduction. First, it is always difficult to begin; second, the introduction has an integrative function and so is not easy to structure. The best place to start is the methods section, because this is least dependent on the other sections for its content or form. As soon as the experiment or data collection is under way, the writer can write this section. It entails only describing the materials and methods, and one can write directly from observation. If procedures are later modified, the preliminary draft can be revised, or annotated for later revision. In a dissertation, if the review of literature is not strongly dependent on the introduction, it may also be written before the introduction. The acknowledgments and biographical sketch can also be written in preliminary form. The abstract should be written after the paper is completed, from a nearly final draft.

3.8.4 Writing for the Reader

In writing a scientific paper, research scientists have two main tasks: (1) to transform their insubstantial and multifaceted thoughts into concrete, linear (i.e., sequential) writing and (2) to make the linear writing transmit their multifaceted thoughts to readers. The first is a familiar task to research scientists, who are constantly putting new ideas and new findings into words. The second task is rarely seen as a separate task. It is assumed that achieving the first accomplishes the second. Yet the difference in readers' interpretations of the same material belies that assumption. The writer must therefore make an effort to translate the thought into words in a form that allows readers to derive the thought exactly as the writer intended it. Furthermore, he or she must make it *easy* for the reader to do this. This last objective in writing is usually called *readability*.

The term is not used here in the usual sense of readability formulas for short familiar words or short simple sentences. Such formulas are not applicable to scientific writing or readers of scientific papers who are well acquainted with most of the scientific terms used in the paper, and with long words and complex structures used in scholarly writing. Readability is used here to mean writing in which readers do not have mentally to bridge gaps, reread to decode, or halt even momentarily to interpret what is written. The writer's objective is immediate comprehension by the reader so that he-she does not have to study the paper.

When research scientists talk to their colleagues about research, their listeners can ask questions when the exposition is confusing. In writing, writers must anticipate how readers will read what they have written and anticipate their questions and confusions. The research scientist and reader both have a conglomeration of concepts and data—thoughts—in their minds, which are very similar, though in different configurations. Both also share the same scientific language. In fact readers have much more information and ideas than are needed. What they lack, however, is the particular conceptual structure that the writer has in mind. Writers must therefore extract from the multitudinous concepts, data, and thoughts in their minds those elements that constitute the conceptual structure that they wish to transmit to readers. They must then incorporate that conceptual structure into language patterns that reconstruct it, so as to elicit from the multitudinous concepts, data, and thoughts in readers' minds, only those details and concepts required for apprehending the conceptual structure that the writer intended.

In developing an argument, the writer must lead readers step by step. Readers expect writers to provide cues or directions; therefore writers must not mislead them, raise false expectations, or confuse them by omitting cues or by supplying implicit or wrong cues. Writers must not surprise them by leaving gaps or changing direction without warning.

Whenever the writer leaves a gap, readers bridging the gap may continue in a different direction. When the writer makes a turn without cueing readers, readers may continue in the original direction and arrive at a different destination, or they may become confused and have to go back to find their direction. Even if they recognize immediately that the writer has not cued them, they must stop to determine the writer's intent. They may choose a different bridge or direction from the writer's. Moreover, readers may not discover that their interpolation or interpretation does not match the writer's. Therefore, leaving readers undirected or misleading them makes for confusion and miscommunication. Such haltings may break readers' train of thought and so distract them or discourage them from reading attentively.

Appendix 3A

Use of Present Tense in Sentences from Introduction and Discussion Sections

a. Statements of Generally Accepted Knowledge Not Requiring Citation of source

1. Many species of insects are associated with fungi.
2. The electron plasma waves produced upon stimulated Raman scattering (SRS) from laser-produced plasmas are known to generate high-energy electrons.
3. Polaris is the Cepheid with the lowest light amplitude.
4. Tse'an Bida Cave (1,460 m elev.) is located in a large canyon where a small perennial spring supports a local riparian plant assemblage.

b. Present Importance, Interest, Usefulness

1. Specifically, Boolean-based algorithms are particularly well suited for the task of holistic comparison.
2. The oxides of nitrogen are of considerable current interest to atmospheric and environmental chemists. . . .

3. Sensitivity analysis is an important numerical tool for the physical investigation and validation of mathematical models.

4. An interesting point of contrast between this and previous work is the selection of molecules from which to derive the pharmacophore.

c. General Comment, Concept, Explanation

1. We can characterize two conceptions of the role of language in the study of theoretical cognitive constructs.

2. Expectations about the effect of alcohol on social behavior are thus present on a cultural level, to be learned by anyone before taking the first drink.

3. The use of simple stencil masks does restrict pattern geometries, since patterns which close on themselves are not permitted.

d. Purpose of Paper

1. This paper considers optimal control problems which arise in the study of aero-associated orbital transfer.

2. The study presented here deals with isolation and further characterization of hamster oncofetal pancreatic antigens.

e. General Statement of Findings

1. Capture efficiency may be affected by aphid, vegetational, environmental, and methodological-machine factors.

2. The results of the present study provide some evidence about the reliabilities and validities that can be achieved with standardized case simulations.

f. Limitations of Study or Method

1. In addition to these limitations, the study is limited in how easily the results can be generalized.

2. The treatment of calcite and siderite as separate minerals rather than a solid solution causes some degree of error, but cannot be overcome at this time.

g. Conclusions or Inferences

1. Of the two net migration models covered by this study, Model III appears to be a better model for rapidly growing areas and Model II for slowly growing or declining areas.

2. Thus it seems that there is, as yet, no consistent explanation for the DeVoe-Brewer data.

4

Introduction and Literature

4.1 INTRODUCTION

The introduction is the beginning of the paper proper. It is a difficult section to begin with in writing the paper. It is recommended, therefore, that the writer begin with a later section—the methods section (see Chapter 5) and then return to this chapter.

4.1.1 Function and Parts

The introduction is the part of the paper that provides readers with the background information for the research reported in the paper. Its purpose is to establish the framework for the research, so that readers can understand how it is related to other research.

The introduction represents *in abstract* the research scientist's thinking about the problem. In writing the introduction, the writer abstracts from his or her past thinking about the problem those details and the logical structuring that represent the final conceptualization of the problem. However the introduction is not a chronicle of the vicissitudes of the research scientist in identifying a research problem and setting it in a framework for research. Moreover, this final conceptualization can only be a refining of the original conceptualization out of which the research developed; it cannot be an ex post facto construction superimposed on the research after it is completed. The writer cannot reconceptualize the problem or rewrite the intro- duction *after* the research to fit the research or findings.* Such a procedure would be a contravention of the systematic character of scientific inquiry and the rigor required in using the scientific method to conduct research.

In the development of the introduction, a general research *problem* is identi- fied, and within this problem a particular research question, or questions, is isolated to address in the research. Out of the broad literature, the material is selected that provides the *background* for these. The answering of the research question consti- tutes the *objective* of the research. In considering the literature and the problem, the research scientist decides how to look at the problem. This establishes the *theoretical framework,* which may be explicit or implicit, and leads to some expectations, that is, *hypotheses,* and to a methodology.

The introduction therefore usually includes at least the following parts,

Ex. 4.1

   a. Derivation and statement of the problem and discussion of the nature of the problem.

   b. Discussion of the background of the problem.

   c. Derivation and statement of the research question or objective(s) of the research.

and may, in addition, include other parts, such as hypotheses, practical importance, outline of the paper, and so on. The immediate focus of the introduction is the objective of the research, that is, the specific question that the research is designed to answer. However, the question arises out of the research problem, which the writer must frame for readers and make meaningful to them.

4.1.2 Research Problem

In structuring the research problem for the reader, the writer is trying to extract from the general background of knowledge on the subject, the problem of interest, that is, the locus of uncertainty, confusion, contradiction, or controversy, or the lacuna in knowledge (*Table 4.1*).

4.1.3 Background and Literature

The background is derived mostly from the literature. In some papers, the develop- ment of a theoretical framework or model is an important part of the background. The background may also include statements of generally accepted knowledge, the importance of the research, and so on.

In framing the problem, the writer must determine how far—both chronologi- cally and conceptually—the introduction must go to provide peer readers with an adequate background for the research. Although in the research scientist's thinking, the problem may have arisen out of the broad literature; for the reader, the problem must arise directly from the immediately pertinent literature. Not all the concepts,

---

*I would like to thank D. B. Zilversmit for drawing this problem to my attention.

TABLE 4.1   Explicit and implicit statements of source of research problems from introduction sections

1. Past research has not examined sufficiently or satisfactorily explained the increase since 1940 in labor force participation of wives.
2. There are two obstacles to such studies: the methodological problem of tracking pollen grains to measure the carryover and the statistical problem of comparable data sets.
3. Unfortunately, we have little comparative research on the ability of local areas to mobilize and deal with problems. . . .
4. . . . According to (*10*) the slow degradation is preceded and facilitated by the chemical interaction of the molten sodium with the solid electrolyte reflected by the coloration of the ceramic in contact with sodium even in the absence of electric current. In the literature (*10–14*) there are some differing opinions concerning the influence of current and/or of the contact with molten sodium or sodium vapor on the processes of degradation and chemical coloration. . . . [J. Sol. St. Chem. 63, 1986, 2]
5. Despite extensive investigations [1]–[4], most techniques used have often been indirect, requiring considerable chemical separation, and impractical for the purposes of hydrochemical exploration for uranium, further understanding of the migration of actinide materials through soil and mineral deposits, and environmental monitoring. [IEEE J. Quant. Electron. QE-22, 1986, 998]

findings, and literature that have presented themselves to the writer are directly pertinent to the problem.

Broad-based introductions such as those shown in *Table 4.2* and *4.3* are superfluous for readers of the journals because they present knowledge or information well known to them. In *Table 4.2a,* the first paragraph is general knowledge for readers. The introduction could have begun more directly with the second paragraph, modified to include reference to chemical pollution. The basic exposition of the central concept of the pharmacophore in *Table 4.2b* seems scarcely necessary for the readers of the specialized journal. In *Table 4.2c* the introduction to a paper on estimating flowing water resources begins by considering estimating, inventorying, and management as evolutionary characteristics "unique to recent man." It is fallacious to descend precipitously from the extraordinarily broad, general universe of discourse of human evolution to the extremely narrow one of estimating flowing water resources of U.S. rivers. In the following example, the first sentence,

Ex. 4.2

Introduction

Chemistry is the science that concerns itself with the structure and properties of pure substances and with the reactions that transform one set of substances into another. In the gas phase, these transformations usually occur in complicated sequences of elementary steps as the consequence of a quasi-least-action principle which states that multiple collisions and/or major bond rearrangements are too improbable and/or too costly in potential energy to provide an efficient reaction path. The elucidation of the sequence of steps, i.e., of the reaction mechanism, has long been and continues to be the subject of much research. In order to characterize the single, elementary steps, it is necessary to measure their rate parameters as directly as possible and to understand their dynamics with the aid of reaction rate theory. That is the subject of this article. [J. Phys. Chem. 88, 1984, 4909]

TABLE 4.2   Introduction with opening (*a, b*) presenting general knowledge familiar to readers
of journal or (*c*) too distant from research problem or question

---

a. *Introduction*

The oxides of nitrogen are of considerable current interest to atmospheric and environmental chemists. They are formed predominantly in the high-temperature or primary combustion zones of both
stationary and mobile power sources and are common atmospheric pollutants. Like hydrocarbons,
oxides of nitrogen contribute to the formation of photochemical smog. Fossil fuel combustion is
the principal source of man-made nitrogen oxides although some chemical processes, such as nitric
acid plants and fertilizers, are responsible for local emissions.
Oxides of nitrogen and nitrate aerosols have been studied extensively[12] because of their importance in
atmospheric chemistry. . . . [J. Phys. Chem. 88, 1984, 5334]

b.

A central concept in medicinal chemistry is that of the pharmacophore, a specific three-dimensional
arrangement of essential chemical groups common to active molecules, that is recognized by a single receptor. Although the concept involves many simplifying assumptions (a single binding mode,
a single set of important interacting groups), it has proven useful in rationalizing pharmacological
data. Much of the effort in modern pharmacology goes into developing and testing hypotheses
about which chemical groups are important for a particular biological activity and what the three-
dimensional arrangement of those groups is in the receptor-bound or "active" conformation of
each molecule. These active conformations may not be the same as conformations in crystals or
global minimum energy conformations either in vacuo or in solution. Since nothing is known about
the atomic-level properties of the receptor associated with most biological activities, one must deduce the pharmacophore from a set of active and inactive ligand molecules, some or all of which
may be conformationally flexible.
The first step in deducing a pharmacophore is to choose the groups essential for activity in each molecule. A group can be an atom, a geometrically defined point within the molecule (e.g., the center
of a phenyl ring), or a "receptor point" with a fixed relationship to specific atoms in the molecule
(e.g., a hydrogen-bond acceptor along a N-H axis in the molecule). For the purpose of discussion
we will assume three such groups A, B, and C. . . . [J. Med. Chem. 29, 1986, 899]

c. *Introduction*

Estimating and inventorying resources so that we know how much we have to manage is a human behavior characteristic that separates us from other organisms. Management is another behavioral
characteristic unique to recent man and shared in only primitive ways by other organisms, such as
nest building by birds and water-level management by beavers. Encyclopedic references provide information, such as length and discharge for selected classes of rivers, e.g., the 100 largest. Statistics
usually are not available for the brooks, streams, runs, and other small waterways that we confront, cross, live next to, and rely on on a daily basis. This paper is the product of an effort to
estimate the flowing water resources of the United States and to provide information on methodology and suggestions as to the use of this information in water quality, fisheries, and other resource
management activities. The system is usable on individual watersheds to account for local variations
in baseline data and thus make the results more accurate for site-specific applications. [Wat. Res.
Bull. 21, 1985, 291]

can serve as an opening sentence to almost *any* paper in chemistry. Most chemists do not need to be told this, and the statement is very distant from the subject of the paper. In fact, the chameleonlike quality of such an introduction serves as a clue that it is starting too far from the focal point of the paper.

Because the function of the introduction is to help to frame the problem for readers, the literature must focus narrowly on the immediate problem:

Ex. 4.3

> Although DDT was banned in the United States in 1972, DDE and DDT[3] continue at high levels in certain parts of the United States (Fleming and O'Shea 1980; Fleming and Cromartie 1981; Clark 1981a; White *et al.* 1983). DDT continues in use in other countries, including wintering areas of birds that breed in the United States. Evidence from the field and from experimental reproductive studies with waterfowl (Haegele and Hudson 1974; Peakall *et al.* 1975; Longcore and Stendell 1977) suggests that residues are lost very slowly, so that regardless of when or where a bird encounters DDE in its travels, it still may have a substantial portion left at the breeding season, a time when small amounts in the body impair reproduction of birds of several species. Controlled studies of loss rates of DDE in wild species, however, are lacking.
>
> Our study was undertaken to (1) extend and broaden the evaluation of lethal residues of DDE with additional species of birds, and (2) determine the loss rate of DDE following high-level exposure. [Arch. Envir. Contam. Toxicol. 13, 1984, 1]

so that only those sources that are needed to outline the problem as logically and concisely as possible are used. Moreover, the writer must select from the cited sources only those details that help to structure the problem and show how the pertinent earlier concepts and findings lead to the question addressed in the present paper. The writer can, however, refer readers to review or key papers, where appropriate, to make a broader view of the problem available to them.

One cannot, however, carry the reduction of references to the literature so far as to appear to stand alone in research on the problem, as in the following very abbreviated introduction:

Ex. 4.4

> This study was initiated to locate precisely the region and the cells that are responsible for the geotropic curvature in primary roots of corn. We estimated cell lengths in the first 4 mm of the epidermis, and one row each of the outer, mid, and inner cortex, and counted the mitotic figures in the above tissues plus the entire cortex, the endodermis, pericycle, and vascular tissues. The correlation of the cell length and cell division data enable us to explain why the curvature response begins at a specific location of the roots. Examinations were at 15-min intervals for 240 min following red light irradiation and geotropic stimulation. [Plant Physiol. 61, 1978, 7]

4.1.4 Research Question and Objective

The research question is the immediate objective that the research addresses. For example, the research problem may be the postnatal effects of intrauterine alcohol on children, but the particular question addressed might be the incidence of anatomical defects, the types of physiological malfunctioning, the kind of mental re-

TABLE 4.3   Excerpt from introduction illustrating identification of research problem
and extraction of research question from the problem

Until recently, little research had been done, both in this country and cross-culturally, on parent sex
differences in the child-rearing process. The focus tended to be solely on mothers in relation to
their children, with little attention paid to the father's role. Research that did include fathers was
generally on children's *perceptions* of parental roles and behavior;. . . .

In the last 10 years, however, widespread changes in family roles and structures . . . have led researchers to focus increasingly on the role of the father in the family, . . . The studies have generally been in two areas. The first has looked at fathers' interactions with infants in the first 2 years
of life, comparing fathers' behaviors with mothers'. . . . the most persistent findings were that
fathers engaged in more active and stimulating interaction than mothers (e.g., Clarke-Stewart, 1978
. . .) and that mothers spent a greater proportion of their time with infants in caretaking functions
(vs. play) than fathers did. . . .

The second area of father-child research has focused on fathers' interactions with preschool-aged children, generally in a structured play, task, or teaching situation. Fathers again were found to be
more active and physical than mothers . . . and to give more functional information and encourage
children's task performance more than mothers (Mazur, 1980). In addition, fathers were again
found to differ more than mothers in their behaviors toward girls and boys. . . .

To understand the different influences fathers and mothers may have on children's social and emotional development, it is important to go beyond the study of parents with infants and preschoolers
and to observe the everyday patterns of parent behavior that emerge in relation to older children
who have entered the complex worlds of school and community. . . . The present article, . . . focuses on mothers' and fathers' interactions with their school-aged children, in the home. It considers whether or not there are substantial differences between fathers' and mothers' behavior toward
children, whether these behaviors correspond to those found in studies of parents in other cultures,
with younger children, and whether parental roles that are traditionally ascribed to the Mexican
culture in fact emerge. . . . [Develop. Psych. 20, 1984, 995-6]

tardation, the presence of hyperactivity, and so on. The development of the research
question can be seen as a successive focusing from the general literature, to the
general problem that forms the background of the research, to the particular research question.

In *Table 4.3* the problem extracted from the literature is the differences between mothers and fathers, particularly fathers, in their interactions with their children. Through a review of the studies with infants and preschoolers, the research
scientist highlights the importance of studying parents interacting with older children, and so arrives at the research question: How do mothers and fathers interact
with their school-aged children? The research question is not usually expressed in
the form of question. It may be expressed as a problem, purpose, proposed solution,
or hypothesis.

The objective, which is commonly called the *purpose* of the research, is to
answer the research question. The term "objective" is used here to avoid the dual
meaning of purpose. As the *what* of research, purpose refers to the particular subject, topic, question that the research scientist plans to address. This purpose (*Ex.
4.5a* and *b*) is here termed the "objective":

Ex. 4.5

  a. The purpose of the research reported here was to determine whether the drug accelerated the metabolism of cholesterol.

  b. The purpose of this research was to determine the molecular size and oligomeric structure of the receptor proteins.

  c. The drug was studied because it had promise of application in treating atherosclerotic conditions.

  d. The purpose in studying the oxides of nitrogen is that they are important in contributing to photo-chemical smog.

As the *why* of research, purpose refers to the reason or justification for the research, as in *Ex. 4.5c and d*. The purpose as objective must be provided in every introduction, but it may be implicitly stated as the subject of the research or the paper or the method (*Table 4.4*); the purpose as reason or justification is provided only if appropriate or pertinent. Logically, the objective, addressing the question arising out of the problem, is the end of the argument in the introduction; however, it may

TABLE 4.4   Research objective (a) stated explicitly or (b) implicitly as subject of paper

a. *Explicit*

1. The question arises as to what role these lipids play in the so-called virus-specific hemolysis and fusion thought to be mediated primarily by viral glycoproteins (3–9).
2. The purpose of the study reported here was to provide evidence on this general question by addressing the following specific ones: . . . .
3. Our objectives in studying *P. ferruginea* were to evaluate its transfer to the Heterobasidiomycetes, to determine if the conidial form was its anomorph, and to characterize the variations between isolates.
4. The purpose of this study was to reduce the dark current of an MSM-PD to a few nanoamperes.
5. It was the purpose of this investigation to assess the significance of the new white and black redistribution patterns taking explicit account of their demographic structures.
6. The purpose of this study was to examine this concept in detail for the case of two-point boundary value problem.

b. *Implicit*

1. In this paper, two fundamental concepts of celestial mechanics are described: instabilities and uncertainties.
2. In this paper we reexamine and clarify the structure of the unemployment-crime relationship.
3. In the present article, the regional distributors of mucus glycoprotein is compared in lamb and sheep stomachs.
4. In this paper, we present exact quantum results using the TDWP method for a collision, where inelastic, reactive, and dissociative scattering occur all simultaneously.
5. In the present study we describe the chain length specificity of the fatty acid oxidizing systems in mitochondria and perioxisomes from the livers of trout fed diets with and without PHFO.
6. In this study the use of a combined fluorescence dye uptake and dye exclusion viability assay permits a direct comparison of the effect of H_2O_2 on promastigotes versus anastigotes.

be stated near the beginning or in the middle, depending on the structure of the introduction.

4.1.5 Other Elements of the Introduction

It is not uncommon for the introduction to begin with the importance, usually the scientific importance, of the problem or of the research:

Ex. 4.6

> a. As is well known, the electrons can degrade laser-fusion target gain; therefore, it is crucially important to gauge the level and spectrum of these plasma oscillations.
> b. Equilibrium data for blood are important because they describe the functional properties of hemo-globin *in vivo* in its physiological environment.
> c. This question is centrally important both for the construction of sociological theories of crime and for the formulation of social policy.

The paper may begin with a general statement that is generally accepted or supported by the literature (*Ex. 4.14, 4.15*). Then the scope of the study may be delimited. In papers in which hypotheses are explicitly formulated, the statement of the question or objective leads to the development of the hypotheses. These may be explicitly labeled and numbered, for example, hypothesis 1, hypothesis 2, and they may be intercalated in the development of the introduction, (*Table 4.5b*) or presented together (*Table 4.5a*). The writer may also point to limitations, emphases, or omissions in the research and account for them. Some writers close the introduction with the results or conclusion, or with a summary of the subsequent parts of the paper (*Table 4.6*).

4.2 WRITING THE INTRODUCTION

In writing the introduction, writers are hampered by an embarrassment of riches and by great latitude in choosing from them. Not only do they have all the possible related literature to choose from, but they also have available the varying approaches as well as the multifarious details of the research in mind, both pertinent and irrelevant. Yet they must reduce these to a concise, tightly structured argument, even though the introduction offers no organizing principle. Being the first section, the introduction has no fixed origin and no inevitable end point. Framing the research question in the most logical and economical manner is therefore the challenge in writing the introduction. Moreover, since writers are likely to write the introduction first, they tend to write expansively and at length. All these conditions predispose the writer to start too far from the subject of the research and make it diffuse, overextended, and ambulatory. It is advisable, therefore, *not* to start by writing the introduction. It is preferable to start by writing a section that is more concrete and

TABLE 4.5   Hypotheses in text of introduction

a. *Grouped*

It has been argued that is is only when dependency relations are not institutionalized that the degree of dependency will predict administrative differentiation. Thus, it is hypothesized:

Hypothesis 1: Dependence on public sources of support will strongly predict the number of administrative offices that manage public-funding relations among private institutions.

Hypothesis 1a: Dependence on public sources of support will not predict the number of public-funding offices among public institutions.

Hypothesis 2: Dependence on private sources of support will strongly predict the number of administrative offices that manage private-funding relations among public institutions.

Hypothesis 2a: Dependence on private sources of support will not predict the number of private-funding offices among private institutions.

If differentiation is linked simply to the management of particular types of dependency relations, there should be no interaction effects of control and dependency as predictors of differentiation. If administrative differentiation is affected by the institutionalization of the relationship, dependence on particular sources of support should show an interactive relationship with control in predicting differentiation. [Admin. Sci. Q. 30, 1985, 4–5]

b. *Intercalated*

We have found that firms that become capital-dependent are more likely to add a financial director to their board, even when the prior existence of financial directors is controlled for. . . . The preceding suggests the following hypothesis:

Hypothesis 1: Broken ties with financial firms are more likely to be reconstituted than ties with nonfinancial firms.

Whether a firm is under family or management control (that is, the type of ownership) may affect the potential of the sending firm to exert power over the receiving firm. Management-controlled firms are generally more vulnerable to external control because their ownership is more easily subject to change; . . . . On the other hand, by controlling the majority of stock or a large minority bloc, family-controlled firms are more likely to have discretion over the board-selection process (Berle and Means, 1968). This suggests that

Hypothesis 2: Management-controlled firms are more likely to reconstitute their broken ties than are family-controlled firms.

Several researchers have argued that a multiple interlock (interlocking firms share two or more directors) is more likely than a single interlock (firms share one director) to indicate control or coordination between two firms . . . however, when other factors were controlled for, the relationship disappeared. We predict that

Hypothesis 3: A broken tie that is part of a multiple interlock is more likely to be reconstituted.

The extent to which an outside director is able to exert influence on a board may depend partly on the size of the board. Other things being equal, the smaller the board, the larger the role played by particular outside directors. Therefore, we expect that

Hypothesis 4: The smaller the firm's board, the more likely the broken tie is to be reconstituted.

Finally, because this study is longitudinal, we can examine. . . . [Reprinted from ''Broken-Tie Reconstitution and the Functions of Interorganizational Interlocks: A Reexamination'' by Stearns & Mizruchi pub. in Admin. Sci. Q. 31(4), 1986, 527 by permission of Administrative Science Quarterly. Copyright © 1986 Cornell Univ.]

TABLE 4.6   Summaries at end of introduction outlining remaining sections of paper

a.

The first part of this paper investigates the relative importance of various institutional forms of problem solving within the total sample and across the different communities. After ascertaining the major patterns, the second section focuses on the social-structural characteristics of communities that generate different techniques. In the later section, we assess various theoretical perspectives on community mobilization. [Amer. Soc. Rev. 49, 1984, 828]

b.

The plan of this paper is as follows. In section 2 we derive the two-fluid equations from our kinetic equations in the low temperature region, obtain explicit expressions for the thermodynamic properties of the superfluid, and show that they satisfy the usual superfluid relations. We also obtain integral equations that when solved give explicit expressions for the transport coefficients. In Section 3 we solve these integral equations approximately for very low temperatures, $na\lambda^2 \gg 1$, by using what would correspond to a first polynomial approximation in classical kinetic theory (for a review see Ref. 5). In Section 4 we consider the moderately low temperature region. Since the calculations are identical in spirit to those of Sections 2 and 3, we merely quote the analogous results. In Section 5 we discuss the results of this paper, compare our results with those of previous workers, and briefly discuss the critical region, near the λ-point, in a dilute superfluid. In the Appendix we show that the approximations made in I are consistent when deriving two-fluid hydrodynamics for a dilute superfluid. [J. Low Temp. Physics 58, 1985, 400]

has a predictable organizing principle, for example, the methods section, which then becomes the end point for the line of development in the introduction.

The introduction should not be too general. It should not be written like the opening of a lecture. Neither should the introduction be a comprehensive development of the problem. The subject of the introduction is not the broad background of the problem but the framework of the particular research for readers who already have a knowledge of the broad background.

4.2.1 Structuring the Introduction

The writer can avoid an overextended and diffuse introduction by focusing on its three essential parts: the research problem, research question or objective, and the background for these. The writer should be able to write down easily the specific question(s) that the research is designed to address.

The writer's task is to structure the development of the problem so that the argument leads logically and inevitably to the question or objective. In doing this, the writer draws upon the literature and provides such other background as is required by readers. This may include accepted general knowledge related to the problem or principles well-known in the discipline or related disciplines.

It is common to start with the background literature to define the problem; from this the question emerges and the objective follows. However the introduction may start with the problem or even begin with the objective or question:

Ex. 4.7

    a. problem → background → question → objective
    b. background → problem → question → objective
    c. question → objective → problem → background

Actually, the background is drawn upon at any point in the introduction, and subsidiary elements, such as methods, hypotheses, the importance of the research, and general knowledge, or explanations are worked into its development as seems functional to the writer:

Ex. 4.8

    a. The electron plasma waves produced upon stimulated Raman scattering (SRS) from laser-produced plasmas are known to produce high-energy electrons. As is well known, these can degrade laser-fusion target gain, and therefore it is critically important to gauge the level and spectrum of these plasma oscillations.
    b. *Escherichia coli* can use nitrate as an effective electron acceptor for anaerobic respiration by reducing it to nitrite.

The lack of a fixed structure allows the writer great freedom in ordering the parts, so that any order that fits the argument can be followed; however, this freedom means that the writer has little guidance. Appendix 4A gives examples showing the variation in the development of the introduction.

If the significance or importance of the problem is included (*Ex. 4.6, 4.18*), its relation to the argument determines its position. If the research is important enough to serve as a major point of interest, the introduction may begin with a statement of this importance.

Some papers include a brief statement of the results or conclusions at the end of the introduction:

Ex. 4.9

    a. The feeding stages of the mycophagous foreign grain beetle were found to exhibit higher chitinase levels than the corresponding stages of the lesser grain borer.
    b. We have shown that these vessels are responsive to vasoactive drugs.

The reason for including the results or conclusions is that they answer the research question when it is raised. The reason for not including the results or conclusions is that they interrupt the logical development of the paper. Since the abstract provides readers with the results and conclusions, there seems little justification for restating them almost immediately afterwards in the introduction, and then a third time in the results and discussion section.

In some papers, the writer closes the introduction by indicating the parts that are to follow (*Table 4.6*). In a paper with the conventional sections, this is unnecessary. In theoretical, analytical or long complex papers, outlining the paper may help to orient the reader (*Table 4.6*). However, if the headings clearly describe the sections, such a preliminary outline of the structure is superfluous.

4.2.2 Choice of Tense

The choice of tense in the introduction depends on whether the writer is discussing the research, the paper, or general scientific knowledge or problems. If the *paper* is being introduced, the present tense is appropriate, since the writer is addressing the reader, with the paper in hand (*Appendix 3A*):

Ex. 4.10

   a. This report *describes* the instability of H_2Ahp in certain conditions and the stabilizing effect of divalent cations.

   b. This paper *examines* the impact of caregiving via an examination of four generic characters of well-being measures. . . .

   c. The main purpose of this article *is* to derive a saddlepoint approximation for the density of f_n of a general statistic V_n.

When the writer is introducing the *research,* the past tense is used:

Ex. 4.11

   a. The objectives of this study *were* to determine differences in reflectance between diseased and healthy watermelon leaves in the photographic and near-infra-red wavelengths.

   b. This study *was designed* to yield descriptive information on reading instruction for upper-elementary mildly handicapped students.

   c. This paper *undertakes* [*describes*] such an investigation. We *develop[ed]* and *test[ed]* hypotheses derived from different theoretical perspectives.

   d. The studies presented here *were carried out* to determine whether $\alpha39$ and $\alpha41$ [*would*] *interact* differently. . . .

The research should not be introduced in the present tense, as in *Ex. 4.11c*. Here the authors are not developing and testing hypotheses as they write. In *Ex. 4.11d* the verb in the subordinate clause should not be in the present tense because it is directed forward to an expected finding in the past.

It is also not appropriate to use the future tense to introduce a report on past research (*Ex. 4.12a*):

Ex. 4.12

   a. The use of high contrast resist processes will permit sharply defined resist patterns with good profiles even with the diffraction limited image shown in Fig. 2(b). However, these high contrast processes have a critical dose which will define where the edges of the resist will be formed. Small variations in the exposure level of a poorly defined image will result in variations in the resulting linewidths. [J. Vac. Sci. Techn. B, 4, 1986, 11]

   b. Theoretically, any variations in amplitude and phase across the face of a disk may be modeled, provided cylindrical symmetry exists. To investigate these effects experimentally, an apparatus designed to study radiated fields of disk transducers to a high resolution has been constructed as will be described in Sec. II. [J. Acoust. Soc. Amer. 80, 1986, 5]

It is preferable not to use the future for reference to the paper itself (*Ex. 4.12b*), since the reader has it in hand at the time of reading.

The introduction may begin directly with a report from the literature, which is frequently in the present perfect tense. The reference may be highly specific:

Ex. 4.13

   a. Aerial deposition of F⁻ *has caused* injuries to vegetation (Bunce, 1984), domestic animals (Shupe et al., 1979), and wildlife (Karstad, 1967).

   b. The problem of excessive adiposity in broilers *has been increased* by *ad libitum* feeding and selection based on growth rate (McCarthy, 1977).

or more general:

Ex. 4.14

   a. Considerable advances in our understanding of the role of motivation *have marked* the last decade (e.g., Atkinson . . .).

   b. Study of the acquisition of gender understanding *has been undertaken* with wide variation in terminology, procedure, and age of children.

   c. Sensitivity analysis *has been* an important numerical tool for the physical investigation and evaluation of mathematical models [1–5].

The literature may be mentioned so generally that it may not actually be cited:

Ex. 4.15

   a. Although the decline in mental function with hypoxia *has been* well *documented* for decades, its pathophysiology is still unknown.

   b. Traditional approaches to the study of absenteeism *have emphasized* the centrality of the work situation.

   c. Simple, highly sensitive and rapid methods of determining trace quantities of uranium *have* long *attracted* the attention of chemists.

Such statements are often followed by more specific statements that cite the literature more specifically. When the literature is reported as part of the development of the argument, the past tense is the appropriate tense to use (*Table 4.7a*). When past research is challenged, the report of the research is in the past, but the challenge may be in the past, present perfect, or even present:

Ex. 4.16

   Green* (1980) *interpreted* this to mean . . . , but DeLacor (1984) *raised* [has raised, or raises] questions that. . . .

When a finding has been widely confirmed or is generally accepted, and can be made without citation, the present tense can be used (*Appendix 3A*):

Ex. 4.17

   a. It is widely accepted that the most important tasks of programming are the design of data structures and of algorithms.

   Green, DeBretor, DeBrion, DeCanet, DeLacor. Fictitious names used in fictitious examples of reference citations. When any of these names appear, the reference is fictitious.

TABLE 4.7   Part of introduction section illustrating discussion of literature as (a) report of background research in the past tense and (b) exposition of background knowledge in the present tense

a. *Report in Past Tense*

Wangensteen *et al.* (1970, 1971), Kutchai and Steen (1971), Lomholt (1976) and Tullett and Board (1976) in studying oxygen permeability of the shell and shell membranes of a variety of bird eggs, concluded that the inner shell membrane was responsible for gas permeability during incubation. A functional change was reported in the inner membrane which resulted in an increase in oxygen diffusion during incubation in turkey eggs, by Tullett and Board (1976) and in chicken eggs by Kayar *et al.* (1981). By the end of incubation, Kayar *et al.* (1981) reported that total oxygen flux resistance for the shell and inner membrane dropped from 88 to 6% in chicken eggs. Tranter *et al.* (1983) reported that a limiting membrane in the chicken egg, visible with a scanning electron microscope, affected oxygen permeability with the occurrence of cracks in the membrane during incubation.

The purpose of this investigation was to determine. . . . [Comp. Biochem. Physiol. 90A, 1988, 421]

b. *Exposition in Present Tense*

1. The flavoenzyme mercuric ion reductase has the unique ability to catalytically reduce Hg(II) to Hg(0): . . . . This function coupled with subsequent volatilization of Hg(0) from the cell constitutes the molecular basis for bacterial resistance to mercuric ions (Summers & Silver, 1978; Robinson & Tuovinen, 1984). Recent structural studies of mercuric reductase (Fox & Walsh, 1982, 1983) and sequencing of the Tn501-encoded structural gene, mer A (Brown et al., 1983), reveal an active site redox-active disulfide and remarkable active site homology with the flavoprotein oxidoreductases, human glutathione reductase, and pig heart lipoamide dehydrogenase (Williams et al., 1982). In particular, the active site disulfide and the intervening four residues are completely conserved (Jones & Williams, 1975; Krohne-Ehrich et al., 1977; Matthews et al., 1974; Brown & Perham, 1972; Burleigh & Williams, 1972). Furthermore, two-electron reduction of all three enzymes results in the formation of an active site dithiolate that is capable of complexing mercuric ions (Massey & Williams, 1965; Casola & Massey, 1966). Mercuric ion reductase, however, is the only enzyme with the capability to catalytically reduce bound Hg(II) to Hg(0). [Biochem. 24, 1985, 6840]

2. The increased preference for city residences among whites has been tied to shifts in family formation patterns, particularly evident among individuals born after 1945, which have resulted in dramatic increases in the formation of single, childless, and two-earner households (Masnick and Bane, 1980; Cherlin, 1981). Such households, it is held, will give greater preference to central residential locations than did their counterparts of the previous generation, who were more heavily engaged in childrearing (Alonso, 1980; Long, 1980). The greater suburban movement of blacks can be linked to rising black social status, some elimination of. . . . This opening of suburban residential opportunities also tends to be most evident for blacks born since 1945 (Pettigrew, 1980). [Amer. Soc. Rev. 49, 1984, 804]

b. The primary source of moisture for Colorado cyclones is air which enters the continent from the Gulf of Mexico.

c. Antisocial aggressive behavior, like many other behaviors, appears to be overdetermined.

The introduction frequently states the importance, significance, or interest of the research or subject (*Appendix 3A*). Such statements constitute present commentary or evaluation and so are in the present tense:

Ex. 4.18

   a. The notion that there might be a relationship between superhelicity and gene expression and that
      the secondary structure of DNA in the promoter region of genes might be involved in their tran-
      scription *is* currently *drawing* increasing attention.

   b. A topic of active concern in the study of confined, current-carrying plasmas *is* the nature of large-
      scale bursts of descriptive magnetohydrodynamic activity to which the plasmas are susceptible.

When the introduction is treated as a state-of-knowledge lecture presentation,
it may assume an essay form and be written in the present tense, as shown in *Table
4.7b,* with the references cited as support for statements made in the present tense.
This kind of treatment is not consonant with rigorous scientific writing. It confers
the status of universal truth on the results of research and converts the research
scientist from an observer or investigator to a commentator or authority.

4.3 LITERATURE AND REFERENCES

4.3.1 Search of Literature

The actual search of the literature is part of the research process. The discussion
here addresses the treatment of sources *after* they have been located and read. The
search should precede the research, in order to:

Ex. 4.19

   a. Avoid repeating research that has already been performed.
   b. Identify and structure the problem and set it within the broader framework of earlier research.
   c. Develop insights that might help in developing hypotheses, methods, et cetera.
   d. Avoid misinterpretation of the results.
   e. Avoid omitting references that peer readers consider pertinent.

Therefore, time spent on reviewing the literature thoroughly before beginning the
research is usually time saved in research.

The literature is not reported in its entirety in the paper. In most scientific
papers, any review of the literature is incorporated in the introduction or the theoretical framework, and it is usually restricted to those sources that pertain directly
to the problem. When the pertinent literature is extensive or complex, it may be
desirable to write a general review in a separate section, but even this must center
on the problem that the research addresses. A review of literature is still included
in most dissertations.

4.3.2 Sources to Cite

Research scientists should cite the source for anything that is not original with them,
if it contributed to the research or the paper. Such sources are usually printed
sources from the published literature, but they may be manuscripts, typescripts,

TABLE 4.8 Different styles for citation of references in text

1. In recent years, optoelectronic integrated circuits (OEICs) have been studied by many workers [1]–[4].

2. Both natural and synthetic compounds are known to induce resistance against virus infection [*1, 4*]. . . .

3. The full set of all such quantities can be used to address a wide variety of questions about the mathematical modeling of physical processes [1–10].

4. Previous light microscopic axoplasmic transport studies in the rat brain have demonstrated that the dorsal striatum projects onto the ventral region of the substantia nigra pars reticulata (SNr)[18,35] which gives origin to the nigrotectal projection.[6,7,17,41]

5. Naismith (4) and Ledermann and Rosso (5) have shown that fat deposition accounts for some of the additional increase in weight in the rat during pregnancy.

6. Ethnographic research has shown that classroom learning is reflexive and interactive and that language in the classroom draws unevenly from the sociolinguistic experience of children at home (Bernstein 1975, 1982; Cook-Gumperez 1973; Heath 1982, 1983; Labor 1972; Diaz, Moll, and Mehan 1986; Mehan and Griffin 1980).

7. Recent detailed correlations of Pi magnetic pulsations with auroral light and riometer absorption fluctuations not only have confirmed that Pi is related to particle precipitation, but that it is produced by local ionospheric currents generated by the particle precipitation [*Arnoldy et al.,* 1982; *Engebretson et al.,* 1983; *Oguti et al.,* 1984; *Oguti and Hayashi,* 1984].

8. In the case of composites the departures from Fickean diffusion were attributed, at least partly, to damage in the form of debunchings at the fiber/matrix interfaces by ASHBEE and WYATT (1969), HEDRICK and WHITESIDE (1977), DEIASI, WHITESIDE and WOLTER (1978), SHIRRELL *et al.* (1979), CROSSMAN, MAURI and WARREN (1979), LEUNG, DYNES and KAELBLE (1979), ANTOON and KOENIG (1981), DRZAL, RICH and KOENIG (1985), and JACKSON and WEITSMAN (1985).

9. . . . 80 taxa are recorded from the rhizoplane, cauloplane, seeds, moribund plants and dead material (Conners, **1967**; Fries, **1944**; Gessner and Lamore, **1978**; Grove, 1913; Jones, **1976**; Klecka and Vukolov, **1937**, . . .

letters, or even conversations. The contributions may include statistical data, experimental measurements, equations, concepts, theories, suggestions, and graphic material such as tables and graphs. Sources are cited (1) to provide the bibliographic information on the sources and (2) to give credit to the contributors.

The assigning of credit is necessary for ethical, legal, and scientific reasons. Ethically, the use of someone else's writing without citing the source presents it as one's own and constitutes plagiarism. If the material borrowed is copyrighted, using it without acknowledgment exposes the borrower to legal action. Scientifically, not citing the source deprives scientists of credit for the originality of their research, which has a high value. Citing their research is therefore a recognition of the value of their scientific contributions.

In selecting references that should be cited, research scientists cannot include all the many papers they have read on the topic. They can cite only papers they have read and only those directly pertinent to the research. Even so, the writer may have numerous references that can be cited for a particular statement in the text. Then some criteria must be established for reducing the number of references, to avoid a series of citations that may be a paragraph long (*Ex. 4.20a, Table 4.8.8*):

Ex. 4.20

a. The enormous and increasing power of banks and banking systems is well documented in the literature (Allen 1978; Baum and Stiles 1965; Bearden 1982; Bunting 1976; Burns 1974; Fitch and Oppenheimer 1970; Glasberg 1981a, 1981b; Gogel 1977; Gogel and Koenig 1981; Knowles 1972; Koenig 1979; Kotz 1978, 1980; Levine 1972; Magdoff 1967; Mariolis 1975; Mintz and Schwartz 1981a, 1981b, 1983a, 1983b; Mizruchi 1982; Mokken and Stokman 1978; Pelton 1970; Ratcliff 1980; Ratcliff et al. 1979a, 1979b; Rifkin and Barber 1978; Schotland 1977; Scott 1979; Sonquist and Koenig 1978; Whitt 1974; Zeitlin 1976). The major function of commercial banks, their raison d'être, is to insure credit to borrowers. [Rur. Soc. 52, 1987, 187]

b. Various factors such as oxygen[83,85,110,189,227,238,243-245] atmospheric contaminents[126,138,139,162,177,180,246] and the presence of moisture,[83,122,124,164,178,189,222,243,244,262-265] influence the degradation in varying degrees. [Polym. Degrad. Stab. 17, 1987, 134]

or a long series of reference numbers. (*Ex. 4.20b,* review article). A citation of more than four references should be justifiable.

When the literature that might be cited is extensive, it is preferable to indicate that many have worked on the problem and then cite only some of the pertinent papers, than to try to make the citation comprehensive. The papers cited can be chosen for their importance, comprehensiveness, or variety, for example, key review articles, classical or seminal papers, earliest papers, representative papers, and so on. Of the references then remaining, some less important references may be deleted, since presumably they are indirectly cited in the review articles, and some may be cited as specific references.

4.3.3 Accuracy of Bibliographic Information

Because sources are cited to help readers retrieve the references, it is essential that the bibliographic information supplied be *absolutely* accurate. It is impossible to overemphasize the need for accuracy in bibliographic citations. Moreover, writers must assume the full responsibility for the accuracy and completeness of their references. Editors cannot be expected to recognize bibliographic errors or omissions.

A complete and scrupulously accurate citation of references is not only a courtesy to other research scientists but a saving of a great deal of human time and energy. When a reference is cited incorrectly, it may waste one to several hours of a scientist's time. Multiplied by the various research scientists interested in a paper internationally and into the future, a great deal of time might be spent searching for *one* paper.

Every effort should be made, therefore, (1) to record the bibliographic data *absolutely* accurately, (2) to copy it as few times as possible, and (3) to copy it, if necessary, as scrupulously accurately as possible. The record should be complete. When a source is first recorded, it should include more information than might ultimately be needed to identify the source unambiguously when cited: full names as printed, with diacritical marks and accents, every fact of publication *exactly* as it is printed, and so on.

Research scientists should record all bibliographic information from the reference book, *in hand*. Sources should not be cited from bibliographic citations. To facilitate subsequent reexamination or verification of the sources, the record should include the library call number. It is worth making a photocopy of the title page. Writers should avoid having to record bibliographic information from notes. If this is unavoidable, one should proofread the bibliographic notes *backward* as well as forward, both in comparing the notes to the original source and to the record. Using a computer program for bibliographic work saves time and effort and ensures accuracy—if the bibliographic data have been entered accurately.

The bibliographic information collected should be complete and for a journal should include at least

Ex. 4.21

a. author	d. volume number	g. full pagination
b. title of the article	e. number or month of issue	h. series (if there is a series)
c. name of the journal	f. date	

and any other facts of publication. For a book, the bibliographic information should include at least the author(s), title, publisher, and place and date of publication. The writer must also consider, if applicable, the subtitle, number of the (later) edition, volume number, and name of the editor or translator. Depending on the book, it may include other information: the book may be a government document, a book in a series, the report of an agency, committee, or institution. The notes should also contain information such as the name of the chairperson or the committee or agency issuing the book, the report number or series, and other special details that might be needed to identify the publication. If the author is not named on the title page, the name may be found in the preface or foreword, or the library catalog. For book reviews, letters to the editor, chapters in a book, notes in a journal, or pamphlets, the writer must note all information that might be needed for retrieval. In sum the record must have any information that might possibly be needed for the final citation in absolutely accurate form.

4.3.4 Citing References in the Text

In most scientific journals references are commonly cited in a conventional abbreviated form in the text, and the complete citation is listed in the reference section at the end of the paper. Every reference cited in the text must be represented in the list of references, and every reference in the reference section must be cited in the text.

For scientific research, the reference cited must be the original or primary source. Rarely, if the primary source is unavailable, as in an obscure or inaccessible periodical or a difficult foreign language, the writer may cite a secondary source. The original source is then cited as being *in* the secondary source (*Ex. 4.29a*). If the work is written in a language that is not commonly known, it may be available in translation; then the translator is included in the citation of the full reference (*Ex. 4.29c*). Journals in languages other than English often include an abstract in English or a common European language (*Ex. 4.30d*). Similarly, conference papers may be published as abstracts in proceedings and may not be later published in full. If the writer is citing an abstract, this should be stated in the full citation (*Ex. 4.30d*). Some sources are reprinted, and the writer may have used the reprint; then both dates must be cited (*Ex. 4.29c–e*).

Numerical system.   Journals follow one of two systems for citing references in the text: the author-year system and the numerical system. In the numerical system, the reference in the text is cited by a number, which corresponds to the number of the reference in the reference section. The references in the text may be numbered in order of first mention. Then the first reference mentioned for the first time is number 1, the fifth reference mentioned for the first time is number 5, even though it may be the eighth reference cited, for example, *1, 2, 3,* 1, *4,* 2, 1, *5,* and so on. The full citations are arranged in the same order and numbered correspondingly.

However, the references in the reference section may be alphabetized and then numbered; then the numbers used for citing references in the text will be in no particular order. In this system, references in tables can be treated like references in the text, but these numbers cannot be used for substantive footnotes to the table; symbols or letters must be used instead. When the references are numbered in the order mentioned in the text, references in a table are cited in full in the table, because their position in the final printing cannot be predicted.

The position and type of numbers used for reference in the text are determined by the style of the journal (*Table 4.8*). The numbers may be placed at the end of the sentence, immediately following the name of the author cited, or immediately following the data or idea referred to. They may be free or enclosed in parentheses or brackets; they may be in reduced type or in type of the same size as the text; they may be in roman, italic, or boldface type; and they may be set on the same line as the text or as superscripts. They are spaced apart in the text line, but not spaced apart if superscripts.

Author-year system.   In the author-year system, the author's surname is used, together with the year in which the pertinent reference was published (*Ex. 4.22*). The author(s) name(s) may be used indirectly and is then parenthetical (*Ex. 4.22b*) or directly as part of the sentence (*Ex. 4.22a*):

Ex. 4.22

   a. Green (1985) found that 4-month-old infants chose more often to listen to human voices.
   b. These findings suggest that they may be particularly well matched to the young infant's perceptual and attentional capabilities (Green, 1985).

Whenever a source is cited again, the name and date are repeated, as in *Ex. 4.22b*. Forms derived from footnote form such as Green (ibid.) or (Green, op. cit.) states that. . . . are not used.

A comma often separates the date from the author, e.g., (Green, 1985), but in many journals this is omitted. A series of references is usually separated by semicolons, e.g., (Green, 1984; DeCanet & DeBrion, 1986; DeLacor, 1986; DeLacor & Green, 1985). For joint authors, journal style may require the ampersand, as in the preceding examples, or and.

If the citations refer to papers by the same author, they are cited, usually chronologically by year, without repeating the name, as (Green, 1983, 1984) or if directly, Green (1983, 1984) showed that. . . . Papers by the same author, published in the same year, are alphabetized in the reference section and cited alphabetically, e.g., (Green, 1985a, b). If the citations refer to different authors, they are cited chronologically or in whatever order is most appropriate. The writer must order such citations intentionally, because an arbitrary order will be read as meaningful.

When reference is made to the author (DeCanet) of a chapter in a work, edited by another person (Green), the reference in the text is to the author of the chapter, that is, (DeCanet, 1987, ch.3), *not* to the editor of the book, e.g., (Green, 1987, ch. 3). The full citation indicates this, as DeCanet, . . . *in* Green, 1987 (see *Ex. 4.29a*). When the reference is to a later reprint or translation of a source, the latest date may be cited in the text, that is, (Weber, 1947), but both must be cited in the reference section (e.g., *Ex. 4.29c–e*). If a paper is not yet published, the date can be replaced by "in press". Many journals will not accept "submitted for publication," or even "accepted for publication;" if the paper is not in press, it must be treated like unpublished material.

The citation in the text may include reference to a page of a book or long article (DeBretor, 1984, 627) or a chapter (DeCanet, 1987, ch. 3), or other parts: equation (DeBrion, 1986, eq. [3.5]), section (DeLacor, 1985, sec. 5.2.3), or note (Green, 1977, review).

When a paper is written by several authors, journal style varies in the number of names that are permitted in the text citation, usually only two or three, (Green and DeLacor, 1983) or (Green, DeBrion, and DeCanet, 1987). When the number of authors exceeds the number permitted, the citation is reduced to the name of the first author followed by "et al." in roman or italics, (Green et al., 1982). Only the *alii* (others) is abbreviated and followed by a period; *et* (and) is not. The abbreviation should not be used for *any* combination of authors beginning with the same author. For example, the papers by Green, DeBrion, and DeBretor (1985) and by Green, DeCanet, and DeLacor (1986) should be cited as in *Ex. 4.23* or *4.24a,* not simply as Green et al.:

Ex. 4.23

> Green and her coworkers (Green, DeBrion, and DeBretor, 1985; Green, DeCanet, and DeLacor, 1986), however, concluded that the liver tryptophan pyrolase converted L-tryptophan to . . .

When there are several references beginning with the same author, and the authors are numerous, one may use the first two authors, if these distinguish the different references (*Ex. 4.24a*), or annotate the text citation (*Ex. 4.24b*):

Ex. 4.24

    a. Green and her coworkers (Green, DeBrion, et al., 1985; Green, DeCanet, et al. 1986). . . .

    b. Green and her coworkers (Green et al., Conversion to L-tryptophan, 1985; L-kynurenine to D-tryptophan, 1986). . . .

Authors cited in the text directly are cited by their surnames, without their titles. When authors are referred to directly, they can subsequently be referred to by a pronoun (*Ex. 4.25a*):

Ex. 4.25

    a. This was described as a bicubic spline surface by Green and DeCanet (1978). *They* stated. . . .

    b. This was originally described (Green and DeCanet, 1984) as a bicubic spline surface. *They* stated that the method for obtaining this surface is liable to produce spurious numerical anomalies.

When authors are cited only indirectly, that is, in parentheses, they should not be referred to by a pronoun subsequently, as in *Ex. 4.25b*; the parentheses set the names outside the structure of the sentence.

    In referring to a source at length, the writer must make the relation of the source to a passage clear. When the greater part of a paragraph refers to earlier work, it is clearer to start the paragraph with a direct reference to the authors:

Ex. 4.26

    Green (1985) and DeBretor (1984) examined strains of *Streptococcus alba* in plastids by electron microscopy and found that they contained . . .

The subject may be reinforced by subsequent references such as "they", "their method," and by citing them again at the end of the paragraph if appropriate. When the parenthetical citation is at the end of the sentence, it usually precedes the period (*Ex. 4.22b*). In the author-year system, references in tables are treated like other references in the text.

    If the source has no author, the citation in the text is usually derived from the full citation in the list of references (*Ex. 4.28*). Where such titles are long, they may be abbreviated. For unpublished material, many editors prefer to avoid the citation "(private communication)"; a footnote is more informative. If a source has not been read, it should not be cited. If a secondary source has been read, for example, a reference to an interpretation by DeBretor in an obscure journal, discussed by Green, whose publication is available, it is cited as "(DeBretor *in* Green, 1985)".

Author-year versus numerical system.   Although the numerical system saves space and cost, the author-year system promotes accuracy. With the author-year system it is easy to insert or delete citations, because changes do not require changes in other references. More important, the author and year may recall the subject of the paper to the writer and reader, whereas a number conveys no information. Also, one is less likely to make an error in writing an author-year citation than a number, and the mistake is more detectable and so correctable.

In the numerical system, whenever a reference number is changed, every number affected by the revision must be changed in the reference section and in the text citations. Such changes are bound to increase the chances of error. Even a careless inversion or transposal of numbers may result in an undetectable error. It is also easy to confuse numerical references with similar numbers in the text (*Ex. 4.32*). Writers should therefore use the author-year form of citation in writing the paper, even if the numerical system is used in the journal chosen. The writer may even annotate the citation, for example, "(Green, 1984, rat)" and "(Green, 1986, rabbit)" to call the paper to mind. Then when the manuscript is ready to be typed for submission to a journal, he or she can number the references. This means that they will be numbered only *once,* and in their *final* form.

4.3.5 List of References

The references cited in the text are cited in full at the end of the body of the paper in a list of references titled, "References," "References Cited," "Literature," or "Literature Cited," depending upon the journal, although the section may be untitled. It may even be titled "Bibliography," but this includes any reference on the subject of the bibliography, as defined by the compiler. The references in a bibliography need not be cited in the paper and the writer need not have read them. The function of a bibliography is to provide readers with a comprehensive or representative list of references on a subject. The function of the list of references is to provide readers with the full reference only to sources cited in the paper. Occasionally, especially in some review papers, a scientific paper may have a bibliography in addition to the reference section or instead of it.

The reference section should include every publication cited in the text of the paper and *only* those actually cited. It should not include papers that are directly pertinent to the research, but which the writer did not cite or did not read, or papers that are generally related to the subject but not *directly* pertinent to the research reported.

Variations among journals.   There are official and semiofficial manuals covering both broad and narrow disciplines that recommend a style for citing references in journals in the discipline: *ACS Style Guide* of the American Chemical Society, the *CBE Style Manual* of the Council of Biology Editors, and the *Publication Manual* of the American Psychological Association. However, many of the journals in a discipline do not uniformly follow the style set forth in such manuals. The style for citing common types of references is usually described in the notice to contributors. For less common types, the writer must consult a manual in the discipline or a more general manual, for example, the *Chicago Manual of Style.*

Style of references.   In examining citations to determine the style, the writer must be prepared to focus on minutiae. He or she may find it helpful to develop a style sheet for the reference section.

The writer must first determine the format for the heading—its title, position, type face, capitalization; then the overall arrangement, whether the entries are alphabetical, numbered, whether the numbers are in parentheses, brackets, indented, et cetera. The numbering may correspond to the alphabetical order of all the references cited or to the order in which the references are first mentioned in the text. In the first instance, the numbers in the text for citing the references will not be in sequence; in the second, the authors' names in the list will not be in any recognizable order. After the first reference to an author, the name may be represented by a dash for subsequent references. In alphabetizing a list of authors, one needs to pay particular attention to names that have prefixes, for example, *de, la, van, von,* or *Mc,* which may be separate or solid, and to non-English names. The pattern of indentation should be noted:

Ex. 4.27

   a. Bottke, W. (1982). Isolation and properties of vitellogenic ferritin from snails. *J. Cell. Sci.* **58,** 225–240. [paragraph]

   b. Newby, H. (1982), Rural Sociology and Its Relevance to the Agricultural Economist: A Review, *Journal of Agricultural Economics,* XXXIII, 125–165. [flush]

   c. Campbell, W. H., Rapid auroral luminosity fluctuations and geomagnetic field pulsations, *J. Geophys. Res.,* 75, 6182, 1970. [hanging indentation]

In determining the style for the individual entries, the writer should first note the elements that are included and their order; then each element can be examined in turn for details of punctuation, capitalization, et cetera. The different styles followed by various journals, together with a fictional paper cited in the style of each journal, are illustrated in *Appendix 4B.* The authorship can be examined for the treatment of the author's first name and surname, initials, and their order and arrangement together with style of capitalization and punctuation. With joint authors, the authors may all be treated in the same way, or the first author is cited with surname first and the remaining authors are cited in normal order. In some journals all authors of a paper are listed; others limit the number of names and use "et al." for authors in excess of the number permitted. Some journals use only initials for first names, even though the full name is used in the original publication.

When a publication has no author specified, the writer has various alternatives. Well-known reference books can be cited by their title, for example, *CBE Style Manual;* well-known series of government publications by government agencies can be referenced under the agency, for example, U.S. Dept. of Labor; and committee or group reports may be listed under the name of the group:

Ex. 4.28

   a. FDIC [Federal Deposit Insurance Corporation]. 1982. Bank Operating Statistics, Washington, DC: FDIC.

   b. National Center for Health Statistics. 1983. Current Population Report, Series P-20, No. 395.

   c. U.S. Bureau of the Census. 1983. County and City Data Book. Washington, DC: Government Printing Office.

   d. ISO 140. Measurement of the sound insulation in buildings and of building elements. July 1978.

e. ILO: Equal opportunities and equal treatment for men and women in employment. Report VII International Labor Conference, 71st. Session, Geneva, 1985.

Other references without known author are listed by title, not as Anon. (i.e., Anonymous) and are alphabetized by the first substantive word.

Editors and translators are cited, but their position relative to the title varies with the style of the journal. If the source cited is included in another source, the author of the chapter and its title are listed in the reference section, followed by the editor and title of the book, together with the necessary bibliographic information (*Ex. 4.29a*). The editor, translator, title, and other bibliographic details of the book are part of the reference cited:

Ex. 4.29

a. Birks, H.J.B., 1981. The use of pollen analysis in the reconstruction of past climates: a review. In T. M. L. Wigley, M. J. Ingram, and G. Farmer (editors), *Climate and History,* Cambridge: Cambridge University Press, pp. 111–138.

b. Fireman Bruce and William Gamson. 1979. "Utilitarian logic in the resource mobilization perspective." Pp 8–14, Mayer Zald and John McCarthy (eds.), *The Dynamics of Social Movements.* Cambridge, Mass.: Winthrop.

c. Weber, Max. 1947. *The Theory of Social and Economic Organization.* Glencoe, Ill.: The Free Press. A. M. Henderson and Talcott Parsons (trans.). Wirtschaft und Gesellschaft. Tubingen: Mohr. 1922.

d. Durkheim, Emile. 1951. *Suicide: A Study of Sociology.* New York: Free Press. (First published in 1897.)

e. Halbwachs, Maurice. 1978. *The Causes of Suicide.* London: Routledge and Kegan Paul (first published in 1930).

If the source is reprinted or translated, this is indicated in the reference cited (*Ex. 4.29c–e*). If the type of publication is not apparent from the form or content of the citation, this should be stated:

Ex. 4.30

a. Koenig, T. 1929. Social networks and role of big business. PhD. dissertation, University of California.

b. Lee, John R. 1977. "Agricultural finance situations and issues." Paper presented at the 1978 Food and Agriculture Outlook Conference. U.S. Department of Agriculture, Washington DC.

c. Tracy, Harry. 1984. "The subtreasury plan." Appendix in James ("Cyclone") Davis, *A Political Revelation.* Independently published in Dallas, Texas.

d. Westell, R. A., R. L. Quaas, and L. D. Van Vleck. 1984. Genetic groups in an animal model. *J. Anim. Sci.* **59** (Suppl. 1): 175 (Abstr.)

Italics and quotation marks are variously used to distinguish book titles from chapter titles, titles of papers, and titles of journals. The title of the reference should be the original title; in languages that are not alphabetical or that use a different alphabet, it may be transliterated or it may be translated. The name of the journal is cited in the original language and is usually abbreviated. Style manuals in a discipline often list the standard or suggested abbreviations for journals in the discipline.

If a manual is not published for the discipline, the reference sections in earlier issues of the journal can be consulted. If the journal to be cited is an obscure one, the writer can consult the publication of the International Organization for Standardization (1985) or one of the various abstract or bibliographical indexes that list standard abbreviations for journals. These indexes are interdisciplinary so that they supply abbreviations for most of the periodicals that a writer might need. They also provide bibliographical guidelines, which are helpful in abbreviating journals that are not listed in the index. The writer should be especially careful about abbreviating the names of non-English journals, because the abbreviation might not distinguish between journals in different languages that have similar abbreviations but different names.

The position of the date varies; it may be near the beginning of the entry sometimes conspicuously positioned, or it may be near the end. The volume number of the journal is often distinguished by the designation *vol.*, *v.*, or typographically by being set in italics or boldface. The inclusion of the number of the issue varies with the style of the journal and with the need to cite it to ensure identifying the issue. Finally, the pagination varies as to whether it is so labeled and whether only the first page or the full pagination is given, and if so, whether the numerals are abbreviated.

After determining the style, the writer can prepare each entry, following the style as accurately and consistently as possible. This task is greatly facilitated if a computer program for bibliographic work is used. The references should then be edited with an eagle eye, verifying each period, comma, each space or indentation, the particular type, et cetera.

4.3.6 Footnotes

In most scientific journals, footnotes are used for notes to the text, not for citing references to scientific papers. In journals in which references are cited by number in order of mention in the text, footnotes are used to cite references in the tables. Generally however, footnotes are discouraged because of the added cost; in fact, in some journals, footnotes are allowed only in tables. Footnotes should therefore be avoided except when they are indispensable.

Footnotes to the text are used for comments, data, or other information that is not directly pertinent to the development of the argument but is important for readers to know. Footnotes are often used with the titles of papers to give information about the article as a whole, for example, to acknowledge grants (*Table 4.9*), or with authors' names, to give the temporary or permanent affiliation, and other information.

Substantive footnotes function essentially as asides, to complement or supplement the text that is footnoted; they are not substitutes for omissions in the text and should not be used for afterthoughts. The footnote may be explanatory, evaluative, or elaborative. For example, it may supply supporting evidence for a statement

TABLE 4.9 Different uses and forms of footnotes

1. [2] We are indebted to Dr. B. Kawohl for his suggestion concerning this method and the reference.
2. [†] Extra pure grade, Wako Chemical Co., Tokyo, Japan.
3. [1] The abbreviations used are: MDCK cells, Madin-Darby canine kidney cells; BHK cells, baby hamster kidney cells.
4. [*] Present address: Code 4040, US Naval Research Laboratory, Washington, D.C. 20375.
5. [1] The concept status group has undergone considerable modification since it was first developed by Weber (. . .). The term is here used to refer to culturally delineated collectivities (e.g., males and females).
6. [*] Corresponding author.
7. [†] Reference to trade names do not imply government endorsement of commercial products.
8. [1] Manuscript received 20 July 1986; accepted 20 May 1987.
9. [2] By acceptance of this article, the publisher or recipient acknowledges the U.S. Government's right to retain a nonexclusive royalty-free license and to any copyright covering the article.
10. [1] The research in this study was conducted under a grant from the RANN Program (Research Applied to National Needs) of the National Science Foundation (G1-32990). Opinions and conclusions expressed in this paper are those of the authors and do not necessarily reflect the view of NSF.
11. Primarily as a display of frugality, we have elected not to include the photographs of the gels illustrating these points. These points were made available to the reviewers, and on request copies will be sent to interested readers. [included in text in results section, but appropriate for footnote]
12. [2] Portions of this paper (including "Experimental Procedures," Table I, and Figs. 1 and 2) are presented in miniprint at the end of this paper. Miniprint is easily read with the aid of a standard magnifying glass. Full size photocopies are available from the Journal of Biological Chemistry, 9650 Rockville Pike, Bethesda, MD 20814. Request Document No. 86 M-4522, cite the authors, and include a check or money order for $2.80 per set of photocopies. Full size photocopies are also included in the microfilm edition of the Journal that is available from Waverly Press. [J. Biol. Chem. 262, 1987, 8483]

in the text or an explanation or justification of the modified statistical procedure. The informational notes may provide more specific or detailed information, such as actual statistics, equations, calculations, definitions of terms or symbols:

Ex. 4.31

a. [18]A trivial algebraic manipulation shows that a positive impact of changes in unemployment levels is equivalent to a negative impact of unemployment lagged one year. Specifically, let Y_t denote C_t or $\Delta^2 C_1$ or their log-transformed equivalents. Then if $\beta_1 < 0$ and $\beta_2 > 0$,

$$Y_t = a + \beta_1 U_t + \beta_2 \Delta U_t + \epsilon_t$$
$$= a + \beta_1 U_t + \beta_2 (U_t - U_{t-1}) + \epsilon_t$$
$$= a + \beta_1 U_t + \beta_2 U_t - \beta_2 U_{t-1} + \epsilon_t$$
$$= a + (\beta_1 + \beta_2) U_t - \beta_2 U_{t-1} + \epsilon_t$$

as claimed. [Amer. Soc. Rev. 50, 1985, 324]

b. **6**

Analyses of variance also were conducted in which the interaction effects of organizational size (total FTE) and organizational control (public-private) were examined. Past literature has suggested that these might be important moderators of main effects in colleges and universities. However, no significant interaction effects result in any of the 14 analyses; hence, these variables are not discussed further in this study. [Administ. Sci. Q. 32, 1987, 233]

A footnote may be used to name the supplier or to acknowledge special strains or equipment supplied to the research scientist. It may be used to define a term as it is used in the paper. In sum, anything that does not fit into the argument or that would intrude on it, but that readers should know, can be included as a footnote. Because they are asides, the writer should make them as brief as possible.

Footnotes to the text are numbered sequentially throughout the paper; in dissertations they are numbered by chapter, that is, 3.1, 3.2, 3.3, et cetera. They are placed as a superscript after the word or end of the sentence to which they refer, usually outside the terminal punctuation. When a footnote to a number is used, the writer must make sure that the number of the footnote is not mistaken for an exponent, especially in mathematical material:

Ex. 4.32

    a. Alloy scattering deformation potential ΔE_L $(EV)^3$. [*text*]

    b. [3]This is obtained from the calculations of energy band alignment $\Delta E_c/x$ [13], combined with data from Ref. [17] and [16]. [footnote]

In *Ex. 4.32a,* the superscript 3, which appears to be an exponent, actually refers to the footnote in *Ex. 4.32b.* It would be preferable to find a *word* in the text to which the footnote can be attached.

When the paper is prepared for publication (see Chapter 9), footnotes are usually typed separately from the text under the heading ''Footnotes'' or ''Notes'' at the end of the paper. For some journals, they are typed in the text, on the line immediately following the line in the text that has the footnote number, and they are set off from the text proper by a line above and below the footnote.

4.3.7 Quotations

Direct quotations are more common in papers in the social sciences than in the natural sciences. This is due partly to the older tradition of scholarly writing in the social sciences and partly to the nature of social science research. Social scientists quote a passage because it is the original *statement* of the idea addressed in the development of the argument, and quoting the passage is more accurate than paraphrasing it. In the natural sciences, the form of the statement is not as much a part of the structure of knowledge. There is, therefore, much less danger of misrepresenting earlier research scientists and so less need for quoting their exact words than in the social sciences. The concepts, findings, conclusions, et cetera can simply be cited, or at most paraphrased.

Quotations are used for various purposes. They may be used to quote well-known or accepted authority or simply to quote earlier research that has a bearing on the present research. The quotation may be limited to *terms* used by earlier writers. Actual speech is always quoted, for example, the questions of interviews and the responses of participants. However, quotations should not be used when the *actual words* are not essential to transmitting the message, as in the following example:

Ex. 4.33

CONCLUSIONS

Bachrach (1983*a,* p. 859) has noted that "the process of becoming a biological parent has been extensively documented . . . since the 1930s (but) other forms of parenthood (such as adoption) have received less attention." Our study was conducted to fill this void. . . .

Bachrach (1983a, p. 863) reported that "the percentage of ever-married women 30 to 40 years of age who had never borne a child increased from 8.3% in 1970 to 13.1% in 1979." [Soc. Biol. 33, 1986, 246]

This report on the extensive literature or the results hardly merits direct quotation. The form of quotations varies with their length and their use in the text, and follows the style of quotations in general writing. A comprehensive treatment can be found in the *Chicago Manual of Style* and other manuals.

In quoting a source directly, the writer must be accurate, both in quoting the exact words and in conveying the original author's (or speaker's) intent. The objective is no less than absolute accuracy. A direct quotation must be an exact quotation, *word for word.* If a word in the quotation is likely to be taken to be a mistake in quoting, the writer can add "[sic]" following the word, but this should not be used simply to correct an error or display the quoting author's punctiliousness. Quotations should therefore be treated like references when they are collected.

Certain *minor* changes are allowed to fit the quotation into the new text. Quotation marks in the original passage are retained within a block quotation, but they are changed to single quotation marks within a quotation enclosed in double quotations. The capitalization of the initial words may be changed. The final punctuation and the punctuation where material is omitted in quotation may also be changed, as well as footnote superscripts; however, the author's parentheses are retained. Therefore, any parenthetical notes by the quoting writer are enclosed in brackets. These may be used to supply an omission or make a correction or comment, but they must be very brief. Original typographical designations are also retained; therefore any italics are considered to be the emphases of the original author. If the quoting writer would like to italicize words in the quotation, a note must be added in brackets, such as, "[italics added]," "[emphasis added]." When parts of the original text are omitted, this must be indicated; ellipsis dots are used to replace the missing text—three within a sentence, four where the text omitted includes one or more sentences or the beginning or end of a sentence.

The need for accuracy applies to implicit meaning as well as explicit statement. The use of quotations must accurately reflect the writer's intent. The quoted author's words should not be used out of context to imply something that he or she did not intend to say.

4.4 SOCIAL SCIENCE PAPERS

The introduction of a paper in the social sciences often requires a more extended development of the argument than the introduction in natural science papers. It would be almost impossible to write an introduction as short as that in *Ex. 4.4* (also

Appendix 4A.10) in the social sciences. It has more discussion of the background and problem, and usually includes the development of the theoretical framework.

4.4.1 Theoretical Framework: Hypotheses and Propositions

In the natural sciences, the theoretical framework for studying a phenomenon is so well established and so much a part of the structure of the discipline that it does not have to be presented explicitly. A biologist might look at lactation within a neurological, biochemical, behavioral, or ecological framework without developing it explicitly. Social scientists must make their theoretical frameworks explicit. They may elect to study adolescent gangs in a theoretical framework derived from developmental, small group, or social alienation theory, but must set the problem explicitly within the framework chosen. If the development of the theoretical framework is simple and brief, it may be part of the introduction; if it requires much development or discussion it may be made a separate section with its own heading, for example, "Theoretical Background," "Theoretical Framework," "Theory." The theoretical framework leads from the development of the research question to the development of an appropriate methodology.

The theoretical framework permits the formulation of particular hypotheses. The great value of hypotheses in scientific research is their usefulness in predicting and guiding research and their susceptibility to testing (see Chapter 1). In the social sciences, hypotheses are explicitly stated and usually so labeled (*Table 4.5*); in the natural sciences, they are not. Hypotheses are not always stated in a predictive form in the social sciences; writers may not distinguish rigorously between the explanatory and predictive character of hypotheses (*Ex. 4.34b, c, e*) and the declarative form of propositions (*Ex. 4.34 a, d*):

Ex. 4.34

    a. The centralization of a state's educational system lowers the degree of administrative complexity of school district organizations.

    b. The more centralized a state's educational system, the lower will the degree of administrative complexity be in the school district organizations.

    c. If a state's educational system is centralized, the school district organizations will show correspondingly less administrative complexity.

    d. Organizations that rely on firm-specific skills and highly trained workers fragment jobs.

    e. If an organization relies on firm-specific skills and highly trained workers, it is more likely to fragment jobs than organizations that do not.

In *Ex. 4.34 a, d,* the statements are assertions or propositions. Nothing in their form indicates a prediction. In *Ex. 4.34 c, e,* the sentences in the conditional are stated as clearly predictive hypotheses, and they state the conditions for the predicted consequences. In *Ex. 4.34b,* the future tense gives the statement a predictive character. An hypothesis cannot logically be stated as a general declarative statement when it is predicting what the research will demonstrate! In an hypothesis derived from a theoretical framework, the predictive meaning is central. The function of the hypothesis is to specify, on the basis of the theoretical framework, the results *expected* under particular conditions. The results of testing the hypothesis can support, not *prove,* the hypothesis (*Chapter 1*).

A proposition is not predictive, and it is not part of scientific inquiry. It is an assertion, which is taken to be true or to be proved true. Its "proof" need only be rational or admissible. Some of the "proof" can come from empirical scientific studies, but the problem is framed and the result is stated *before* the proposition is proved to be true. The research question or objective is often addressed by propositions, which are so labelled, even though the study is an empirical study. The propositions then function implicitly as hypotheses, and they may even be explicitly labeled as hypotheses; however, they lack the rigor of hypotheses stated in explicitly predictive form.

Hypotheses should be expressed in a conditional or predictive form. If stated as propositions, they should at least be in the conditional or future tense (*Ex. 4.35*):

Ex. 4.35

> Feeding rats a normal diet without vitamin E should [would, will] result in deleterious effects on the animals.

If several hypotheses are stated, they should be enumerated so that reference can be made to them subsequently by number (*Table 4.5*).

4.5 DISSERTATION

Because a dissertation is more comprehensive than a paper, it must demonstrate a good understanding of the research problem and a thorough knowledge and understanding of the literature. The student must be able to make order of the multitude of details of the literature and demonstrate a grasp of the many elements and their interrelationships.

Graduate students must understand the position of their limited research question as part of the larger research problem and the theoretical framework in which it is embedded. Their competence as future research scientists will be based partly on an understanding of the theoretical structure of the discipline. Indeed one of the most important functions of the introduction and literature review in a dissertation is to present a clear analysis of the problem and a thorough elaboration of the theoretical framework, so as to demonstrate the student's thorough understanding of the theoretical structure of the problem. Both the introduction and the review of literature are therefore more comprehensive than they are in a scientific paper. In fact, they are usually separate chapters, and the added details are not simply a packing in of more details, but a thorough structuring of the network of relations in the problem.

4.5.1 Introduction

The objective of the introduction in the dissertation is to set the problem into a broad perspective and to develop the perspective fully down to the narrow perspective of the actual research. The introduction reaches farther back from the problem and explores more of its implications than in a scientific paper. It includes more details on the problem and its background, and these are developed more fully. The underpinning for the research question is therefore a comprehensive rather than skeletal discussion of the problem and its background. Any discussion relative to the theoretical background or analysis is also more ample and detailed. The background for each hypothesis may be developed more formally in detailed logical steps, and the research objective may be derived from an extended discussion. The scope of the research, and its importance or application may be given more space than in a paper.

4.5.2 Review of Literature

In the dissertation, the review of the literature is a comprehensive and critical review. Faculty advisors expect the student to have a broad acquaintance with the literature and to be able to evaluate it critically and put it in a meaningful perspective. For the advisors, this evaluation and interpretation is a measure of the student's broad understanding of the principles and theoretical structure of the discipline and insight into the particular problem in relation to the broader problems in the discipline.

Historical versus topical review.   There are two general methods of reviewing the literature—historically or topically. In the *historical* or chronological method, the review of the pertinent literature starts with an early publication—presumably a publication that has a central point of reference for the research. The studies following the initial study are then reviewed, more or less in the order of their publication, primarily as they relate to the present research broadly conceived. As the writer progresses from reference to reference, he or she must evaluate the contribution of the new research to the development of a knowledge and understanding of the problem. The results of this process should be a description of the historical development of the research problem and how this development led to the particular question studied and the research reported in the paper. Although the historical review is more or less chronological, the dates of the papers are rarely important in themselves. One does not usually start with "In 1985, Green found that. . . . " because the dates are not usually in themselves significant.

The *topical* method of reviewing the literature is to take the research problem—as a problem—and then draw on the literature to develop an integrated view of it. This method is really a presentation of the *structure* of the problem at the present stage of knowledge. In the topical review, the writer starts with some problem, question, or point that is central and basic to the research problem. This is not necessarily associated with the earliest papers nor with any particular period in the research. The writer focuses on the central problem and selects papers from the literature to develop the conceptual structure of the problem. Because progress on a problem is rarely linear and because the literature is drawn upon as necessary to construct the framework of the problem, the papers discussed will not necessarily be discussed in their chronological order.

In the historical review, the significance or importance of the different research studies is related to the development of the problem over time; the review constructs a historical structure of the problem. In the topical method, the chronological development is subordinate and even irrelevant to the structure of the problem. The review is in the nature of a critical examination of the state of knowledge at a particular time, after the fact. It is not a summary, but a state-of-the-art synthesis of all the pertinent research on the problem at that time. Where the historical review is cumulative and the perspectives at successive stages are more or less additive and linear, the topical review is abstractive, so that the perspectives may be very different at different stages in the development of the problem. One cannot simply add to a topical review at a later time to make it current.

The pitfall in writing a historical review of literature is that it will become a series of loosely related summaries or a series of brief abstracts, and that one will be tempted to digress to interesting but irrelevant parts of the studies reported. The pitfall in the topical review is that one will fail to integrate the literature logically and coherently, and so will not achieve a closely structured synthesis.

Although the review of literature should be a critical review, "critical" here is neutral, meaning "discriminating" and "judging carefully," not fault-finding. One does not stop being a scientist in assuming the role of critic. The writer needs to read the literature thoughtfully judging carefully the effectiveness and contribution of the various research studies to elucidating the research problem. Such a reading precludes evaluative labels and focuses on making substantive specific, descriptive, or analytical comments about the research.

Extent of literature review.   Where a great deal of research has been conducted on a problem, a review of the literature may be an extensive treatment. But the criterion for an effective review of literature is not its length or its comprehensiveness but the creative insight into the earlier research and a synthesis of the literature into a conceptualization of the problem that illuminates it.

The collecting and integrating of the literature may serve as an especially useful function in the theoretical development of the problem. If little research has been conducted on the problem directly or the literature is widely dispersed, the review of literature is an opportunity to bring the dispersed references together and organize them into a meaningful conceptual structure. For example, in a problem of infant nutrition and growth, a research scientist found that the biological concomitants had been intensively and widely studied; however the biological model did not account for the epidemiological findings. The discrepancy raised the question of whether social-cultural factors might help to explain the epidemiological findings. However, the social-cultural influences on infant nutrition and growth had received little attention in the literature. Such influences had been hinted at or only briefly mentioned in papers on the biological factors, and only indirectly addressed in sociocultural studies; the problem was not considered central to either discipline. By drawing together the fragments of social-cultural data from the widely disparate sources, the research scientist was able to bring the available literature together into a conceptual structure of the social-cultural determinants of infant growth, which addressed the problem of the inadequacies in the biological model. This example illustrates how the review of literature allows one to develop insight into a problem, even when the literature is not readily or immediately seen as addressing it.

Tense.   The tense appropriate to reporting literature is the past tense. A review of the literature in the present is not a review of empirical literature but an essay on the research problem or a summary of present knowledge on the problem. Tense is more likely to be a pitfall in the topical review because the writer tends to take the results of the various studies as givens rather than simply as findings.

Appendix 4A

Order and Development of Introduction

The parts of the introduction may be variously ordered, depending on their importance and relation to one another. In the following examples, the various parts are labeled to show their order and the structure of the introductions; some of the examples have been abbreviated and details omitted where they did not seem necessary. The characterization of the part is enclosed in brackets, and it applies to all of the material between it and the preceding bracket.

1. Species of *Balantidium* have been reported from at least six species of freshwater fishes (1, 2, 6–8) and one salmonid (*Salvelinus fontinalis* Mitch.) also from a freshwater environment (4). In each, the *Balantidium* is an intestinal symbiont and might be a parasite (7, 8). [*literature*] Reported here is the occurrence of *Balantidium prionurium* n. sp. from the intestinal lumen of the herbivorous saltwater surgeonfish, *Prionurus punctatus*. [*findings*] [J. Protozool, 32, 1985, 587]

2. Since 1975, ion transport across dog tracheal epithelium has been studied extensively by using tissue sheets mounted in Ussing chambers (20). [*literature*] In such a preparation, the basolateral membranes of the cells are separated from the bathing medium by a layer of connective tissue 500–1,000 μm thick. [*past method*] The size of this collagenous dead space makes it impractical to study the exchange of ions across the basolateral side of the tissue. [*problem*] To overcome this problem, we decided to develop an isolated cell preparation in which both apical and basolateral

membranes would be in direct contact with the bathing medium. [*objective, new method*] Such a preparation should allow us to study how Na and Cl enter the cells from the serosal side of the tissue. [*advantage of new method*]

This paper describes a method for obtaining dispersed isolated cells, the viability of which is assessed by a number of methods. [*subject of paper*]

Clearly isolated cell preparations offer several advantages, and our preparation will, we hope, prove suitable for a variety of studies on the biochemistry of the tracheal mucosa. [*importance*] [Amer. J. Physiol. 241, 1981, C184]

3. **1. Introduction**

In this paper we deal with the linear approximation of the nonsymmetric Kaluza Klein theory (Kalinowski 1982, 1983a, b). [*subject of paper*] We consider separately the electromagnetic (five-dimensional) and nonabelian ($(n + 4)$-dimensional) cases. We find electromagnetic and Yang-Mills field Lagrangians in the nonsymmetric Kaluza-Klein theory up to the second order of approximation with respect to. . . . [*method*] The interesting point for the electromagnetic case is that to the first order of approximation with respect to $h_{\mu\nu}$ there is no coupling between $h_{[\mu\nu]}$ and the electromagnetic field. [*scientific interest*] This is the same as in the nonsymmetric theory of gravitation (see Moffat 1982). [*literature*] In the nonabelian case this is in general not true. The skewon field couples to the Yang-Mills (gauge) field in the first order of approximation and this term is proportional to the dimensionless constant μ. This constant in the nonabelian-nonsymmetric Kaluza-Klein theory is connected to the cosmological constant (Kalinowski 1983a). [*discussion*] Thus we get an interesting connection between the cosmological constant and a coupling constant between skewon and Yang-Mills fields. [*conclusion*]

The paper is organised as follows. In the second section we describe some elements. . . . [*structure of paper*] [Class. Quant. Grav. 1, 1984, 157].

4. **INTRODUCTION**

Much current practical interest exists concerning the diffraction of elastic waves by a random distribution of elastic inclusions in a homogeneous matrix. [*scientific interest*] Determination of static elastic properties of particulate and fibrous composites by measuring effective ultrasonic wave speeds proves very effective.[1-4] Several theoretical studies[5-9] show that for long wavelengths one can calculate the effective wave speeds of plane-longitudinal and plane-shear waves moving through such a composite material. [*literature*] One makes two simplifying assumptions: (1) the wavelength is long compared with the dimensions of individual inclusions and (2) the Lax[10] quasi-crystalline approximation. [*method*]

In the present study, we consider the propagation of plane waves through a composite containing a "sea" of a random homogeneous distribution of prolate ellipsoidal inclusions and "islands" of another phase represented as aligned oblate ellipsoidal inclusions. This was done to model the properties of the composite shown in Fig. 1. [*problem, objective*] The aligned oblate-spheroid inclusions induce a wave-speed anisotropy. [*comment*] Aligned inclusions were considered by Chow[11] . . . and by Willis[9]. . . . A multiple-scattering approach similar to that used in Refs. 5-8 was used in the present study. Earlier, Twersky[12-16] used a multiple-scattering approach . . . [*literature, method*] The present study is unique in considering a nonhomogeneous particle-reinforced composite as a random distribution of aligned ellipsoids in a matrix containing a random distribution of nonaligned particles. [*unique method*] [Acoust. Soc. Amer. 79, 1986, 239]

5.   **1. Introduction**

The extended capabilities of electron-microscopic imaging and electron diffraction methods available today require a re-thinking of some of the methods of theoretical interpretation. [*general problem*] Particular attention has to be paid to usually adopted approximations and to the limits of their validity. [*specific objective*]

One of the very common approximations neglects the usually small difference between the symmetrical and the general Laue cases. [*deficiency in earlier research*] This means that one neglects the fact that the inclination of an external or internal crystal boundary affects the boundary conditions of the electron waves. . . . Hence it affects the excitation errors which enter the fundamental (eigenvalue) equation and modifies its solutions, *i.e.* the eigenvalues and eigenvectors (amplitude ratios) of the individual Bloch waves, and the correlated extinction lengths. [*discussion of deficiency*] The consequences of inclining a surface to the fine structure of diffracted beams were treated in early papers, *e.g.* von Laue (1948), . . . As argued more recently the sensitivity to boundary inclination increases if diffraction arises from regions of stronger local curvature or of larger inclination of the dispersion surface (Sheinin & Jap, 1979). [*literature*] Related situations occur, for example, on working with weak beams or with many beams (high-resolution electron-microscopic imaging, convergent-beam electron diffraction). In such cases, modifications due to reflections from non-zero-order Laue zones have to be taken into account by suitable approximations, too. [*discussion of problem*]

Various attempts have been made to treat the eigenvalue equation of the Bloch waves in these respects. Niehrs & Wagner (1955) first showed that in the general Laue case, also, the fundamental equation can be formulated as an eigenvalue equation and that it can be transformed to a Hermitian form of the dynamical matrix. This work and further studies (Whelan & Hirsch, 1957; . . .) confirm that the influence of an inclined boundary is usually very small. [*discussion of literature*] However, owing to unsuitable choices of angles or of coordinate systems, conveniently small terms to account for such influences have not been obtained. [*deficiency of earlier methods*]

The present paper shows how the influence of boundary inclination as well as the influence of reflections from non-zero-order Laue zones can be described by convenient small correction terms. [*research objective*] They will be treated as perturbations, which method is more clearly justified than in the work of Gjønnes & Gjønnes (1985). Non-centrosymmetry is included. [*method vs. earlier method*] [Act. Crystal. A43, 1987, 684]

6.   **I. Introduction**

Optical surveys of Galactic supernova remnants, such as those by van den Bergh . . . are intrinsically limited by the low quantum efficiency of photographic plates. Furthermore image rotation and telescope flexure place an effective limit of about three hours on individual exposures. [*limitations of earlier method*] As a result of these limitations it has so far been possible to observe only 40 (van den Bergh 1983) optical supernova remnants in the Galaxy. [*gap due to limitations*] The presently known optical remnants are distributed more-or-less uniformly in Galactic longitude and are typically located at distances of a few kpe. The effects of interstellar extinction are particularly severe in the blue and green regions of the spectrum—where photographic plates are most efficient. [*general knowledge*] As a result it has not been possible to observe the majority of known radio supernova remnants, most of which are located in the general direction of the Galactic center, at optical wavelengths. [*problem*] [Publ. Astron. Soc. Pacific 98, 1986, 448]

7.   The theory and application of latent-variable structural equation models and their relevance to developmental issues have become well established. [*established theory*] Reviews of the field by, . . . texts by authors such as. . . . [*literature*] In many of these treatments, the issue of sample size is invariably raised but typically not treated in sufficient detail to provide useful information to users of these methods. [*problem*] This paper will attempt to provide some guidelines and alternatives regarding sample-size and goodness-of-fit issues in latent-variable structural equation models. [*objective*] The developmental researcher is probably most acutely aware of sample-size problems since models of human development are often complex and involve many variables. [*comment on problem*] However, the number of subjects available to test such models is often small or, at least, small relative to the size and complexity of the assumed model for the data. [*problem*] For example, Bornstein and Benasich (1986) recently tested a latent-variable model of habituation in infants using 35 subjects. [*literature*]

The question of how many subjects are needed before estimating and testing a latent-variable structural equation model has plagued researchers who are forewarned about the necessity of large samples for appropriate statistical conclusions. [*problem*] While the statisticians can find solace in asymptotic statistical theory . . . the developmental researcher using these methods is often left wondering about the relevance of such theory for finite samples. [*discussion of problem*] This article will review results from previous work that has attempted to address the sample-size and goodness-of-fit issues and will introduce a new method of estimation especially designed for small samples. [*objective*] [Child Devel. 58, 1987, 134]

8. **1. Introduction**

The isomerization via intramolecular hydrogen atom migration forms an important class of alkyl and alkenyl radical reactions [1,2] and can be quite often encountered in the postulation of mechanisms of the processes involving radical intermediates. In our previous paper [3] the bond-energy–bond-order (BEBO) method of Johnston and Parr [4] was extended for the activation energy calculations of intraradical 1,2-hydrogen shifts. [*literature on process and method*] The main features of the extended method are the resolution of the triplet interaction of non-collinear atomic orbitals into σ and π contributions and the consideration of bond strain in the three-membered cyclic transition state. [*characterization of earlier method*]

The aim of this paper is to present a further extension of the BEBO method for the calculation of activation energies of 1,3- and 1,4-hydrogen shift reactions and to make an estimation of activation energies of 1,5-hydrogen migration. [*objectives*] The calculated activation energies are compared with the experimental results for a set of 11 reactions listed in table 1. [*method*] [Chem. Phys. 117, 1987, 219]

9. Although there have been various reports of multiple sclerosis (MS) in children (Low and Carter 1956, . . .), the onset of this disease occurs most frequently during adulthood and its manifestation during childhood is rare. [*literature*] For this reason the occurrence of MS in children often is not even considered, which may mean that the correct diagnosis is not made for young patients with a relapsing neurological disease, although the clinical condition may be indicative. [*problem*]

A diagnosis of MS basically depends on clinical criteria, *i.e.* neurological signs suggesting multiple lesions of the nervous system and a relapsing course (Poser *et al.* 1983). Findings from laboratory investigations, such as delayed latencies of evoked responses and the presence of oligoclonal antibodies in the cerebrospinal fluid (CSF), may lend diagnostic support to clinically suspected MS. Moreover, the detection of cerebral plaques by means of computer tomography (CT) is of particular importance and is considered to be of pathognomonic value (Bye *et al.* 1985). [*discussion of past methods*] However, there is no specific test to confirm this diagnosis. [*problem*] Since the introduction of magnetic resonance imaging (MRI), it has been expected that this method would increase diagnostic certainty in cases of suspected MS. Previous reports on MRI findings for adults suffering from MS confirm this expectation (Gebarski *et al.* 1985) . . . [*discussion and literature*]

In the present paper we report and contrast MRI and CT findings for three children who were diagnosed as definite MS cases according to the criteria of Poser *et al.* (1983) and who were treated in our hospital between 1980 and 1985. [*objective, methods*] [Dev. Med. Child. Neur. 29, 1987, 586]

10.   INTRODUCTION

Analysis methods for estimating age-specific survival rates from the ringing of young birds have generally been based on a single underlying model, or special case of this model. [*standard method*] Our objective is to review this model that we will call the life table model, its assumptions, approaches to estimation of the model's unknown parameters and problems in making inference concerning age-specific survival rates. [*objectives*] Our work was motivated, to some extent, by the recent paper by Lakhani & Newton (1983). [*literature*] [J. Animal Ecol. 54, 1985, 89]

Appendix 4B

Citations of Sample Articles from Different Journals to Show Variations and Corresponding Citations of Fictitious Article to Highlight Style Differences

Citations from Sample Journals	Citations of Fictitious Article

The Journal of Astronautical Sciences

[11] CORLESS, M. and LEITMANN, G. "Adaptive Control of Systems Containing Uncertain Functions and Unknown Functions with Uncertain Bounds," *Journal of Optimization Theory and Application,* Vol. 41, September 1983, pp. 155–168.

[13] GREEN, M. A., and DEBRETOR, E. "Present Tense and Scientific Rigor in Reporting Research," *Journal of Scientific Inquiry,* Vol. 23, March 1988, pp. 545–550.

Applied Acoustics

6. P. G. Craven and B. M. Gibbs, Sound transmission and mode coupling at junctions of thin plates, part I, part II, *Journal of Sound & Vibration,* **77** (1981), 417–35.

13. M. A. Green and E. DeBroter, Present tense and scientific rigor in reporting research, *Journal of Scientific Inquiry,* **23** (1981), 545–50.

Journal of Non-Equilibrium Thermodynamics

[12] Eber, N., Janossy, I., An experiment on thermo-mechanical coupling in cholesterics. Mol. Cryst. Liq. Cryst., 72 (1982), 233.

[13] Green, M. A., DeBretor, E., Present tense and scientific rigor in reporting research. J. Sci. Inq., 23 (1988), 545.

Journal of Nutrition

13. Heth, D. A. & Hoekstra, W. G. (1965) Zinc-65 absorption and turnover in rats. I. A. procedure to determine zinc-65 absorption and the antagonistic effect of calcium in a practical diet. J. Nutr. *85,* 367–374.

13. Green, M. A., & DeBretor, E. (1988) Present tense and scientific rigor in reporting research. J. Sci. Inq. *23,* 545–550.

Metabolism

14. Irsigler K, Kritz H: Long-term continuous intravenous insulin therapy with a portable insulin dosage-regulating apparatus. Diabetes 28:196–203, 1979.

13. Green, M. A., DeBretor E: Present tense and scientific rigor in reporting research. J Scien Inq 23:545–550, 1988

The New Phytologist

COUGHTREY, P. J. & MARTIN, M. H. (1977). Cadmium tolerance of *Holcus lanatus* from a site contaminated by aerial fallout. *New Phytologist,* **79,** 273–280.

GREEN, M. A. & DEBRETOR, E. (1988). Present tense and scientific rigor in reporting research. *Journal of Scientific Inquiry,* **23,** 545–550.

Rural Sociology

Flora, Jan L., and John M. Stitz
  1985 "Ethnicity, persistence, and capitalization of agriculture in the Great Plains during the settlement period: wheat production and risk avoidance." Rural Sociology 50 (3):341–60.

Green, M. A., and E. DeBretor
  1988 "Present tense and scientific rigor in reporting research." Journal of Scientific Inquiry 23(3): 545–50.

Behaviour

CHRISTENSON, T. E. & LEBOEUF, B.J. (1978). Aggression in the female northern elephant seal, *Mirounga angustirostris.*—Behaviour 64, p. 158–172.

GREEN, M. A. & DEBRETOR, E. (1988). Present tense and scientific rigor in reporting research.—J. Sci. Inq. 23, p. 545–550.

Social Biology

FRIEDLANDER, S., and M. SILVER. 1967. A quantitative study of the determinants of fertility behavior. Demography **4**(1):30–70.

GREEN, M. A., and E. DEBRETOR. 1988. Present tense and scientific rigor in reporting research. J. Sci. Inq. **23**(3):545–550.

Journal of Mammology

TIMM, R. M. and R. D. FRICE 1980. The taxonomy of *Geomydoecus* (Mallophaga: Trichodectidae) from the *Geomys bursarius* complex (Rodentia: Geomyidae). J. Med. Entomol., 17:126–145.

GREEN, M. A. and E. DEBRETOR. 1988. Present tense and scientific rigor in reporting research. J. Sci. Inq., 23:545–550.

Journal of Sound and Vibration

32. C. S. KIM and S. M. DICKINSON 1987 *Journal of Sound and Vibration* **114**, 129–142. The flexural vibration of line supported rectangular plate systems.

13. M. A. GREEN and E. DEBRETOR 1988 *Journal of Scientific Inquiry* 23, 540–550. Present tense and scientific rigor in reporting research.

Journal of General Physiology

Bell, J., and C. Miller. 1984. Effects of phospholipid surface charge on ion conduction in the K^+ channel of sarcoplasmic reticulum. *Biophysical Journal.* 45:279–287.

Green, M. A., and E. DeBretor. 1988. Present tense and scientific rigor in reporting research. *Journal of Scientific Inquiry.* 23:545–550.

Geophysics

White, J. E., and Sengbush, R. L., 1963, Shear waves from explosive sources: Geophysics, v. 28, p. 1001–1019.

Green, M. A., and DeBretor, E., 1988, Present tense and scientific rigor in reporting results: J. Sci. Inq., v. 23, p. 545–550.

Comparative Biochemistry and Physiology

Bettke W. and Crichton R. R. (1984) Vitellogenic ferritin of *Lymnaea stagnalis* L. (Mollusca. Gastropoda) differs in structure from soma cell type ferritin *Comp. Biochem. Physiol.* **77B**, 57–61.

Green M. A. and DeBretor E. (1988) Present tense and scientific rigor in reporting research *J. Sci. Inq.* **23**, 545–550.

Journal of Geophysical Research

Coroniti, F. V., and C. F. Kennel, Auroral micropulsation instability, *J. Geophys. Res., 75,* 1863, 1970.

Green, M. A., and E. DeBretor, Present tense and scientific rigor in reporting research, *J. Sci. Inq., 23,* 545, 1988.

Paleogeography, Paleoclimatology, Palaeoecology

Balsam, W. L. and Heusser, L. E., 1976. Direct correlation of sea surface paleotemperatures, deep circulation, and terrestrial paleoclimates: foraminiferal and palynological evidence from two cores off Chesapeake Bay. Mar. Geol., 21:121–147.

Green, M. A. and DeBretor, E., 1988. Present tense and scientific rigor in reporting research. J. Sci. Inq., 23:545–550.

Contributions to Mineralogy and Petrology

Verpaelst P, Brooks C, Franconi A (1980) The 2.5 Ga Duxbury massif, Quebec: a remobilized piece of pre-2.0 Ga sialic basement(?). Can J Earth Sci 17:1–18

Green, M A, DeBretor E (1988) Present tense and scientific rigor in reporting research. J Sci Inq 23:545–550.

Physica Status Solidi (B) Basic Research

[4c] C. H. PERRY and D. K. AGRAWAL, Solid State Commun. **8.** 225(1970).

[13] M. A. GREEN and E. DEBRETOR, J. Sci. Inq. **23,** 545(1988).

International Journal of Quantum Chemistry

[13] A. S. Davydov and A. V. Zolotariek Phys. Lett. A **94**, 49 (1983).

[13] M. A. Green and E. DeBretor J. Sci. Inq. **23**, 545 (1988).

Phillips Journal of Research

[12]) T. H. DiStefano and M. Shatzkes, Appl. Phys. Lett. 25, 685 (1974).

[13]) M. A. Green and E. DeBretor, J. Sci. Inq. 23, 545 (1988).

Biochimica et Biophysica Acta (N)

31 Spieckermann, P. G. and Piper, H. M. (1985) Basic Res. Cardiol. 80-2, 71–74.

13 Green, M. A. and DeBretor, E. (1988) J. Sci. Inq. 23-3, 545–550.

Journal of Molecular Liquids

10 P. H. Yang and J . A. Rupley, Biochemistry 18 (1979) 2654-2661.

13 M. A. Green and E. DeBretor, Journal of Scientific Inquiry 23 (1988) 545–550.

Molecule Chemistry

25. G. A. Huff and C. N. Satterfield, *J. Catal.* **85**, 370-379 (1984).

13. M. A. Green and E. DeBretor, *J. Sci. Inq.* **23**, 545–550 (1988).

Journal of Rheology

8. W. K. George and J. L. Lumley, *J. Fluid Mech.*, **60**, 321 (1973).

13. M. A. Green and E. DeBretor, *J. Sci. Inq.*, **23**, 545 (1988).

Inorganica Chimica Acta

8 B. King and F. G. Stone, *Inorg. Synth.*, 7, 193 (1963).

13. M. A. Green and E. DeBretor, *J. Sci. Inq.*, **23**, 545 (1988).

Journal of Plasma Physics

STEVERDING, B. & BIBSON, F. P. 1977 *J. Phys. D,* **10**, 1683.

GREEN, M. A. & DEBRETOR, E. 1988 *J. Sci. Inq.,* **23**, 545.

The Biochemical Journal

Gillam, S. & Smith, M. (1979) Gene **8**, 81–97

Green, M. A. & DeBretor, E. (1988) J. Sci. Inq. **23**, 545–550

5

Materials and Methods

It is with the description of materials and methods that the paper begins the report of the actual research that was introduced at the beginning of the paper. The methods are central to scientific research. Although most readers are interested primarily in the results, these are of little scientific value unless the methods have been carefully planned and accurately and precisely executed.

The methods section is largely descriptive. It is, in fact, the chief descriptive section of the paper; it is also one of the most factual sections. Its purpose is to describe accurately the materials, apparatus, and equipment used, and the procedures performed.

5.1 GENERAL CHARACTER OF SECTION

5.1.1 Theoretical and Practical Objectives

The research scientist's chief concern is accuracy, which in the methods is important for both theory and practice. From a *theoretical* point of view, a clear and accurate description of the materials and methods is an essential prerequisite for validating the results, because the interpretation, explanation, and meaningfulness of the results are directly dependent upon an accurate and adequate report of the materials and methods used in the research. For example, if the method of collecting the sample is not carefully devised and described, the results cannot be generalized to similar samples; in fact it may not be possible to determine what constitutes a similar sample. The methods described in a paper are therefore subjected to severe scrutiny and are expected to conform to strict scientific practice.

Methods may influence the development of a discipline. It may become clear that an earlier, or even a standard, procedure is no longer adequate, and a new method, research technique, or instrument may be developed that overcomes the deficiency or allows research scientists to address problems that could not be addressed before. The new method or instrument may make such a fundamental change in the kind of data collected that it forms the nexus of a new branch of research in the discipline. This emergence of a new field of study out of the fundamental inadequacy of a method constitutes part of the scientific development of a discipline. In fact, one test of a research scientist's creativity is often the ability to devise a method that can overcome the limitations of earlier methods. The methods are therefore significantly associated with the theoretical development of a discipline.

From the point of view of the *practice* of science, an accurate description of materials and methods is required, so that peers can (1) replicate the experiments described, (2) modify them with the assurance that the modification is different from the original in a particular way, (3) apply them under different conditions, or (4) compare the research reported with other research and so extend the research further.

5.1.2 Heading and Location

Usually the section describing the materials and methods is set off under its own heading, although sometimes it has no heading. In the natural sciences, the methods section is variously called "Materials and Methods," "Procedures," "Experimental Methods," and so on (*Table 5.1a*).

The methods section usually follows immediately after the introduction, but may follow the review of literature, or the theoretical framework. The methods section is in fact a direct outgrowth of these introductory sections, which should be so structured that they lead into the description of the methods. In some journals, the methods section is placed at the end of the body of the paper, often set in reduced type or miniprint (*Table 5.2a*). It may be set in reduced type in its normal position, thus allowing readers to skim the methods in passing. The reduced type saves space and reduces the cost of publication, while making necessary information available to readers. In some journals, the typescript itself is reduced photographically and is reproduced directly at the end of the paper (*Table 5.2b*).

5.1.3 Audience and Extent of Description

In describing the materials and methods, the research scientist must decide what to include in the description and how detailed it must be. This decision depends upon

TABLE 5.1   Headings of methods sections in experimental papers: (a) main headings, (b) main headings with specific subheadings

a. *Main Headings*

	Experimental Section	Data Sources and Methods
Experimental	Basic Model and Experimental	Patients and Methods
Experimental Apparatus and	Procedures	Patients and Procedures
Procedures	Proposed Reaction Scheme	Study Areas and Methods
Experimental Approach	Materials and Occurrence	Subjects and Methods
Experimental Design	Procedures	Methods
Experimental Methods	Test Method	Method of Study
Experimental Procedures		

b. *With Subheadings*

Methods	*Methods*	*Experimental Method*
Subjects	Preparations and Analysis of	Facility
Stimuli and Apparatus	Fe(OH)$_3$ Powder	Models
Experimental Design	Density	Instrumentation
Procedures	Thermogravimetry	Test Conditions
Measurements	X-ray Diffraction	Data Reduction
	Mössbauer Spectroscopy	Uncertainties
	Bulk Magnetic Measurements	

the audience and upon the method. For a scientific paper, the readers of the journal are mostly peers who are familiar with procedures commonly used in the discipline or with related or similar procedures. Such readers do not require a detailed description, but the description must be detailed enough for them to evaluate the methods and replicate the research. In *Ex. 5.1*:

Ex. 5.1

> (iii) *Restriction endonuclease digestion*
>
> DNA was digested with restriction endonucleases at 37° in conditions recommended by the supplier (Bethesda Research Laboratories). Eight restriction endonucleases all specific for different hexanucleotide sequences were used. The enzymes were *Bam*HI, *Eco*RI, *Hind*III, *Hpa*-I, *Pst*-I, *Sal*-I, *Xba*-I, *Xho*-I. Electrophoresis of single and double endonuclease digests was carried out in 1 per cent agarose gels in a BRL, H4 gel apparatus. The continuously circulated buffer was standard 1 × TBE; Tris (89 mM)-boric acid (89 mM)-EDTA (2·5 mM) and low voltage (1·5 volts cm^{-1}) gradients were used (McDonnell, Simon and Studier, 1977) to separate the DNA fragments. [Hered. 52, 1984, 104]

the writer is addressing readers who are familiar with both materials and methods. Details are provided only for the exact concentration, molarity, voltage, and so on. In the following example,

Ex. 5.2

> An appropriate volume of enzyme solution was added to 1 to 10 vol of a 50% sucrose-0.1% bromphenol blue (dye marker) solution, so as to obtain the adequate volume (generally

TABLE 5.2   Description of methods (a) after discussion section and (b) at end of paper

a. *Methods after discussion*

b. *Methods in miniprint at end of paper*

DISCUSSION

Of the seven known anthraquinones isolated in this study six have also been isolated
from calli of *C. ledgeriana* [6], with the exception of alizarin-2-methylether.

In a previous study on the alkaloids and
anthraquinones from callus material of *C. pubescens* [7] . . .

EXPERIMENTAL

Biological material. Callus cultures of *C.
pubescens* were grown on medium H at 28° with
a 12 hr day period.

Extraction. Ground, freeze dried, calli
were mixed with 10% aq. NaHCO₃ soln and extracted with CHCl₃. After evaporation of the
solvent the residue was redissolved in Et₂O. The
ethereal soln was extracted with 1 M NaOH,
which was then acidified with HCl and extracted
with Et₂O. . . .

Prep. TLC. The plates used and the solvent systems employed were the same as in ref.
[6]. After detection in daylight, the various
bands were scraped off and the anthraquinones
eluted with mixtures of CHCl₃ and MeOH. . . .
[Phytochem. 25, 1986, 1125]

Miniprint Section.
TRANSFER RNA(5-methylaminomethyl-2-thioridine)
METHYLTRANSFERASE FROM ESCHERICHIA COLI K-12 HAS TWO
ENZYMATIC ACTIVITIES by T. Hagervall, C. Edmonds, J. McCloskey and G.
Bjork

EXPERIMENTAL PROCEDURES

Preparation of tRNA.
Transfer RNA was prepared essentially according to Avital and Elson
(1969), except that the deacylation step was omitted.

Hydrolysis of tRNA for LC/MS analysis
Quantities of 100μg of tRNA were hydrolyzed in 50μl of 0.01M NH₄ OAc,
pH 5.3, by nuclease P1 (2 units/100μg tRNA), for 8 hours at 37°C. Two
μl IN NH₄OH and alkaline phosphatase (0.5 units/100μg tRNA) were
added and incubation continued at 37°C for 12 hours. Aliquots were
examined without further manipulation, by HPLC or combined LC/MS.

Directly-Combined Liquid Chromatography-Mass Spectrometry
Chromatography was carried out with a Beckman 322M liquid chromato-
graph, using two chromatographic systems. System A: two 4.6 mm × 250
mm supelcosil cyanodecyl columns in series, using a 0.25 M aques am-
monium acetate mobile phase (pH 6) run isocratically at 2 ml/minutes.
System B: 4.6 × 250 mm Supelcosil LC-18 DB column with a 3 cm Brown-
lee (Cotati, CA) Spheri-5 C18 precolumn thermostated at 31°C. Mobile
phase. . . . [J. Biol. Chem. 262, 1987, 8494]

---

from 10 to 40 μl) and activity (from ~3 to ~400 fkat or from ~2 to ~300 NF mU) to be applied
to the gel slot. Electrophoresis was run at 300 volts and 0°C for 2.5 h. At the end of the run the
dye marker band was at about 14 cm from the slot bottom. [Anal. Biochem. 143, 1984, 77]

the research scientist could assume that the reader could determine the *"appropriate
volume of enzyme solution"* to use "to obtain the *adequate* volume . . . and
activity. . . ."* Also, it is clear from *Fig. 5.1* that the reader and writer share a sym-

1-Phenyl-1-heptenes (4b and 5b), prepared from a 1.3:1 **4a–5a** mixture, phenylmagnesium bromide, and (tpp)₂-
NiCl₂: liquid olefin mixture (79%),³¹ ¹H NMR (CCl₄) δ 0.7–1.1 (m, 3, Me), 1.1–1.7 (m, 6, methylenes), 2.0–2.4
(m, 2, allyl Hs), 5.4–6.5 (m, 2, olefinic Hs), 6.9–7.4 (m, 5, aromatic Hs). **4b** (63% of mixture): ¹³C NMR δ
14.0 (C-7), 22.6 (C-6), 29.0 (C-4), 31.4 (C-5), 33.0 (C-3), 125.7 (ortho C), 126.5 (para C), 128.2 (meta C), 129.5
(C-2), 130.9 (C-1), 137.7 (ipso C). **5b** (37% of mixture): ¹³C NMR δ 14.0 (C-7), 22.6 (C-6), 28.6 (C-3), 29.7 (C-
4), 31.4 (C-5), 126.2 (para C), 127.9 (ortho C), 128.5 (meta C), 128.5 (C-1), 133.0 (C-2), 137.6 (ipso C). Heating
of the olefin mixture and a catalytic quantity of thiophenol and azobis (isobutyronitrile) under argon in a sealed
tube at 130°C for 3 h, followed by standard work-up, led to liquid olefin **4b**.
2-Octenes (4c and 5c), prepared from a 1.3:1 **4a–5a** mixture, methylmagnesium bromide, and (tpp)₂NiCl₂: liquid
olefin mixture (70%, based on NMR analysis); ¹H NMR (CCl₄) δ 0.86 (t, 3, J = 7 Hz, Me), 1.1–1.7 (m, 9,
methylenes, olefinic Me), 1.7–2.2 (m, 2, allyl Hs), 5.3–5.5 (m, 2 olefinic Hs). **4c** (58% of mixture): . . .

Figure 5.1.   Description of procedure consisting largely of chemical names, formulas, and numbers and symbols representing chemical compounds. [J. Org. Chem.
49, 1984, 4897]

bolic language to such an extent that the preparation of the compounds consists of a highly abbreviated, largely symbolic presentation. The excerpts in *Table 5.3a* indicate the extent to which accepted or known procedures are mentioned without description, comment, explanation, or reference. Familiar methods, such as the following,

Ex. 5.3

dialysis	extraction	hydrolyzation	fluorography	electrophoresis
elution	evaporation	fractionation	methylation	titration

are mentioned or referred to without description, or they may simply be referred to as a standard method, as in *Table 5.3b*. In methods closely associated with a com-

TABLE 5.3   Statements from methods sections (a) reporting generally known procedures without comment, description, explanation, or reference, (b) referring to standard procedure, or (c) instrument

a. *Report of Procedures Without Description or Reference*

1. Glucose concentrations in whole blood were determined colorimetrically with glucose oxidase (*EC* 1.1.3.4; Sigma, St. Louis, MO).
2. Metabolic concentrations were measured fluorimetrically.
3. The unreacted cyclohexanone was extracted into methylene chloride.
4. The organic phase was dried ($MgSO_4$), filtered, and evaporated to dryness. The residue was crystallized from appropriate solvents or purified by column chromatography.
5. Eight palladium gages, each approximately 1000 Å thick, were deposited on the polished surface of each substrate.
6. The photolysis light beam was introduced by dielectric mirrors into the cylindrical photolysis cell.
7. Radioactivity in the fibers was measured by liquid scintillation counting.
8. The autoclave was purged twice with 250 psi of ethylene and then filled to 250 psi.
9. The mixture was evaporated in vacuo and the residual oil partitioned between Et_2O and dilute aqueous NaCl.
10. Serum LH, FSH, and progesterone were measured by radioimmunoassays validated for the respective bovine hormones.

b. *Reference to Standard Procedures*

1. The usual workup gave the acetate of **10** as white crystals.
2. Derivatives . . . were synthesized under inert atmosphere in carefully degassed ethanol (absolute) with the use of standard Schlenk techniques.
3. The material was coupled to aminomethyl-resin by the standard DCC coupling procedure.

c. *Reference to Instrument Used*

1. Cell growth was monitored daily using in vivo fluorescence using a Model 10 Turner Designs fluorometer equipped with a red sensitive photomultiplier tube and large cuvette holder.
2. The sample was centrifuged at room temperature in a Beckman Model B Microfuge for 15 s.
3. Absorption spectra were measured on a Perkin-Elmer double-beam spectrophotometer.

mon experimental instrument the writer may simply refer to taking measurements with the instrument (*Table 5.3c*). It is abundantly clear from these examples that the procedures are written for highly specialized and knowledgeable readers.

Sometimes a professional standard is established:

Ex. 5.4

METHODS AND MATERIALS

As a result of concerns about the difficulty of comparing results from much of the published literature on P adsorption, a group of researchers met in Fort Collins, CO in 1979 to propose a standard P adsorption procedure for soils and sediments.[3] After revision, the following procedure was produced:

1) *Weight of Soil/Sediment*—0.5 to 1.0 g
2) *Soil/Solution Ratio*—1:25
3) *Extraction Time*—24 h
4) *Electrolyte*—0.01 mol L^{-1} $CaCl_2$, unbuffered
5) *Initial Dissolved Inorganic P Concentrations*—0, 6.45, 16.13, 32.26, 161.3, and 323 μmol P L^{-1} as KH_2PO_4 or NaH_2PO_4
6) *Temperature*—24 to 26°C
7) *Microbial Inhibition*—20 g L^{-1} chloroform
8) *Equilibration Vessel*—50 mL or other size to provide at least 50% head space
9) *Shaking*—End-over shaker if available
10) *Separation*—Filter through 0.45 -μm pore diameter filter (0.2 μm for clays)
11) *Analysis*—Any procedure for determination of orthophosphate, manual or automatic, capable of detecting > 750 μmol P
12) *Replication*—Duplicate equilibrations, and single analysis of orthophosphate in solution [J. Envir. Qual. 13, 1984, 591]

Such a standard need only be referenced:

Ex. 5.5

Water samples were analyzed for dissolved oxygen (DO), 5-day biochemical oxygen demand (BOD), electrical conductivity (EC_w), total N (TN) nitrate . . . in accordance with "Standard Methods for the Examination of Water and Waste Water" (American Public Health Association, 1971).

Procedures adopted from the literature that are not standard must be ascribed to their originator by citing them:

Ex. 5.6

a. Protein concentration was determined by using a microscale version of the biuret assay described by Layne (16).
b. First-strand synthesis was performed via reverse transcription (Miller & McCarthy, 1979).
c. Details of the apparatus and the chemical transport method can be found in our previous papers.[10,42,13]
d. Therefore Allinger's well-known MM2 program[30] was used as the starting point for building such a force field.

When such procedures are very unfamiliar to readers, it may help readers to characterize them briefly.

A research scientist who modifies a method reported in the literature cites the paper and characterizes the method, if necessary, then describes the modification, as would be required for a new method:

Ex. 5.7

> Root apical cells from seedlings of onion *Allium cepa* L. were used in this study. All procedures were identical with those described previously (Wick and Dunec, 1983), but some preparations were subjected to prolonged wall digestion; also, 0.1% Triton X-100 was included in the digestion solution.

When the method used is originally described in a publication that is not readily available, it may be helpful to readers to give an abbreviated description of the methods. For methods in a journal in a language that is not widely known outside the source country, for example, Japanese, one may have to have an accurate translation made of the paper—or at least of the methods section. A condensed version of the translation might then be included in the methods section of the paper, and a note about the translation included in the reference citation. This secondary publication saves repeated translation of the methods section of the original paper. For papers in Russian, government-sponsored translation of some journals, for example, *Soviet Journal of Quantum Electronics,* obviates repeated translations.

If the method used is derived from an unpublished dissertation, it can simply be cited, since dissertations are usually filed in university libraries and so can be borrowed. The full dissertation may be purchased from University Microfilm International at Ann Arbor, Michigan, if it has been filed there. Dissertations are not as readily available as journals; therefore it may be appropriate to characterize or summarize the methods. If the dissertation cited is the writer's own dissertation, then decidedly the method should be described.

5.2 PARTS OF METHODS SECTION

The number of different parts included in the methods section depends upon the type of research and its complexity and extensiveness. The various parts of the methods section may be incorporated into one section under a general heading, such as "Methods" or "Materials and Methods," but when these two major parts are extensive or complex, they may be described under the separate subheadings or equivalent specific headings. The section may be further subdivided, and the subheadings for the subdivisions may be general or specific (*Tables 5.1b, 5.4*). For example, papers may describe equipment separately from materials; or the materials section may include several different kinds of materials, which merit separate treatments:

Ex. 5.8

Reagents	Cultures	Equipment	Probe Design
Solutions	Bacteria	Study Site	Virus and Cells
Drugs	Animals	Soil Properties	Hosts and Parasites
Vaccines	Corals	Standard Solutions	Irrigation Equipment
Tissues	Apparatus	Reagents and Media	Electrochemical Apparatus

TABLE 5.4   Methods section divided into subsections with headings

Method

    Subjects. Six experimentally naive subjects, three males and three females, were paid for participating. Their ages ranged from 19 yr. to 23 yr.

    Stimuli and apparatus. The stimulus items consisted of two identical sets of 11 abrasive papers (prepared as before): grit values 36, 40, 50, 60, 80, 100, 120, 150, 180, 220, and 320. The arrangement of the stimuli and the apparatus is described in Experiment I.

    Experimental design. A five-factor, completely crossed design was used: Subjects × Days × Replication × Modality × Grit Value. Each subject judged the roughness of all 11 abrasive papers in blocks, once by touch alone, once by vision, and once by both touch and vision. This was performed twice on each of six days. The order per day of presentation of the surfaces within a modality (T, V, and T + V) was randomized; the six possible orders in which the modalities were used were randomly assigned across subjects within a day. Subjects were given 1 full day of practice.

    Procedure. The subject was given a set of instructions to read, which outlined a magnitude estimation procedure for judging the roughness of a set of surfaces by touch alone, by vision alone, or by both touch and vision together. The subject was told that the first stimulus (grit = 100) in each block should be called 10. This surface served as a standard, and it appeared at the beginning of each block of exposure conditions (T, V, and T + V); it also appeared as one of the subsequent 11 stimulus surfaces. Any positive, nonzero numbers could be used, that is, decimals, fractions, or whole numbers. The subject assigned numbers to represent the roughness of each surface in proportion to that assigned to the standard. On the V and T + V trials, the subject was encouraged to move his or her head to maximize the visual cues to roughness. No time limit was imposed. [J. Exp. Psych. 7, 1981, 909; with permission]

    Similarly, if the methods consist of different procedures, this section of the paper may be subdivided, for example, into the preparatory and the main experimental procedures (*Table 5.5*). The preparatory or preliminary procedures may be included under the materials section, if they describe procedures to prepare materials for the main experimental procedures. They are then separated from procedures for the experimental methods, which are included in the section describing the methods and procedures. Such preliminary procedures may, however, be distinct enough, complex enough, or detailed enough to be treated under separate headings (*Table 5.4–5.6*).

    In a paper in which a theoretical model is developed and applied, there may be a separate section describing the development of the model and a separate section for describing the method of applying the model. The methods section may also include a section describing the experimental design. These sections precede the descriptions of the materials and methods. When the research includes a statistical analysis, this is included at the end of the methods section and is variously labeled:

Ex. 5.9

Analytical Methods	Calculations and	Data Organization and Treatment
Analytic Techniques	Theoretical Aspects	Processing of Analytic Data
Mathematical Analysis	Computations	Statistical Analysis
Mathematical Methods	Computational Details	Statistical Methods
Calculational Procedure	Data Reduction	

TABLE 5.5   General and specific headings for sections on preparatory and main experimental procedures of methods section

Preparatory Procedures

Sampling Procedures	Preparation of Rat Brain Membranes
Surgical Preparation	Preparation of Samples for Polyacrylamide
Isolation of Flagella	Analysis
Capture, Sampling and Marking	Reduction of Trytophan in Proteins Prior to
Preparation of Crude Extract	Acid Hydrolysis

Experimental Procedures

Synthesis	Measurement of Microelectrode Tip Diameter
X-ray Analysis	Coulometric Generation of Divalent Copper
Crystallographic Studies	Reaction of $W(CO_4)$ (NO)Cl with Cl(NO)
Genetic Studies	Infusion and Blood Collection Procedures
Adsorption Studies	Collection and Reduction of X-Ray Data
Tissue Analysis	Immunoprecipitation of Phosphoproteins
Spectrochemical Analysis	Measurement and Analysis of Spin-Relaxation
Laboratory Methods	Transients of Metal Atoms
Field Methods	

5.2.1 Research Design*

For a complex experiment or protocol, the writer may describe the design of the research to make it possible for the reader to understand the relation of the separate procedures to the various sections of the methods and to the research overall. The research design is usually presented at the beginning of the methods section. It may be only briefly outlined, but when the methods are extensive or complex the description may serve as a framework for the methods and may then be under a separate heading (*Table 5.4, 5.11*).

A separate research design section or subsection is particularly useful (1) when preparatory procedures or preliminary experiments precede the main experiments, (2) when the research is very complex, (3) when the research consists of several experiments or different types of studies, (4) when a particular configuration of experiments is planned, or (5) when it is necessary to discuss the reasons for the design or its particular objectives, advantages, or disadvantages in some detail.

Describing complex protocol.   In some papers the research design is very complex and the protocol allows for studying various treatments and their interrelationships simultaneously. Such research is usually prestructured on a formal experi-

---

*This term is used here in its broad sense of the plan of the research. It thus includes the narrower meaning of the term, i.e., a statistical research design.

mental design with a statistical base, for example, a factorial design, a randomized block design, a 4 × 4 Latin square, and so on (*Tables 5.4, 5.11*). The various parts are not segregated into individual experiments that can be isolated and separately reported. In such research the description of the research design is presented; then the procedures are described independently. For example, in a study of the interrelationship of diets and drugs to determine the concentration of the drugs and cholesterol in the blood and their effect on liver and kidney tissue, some procedures would be the same for all animals, and some would be different. In such a protocol, it is not possible to differentiate separate experiments; therefore, the methods section would describe the experimental design and the various procedures:

Ex. 5.10

General Procedures
Preparatory treatment of animals
Diet of treatment groups
Drug treatment
Experimental treatment of animals
Killing of animals

Collection of Samples
Blood—drugs, cholesterol
Tissue—hepatic, renal

Analysis of Samples
Blood—drugs, cholesterol
Tissues—hepatic, renal

5.2.2 Description of Materials

Accurate descriptions of the materials are essential for establishing the ground substances or ground structures for the observations and measurements in the experimental procedures:

Ex. 5.11

Experimental Section

All the chemicals used were reagent grade. Solvents were dried prior to use:. . . . pentoxide, THF was dried over sodium metal/benzophenone, and. . . . Melting points were taken with a Gallenkamp apparatus and are uncorrected. Nuclear magnetic resonance spectra were taken with a Varlan T-60 or XL-200 spectrometer using tetramethylallane as internal standard. . . . All reactions were usually run in a nitrogen atmosphere and all equipment dried in an oven. Purifications involving column chromatography were performed with Merck silica gel 60 (230–400 mesh) using flash chromatography.[15] Microanalyses were performed by Guelph Chemical Laboratories Ltd.

Triacetic ester 5 was prepared according to literature procedure[9] and its identity confirmed by NMR, IR, and MS analysis. . . . [J. Org. Chem. 51, 1986, 2426]

Readers must know their exact form or condition, their purity, consistency, or standardization. The breeding of lines or races of organisms developed especially for experimental studies, the analysis, purification, or synthesis of chemicals before they are used, and the standardization of solutions in preparation for experimental use all reflect the importance of standardizing materials and conditions.

The materials are described incidentally as part of the methods when they are few and simple or when they are commonly used and so do not require detailed description:

Ex. 5.12

Inoculation

Charleston Gray watermelon (*C. vulgaris* Schrad.) seeds were planted in four replicated and randomised plots infested with Fusarium wilt. A normal spacing of 91 cm between plants and 3·04 m between plots was used. *Fusarium* colonies were started from hyphal tips of the fungus grown on potato dextrose agar (PDA). Inoculum was prepared by washing the culture plates with sterile deionised water to form a conidial/mycelial suspension. The inoculum suspension was watered into separate beds of previously fumigated soil (methyl bromide at 2·04 kg/100 flats (45 cm × 60 cm)). [Ann. Appl. Biol. 108, 1986, 244]

They are described under separate headings (1) when they are numerous or include different kinds of materials, for example, chemicals, equipment, and organisms (*Ex. 5.8*), (2) when they are too complex to be introduced incidentally into the description of the methods, (*Fig. 5.1, Ex. 5.11*) (3) when they must be treated or modified in preparation for the research (*Table 5.2, 5.6*), or (4) when they are discussed because the reader must have a full understanding of the nature of the materials before reading the procedures (*Table 5.6, Ex. 5.11*). If the paper is primarily a description of a new method or instrument, then of course, their description constitutes a major part of the methods section. Equipment or apparatus that is new and requires a separate description may be included under a separate general heading, "Apparatus" or "Equipment," or a more specific heading.

Kinds of materials.   Materials may range from chemicals to equipment to organisms. For chemical substances, the formula or name must be given in a form useful to readers. They may be designated by their chemical names, generic names, or trade names or by their chemical or structural formulas. In organic chemistry very complex structural formulas are numbered and thereafter referred to by number in boldface type (*Fig. 5.1, Ex. 7.27*). The source of the substances should be indicated, as well as the degree of purity. This is especially important for compounds that may vary from one supplier to another, as in biochemical substances. The source may be included in parentheses (*Table 5.7*) or in a footnote. The name of the supplier and often the city is given. If the trade name used is registered, the trademark must be used; wherever possible, however, it is preferable to use a generic form of the substance. For solutions, the molarity, pH, dilution series, or other pertinent values are reported. Equipment is described to the extent determined as required for readers. For purchased equipment the trade name, model, and source (usually manufacturer) are provided (*Table 5.7b*).

For equipment designed and built expressly for the particular experiment, a very clear and accurate description is essential, since the paper is the only permanent source for the design. When the apparatus includes both new and standard parts, the configuration of the parts is described, then the newly designed parts are described in detail. These new parts are not described in minute detail, however, because they are described in conjunction with illustrations, which provide detail more effectively than a verbal description (see Chapter 6, especially Fig. 6.6, 6.20–6.22).

TABLE 5.6 Subsections under methods section illustrating separate subsections
for different materials

METHODS

Experimental Materials

The experiments were done using living animals of the species of *Loligo pealei* collected at the Marine Biological Laboratory, Woods Hole, MA during May–June, 1984. After killing the squid, an axon was rapidly dissected and cleaned so that it was ready for an injection of dye generally within 1 h of the death of the animal.

Microinjection

Stock solutions of arsenazo III, phenol red or the inhibitor orthovanadate, were made up in 330 mM K TES buffer, pH 7.3. Small volumes, 15–20 μl, of these were microinjected over a length of 15–20 mm. The usual final concentrations within the axoplasm were 500 μM for arsenzao, 300 μM for phenol red, and 1 mM for orthovanadate.

Solutions

Sea water had the following composition (in millimolars) Na, 455; K, 10; Mg, 55; Ca, 3; Tes (pH 7.8), 10; Cl, 571; EDTA, 0.1 (to protect against heavy metal contamination). For solutions with 112 mM Ca, Mg was omitted and NaCl reduced so that the solution was isosmotic with 1,000 mosml/L. High K solutions and Na-free solutions were made by replacing all Na with either K or Li, respectively. Cyanide sea water had 2 mM NaCN at pH 7.8 added to normal sea water. All solutions were adjusted to 1,000 $\pm$ 10 mosmol/L and pH 7.8. Temperature was controlled at 13°C except when another temperature is given.

Hydrocarbon Derivatives

The chemical reagents used for this study (alcohols, paraffins) were of analytical grade and were obtained from Eastman Kodak. For the relatively insoluble compounds. . . . Br-hexane, solutions of these substances were prepared in sea water by adding sufficient of the chemical reagent so that a saturated solution was obtained. Dilutions of this solution were then used to obtain the desired final concentrations. For the more water soluble compounds . . . an appropriate amount of the reagent was added to sea water to produce the desired final concentration. [Reproduced from Biophys. J. 50, 1986, 12, by © Biophysical Soc., with permission]

The methods may include a discussion of the reasons for designing the particular equipment, advantages over other designs, how it corrects faults in equipment used in earlier studies, and so on.

Experimental animals should be clearly specified and characterized as to species, sex, age, strain, source, and any other characteristics pertinent to the research. A description of experimental rats might read

Ex. 5.13

   a. Male weanling (28-day-old) Fisher-344 rats were purchased from Charles River, Inc., Wilmington, MA.

   b. A total of 216 hairless mice of the Philadelphia Skin and Cancer Hospital SKH/m-1 strain ranging in age from 5 to 7 weeks were used in these experiments.

TABLE 5.7   Excerpts from materials and methods sections with references to sources
of (a) materials and (b) equipment

---

a. *Materials and Methods*

Porcine MBP (p-MBP) was kindly provided by Dr. Max Marsh of Eli Lilly & Co. Peptides were prepared from rabbit MBP described. . . . The fatty acid spin-label were purchased from Aldrich Chemical Co.. . . .

The proteins were lyophilized twice from 99.5% D_2O and dissolved in 99.96% D_2O (Merck Sharp & Dohme) at a concentration of 0.50–1.0 mM [Biochem. 23, 1984, 6041]

b. *Instrumentation*

. . . For the decay-time measurements on $CsMgCl_3$-Bi^{3+} a photon-counting system (EG&G) was used. Details of this system have been described elsewhere (*12*). The emission wavelength was selected by a double monochromator (Jobin Yvon, HRD 1), equipped with a photomultiplier tube (RCA C31034). Measurements of the time-dependent behavior of the emission bands in $CsCdBr_3$-Pb^{2+} were performed using a nitrogen laser (Molectron, Model UV 14). The emission was analyzed with a Spex 1704 X 1-m monochromator and detected by a photomultiplier tube (RCA C31034). . . . [J. Sol. St. Chem. 55, 1984, 345–4]

Experimental plants should be described as to species, stage of development, condition, parts of the plant, and other characteristics pertinent to the research. The treatment of animals or plants in preparation for the research is described in the materials section if brief and simple, or under a separate heading, if extensive or complex.

Scientific names of organisms.   For organisms, the full, correct, scientific, i.e., Latin name (epithet) is given. This makes absolutely clear exactly which species is being studied, and the name is recognized by scientists throughout the world. To facilitate retrieval of the research, the writer should make an effort to include the scientific name in the title:

Ex. 5.14

a.   **Balantidium prionurium n. sp., Symbiont in the Intestine of the
Surgeonfish, Prionurus punctatus**[1]

b.   Homing and Site Fidelity in a Neotropical Frog,
Atelopus varius (Bufonidae)

In studies of domestic animals, however, as in the veterinary sciences, the common name is usually used throughout, for example, horse, Hereford cattle, Berkshire swine.

When the full scientific name is given, it includes the genus and species, any inferior taxon (e.g., subspecies, form, variety, cultivar) if pertinent, and the author of the epithet, e.g., *Coccinella novemnotata* Herbst, *Actias luna* (L.), *Heraclides andraemon* Heubner subsp. *bonholei* (Sharp). A larger taxon may be included, in parentheses, such as the family or order (*Ex. 5.14b*), but it is not essential. The genus and species are set in italic type, but other parts of the epithet are not italicized, for example, *Rosa damascena* Mill. cv. 'Versicolor.' The generic name is always capitalized, as in *Rosa, Cocinella, Mus,* as are cultivar names, for example, 'Versicolor' (York and Lancaster rose), *Phaseolus vulgaris* L. 'Aurora.' Cultivars are often referred to simply by the common or scientific cultivar name, for example, 'cv. Aconitifolium' or just 'President Lincoln.' The name used should be the *correct* scientific name taxonomically. This can be verified by consulting a taxonomist.

After the full name has been given at the first use in the text, the name is thereafter shortened by abbreviating the genus to the first letter and omitting the author's name, but the name is still italicized, for example, *Falco sparverius* becomes *F. sparverius*. In the abbreviation of the generic name, each name is treated independently, whether in the same or a different genus, that is, full name when first mentioned and genus abbreviated thereafter. In a dissertation, the full name is again used at its first appearance in each chapter.

Whenever two genera begin with the same letter, however, the generic name of one species cannot be abbreviated independently of the other without causing confusion. For example, if the paper begins by discussing *Streptococcus cremoris,* the generic name is written out at the first mention, and thereafter abbreviated for any species of the genus, for example, *S. lactis*. Now, if *Staphylococcus aureus* is introduced, it is written out in full when it is first used, as is customary, because it is a different genus. If the succeeding references continue to be to species of *Staphylococcus,* then the generic name is abbreviated, for example, *S. aureus*. If, now however, another species of *Streptococcus* is introduced, for example, *Streptococcus lactis,* it cannot be abbreviated to *S. lactis* because it would be interpreted to refer to *Staphyloccus lactis,* the last genus written out in full. The genus must be again written out in full, for example, *Streptococcus lactis,* after which any species with the abbreviated generic name represents a species of *Streptococcus* until a species of *Staphylococcus* is introduced. This must again be written out in full, even though it has been written out in full earlier. Thus with each change in genus, the generic name must be written out in full. When there are several genera beginning with the same letter, abbreviations using several letters may be used to differentiate them, for example, *Staph. (Staphylococcus), Sh. (Shigella), Sal. (Salmonella)*.

5.2.3 Description of Procedures

The procedures described include any operations or measurements performed before or during an experimental study. The central or experimental procedures are those that test the hypotheses and answer the research question. Procedures may range from very simple collection and testing techniques to highly specialized analytical procedures such as are used in molecular biology and atomic and nuclear studies. In more complex studies, there may be preliminary or preparatory procedures, as well as the central procedures.

The focus of the methods section is the description of the operations or procedures performed. The new method may consist largely of standard assays, or it may be an entirely new procedure. For example, there are no standard procedures for setting up plots of natural vegetation or cultivated plants for observation and experimentation. Such procedures depend on the objectives and design of the experiment, as well as on the site available; therefore they have to be described in full. On the other hand, assays of material collected from the plot might be standard assays and would not be described, for example, obtaining the dry weights of the samples or determining the protein content of the leaves. Similarly, the treatment protocol for animals is directly tied to the experimental question and so must be described, but any standard procedures for assaying blood plasma or preparing histological sections may only be mentioned. Likewise, the procedures for preparing chemical compounds for the synthesis or analysis of a compound may simply be named, if familiar to readers. These various experimental methods and procedures are often described under their own headings or subheadings (*Table 5.5*):

Ex. 5.15

METHODS

a. Materials
b. Viruses
c. Irradiation of virus
d. Cells and infection
e. Preparation and anion-exchange chromatography of cellular postribosomal supernatant

f. Assay of protein kinase activity
g. Assay of DNA polymerase
h. Determination of minimum concentration of cycloheximide for inhibition of protein biosynthesis

Even though different procedures are interrelated and carried out concurrently, they are described separately (*Ex. 5.15*). Field procedures are described separately from laboratory procedures. Analyses by different procedures, for example, chemical, X-ray diffraction, are described separately. These separate procedures may be described under separate headings depending on how distinct, complex, or extensive they are (*Table 5.1, 5.5*).

Procedures preparatory to the main experiment (*Table 5.6*) or sample-collecting procedures must be described because they establish the point of departure for the research. Such procedures may have a systematizing, stabilizing, analytical, or synthesizing function in the research and may even be the central focus of the research. It is important to collect samples systematically. The procedure for collecting samples may be so crucial, that if it is not accurate or carefully controlled, it may vitiate the research. After samples are collected, they may have to be treated—to preserve them or to prevent contamination. Animals and plants may have to be treated to stabilize their condition before they are subjected to experimental procedures. Such preparatory procedures are usually not standard and need to be described. For research with chemical substances, the need to establish an initial level of purity, concentration, or chemical or physical structure may require a preliminary analysis or synthesis of the compounds (*Ex. 5.11*).

5.2.4 Analysis of Data

Any statistical, computational, or other mathematical methods used to manipulate or analyze the data should be described in the methods section. Such analyses are as much part of the methods as the physical, chemical, or biological analyses. The analysis may be described at the end of the methods section or experimental design section, or elsewhere if appropriate. It may be described in the results section if it is brief and not known before the results that the analysis is required.

Many analytical methods are standard procedures and so need only be named, for example, linear regression, Mann-Whitney U test, Pearson product-moment-correlation, Student's t-test, ANOVA (*Table 5.8*). If the research scientist modifies a standard analysis or an analysis presented in the literature, the source is cited, and the modification of the analytical procedure is described. The analysis is described in full, if a new method of analyzing the results, or if the method of applying the analytical procedures is unusual. When the analysis is extensive or includes discussion as well, it is treated as a separate section (*Table 5.8*) with its own heading (*Ex. 5.9*). The discussion of the analytical method is included in the analytical section even when it is very extensive (*Table 5.8c*). Indeed the analytical section may be largely an explanation and justification of the analytical methods.

5.2.5 Discussion of Methods

Placement of discussion.   Although the methods section is preeminently a descriptive section, it may include discussion. Discussion of the methods may not necessarily belong in the methods section, however.

Any discussion of the methods that is central to the investigation as a whole, that is, to the *design of the study,* may be included in the introductory sections *before* the main methods section. Such a preliminary discussion may be an explanation of the methodology, or a justification for using the particular materials, equipment, procedures, or analyses, relative to earlier research studies. Discussion of the methods in such contexts usually belongs in an introductory section but, if extensive, may be in the opening section in the methods section.

Any discussion of the methods relative to the *findings,* that is, the results, belongs in the discussion section of the paper. However, any discussion of materials, equipment, and procedures in themselves, *as materials or methods,* belongs in the methods section proper, not in the discussion section, since the methods are best validated at the time that they are described. In other words, any discussion that contributes to a full understanding of the methods, and that is not dependent on the results obtained, is included in the methods section.

Any discussion of the statistical analysis, *as analysis* is included in the methods section, because the validity of the analytical methods used must be established before the results are presented. It is not possible to evaluate statistical results without a thorough understanding of the statistical methods used.

TABLE 5.8   Examples of (a) references to standard statistical and other mathematical procedures, (b) description of computation, and (c) discussion of analysis

a. *Statistical Analyses*

1. Chemical analysis results were treated statistically using the SPSS statistical package (Nie et al., 1975), particularly the Oneway, Scattergram, Pearson correlations, and Regression programs.
2. The analysis of the spectra was made by a nonlinear iterative least-squares-fitting algorithm.
3. The matrix inversion method used in step 5 was based on Cramer's rule.
4. Bartlett's (1950) test was used for the orthogonality of the determinant of a sample zero-order correlation matrix.
5. Table 5 presents the results of a two-stage least-squares multiple regimen analysis in which 1983 salary is the dependent variable and the independent variable is 1978 salary.

b. *Computation*

V. COMPUTATIONAL DETAILS

A. Potential energy curves

Two different contracted Gaussian basis sets were employed in the present study. One, referred to as the smaller basis, is a $(4s,2p)$ basis which is an uncontracted version of the basis of Dunning and Hay.[33] The other, the larger basis, consists of the former augmented with a single polarization function (with an exponent of 0.15). The calculation of the propagator proceeds in two steps. First, an AGP reference state, here representing the $X^1\Sigma_g^+$ ground state of Li_2, is obtained variationally.[30] Then the propagator matrix is constructed and diagonalized to give vertical excitation energies. The associated eigenvectors give the excitation operators and can be used to calculate electron transition moments and oscillator strengths which are needed to obtain radiative lifetimes. . . . [J. Chem. Phys. 81, 1984, 3979]

c. *Discussion of Analysis*

In each case, the dependent variable can assume one of only four possible values (0–3). Because multivariate linear analyses are based on assumptions requiring an interval level of measurement for the dependent variable (Maddala, 1983), the use of ordinary-least-squares techniques is inappropriate.[2] Instead, an ordered probit model was employed. This is an extension of a dichotomous probit model that is applicable to analyses involving ordinal dependent variables (McKelvey and Zavoina, 1975; Winship and Mare, 1984). With this analytic procedure, a distinction is drawn between an underlying theoretical dependent variable, which has an interval scale of measurement, and the observed dependent variable, which is ordinal. In the present case, the latent theoretical variable may be thought of as the amount of pressure for administrative differentiation; the observed variable is the presence of zero, one, two, or three offices. . . . [Admin. Sci. Q. 30, 1985, 7–8]

Discussion of materials and methods.   The discussion of particular materials depends on the reader's requirements and accompanies the description of the materials. For example, it may be necessary to give the reasons for choosing a particular form of a chemical compound, to explain the choice of particular conditions, dosages, pesticides, and so on, or to justify the use of a particular kind of instrument or configuration of the apparatus.

Procedures may be discussed from various points of view. When preliminary procedures are performed, it may be necessary to explain how they are related to the central procedures. These are usually discussed before the description of the main experiments. The discussion of central experimental procedures may include explanations of particular parts of the procedures and the relationship of procedures to testing the hypothesis.

Discussion of the analytical procedures may be necessary to explain or justify the analysis used or to compare it with other analyses that might have been used, or to discuss its merits or any disadvantages in the particular study. The discussion of analytical procedures is usually integrated into the description of the analysis because of the very close relationship between the two. Detailed discussion of the analytical methods is more common in papers in the social sciences.

The treatment of any discussion relative to the description of the methods depends on its length and complexity. It is made an integral part of the description if it entails just a reason or a relatively brief explanation. It is then often effectively placed after the description. If the discussion is more extensive or complex, then it is conducive to clarity to keep it separate from the description. It may precede the pertinent description or follow it, the position depending on its generality, its complexity, and readers' need to understand the explanation *before* the materials or the methods are described. For example, a general discussion or an explanation of a complex procedure is more effective placed before the description of the pertinent materials or methods. If the discussion is needed to justify or support the particular materials or methods used, and readers need to know the method to understand the explanation, it is placed after the description. If it is designed to raise general methodological problems that need to be resolved before the materials or methods can be presented, it belongs before the description. When the discussion is in the nature of a summary explanation, it is more appropriately placed after the description.

5.2.6 Methods in Descriptive Papers

Descriptive or observational studies in the natural sciences range from naked-eye observations to those made with highly developed instruments.

Methods used in naked-eye observations must be made systematic, and their systematic character must be described. For example in a study of maternal behavior in horses, the writer must define nursing behavior, before the time spent in nursing can be measured. The writer must also describe the conditions under which the observations were made, that is, time of day, period of observation, frequency, position of observer, and so on. The methods of measuring time and distance must also be described. Even the method of recording observations must be described. For example, if one person was both observer and recorder, the procedure designed to avoid missing observations while recording or to regularize them must be described.

Descriptive studies also include studies of natural entities or phenomena that are too large or too small for naked-eye observation. The research scientist is then limited to indirect observation and to methods based on highly developed instrumentation and theoretical models. The materials in such studies and the highly developed instruments designed to study them are often difficult to describe. The writer does not describe an atom or Mars, the electron microscope, or the Arecibo ionospheric dish. The instrument is taken as a given and is not described, and the "material" is described only in relation to the elements known and to be studied. The methods are largely oriented to describing (1) the procedure for using the instrument to collect the data and (2) the calculations and computations or mathematical analyses used to "reduce" the data, that is, to determine what has been observed.

5.3 WRITING THE METHODS SECTION

The methods section is the part of the paper that should be written first. It is the best part of the paper to begin with, because (1) it is largely reportorial and descriptive and so easier to write than the introduction, (2) it can be written from direct observation, and (3) the methods usually remain essentially the same once data collection begins, and so the section can be revised to an advanced draft before the research is completed. Therefore, as soon as the research is actually under way, the research scientist can write this major section of the paper completely. This first description of the methods should be comprehensive to allow for later condensation.

5.3.1 Descriptive Character

The methods section is essentially a descriptive section, and an accurate, clear description of the methods is the central scientific requirement. The form of descriptive writing discussed in Chapter 3 is particularly applicable to the methods section. The descriptive function of the methods section determines the universe of discourse. The focus is on materials and on past systematic activities and procedures; therefore this section is written in the third person and reported in the past tense.

The process of writing a scientific description for materials is different from the process of writing a description for methods. For methods, the objective is to convert the sequential (i.e., linear) order of activities into the linear form characteristic of writing. The linear order is inherent in the sequence of activities, and the translation from the actual events to writing the description is easily made; it requires only ordering of the activities and procedures. The sequential character of methods is therefore an inherent integrating device for describing them; however, materials have no such inherent linear relationships for the writer to draw upon. For physical entities such as materials, the writer must reduce them to two-dimensional structures that permit describing them in a linear order. The writer may have graphic representations of the material as the basis for a description, for example, a diagram of the instrument. This reduces the three-dimensional object to a two-dimensional structure, which makes it easy to identify the elements and their relationship and to draw a flow diagram that reflects the relationships linearly. For entities that are not illustrated, or that cannot be illustrated, the writer can lay out the elements of interest in relation to one another in a sketch or network and then order them into a path diagram sequentially, and so into a linear order for writing.

The description cannot be treated as a list of parts, except in highly formalized description, as in specifications for an instrument or scientific descriptions of organisms (*Ex. 3.22, 7.4*). The parts of the structure must be integrated. One cannot simply list each part and describe it:

Ex. 5.16

> The platform is cast aluminum, with a finely ridged edge and. . . . The tube tapers to the base and rests. . . . The tube is ringed inside with. . . . The sensing mechanism has two electrodes, which. . . .

If an instrument is complex and the description of the parts is very detailed, the relation of the parts overall can be described at the beginning of the description, then the separate parts are described in detail. However even then, the description must be written to show the arrangement and interrelationship of the parts; the linear order in which they are described must translate back to the multidimensional structure.

5.3.2 Procedural versus Chronological Order

The sequential steps of the procedures have a chronological character that tends to elicit a descriptive narrative style of writing, but the methods, though sequential, are not a story. The methods section is not simply a list of activities described in their chronological order. It is a structuring of the various activities into a meaningful, scientific procedure.

A method often consists of several procedures; some of these various procedures are performed more or less concurrently, and some of the steps of the different procedures may be interwoven. For example, if rats are given a series of injections of an experimental drug over a period of time, and blood samples are then drawn and analyzed between injections, and finally the animals are killed and the livers removed and examined histologically, a chronological description might be like the numbered steps shown in *Table 5.9a*. A reader would have difficulty understanding the methods, because the interweaving of the different procedures would obscure the separate methods that constitute the methods as a whole. If instead, the various procedures are separated out, and the different procedures for blood analysis, removal and analysis of the liver, and so on are described separately as shown in *Table 5.9b*, the reader would have a much clearer understanding of the methods. A description of the research design preceding this description of the procedures would help readers to understand the relationships.

Chronology is important, however, in ordering the steps within a procedure, so that later activities are not described before earlier ones on which they depend. For example, the writer should not describe a newly developed assay before describ-

TABLE 5.9   Comparison of chronological and procedural ordering of parts of methods section showing that procedural order makes structure of the methods more readily understood than chronological order

a. *Chronological Order*	**b.** *Procedural Order*
1. Injection of drug	1. Protocol and method for injection of drug
2. Collection of blood sample	2. Serum studies
3. Analysis or storage of blood sample	a. Collection and storage of blood samples
4. Injection of drug	b. Analysis of blood samples
5. Collection of blood sample	3. Tissue studies
6. Analysis or storage of blood samples	a. Killing of rats
7. Injection of drug	b. Removal and fixation of liver
8. Collection of blood sample	c. Biochemical assays
9. Analysis or storage of blood sample	d. Histological preparation of liver tissue
10. Killing of rats	
11. Removal and fixation of liver tissue	
12. Preparation of liver for histological study	
13. Analysis of stored blood samples	
14. Biochemical assays of tissue	

ing the methods used in collecting or preparing the samples for the assay. Similarly direct chronology is to be preferred to indirect chronology:

Ex. 5.17

   a. Sulfate was analyzed by atomic absorption of barium, *after* samples were precipitated with barium chloride to form barium sulfate.

   b. Each vacuole initially contained approximately 0.5–1.0 mL cryoprotectant solution and culture material *before* lyophilization. *After* the samples of culture material were frozen at $-30°C$, they were diced at pressures between 10–20 μHg, then brought to room temperature and then sealed under vacuum.

In these examples, a procedure is described, then conditions or procedures that preceded the procedure are described. Readers must then reorder the steps mentally to establish the actual sequence of activities.

5.4 METHODS SECTION IN SOCIAL SCIENCE PAPERS

Research papers in the social sciences reflect two major methodological problems, (1) controlling conditions and (2) developing and standardizing methods of measurement (see Chapter 1). These make methodology a central concern for social scientists. They also make descriptive, analytical, or survey studies more common than experimental studies.

5.4.1 Parts of Section

Heading and parts. The methods section in social science papers is often called "Methodology," the name possibly reflecting better than "Methods" the integral relation between methods and theory; however it is also referred to as "Methods" and by other names (*Table 5.10*). In many analytical and survey papers, the methods section follows the section on the development of the theoretical framework.

Social scientists may describe three "materials" in the methods section: participants (sample or subjects), site, and instruments, as well as the procedures, variables, and analyses (*Table 5.10, 5.11*).

TABLE 5.10   Examples of types of headings for sections and subsections in social science papers

General Heading

Method(s)	Data and Methodology	Design and Procedures
Methodology	Material and Methods	Method of Inquiry

Theoretical and Experimental

Research Design	Research Design and Method	A Typological Model
Basic Model	Theoretical Framework	Spatial Allocation Model
Theoretical Model	Experimental Design	Generalized Spillover Model
Hypotheses	Labor-Supply Framework	

Participants, Sample, and Site

Sample	Passive Observers	Study Region
Subjects	Active Observers	Study Area
Participants	Sample Selection and Data	Study Site

Sample Characteristics, Variables

Sibling Placement	Demographic Characteristics	Family Stakes and Student Ability
Exogenus Variables	Types of Opinion Leaders	
Community Problems	Educational Expectations	Parent-Child Consensus Variables
Independent and Dependent Variables	Task Characteristics	
	Historical Birth Cohort	

Procedures

Measurements	Testing Procedures	Testing for Transmission Inflation
Sample Survey	Measurement Procedures	
Empirical Tests	Methods of Data Collection	Scoring of Behavior and Recording Data
Estimation Method	Cluster Definition Method	

Participants, site, instrument.   The chief "material" in social science research is human beings. Most papers in the social sciences have a section describing the "subjects" or "sample," but the heading may be more specific, for example, "Patients" or "Residents" (*Table 5.10*).

The characteristics of human subjects must be carefully selected and clearly specified, since the attributes of a sample are important for determining the reliability of the research (*Table 5.11*). Some participants are chosen because they are members of a group of interest, such as, smokers, members of a particular craft, profession, type of organization; some may be chosen simply because they are available, such as, students in a professor's class. Criteria for selection are reported, and special attributes of participants required for the research or differences among participants are described, as for example, age, sex, and occupation. The use of human subjects in research raises ethical questions about their treatment. Participants must be informed and kept anonymous. Ethical standards for treating such subjects are to be found in a publication of the American Psychological Association: *Ethical Principles of Psychologists*. The paper should include a statement about conformance to requirements of institutional committees, review of research proposal by such a committee, written consent of participants, and so on:

Ex. 5.18

   a. The mice were cared for according to Department of Health and Human Services guidelines for animal care.

   b. Forty-three severely malnourished children (22 female, 21 males, ages 18–60 mo.) were recruited in the outpatient clinic and emergency ward of. . . . Written informed consent was obtained from the parent or guardian of each child.

Social scientists must often describe the "site," the place where the participants were studied (*Table 5.10, 5.11*) in some detail, if the organization or institution is pertinent to the study. Banks are a different social environment from government agencies, and factories different from military units. Where the attributes of a site are important for the study, the characteristics desired and the selection process are described. The instrument is usually a questionnaire or interview schedule (*Table 8.11*). It may be a standard instrument, a modification of such an instrument, or a new instrument designed specifically for the study. The objectives and procedures for developing the instrument are described and discussed:

Ex. 5.19

   Instruments.—Three instruments were used in the study: the LTCIS (Falcone, 1979), a Telephone Survey Guide, and a Chart Audit. All were deemed to have face validity. The LTCIS was initiated by Jones . . . the first part elicited general information, including a person's health and functional status. The person's functional status score could range from zero to seven and reflected the number of dependent activities of daily living (ADLs). In part two of the LTCIS translated are service needs using 20 assessment items to systematically project whether any of the 11 identified services are needed. Services include audiology, dental . . . shopping. Prior to the initiation of the study, the GCNS received formalized training in the use of the LTCIS . . . In this study, consistency was maintained by limiting the use of the protocol to one clinical nurse specialist.

The Telephone Survey Guide was developed to assist in gathering data regarding patient outcomes at 2 and 4 weeks post-discharge. Questions regarding the patients' present placement, resources utilized, changes in the original plan, and assessment of the patients' perceived needs for relocation or transfer from their present placement were included. The Chart Audit Guide aided in retrieving data from hospital records (principal and secondary diagnoses discharge date, and number and dates of readmissions). [Geront. 27, 1987, 578]

If a standard questionnaire is used, the writer may only need to name it, e.g., the Mood Assessment Scale, or Cornell Medical Index. The development of a new questionnaire presents a writing problem as well as a scientific problem. The measures chosen must be appropriate to ensure validity, and accurate and neutral to ensure reliability. Both these goals depend on precise language. The research scientist must denote variables clearly to measure them accurately. Even with clear, conceptual delimitation, it is not easy to achieve clear, sharp denotation in writing the instrument. In framing questions or statements for a questionnaire, therefore, the research scientist must phrase them so that the respondents understand them to mean exactly what the research scientist means; otherwise, the instrument will not only be measuring some other variable, but an unknown variable at that. Besides communicating accurately, the questions must be written without bias so that they do not bias responses. The questionnaire is not usually included in the paper, but it is usually described in enough detail and with examples of questions and method of scoring to make its structure, form, and coverage clear.

Methods.   The section describing procedures for testing the hypothesis may not be as dominant a part of the paper as other procedures, such as selection of the sample and standardization of measurement techniques, which are often important enough to warrant separate headings. Since replication is extremely difficult in the social sciences, statistical analyses are used to control for intervening variables, and in a sense substitute for replication. Papers in the social sciences, therefore, usually include a section on the analytical procedures, which are described and often discussed. Experimental procedures are treated as in other sciences.

5.5 DISSERTATION

The methods section in the dissertation is similar to that of a comparable paper, but it is written in more detail. It must serve as a comprehensive report of the methods to faculty advisors, and it may have to serve subsequent students who work in the same laboratory. In fact procedures, analyses, or computations may be so detailed that they are set off in appendices. Moreover, the dissertation includes more discussion of the methods, to explain or justify them. Such discussion is appropriate to the more comprehensive character of the dissertation and to its function in demonstrating that the student is innovative and has a full understanding of particular techniques and procedures relative to the theoretical structure of the research. In an experimental study, the research design is discussed in some detail.

TABLE 5.11 Subsections of methods section describing (a) research design, (b) persons participating as subjects, (c) site of research

a. *Research Design*

1. *Study Design:* A two-period cross-over design with a two-week initial baseline period, two four-week treatment periods and a one-week recovery period was used.

2. *Design.* Two different procedures were developed to determine whether children recognized equivocal situations. The first procedure required children to indicate whether the situation presented was associated with only one emotion or whether it might be associated with both positive and negative emotions. When an equivocal situation failed to elicit a two-emotion response, the child was asked a second question. This question was designed to prompt the children to think about possibile individual differences in their peer group. This question required the children to indicate whether "almost everybody" felt the same way or whether "some kids do and some kids don't" by using the pictures described above. Thus, there are two measures of recognition of the equivocal nature of a situation ("mixed responses"), unprompted and prompted.

In the present study, the prompting question was not asked immediately after the first question. Instead, the experimenter kept a written record of responses to the first question and asked the prompting question, where called for, in a separate procedure after all of the situations were presented once. This was done for two reasons. [*Dev. Psych.* 23, 1987, 115]

b. *Participants*

1. *Subjects*
Five age-groups of male children participated in this experiment (Table I). All the children were in regular classrooms and wrote with their right hand. [*Developm. Med. Child Neuro.* 29, 1987, 727]

2. *Subjects.* Informed written consent was obtained from nine healthy volunteers (four female, five male), who were 27.7 ± 2.6 yr. old and weight 74.4 ± 5.6 kg. All were nonobese (ideal body weight, $110 \pm 4\%$, Metropolitan Life Insurance Co. tables, 1985) and had no family history of diabetes mellitus. Six of the nine subjects exercised two to four times a week on a noncompetitive basis, but none participated in strenuous physical activity for at least 24 h before study. [*J. Clin. Invest.* 81, 1988, 1564]

c. *Site*

1. *Study Sites*
The study was carried out in four streams in the Ellis Hollow area of Tompkins County, New York (Fig. 1). The streams flow through eastern deciduous forest and have similar heterogeneous substrates, with components ranging in size from silt to boulders. Thirteen riffle sites (length 8–26 m) and four pools (depth 0.34–1.1 m and diameter 1.9–4.15 m) were sampled. The riffle sites were divided into six categories based on flow regime and stream size (Table 1). All pools were permanent and located between two adjacent riffle areas. One pool was between the two permanent large (PL) riffle sites, one between the two intermittent small (IS) riffle sites, and each of the other two between an intermittent medium (IM) riffle site (upstream from the pool) and an permanent medium (PM) riffle site (downstream from the pool). [*Can. J. Zool.* 66, 1988, 5479-80]

2. *Site*
The site of this study was the Oakwood Village Health Care and Retirement Community in Madison. Site selection was dependent upon its convenience of location, sufficient complexity to "challenge" the simulation technique, and permission from the administrator of the facility. Oakwood is composed of a 213-unit retirement center, a 137-bed skilled nursing facility, and a 15-unit community-based residential facility; and has approximately 400 residents. [*Geron.* 27, 1987, 170]

In the social sciences, the discussion of the methods may be a more important part of the methods section than in the natural sciences. It must demonstrate the student's understanding of complex methodological problems and issues, and so may be more extensive than in a paper. The site and participants are described and discussed in greater detail, as well as the procedures for choosing them and for ensuring reliable and accurate response to the questionnaire. Objectives of questionnaires and the reasons for various questions are explained specifically and in greater detail, and the complete questionnaire, as well as other documents, are usually included in appendices.

6

Tables and Illustrations

6.1 INTRODUCTION

Some parts of the research can be reported more effectively and more clearly by complementing the writing with tables and illustrations. Tables are visual displays consisting of sets of numerical or verbal elements arranged in one or more lists or in a grid of columns and rows; illustrations consist of pictorial elements. In a scientific paper, any visual representation that is not a table is called a figure, which may consist of a single illustration or several. Tables and illustrations are used in scientific papers whenever they can more clearly communicate the writer's message than words alone. They are usually found in the sections that present new information, the results and the methods section.

6.1.1 Uses of Graphics

Graphics are used for various purposes: (1) to illustrate natural objects, structures, or phenomena, (2) to present observations or measurements graphically, (3) to illustrate interrelationships, (4) to illuminate and clarify concepts, (5) to condense information and save space, and (6) to record new scientific data. Visual conceptualization, for example of instruments or physical structure, would be extremely difficult without illustrations to supplement the written text. Graphics are a necessity when the actual data or measurements are indirect and already in graphic form, as in oscillograms, chromatographs, computer-generated graphics, and so on. The presentation of a large body of numerical data would be unreadable, and almost incomprehensible even for only a few variables if presented in written form. Yet the same data arranged in a table fall readily into patterns of relationships and become comprehensible.

Graphics are generally more economical of space than is written text presenting the same data. The description of newly designed equipment would be very detailed and lengthy if not accompanied by a diagram. Graphics also serve as reservoirs for details, which the writer can draw upon in writing. Moreover, by thus removing the additional details from the path of the written text, graphics allow the structure of the argument to stand out. They thus not only provide economy of communication but also contribute to achieving a clear, concise, coherent presentation.

6.1.2 Visual Effectiveness

Graphics must meet (1) documentary or archival objectives, to record graphically new methods, findings, concepts, and so on, and (2) visual and semantic objectives, to give these new elements meaning visually. The documentary objective is easy for research scientists to realize by simply presenting the observations and data. This direct reporting is not enough; the graphic must also make a coherent statement. It is as necessary to organize the graphic presentation visually as it is to organize the written presentations verbally.

Making graphics *visually* meaningful is important because form communicates meaning. A graphic presentation, organized visually for meaning, can be understood immediately; the reader sees the patterns or relationships holistically as a unit. *Tables 6.1 and 6.2* illustrate the difference between a table prepared for documentary or archival purposes and one designed for visual effectiveness. *Table 6.1*, the original table, includes all the information required to interpret the data; the revision in *Table 6.2* has exactly the same information but is visually much more effective. Simply by looking at it, readers can immediately see the components of each model, information that they cannot get from *Table 6.1* without studying it attentively. This requirement for effective visual communication applies to all graphic forms in scientific papers, unless they are intended mainly for archival purposes, as in astronomical measurements and calculations (*Fig. 6.1*) and tabulations of economic or census data. The main function of such tables is to order a large array of data in convenient form, so that they can be readily retrieved; however, even these should be made as effective as archival and other constraints allow.

TABLE 6.1 Table suited to documentary or archival purposes

Variables in Model	Adjusted R²	Predicted Mean Square Error
TAPESCRE INTEXPER MARSTAT MARSTATI PREPARED	.2436	120.3587
DAYS TAPESCRE AGE MARSTAT PREPARED	.2388	121.3629
DAYS TAPESCRE MARSTAT MARSTATI PREPARED	.2388	120.8587
TAPESCRE AGE MARSTAT RATE PREPARED	.2387	122.4494
TAPESCRE INTEXPER AGE MARSTAT MARSTATI PREPARED	.2379	122.0762
TAPESCRE INTEXPER MARSTAT MARSTATI RATE PREPARED	.2412	121.8962

TABLE 6.2 Table 6.1 revised to make it visually effective

Model	Variables in Model								Adjusted R²	Predicted Mean Square Error
	TAPESCRE	INTEXPER	MARSTAT	MARSTATI	PREPARED	DAYS	RATE	AGE		
1	×	×	×	×	×	—	—	—	0.2436	120.3587
2	×	—	×	—	×	×	—	×	0.2388	121.3629
3	×	—	×	×	×	×	—	—	0.2388	120.8587
4	×	—	×	—	×	—	×	×	0.2387	122.4494
5	×	×	×	×	×	—	—	×	0.2379	122.0762
6	×	×	×	×	×	—	×	—	0.2412	121.8962

[Courtesy of W. Graham]

TABLE III. Radial velocities for stars in the U direction.

NAME	R.A.	DEC.	ST	VIS	PHO	VRAD
SAO76948	50147	295427	K0	7.8	9.2	-40.4
SAO76958	50302	294156	F0	8.0	8.5	19.3
SAO76983	50608	293200	G5	8.9	9.8	6.4
SAO76988	50632	290453	G0	8.6	9.7	-20.5
SAO76989	50633	294411	F8	6.6		6.0
SAO57722	50750	300931	K0	9.5	11.0	23.7
SAO57724	50753	301429	K0	8.6	9.5	-30.9
SAO77013	50756	294326	F5	8.3	9.2	49.1
SAO77018	50809	293323	F8	9.4	10.7	-48.7
SAO57740	50915	300933	F8	8.7	9.3	-65.8

Figure 6.1. Part of large table illustrating use of graphic for documentary or archival purposes. [Astron. J. 93, 1987, 595, with permission]

6.2 TREATMENT OF GRAPHICS

6.2.1 Planning and Preparing Graphics

The tables and illustrations should be planned as the paper is being planned, and even more than in the writing, they should be prepared while the research is in progress because drawing, drafting, and processing photographs are likely to cause delay. Planning and preparing graphics early helps writers to integrate them into the text, because the graphics and the written text can then evolve together, and also save rewriting and revising. With a sketch or table at hand while writing, writers can revise the graphic to make it visually effective, avoid including too much information in the text, and make better decisions about including or excluding a graphic. Such parallel development of graphics and text also ensures their being consistent with each other in terminology and in content.

Planning the graphics can begin with the outline, where the writer can note points at which a table or illustration would be desirable. Or early in the research, the writer can tentatively list tables and illustrations to include in the paper. During the research, the research scientist can set up the headings for the tables and draw preliminary sketches.

6.2.2 Scientific and Practical Criteria for Selection

In planning the graphics for the paper, the writer must consider (1) graphics essential for reporting the research, (2) material especially suited to graphic presentation, (3) the type of graphic best suited for presenting the information, and eventually, (4) the structure of each graphic for visual effectiveness. The chief scientific criterion for graphic presentation is whether the data must be presented in graphic form. The keynote is the *necessity*, because the scientific criterion is constrained by the practical criterion, the high cost of reproduction. Most journal editors do not have the funds to publish tables and illustrations freely, and some of the cost may be passed on to authors:

Ex. 6.1

> For Wistar Institute Press journals, it is the current policy to underwrite all normal black and white tabular and illustration costs. However, to help defray abnormal costs, authors will be requested to pay $75 per page for tabular and illustration pages that exceed 50% of the total number of printed text pages. Such payments . . . are not a prerequisite to the publication of any article. However, because of the very high cost of color work, such work will be initiated only at the author's request and expense. [J. Expt. Zool. 242, 1987, back cover]

Planning the graphics carefully can reduce the number of tables and figures as well as make effective use of space.

One chooses different types of graphics for different purposes. The same material can be illustrated in different ways, depending on scientific objectives. For example, to illustrate the appearance of the lesions on the stems of diseased plants, a photograph may be the most suitable graphic form. To illustrate the superficial distribution and structure of the lesions, a drawing might be more informative. For an understanding of the morphological or anatomical changes, a diagram of a cross section of the stem would show changes in the tissue layer. For histological or cellular changes in structure, a photomicrograph might be most appropriate. Although the subject or content of a graphic largely determine the *type* of graphic to be chosen, the purpose determines how the data are to be displayed.

Graphics must also be adapted to the size and format of the journal. In planning the graphics for a paper, research scientists should plan in terms of printed page widths and column widths to make economical use of space.

6.2.3 Graphics and Text

Depending on the research, either the text or the graphic may be dominant. Most of the information or data may be in the graphic, as in tables of extensive data or the diagram of experimental equipment; then the graphic serves as the substantive record, and the text presents only the salient elements. Or the text is the substantive record, as in the development of a theoretical model, and the graphic presents a schema to facilitate an understanding of it.

Because of their dual function, graphics create a dilemma for writers. By their graphic character, tables and figures stand outside the text. They are independent of it physically, and they must be self-contained for readers. They provide the evidence and examples for developing the text argument, but they do not constitute the argument. They are only the means, and so must be integrated into written text (see Chapter 7).

The objective in integrating graphics into the text is to draw them into the development of the argument, as appropriate. The text should highlight, summarize, or interpret details in the graphics, so that the reader can understand them. The writer refers explicitly to a table or figure and to the particular elements that are to be noted, and presents in the written text the ideas that he or she wants readers to derive from the graphic. In this way, readers are not dependent only on the graphics for their information and interpretation.

In integrating graphics into the text, the writer must make the text and graphics consistent with each other. If the text states that the structure shows a long projection, the projection in the illustration should be definitely long; if the text gives a value of 117.78, the table should not show the value to be 117.84. References to the title, notes, headings, or labels of a graphic should be consistent in content and terminology.

Although graphics must be integrated into the text, each table and figure must stand complete and independent of the written text. It should have all the information that readers require to understand it without reference to the text. This independence of the text accounts for the lengthy summary description of methods or protocol in the headnotes or footnotes of some graphics (*Table 6.16*) and for extensive legends.

The text must refer explicitly to every table and figure at least once, and may refer to it more than once. References made to tables or figures may be general or specific. When the table has extensive columns and rows of values, or groups of similar variables, it may be helpful to readers to include column number along with the variable named. In the text, references to graphics may be part of the structure of the sentence (*Ex. 6.2a*), or they may be set off parenthetically (*Ex. 6.2b–d*):

Ex. 6.2

    a. Omitting the real interest rate for the equation in *Table 7,* as expected, had little effect on the other estimated parameters of the equation, as is shown in *Table 8.*

    b. This difference is clearly seen in the 2D-J spectra (*Fig. 5*).

    c. This difference is already seen in the 2D-J spectra (*see Fig. 5*).

    d. This difference is already seen in the 2D-J spectra. (See *Fig. 5.*)

When parenthetical, they are usually included as part of the sentence, (*Ex. 6.2b,c*) and less effectively as a separate sentence (*Ex. 6.2d*), with the first word capitalized and the period falling within the parentheses.

6.2.4 Enumeration and Titles

Because tables and figures must be integrated into the text, they must be numbered; because they must stand independent of the text, they must be given a title and any additional information required to explain them. Even a minor table or figure should be numbered, to facilitate reference and discussion, and to allow the compositor more latitude in making up the page.

Enumeration.   Tables and figures are usually numbered and designated by label and number, as in, Table 3, Figure 3, and so on. The labels may be capitalized or italicized, and "figure" is often abbreviated (*Appendix 6B*), especially in the text. They are numbered consecutively in two separate series, that is, Table 1, 2, 3, . . . and Fig. 1, 2, 3,. . . . . Each table or figure is numbered in the order of its primary reference in the text; however, secondary references may interrupt this order, and the order of *references* to them might be *Table 1, 5, 2, 3, 5, 4, 5, 6.* In the manuscript the primary reference is designated by a displayed centered note: "Insert *Table 5* [or *Fig. 5*] about here," set with extra space above and below it and often with a line across the page between the text and the space.

Titles.   Each table and figure should have a title to describe the subject of the graphic and the data included. Because it lacks a title, *Table 6.3* cannot even be identified as to subject without reference to the text. Titles follow the numerical designation of the table or figure. Titles of tables, with their number, are regularly placed above the tables; titles of figures are usually placed below the figures, but may occasionally be above the figure.

The function of the title is to describe the contents of a table or the subject of a figure fully and accurately; therefore it must be specific, informative, clear, and complete. As titles are analogous to headings or labels, they must be brief—as short as is consistent with describing the contents. Because tables and figures must be understood without reference to the text, titles may become excessively long, unless efforts are directed to making them succinct. Titles should include the significant elements of the graphic; then if supplementary or explanatory material is required, it can be included in headnotes, footnotes, legends, or in the table or figure itself, as appropriate. If notes are extensive and apply to a series of similar tables, or figures, the title and notes of the first graphic can provide the full details, and reference is made in subsequent members of the series to the graphic with the complete information.

6.2.5 Dissertation

For a dissertation, graphics do not have to be of professional quality for photoengraving and publishing, and the space available for graphics is not as limited. A

TABLE 6.3   Table without title or explanatory notes

TABLE I

Xo (μm)	Ax (μm)	Ay (μm)	d (Å)	w (μm)	g (μm)	r_{exp} (%)	r_{IM} (%)	r_R (%)	r_{FE} (%)
7	4.93	2.91	2300	30	3	11	10.8	13.9	10.5
7	5.44	3.71	3200	30	7	22	19.2	27.2	18.8
7	6.05	4.11	2400	30	3	8	8.8	11.4	8.9

[C. Sabatier & E. Caquot. IEEE J. Quant. Electr. 22, 1986, 36, © 1986 and with permission of IEEE]

table or figure can occupy a full page, and there are few restrictions on the number of tables or figures included.

Nevertheless they should be prepared as carefully as possible. In general, where meaning is conveyed, the graphic should come as close as possible to the quality required for a paper. Where only form is affected, for example, in the professional evenness of lines, spacing, and positioning, a less professionally prepared graphic is permissible. However, preparing graphics of a quality suitable for publication gives graduate students experience while they still have advisors available for consultation.

TABLES

6.3 FORM

6.3.1 Basic Structure

Research often generates numerous observations, measurements, and quantities that must be reported. When such elements are numerous or detailed, they are too confusing to follow in the written text and are more effectively presented in tabular form. The elements in a table may consist of measurements, values derived by various types of calculations, or words, phrases, and even sentences (*Table 6.47*). The discussion here focuses on tables with numerical data.

A table may consist simply of a series of items, which are arranged vertically in one or more lists or columns (*Tables 6.4, 6.45b*). The items are related only to the column heading or title, and any ordering of the items is within the list, not across rows. Such lists may consist of questions asked in a questionnaire, lists of different types of plants, animals, soils, the composition of a solution or diet, the demographic or other characteristics of a group, and so on. Usually, however, tables are correlational; there are usually two sets of elements. One of these, the column at the left constitutes the row headings for the table. The others constitute the columns of quantities or values that correspond to the row headings.

At a minimum, such a table consists of a list of items with a corresponding list of values (*Tables 6.26a, 6.35a, 6.36*), but more often, it consists of a list of items which correspond to several lists (columns) of values. Each value represents the intersection of the variable in the column heading and row heading and thus expresses the relation between the two, but its relation to other values in its column and row may also be of interest. Therefore, although the table presents discrete values, it displays the relationship of the various variables.

Fig. 6.2 illustrates the parts of a table. These elements may be variously set off by "rules," that is lines. Lines are usually used to set off headings or sets of data. In a ruled table, the columns may be separated by vertical rules and one or

TABLE 6.4 Table consisting of a list of items

TABLE 1

Summary of the Variables Investigated in This Report

POSSIBLE ARTIFACTS

Task Variables

Horizontal versus vertical paper orientation (Experiment 2)
Time course—when does the difference begin?
Type of reading—proofreading versus comprehension (Experiment 3)
Real-life reading distances (Experiment 4)
Visual angle (Experiment 5)
Eye movements (Experiment 6)

Display Variables

Dynamic characteristics of CRT displays (Experiment 7)
Different CRT displays (Experiment 8)
Self-luminous versus reflective displays
Display contrast
Scale of layout
Character height/line-width ratio
Polarity (Experiment 9)
Aspect ratio
Same font (Experiment 10)

Personal Variables

Experience at reading from CRT displays Age

[Hum. Fact. 29, 1987, 1, © 1987 Human Factors Society, Inc., with permission]

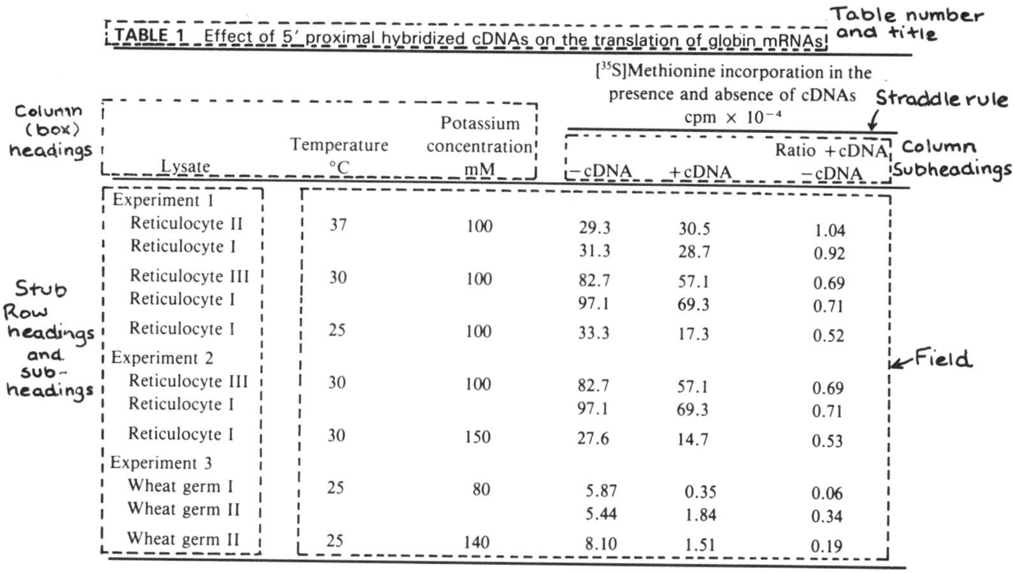

TABLE 1   Effect of 5′ proximal hybridized cDNAs on the translation of globin mRNAs   *Table number and title*

Lysate	Temperature °C	Potassium concentration mM	[35S]Methionine incorporation in the presence and absence of cDNAs cpm × 10⁻⁴		
			−cDNA	+cDNA	Ratio +cDNA −cDNA
Experiment 1					
Reticulocyte II	37	100	29.3	30.5	1.04
Reticulocyte I			31.3	28.7	0.92
Reticulocyte III	30	100	82.7	57.1	0.69
Reticulocyte I			97.1	69.3	0.71
Reticulocyte I	25	100	33.3	17.3	0.52
Experiment 2					
Reticulocyte III	30	100	82.7	57.1	0.69
Reticulocyte I			97.1	69.3	0.71
Reticulocyte I	30	150	27.6	14.7	0.53
Experiment 3					
Wheat germ I	25	80	5.87	0.35	0.06
Wheat germ II			5.44	1.84	0.34
Wheat germ II	25	140	8.10	1.51	0.19

(Annotations: Column (box) headings; Stub Row headings and sub-headings; Straddle rule; Column Subheadings; Field)

Figure 6.2. Table showing parts. [Adapted from J. Biol. Chem. 261, 1986, 13982]

TABLE 6.5   Table with box headings and vertical rules

TABLE III

C/E Ratios of Gamma-Ray Energy Deposition Rate in Stainless Steel with TLD-200
Measurements in Blanket 2Al

Experiment Position	r (m)	C_n	f	TLD-200 Measurement		
				Statistical Error (%)	Systematic Error (%)	C/E
1	0.237	0.88	1.06	2.3	7.9	0.95
2	0.267	0.89	1.06	2.5	7.6	1.01
3	0.296	0.91	1.06	2.6	7.2	1.02
5	0.355	0.93	1.06	2.7	6.6	0.94
6	0.385	0.93	1.06	2.9	6.4	0.92
12	0.563	0.96	1.07	4.4	5.9	0.82
15	0.651	0.96	1.07	5.6	5.8	0.75
17	0.711	0.96	1.07	7.3	5.8	0.70

[Adapted from Tab. 3, Wang et al. Nuc. Sci. Engr. 93, 1986, 269, © 1986 Amer. Nuclear Soc., La Grange Park, IL, with permission]

more of the rows by horizontal rules. In *Table 6.5*, the column headings are completely enclosed in rules, and the individual headings are called "box headings." When box headings are divided horizontally by a rule to separate headings from subheadings, the separate boxes are called "decks," with the heading being in the "upper deck" and the subheadings in the "lower deck."

Ruled tables are costly to set into type and are not as commonly used as tables with only horizontal rules. In most scientific journals, tables have three main horizontal rules: one across the top, separating the body of the table from its title and the main text; one below the column headings, setting them off from the title and the rows below them, and one across the bottom, separating the rows of data from the footnotes and the main text. If a column heading has subheadings, they are included under a horizontal spanning rule (*Fig. 6.2*).

There are many variations on this basic table. The table may lack any of the three rules or all of them (*Tables 6.6, 6.38*), or it may have additional horizontal rules (*Table 6.7*). When data are voluminous, as in census-type tables, horizontal and vertical rules help guide readers through the innumerable numbers in the rows and columns. In such tables the column headings may even be set vertically either perpendicular to the horizontal rule or diagonally from it to save space as in *Table 6.11c*. The wide variation in the use of rules is seen in the various tables in this section. A few journals use rows of dots ("leaders") from the end of the row headings to the corresponding values in the first column (*Table 6.25b*). Tables may include illustrations; then the drawings must be reproduced photographically and combined with the table in printing. Many journals have a standard format for their

TABLE 6.6   Table with no rules

TABLE 2.   Two-way contingency tables for *Bromus tectorum* cohorts during the 1980–81 growing season. Dashes in the chi-squared columns indicate that expected values were not sufficiently large to validate the test. See Table 1 for explanation.

Treatment and cohort	Alive		Dead		χ^2_{homo}	χ^2
	Observed	Expected	Observed	Expected		
Protected						
2 Jan	24		131			
7 Mar	0		11			
Autumn-winter						
22 Nov	45	41·6	190	193	0·86	1·23
2 Jan	11	6·60	25	29·4	4·43*	7·59*
7 Mar	2	0·84	6	7·16	3·07	4·88*
Autumn-spring						
22 Nov	31	33·5	210	208	0·54	1·59
2 Jan	5	5·71	33	32·3	0·13	0·69
7 Mar	3	1·57	9	10·4	3·16	0·80
Winter-summer						
2 Jan	13	22.0	214	205	10.0*	4.26*
7 Mar	0	0·00	2	2	—	—

[Adapted from J. Ecol. 74, 1987, 748, with permission]

tables. Writers should examine tables in recent issues of the journal of choice before designing their tables. Each table should be typed double space beginning on a separate page.

6.3.2 Columns and Rows

The matrix character of tables results in a grid of cells and so sets limits on the space available for the various entries. Space in a table is particularly limited for headings and other verbal entries. Economy of means thus becomes of central importance in constructing tables, and strategies must be devised to make form substitute for words. Whether a set of data is set in a column or a row depends on the relation of the various variables, the amount of space taken up by headings and the values in the field, and the size of the printed page of the journal.

Because a table makes its scientific statement by its layout, the various elements should be planned and ordered so that the relationships are accurate and clear. Any ordering of elements will help readers, even ordering on extrinsic criteria, but ordering on intrinsic criteria reinforces the message. For comparing sets of data, the related columns or rows can be placed adjacent to one another, rows one above the other or columns side by side, depending on the objective. For example, percentages can be set adjacent to totals and calculated values adjacent to measured quantities. It is usually easier to compare values in a column than in a row. For example,

TABLE 6.7 Table with excessive rules

Table III. Mean oxygen consumption (VO_2) and gross energy expenditure (E) before and after 2-MJ test meals of dietary fats and thermic responses (ΔE), and urinary nitrogen excretion (N) after the test meals in 5 or 4 normal-weight men (means ± SE; significance p)

	Sunflower oil (n = 5)		Fasting (n = 5)		Butter (n = 4)		Fasting (n = 4)	
	VO_2 ml/min	E kl/6 h	VO_2 ml/min	E kl/6 h	VO_2 ml/min	E kl/6 h	VO_2 ml/min	E kl/6 h
Before test meals	276±8	1,993±59	283±12	2,033±84	284±11	2,052±73	292±10	2,099±66
After test meals	282±9	2,027±63	284±17	2,037±120	294±10	2,118±65	297±14	2,133±93
p		NS		NS		0.05		NS
ΔE, kl/6 h		34±37		5±37		66±16		34±29
N, g/6 h	2.2±0.1		2.8±0.3		2.7±0.3		2.8±0.3	
p	0.05		0.05					

[Ann. Nutr. Metab. 28, 1984, 248, with permission of S. Karger AG, Basel]

if one wishes to compare mean values accompanied by error values, relative to the column heading (i.e., within the column), then the error values are best placed beside the means. One can then read down the column, without having the error values intrude, as they would if the error values were placed in the column below the mean. However if one wishes to compare the mean values relative to the row headings (that is, across the row), then placing the error values below the means allows readers to make a comparison of the means across the row without the intrusion of the error values between the mean values.

Tables should be adapted to the printed page if possible. A table with numerous columns may not fit across the page of a journal with a narrow page width. The table may then have to be set broadside on the page or extended across two pages, both undesirable alternatives. However, it can be revised so that the columns and rows are reversed. A table may be too narrow for the width of the page; then the excess space may be variously distributed (1) outside the table, (2) between the columns, or (3) between the stub and the columns, although this space should not be excessive. Within the table, there should not be large areas of blank space, especially when other parts are crowded (*Table 6.39*). When a table is too long for a page, a doubling of the table with the two parts side by side may be considered, if the table is narrow enough (*Tables 6.40, 6.41*).

The minimum width of columns is usually determined by the longest element in the heading. In numerical tables, the widest element is likely to be a word in the heading; in verbal tables, it is likely to be a word, or series of words, in the column (*Table 6.47*). The maximum width of columns is determined by the number of columns, their minimum widths, and the width of the printed page. The space between columns is kept equal, and each column can be considered as including half the space on each side of it.

6.3.3 Headings

Each column and row should be given a heading, that is, essentially a label that identifies the values or elements in that column or row. A column heading may extend over several columns, each having a subheading (*Tables 6.6, 6.25, 6.45a*); and a row heading may include several subheadings, each in a separate row (*Table 6.13, 6.19b, 6.30*). The heading at the top of a column or group of columns labels the corresponding column or the group of columns below it. The column heading at the top of the row headings labels the column of row headings; it is not be the label for the column headings, as in *Table 6.8a, 6.9a*. In *Table 6.8a*, the heading, "Temperature K," applies to the column headings, 681, 735, 803, not to the physical quantities in the stub row headings, as it should. The table should be revised as in *Table 6.8b*. In *Table 6.9a*, "Bombarding Energy and Average R_1," apply to four sets of paired numerals (bombarding energy) and (average R_1), not to the column headings "Revealed ordered planes" and "β (deg)." The headings should therefore be revised as shown in *Table 6.9b*.

Headings should be provided for all columns, even the column of row head-

TABLE 6.8 Table showing (a) column heading of row headings applying to the column headings instead of row heading; (b) revised to make it apply to row headings

a.

Table IV. The Values of Physical Properties

Temperature, K	681	735	803
C_s, wt pct[25,52]	0.54	0.80	1.20
ρ_s, kg $\cdot$ m^{-3} [43]	8.87×10^3	8.87×10^3	8.87×10^3
ρ_{ep}, kg $\cdot$ m^{-3} [38]	$6.86_7 \times 10^3$	$6.83_0 \times 10^3$	$6.87_7 \times 10^3$
η, kg m$^{-1} \cdot$ s^{-1} [50]	$1.28_1 \times 10^{-3}$	$1.18_7 \times 10^{-3}$	$1.09_5 \times 10^{-3}$
D, m$^2 \cdot$ s^{-1} [51]	0.81×10^{-9}	1.04×10^{-9}	1.35×10^{-9}

[Metall. Trans. B 17B, 1986, 287; with permission]

b.

Physical properties	Temperature, K		
	681	735	803

TABLE 6.9 Table showing (a) column heading of row headings applying to the column headings instead of row headings; (b) revised to make it apply to row headings

a.

Table VI: Summary of results of damage depth P_{II} for the four irradiations.

Bombarding Energy (keV) Average R_1 (Å)		0.5 607		1.0 624		1.5 652		2.0 760	
Revealed ordered planes	β (deg)	$n \cdot d_{002}$	P_{II}	$n \cdot d_{002}$	P_{II}	$n \cdot d_{002}$	P_{II}	nd_{002}	P_{II}

b.

Revealed ordered planes	β (deg)	Bombarding energy (keV) Average R_1 (Å)							
		0.5 607		1.0 624		1.5 652		2.0 760	
		$n \cdot d_{002}$	P_{II}	$n \cdot d_{002}$	P_{II}	$n \cdot d_{002}$	P_{II}	$n \cdot d_{002}$	P_{II}

ings. This heading should not be omitted (*Tables 6.17, 6.30b, 6.39*) simply because it seems self-evident or because it is difficult to assign a label. For example the row headings in *Ex. 6.3b*, comprising the row headings in a table:

Ex. 6.3

a. Black	Spanish	Italian	German
White	Asian	Polish	Irish

b. Officials	Shopkeepers	Capitalists	Smallholders
Teachers	Craftsmen	Landed	Servants
Minor Officials	Seamen	Proprietors	Paupers
Tradesmen	Fishermen	Farmers	Others

had no column heading, yet they could easily be labeled "Occupations." The column heading for the row headings in *Ex. 6.3a* is embedded in the title "Employment Measures for Selected Ancestry and *Ethnic Categories*" [*italics added*].

Space constraints are usually not as great for row headings as they are for column headings. One should therefore consider using row headings when a table must accommodate many or long headings or subheadings. Also, subheadings can be accommodated more easily in row headings than in column headings. The subheadings of a column heading are usually delimited by a spanning horizontal rule that extends from the beginning of the first subheading to the end of the last subheading (*Table 6.7b, 6.24a, 6.25, 6.45a*). Headings may be set over subheadings or across part of the field without rules (*Table 6.6*), but then it may not be clear which subheadings or columns fall under a heading.

Headings should start with the head word. Because of the limited space, they should be telescopic, terse even. Abbreviations can be used to fit the heading to the space; when they are unfamiliar or are used for the purpose of the table, they are defined in footnotes. An abbreviated form (*Table 6.11b*) is preferable to an arbitrary symbol, number, or letter (*Tables 6.10, 6.11a*), which has little association for readers. Even if it must be explained, it serves as a mnemonic device that can recall the

TABLE 6.10   Table with column headings numbered to save space; numbers replaceable by mnemonic abbreviations of accompanying explanatory footnote (abbreviations not in original table)

Table 13. People killed in the earthquake of 3rd March 1985, by cause of death, age and sex of 145 victims

Ages	Sex	Causes										Total
		1	2	3	4	5	6	7	8	9	10	

One death corresponding to a cerebrovascular accident in a 65-year-old female is not included in the Table for reasons of space. Total deaths–146.

1. Multiple trauma. [*MT*]
2. Head injury (closed). [*HI*]
3. Crush injury of throax. [*CT*]
4. Myocardiac infarction. [*MI*]
5. Head injury (compound fracture). [*HF*]

6. Dorsal and/or cervical fracture. [*DCF*]
7. Skull fracture [*SF*]
8. Subterranean asphixiation. [*SA*]
9. Fractured pelvis. [*FP*]
10. Internal injury. [*II*]

[Disast. 10, 1986, 138, with permission]

TABLE 6.11 Column headings designed to save space by (a) letters or symbols, (b) abbreviated forms, or (c) vertical orientation of type

a.

Table 2
Diagnostic checks for the daily test equations.[a]

Currency data	L	G	$Q(10)$	S	K	I	$Q*(10)$

b.

TABLE 1. The mean July temperature at Des Moines, Iowa (DSM) and two neighboring stations with identical normalized departures from the mean (Z-scores) and correlation with DSM (correlation = 0.81), but unlike year-to-year variances. A discontinuity (+0.5°C) is introduced into the record DSM between 1965 and 1966; RFD is the station identifier for Rockford, Illinois and GRI for Grand Island, Nebraska.

	Temperature (°C)								
	Monthly means			Differences		+0.5 Discontinuity differences		−0.5 Discontinuity differences	
Year	DSM	GRI	RFD	GRI-DSM	RFD-DSM	GRI-DSM	RFD-DSM	GRI-DSM	RFD-DSM

c.

Table 2
Number of scientific publications per country and field

ARGENTINA BOLIVIA BRAZIL CHILE COLOMBIA COSTA RICA CUBA ECUADOR EL SALVADOR GUATEMALA

heading to readers. In *Table 6.10* the numerals, representing the ten causes of death listed in the footnote to the original table can be replaced by the associated mnemonic abbreviations (not in the original). Numbers are regularly used for the column headings of cross-correlation tables to save space, but these usually represent the variables written out in the row headings (*Table 6.14*). *Table 6.11a* has enough space for the letters to be replaced by more informative headings: Log, Godf. (Godfrey test), Skew (skewness), and Kurt (kurtosis). And the vertically printed column headings in *Table 6.11c* could easily be replaced by a two- to three-letter abbreviation set across the column.

Headings that are too long or too wide can be shortened if they share a common element. This can be converted to a main heading and the remaining elements

TABLE 6.12   Original headings with repeated elements revised to eliminate repetition by making repeated elements main heading and other elements subheadings

a. *Original*

Decrease in satisfaction after migration	Increase in satisfaction after migration	Change in satisfaction after migration

b. *Revised*

Satisfaction after migration		
Decrease	Increase	Change

c. *Original*

Plasma urea mg/24 hr	Urine urea mg/100

d. *Revised*

Urea	
Plasma mg/24 hr	Urine mg/100

made subheadings, (*Table 6.12, 6.31*). The revision shortens the original column headings, eliminates the repetition, and highlights the variables. When headings are long but do not share a common element, it may be possible to make them row headings.

Because headings are condensed, it is especially important that they be clear. Consistency and a strict parallelism can contribute markedly to clarity. Coordinate headings should be parallel in form. The terminology used in the headings should be consistent with that in corresponding headings and with the terminology in the title. Capitalization should be consistent. It is desirable to capitalize only the first word of a heading. Besides saving space, this makes it possible to distinguish capitalization for other purposes. The need for consistency extends across the tables in a paper; tables presenting similar data should have a comparable pattern of headings. In *Table 6.13a*, the babies and conditions are the column headings; in *Table 6.13b*, they are the row headings. Therefore though the tables are set one above the other in the journal, the corresponding values are not readily compared. With the column and row headings made consistent, the two tables would be readily compared. This would then suggest combining the two by adding the rectal temperatures in *Table 6.13b*, as a column parallel with "Right eye" and "Left eye."

TABLE 6.13   Related tables with same headings treated as (a) column headings in one table and (b) row headings in the other table

a.

TABLE I

Mean rectal temperatures (°C) with standard deviations for incubated and non-incubated premature babies and fullterm controls

| Premature babies | | Normal controls | |
Incubated (N = 23)	Non-incubated (N = 14)	Incubated (N = 10)	Non-incubated (N = 14)
36·76 ±0·41	NS	36·81 ±0·36	37·03 ±0·30	NS	37·03 ±0·56
	36·78	p = 0·025	37·03	

b.

TABLE II

Ocular temperatures (°C) for incubated and non-incubated premature babies and fullterm controls

| | | Right eye | | | | Left eye | | |
	N	Mean	SD	p	N	Mean	SD	p
Premature babies								
Incubated	23	36·93	0·64	<0.0001	2?	36·73	0·53	<0·001
Non-incubated	14	35·89	0·63		14	35·84	0·63	
Normal controls								
Incubated	10	36·86	0·55	<0·01	10	36·82	0·53	<0·01
Non-incubated	13	36·05	0·75		14	35·84	0·94	

[Dev. Med. Child Neur. 28, 1986, 281, with permission]

Row headings and subheadings, even column headings are sometimes numbered or lettered sequentially. Such enumeration takes up space unnecessarily. It also adds another set of numerals for readers to attend to and so should be avoided. In cross-correlation tables (*Table 6.14*), the numbers substitute for corresponding headings to save space.

Units of measure must be specified and are usually abbreviated. They are included with the headings, not as footnotes to values. In the rows, they usually follow immediately after the heading or subheading, separated by a comma, or enclosed in parentheses. In the columns, since space is limited, they are set on a separate line, the last in the box, sometimes in parentheses (*Table 6.3, 6.5, 6.15, 6.43a*). They may be placed below the horizontal rule that sets off the column headings (*Table 6.25a, 6.33*), but unless this saves space (*Table 6.15b*), this overemphasizes them. When the same unit of measure applies to all the values in the table, the unit can be specified in a footnote, or made part of the title. There, it is more effective within

TABLE 6.14   Cross-correlation table showing substitution of numbers in column headings for corresponding numbers of variables in row headings.

Table 1

Descriptive Statistics and Zero-Order Correlations for Study Variables

Variable	Mean	S.D.	2	3	4	5	6	7	8	9	10
1. No. of nonsupervisory employees*	2.54	0.32	.00	−.03	.05	.41**	−.01	−.07	.04	.01	−.38**
2. Decentralization	86.51	78.36	—	−.27*	−.05	−.01	−.08	−.03	−.09	−.12	.31*
3. Production-process complexity (Woodward)	4.92	1.55		—	.14	−.10	−.09	−.05	.04	−.09	.13
4. Quality-control complexity†	55.65	37.29			—	.19	.13	.18	.14	.10	.37**
5. Structural complexity	5.14	2.00				—	.16	.15	.11	.09	.05
6. Decline (1967–72)†	−1.40	29.04					—	.96**	.86**	.95**	.01
7. Production-process complexity × decline	−10.67	149.57						—	.81**	.89**	.04
8. Quality-control complexity × decline	57.87	1741.77							—	.83**	.03
9. Structural complexity × decline	1.88	153.02								—	−.10
10. Administrative intensity†	10.60	5.06									—

[Reprinted from *Complexity and Administrative Intensity: The Case of Declining Organizations* by McKinley. Admin. Sci. Q. 32, 1987, 96, © 1987 Cornell Univ. with permission]

(*Ex. 6.4b, Table 6.13*) or at the end (*Ex. 6.4c–e, Table 6.17b, 6.22a*), than at the beginning (*Ex 6.4a, Table 6.30b*), where it displaces the more important elements:

Ex. 6.4

   a. *Percentage* of Tenmile Creek Channeled by Sections
   b. Mortality (*in percent*) and analysis of variance between half-sib families of 5 Scots pine. . . .
   c. Comparison of Fourier Series Fit to Sample Air Temperature (*data in degrees Celsius*)
   d. Respondent Characteristics by Respondent Source (*%*)
   e. Measured Values of the Membrane Shear Modules (*dyn/csn*)

Large numerical quantities should be expressed in smaller units to save space, but comparable values should be expressed in comparable units to facilitate comparison. Exponential terms for units, such as 10^{-3} or 10^3, should be avoided in headings, because readers cannot be certain whether the mathematical operation has been carried out in obtaining the values in the table or remains to be carried out. If used, they should be explained. If the sample number (N) follows most of the row headings, and there is space in the width of the table, a separate column for N should be considered.

6.3.4 Field

The values or words in the field are labeled by the headings and title and explained or supplemented, if necessary, by headnotes or footnotes. These values and their relationships are the elements of interest to readers; the headings are just the means of designating the relationship. A column or row should not be wasted for totals, when the values to be added are so few or so easily added that the writer can sum them mentally. If the percentage columns are not easily summed, a footnote can indicate that each percentage column totals 100 percent. Indeed percentages should be used with care to avoid misrepresentation.

Calculated values should be consistent with the accuracy of the original measurements. The standard deviation or error must be specified and accompany the values. Error values may follow the mean values in the same row in separate columns (*Table 6.19a, 6.45a*), or the same column (*Table 6.15b, 6.17a*), or below the mean (*Table 6.15a*). They may be distinguished by parentheses (*Table 6.15a*) or by the ± sign. When means and error values are in the same column, the ± sign is separated from them by a space on each side (*Table 6.15b, 6.33a*). When space is not limiting, they can be positioned to allow the comparisons of interest. Levels of significance can be indicated by asterisks or bullets (•) which are explained in a footnote. Space should not be wasted by devoting a column or row to a few significant values.

Numbers are centered in the column and aligned with the last line of the corresponding row heading. Whole numbers are aligned from right to left; decimals are aligned with the decimal point; values with ± are aligned with the sign. Verbal elements are aligned from the left, with the first word often capitalized; when the text runs over, the succeeding lines are usually flush left. If the entries consist of several lines of text, the first line may be indented to improve readability, or the entries can be separated by a space.

The empty cell is difficult for readers to interpret: A dash is no more helpful than the empty space, and "ND" (no data) or "NA" (not applicable, not available) are not much more explanatory; whatever the indication, it should be explained in a footnote.

6.3.5 Footnotes and Headnotes

Research scientists often have notes or comments to add to the tables. Such notes cannot be inserted in the title without making it too long, or in the column or row without disturbing the structure of the table; therefore they are usually placed in footnotes or less commonly, in headnotes. Headnotes are inserted after the title (*Table 6.15a, b, 6.24a, 6.33*) and are usually reserved for describing the protocol or method. Footnotes may include the same type of notes as headnotes (*Table 6.16*),

TABLE 6.15 Tables with notes on methods (a) in caption as part of title, (b) as headnote separate from the title

a. *Caption*

Table 4. Effect of increased humidity on growth rate and effective turgor (P_e) of bean leaves. Leaves were placed into a humid box (RH %) 12 h before measurements were made. Growth rate was determined as in Table 3 but averaged over 12 h. P, Y and P_e were determined as described in Table 3. (SE/n indicated)

Leaf age (d)	RH	Growth rate ($cm^2 \cdot cm^{-2} \cdot h^{-1}$)	P (bar)	Y (bar)	P_e (bar)
11	60	3.5 (0.5/5)	3.4 (0.3/7)	3.0 (0.1/3)	0.4
	100	3.6 (0.8/5)	4.2 (0.1/7)	3.1 (0.5/3)	1.1
12	60	1.4 (0.4/4)	3.2 (0.1/9)	2.1 (0.2/7)	1.1
	100	1.7 (0.2/4)	3.7 (0.2/5)	2.8 (0.2/3)	0.9

[Planta 167, 1986, 41, with permission]

b. *Headnote*

TABLE II

Effects of cholera toxin and pertussis toxin treatments on PGE_2-induced GTPase activity of canine renal medullary membranes

Canine renal medullary membranes were isolated and subjected to the indicated treatments and then assayed in quadruplicate for GTPase activity in the presence and absence of 10^{-7} M PGE_2 as described in the text. This experiment was performed three times with similar results.

Treatment	GTPase activity	
	$-PGE_2$	$+PGE_2$
	pmol/min/mg protein	
None	18.10 ± 0.31	22.29 ± 0.12*
NAD	19.93 ± 0.14	24.80 ± 0.03*
Cholera toxin	18.89 ± 0.13*	23.83 ± 0.19*
Pertussis toxin	16.49 ± 0.53	20.81 ± 0.34*
Cholera toxin plus NAD	20.29 ± 0.09	24.76 ± 0.37*
Pertussis toxin plus NAD	11.50 ± 0.55	11.69 ± 0.29

*Values were significantly different from control values (values in the absence of PGE_2) ($p < 0.05$).
[J. Biol. Chem. 261, 1988, 13433, with permission]

TABLE 6.16   Table with notes on method in footnote

TABLE 1. Effect of aphidicolin and PAA on the plating
efficiencies of wild-type, Aph[r] mutant, and progeny Aph[r]
mutant viruses

| Virus | Plating efficiency with[a]: | |
	Aphidicolin (2 μg/ml)	PAA (25 μg ml)
Wild type	<0.05	0.60
Aph[r] mutant	0.83	<0.05
Progeny Aph[r]	0.75	<0.05

[a]To determine the plating efficiency, confluent monolayers of human embryonic fibroblasts were infected with approximately 100 PFU of each virus. After a 1-h adsorption period, the cultures were overlaid with 0.5% agarose in Eagle minimal essential medium containing 2 μg of aphidicolin per ml, 25 μg of PAA per ml, or neither compound. The number of plaques was counted at 2 days postinfection, as described previously (22).
[Virol. 61, 1987, 389, with permission]

but they include all other types of notes as well (*Appendix 6A*). They may apply to all the table, to just one column or row, or to a single entry in the table. They usually supply explanatory or additional details, which may qualify, limit, or expand upon the element noted. Such notes should be brief; they should not be so long or so numerous as to be too prominent. *Table 6.26b,* which has only five values, has five footnotes and a source note that occupy more than twice the space of the table. If a column or row has the same entry for every cell (*Table 6.29*), it can be transferred to a headnote or footnote to save space.

Footnotes are designated by superscripts, symbols, numbers, or letters, depending on journal style (*Appendix 6A*). They may be attached to a word or phrase in the title, or to the whole title, and then apply to the whole table. When attached to the column or row heading, they apply to the entries in the column or row; they may be attached to single value in the field. Generally, it is less confusing to attach letters to numerals and numerals to letters. However, when letters are used for footnotes, they may be confused with letters used for other purposes. Symbols avoid this problem, but cause confusion when levels of significance are indicated by asterisks (*Appendix 6A*).

Text footnotes are ordered numerically, alphabetically, and symbolically by convention (*, †, §, #, **, ††, §§, ##, etc.). Footnotes to tables are ordered in a series separate from the footnotes to the text and to other tables. In each table the footnotes are ordered from the very beginning of the title, reading in order from left to right, and from top to bottom, in normal reading order. In the table, the footnote symbols follow the items footnoted; in the footnotes, they precede the note, which is set in reduced type and may be set flush or indented. Footnotes may be a separate paragraph, or run in together in a block (*Appendix 6A*).

In some journals the general notes that apply to the whole table are treated separately without designation, or following the heading "Note" (*Appendix 6A*). Such general notes usually precede the enumerated footnotes. Some tables include the source for the data in the table, which is usually listed separately as the last footnote and labeled "Source." All footnotes are typed in the table. Footnotes may cite references to the literature by author-year; when references cited in the text are numbered in order of mention, then the footnotes in the tables are cited in full.

6.4 TITLES

6.4.1 Form and Substance

The table title is preceded by the table designation and number, which varies in form in different journals. The table designation may be on a separate line from the title, and then centered (*Table 6.17b, 6.18, 6.24*), or flush left (*Table 6.4, 6.32*), or indented on the left-hand margin. It may be on the same line with the title and then similarly the combined table number and title may be centered (*Table 6.6, 6.16, 6.28, 6.42*), flush with the left margin (*Table 6.27, 6.36, 6.40*), or indented (*Table 6.7, 6.11b*). The table designation is capitalized, but it may be all in capitals (*Table 6.13, 6.27, 6.43*), or in capitals and small capitals (*Table 6.19a, 6.24a, 6.33*). It is usually in roman or boldface, rarely in italic type (*Table 6.26c*). The table numbers are usually arabic numerals, but may be roman numerals. Roman numerals are often used for table numbers in dissertations, but arabic numerals designating the chapter and table numbers, as in "Table 5.3," avoid the cumbersomeness of the roman numerals. When the title follows the table numbers on the same line, the table number is usually followed by a period, occasionally by a dash, period and dash, or colon. The title is usually in roman type, but may be in italic (*Table 6.25a, 6.26, 6.33, 6.45a*) or less frequently in boldface type (*Table 6.28, 6.38, 6.39*). In many journals only the first word of the title is capitalized, except for words requiring capitals, but in some journals all the important words are capitalized (*Table 6.28, 6.36, 6.37, 6.39*). The title may be in capitals and small capitals (*Table 6.8, 6.20a, b, 6.35a*), and much less frequently, it is all in capitals (*Table 6.20c, 6.45b*). It usually lacks end punctuation but may be followed by a period (*Table 6.7a, 6.40*). When the title is longer than one line, the succeeding lines may be centered (*Table 6.17b, 6.33a, 6.21d*), flush with the left margin (*Table 6.32, 6.36, 6.40*), rarely indented. Where there is a head note it usually follows the title on a separate line (*Table 6.15, 6.21d, 6.33*), but it may follow directly after the title (*Table 6.15a*), or the same material may be included in a footnote (*Table 6.15c*).

The title must describe the contents of the table; it must be specific, descriptive, informative, and complete, yet concise. The title is a label; it must describe, not interpret or characterize the data presented. A descriptive title is easily formed by combining the row headings in relation to the column headings, as in *Table 6.17, 6.7*. These titles are consistent with the headings, and complete, and so accurately describe the tables. In tables with numerous column or row headings, it may not be possible to include all the headings in the title; then general terms can be used in the title to describe groups of specific headings. For example, in a table in which eight catalysts are listed, the title can refer to "catalysts" or "eight catalysts" rather than name them. The row headings should have a column heading. One may use "variables," but it is preferable to characterize them, for example, "attitudinal variables," "family status," and so on.

6.4.2 Ineffective Titles

The title must reflect the headings that describe the table; it cannot replace or substitute for them. For example in the following title:

Ex. 6.5

TABLE 11

HOUSEHOLD CHARACTERISTICS, SECOND-GENERATION COGNITIVE COMPETENCIES, ATTITUDES,AND SCHOOL AVAILABILITY AS DETERMINANTS OF THIRD-GENERATION SCHOOL PARTICIPATION

INDEPENDENT VARIABLES	ALTERNATIVE SPECIFICATIONS			
	(1)	(2)	(3)	(4)

none of the variables in the headings are identifiable in the title. If they are represented by the numbers used as column headings, then the numbers should be set beside the variables in the title in parentheses, explained in a footnote, or be replaced by abbreviations.

Long titles should be shortened in form rather than substance, for example, by eliminating unnecessary articles and wordy phrases. The initial article usually has little meaning and stands in the way of a more substantive word:

Ex. 6.6

a. *The* Confined Rotator Model: Experimental and Fitting Parameters

b. *The* Relative Activities of Different Cationic Activators

c. *An* Analysis of Entry and Exit in Canada's Principal Trader Lines 1976–1979

and so can be deleted. Similarly wordy phrases or excessive detail should be eliminated; the title in *Ex. 6.7b* is a more concise version of *Ex. 6.7a:*

Ex. 6.7

a. Percent of one-child certificate recipients in 1982 who have renounced the certificate by having second child.

b. Recipients of one-child certificate (1982) renouncing certificate with second child (in percent).

TABLE 6.17   Descriptive titles formed by combining column and row headings

a.

TABLE 3

Effect of increasing the level of sodium in the diet on energy intake, urinary excretion of sodium, potassium, calcium, phosphorus, nitrogen, and hydroxyproline, expressed per mmol creatinine excreted, and blood pressure in normotensive man on low and high calcium diets

	Low calcium diet (n = 6)		High calcium diet (n = 6)	
	Low sodium (supplement 22 mmol sodium/day)	High sodium (supplement 178 mmol sodium/day)	Low sodium (supplement 22 mmol sodium/day)	High sodium (supplement 178 mmol sodium/day)
Energy intake (MJ)*	13.7 ± 22.6	13.5 ± 2.8	13.0 ± 3.1	12.9 ± 2.9
Creatinine (mmol)	14.97 ± 1.52	15.08 ± 1.71	15.06 ± 2.75	14.78 ± 2.34
Sodium (mmol)[†]	8.19 ± 0.91	18.51 ± 1.98	8.53 ± 1.76	18.39 ± 3.17
Potassium (mmol)[†¶]	6.82 ± 1.10	7.47 ± 0.96	7.92 ± 1.09	7.95 ± 1.14
Calcium (mmol)[§]	0.250 ± 0.064	0.309 ± 0.138	0.281 ± 0.163	0.334 ± 0.161
Phosphorus (mmol)[‖]	2.31 ± 0.27	2.44 ± 1.56	2.44 ± 0.44	2.52 ± 0.28
Nitrogen (mmol)	6.41 ± 0.75	6.72 ± 0.80	6.95 ± 1.13	7.25 ± 1.29
Hydroxyproline (mg)	2.35 ± 0.26	2.44 ± 0.34	2.08 ± 0.32	2.08 ± 0.41
Blood pressure (mm Hg)				
Diastolic	67 ± 8	64 ± 9	62 ± 5	63 ± 5
Systolic**	116 ± 9	116 ± 7	110 ± 5	109 ± 8

[Amer. J. Clin. Nutr. 41, 1985, 56, © Am. J. Clin. Nutr., with permission]

b.

TABLE 4

Growth Rates of Labor Productivity, the Capital-Labor Ratio, the Money-Labor Ratio, the Composition of the Capital Stock, and Total Factor Productivity, The U.S. Manufacturing Sector, 1955–81 Annual Averages, in Percent

	1955–65	1965–73	1973–78	1978–81
Labor				
Productivity	2.973709	2.799089	2.232467	1.135900
Capital-Labor				
Ratio	0.362946	0.623719	0.495698	1.097620
Capital Stock				
Composition	0.043499	0.080242	0.024286	0.094745
Money-Labor				
Ratio	0.001746	0.007300	−0.045049	−0.024240
Total Factor				
Productivity	2.565518	2.087828	1.757532	−0.032226

[Atlanta Econ. J. 25, 1987, 49, with permission]

When the title remains long despite trimming, secondary, explanatory, supplementary, or detailed information can be transferred to footnotes.

In working for conciseness writers may write titles that are too short, broad, or general. Although acceptable in chart presentations where they are supplemented by the speaker's comments, such titles are not accurate or adequately descriptive for scientific papers. They encompass more than can be reported in the single table. in *Table 6.18,* the title "Public Attitudes" might be used for any table reporting public attitudes on any topic, at any time, for any population; yet it is restricted to three attitudes scaled on five technologies! In *Table 6.18b* "Race and Class" repre-

TABLE 6.18   Titles too broad or general (a-c), not reflecting headings

a.

TABLE IV
Public Attitudes

Technology	Probability of Public Attitude Scale Level		
	Neutrality 0	Controversy −1	Action-oriented opposition −2
Conventional, scrubber	0.10	0.80	0.10
Conventional, nonscrubber	0.85	0.10	0.05
Fluidized bed	0.25	0.65	0.10
Texaco IGCC	0.40	0.50	0.10
KILnGAS IGCC	0.40	0.50	0.10

[Oper. Res. 34, 1986, 26, © 1986 Oper. Res. Soc. of Amer. with permission. No further repro. permitted without consent of © owner.]

b.

TABLE 5
RACE AND CLASS

SUBJECT	REVIEWS			
	West Indian (%)	British (%)	American (%)	N

c.

TABLE 1
Diet formulation

Ingredient	Diet					
	1	2	3	4	5	6

sents only the percentage of times that race or class was mentioned in three reviews. The title in *Table 6.18c* could be used for any table reporting ingredients in diets. Titles that designate the type of data in a table are also too widely applicable to be specific enough for the particular table (*Tables 6.19, 6.20*). The title of *Table 6.19a* can be used for any table that presents statistical data. The important subject

TABLE 6.19   Tables with titles describing statistical analyses, not reflecting row and column headings

a.

TABLE 1

DESCRIPTIVE STATISTICS

| | Years of school(S) | | Years of experience | | | | Ability (*IQ*) | |
| | | | (*EX*) | | *EX²* | | | |
Father's occupation	Mean	Stand. dev.	Mean	Stand. dev.	Mean	Stand. dev.	Mean	Stand. dev.
Labourers and service workers, farm workers	11·4	2·81	17·7	12·91	480·3	580·06	9·4	1·98
Farmers and farm managers	11·7	3·02	24·0	14·05	772·8	716·18	9·6	2·04
Operatives and kindred	12·4	2·42	16·3	11·56	397·9	496·38	9·8	2·00
Craftsmen, foremen and kindred	12·5	2·49	16·9	12·88	451·6	580·25	9·9	1·94

[Econ. 53, 1986, 499, with permission]

b.

Table 2 Multiple Regression Coefficients

| | Dependent Variables | | |
| | (Residualized Accuracy Scores) | | |
Independent Variables	Grades	Reading	Intelligence
Statistical Stability Model			
Level of Self-Esteem	.223***	.177***	.282***
Stability of Self-Esteem	.044	−.021	.040
Level × Stability Interaction	.039	.061*	.070**
Overall R²	.060	.031	.092
F	31.83***	16.13***	50.62***
Phenomenal Stability Model			
Level of Self-Esteem	.235***	.180***	.300***
Stability of Self-Esteem	−.099	−.088***	−.062*
Level × Stability Interaction	.032	.051*	.063**
Overall R²	.057	.035	.092
F	30.65***	18.40***	50.82***

[Soc. Psych. Q. 49, 1986, 6, with permission]

TABLE 6.20 Titles of tables describing measures or variables, not reflecting headings

a.

TABLE 2

MEANS AND STANDARD DEVIATIONS FOR VARIABLES

| | CIVILIAN SAMPLE (N) | | | | MILITARY SAMPLE (N) | | | |
| | Women (738) | | Men (832) | | Women (309) | | Men (344) | |
VARIABLE	Mean	SD	Mean	SD	Mean	SD	Mean	SD

b.

TABLE 2

ESTIMATED LATENT-CLASS PROPORTIONS

Stable Liberal (%)(1)	"Soft Liberal" (%)(2)	Stable Equivocal (%)(3)	"Soft Conservative" (%)(4)	Stable Conservative (%)(5)	Nonopinion (%)(6)	N

c.

Table 3. Standardized Parameter Estimates and Standard Errors for Social Process Model

| | Independent Variables | | | | | | | | |
Dependent Variables	SES	Family Influences	Attach to Parents	Attach to Peers	TI Smoking	Commitment	Belief	Diff. Assoc.	R^2

of the table, schooling, experience, ability, and father's occupations, are ignored. In *Table 6.19b* neither the heading nor the title mention the actual subject of the table. Generally, if a title can be used for various other tables, this is a clue that the title is not specific enough. If the title does not reflect the row and column headings, it is likely to be too broad or general. The headings of the columns and rows in *Tables 6.18 and 6.19* can be used to develop more specific titles:

Ex. 6.8

    a. Probability of Neutrality, Controversy, and Opposition in Public Attitudes Toward Types of Construction Technologies (*Table 6.18a*)

    b. Relation of Years of School, Years of Experience, and Ability to Father's Occupation (*Table 6.19a*)

    c. Grades, Reading, and Intelligence in Two Stability Models (*Table 6.19b*)

       Titles may be specific but incomplete because of omission of important information. *Table 6.21a* does not refer to the four different studies listed. The title in *Table 6.21b* gives no indication that the distribution is given by type of injury (see *Table 6.10*). In *Table 6.21c,* the title gives little indication that it deals with a cost measure of traffic conditions in four Australian cities. The title in *Table 6.21d* does

TABLE 6.21   Titles of tables incomplete, not fully consistent with headings

a.

TABLE 2

Amino Acid Composition of Human Neutrophil Cathepsin G Expressed as Residues/Molecule

Amino acid	This paper	Feinstein and Janoff 1975 (3)	Twumasi and Liener 1977 (9)	Travis, 1978 (6)

b.

Table 9. Distribution of those injured by the earthquake of 3rd March 1985, by age group and sex

[Type of injury]	0–4			5–9			10–14			15–19			20–24			25–34		
	M	F	T	M	F	T	M	F	T	M	F	T	M	F	T	M	F	T

c.

TABLE 3
Marginal Congestion Costs in $, 1982–83 Prices

Peak-period traffic conditions	Sydney	Melbourne	Brisbane	Adelaide

d.

TABLE 1

Growth indices, total food and fluid intake, total caloric intake, 24-h urine output and plasma glucose of nondiabetic C57BL / KsJ-db + / + m and + m/ + m in mice fed nonpurified diets plus either tap water or tap water with 5% glucose

	Heterozygotes		Homozygotes	
Measure	Water	5% Glucose	Water	5% Glucose

not mention the main column headings, heterozygotes and homozygotes. In the following title:

Ex. 6.9

TABLE 1   Characteristics of Low-Vision Subjects

Subject[a]	Age	Diagnosis	Snellen Acuity	M Acuity	Ocular Media	Visual Field	Enlarged D-15 Color Test

the general term "Characteristics" omits much important information and can be improved by grouping or categorizing the seven characteristics that constitute the column headings.

Titles may be inaccurate because they are not consistent with the data presented. This may be due to the use of an inaccurate, general or class term for some of the variables or measures. In *Table 6.22a* "decedents" does not accurately cate-

TABLE 6.22 Titles of tables not consistent with column headings

a.

TABLE 5. Blood alcohol levels of decedents, October 1981–September 1984, in percent[a]

Blood alcohol levels	Alcoholics ($N = 333$)	Narcotics abusers (126)	Both conditions (989)	Non-abusers (722)

b.

Table 1. Visible Foliar Injury Symptoms and H_2S Emission Resulting from SO_2 Fumigation. Visible injury was measured five days after fumigation. (Data represent averages of four separate measurements.)

Days from sowing	Nodal position	H_2S emission (μmole H_2S/μmole SO_2)	LAN* %

c.

Table 2. Volatile fatty acids produced with hay-conc. as the substrate, in the presence of various inhibitors.

Drug	HAc[a]	Prop.[a]	But.[a]	Total[a]	HAc:Prop 0hr[b]	2hr[b]

d.

TABLE 1. Reversibility of hCGRP inhibition of bone resorption

Preincubation (0–48 h)	Treatment (48–120 h)	^{45}Ca release (%) (48–120 h)

gorize the specific population studied. The research was not a study of decedents, who happened to be alcoholics and narcotics abusers, but a study of alcoholics and narcotics abusers who had died. In *Table 6.22b* the phenomenon studied, "Visible Foliar Injury Symptoms," is in the title but not in the headings. The injury was measured by the amount of "leaf area necrosis," which appears as "LAN" in the column heading. To be accurate, one can replace "Visible Foliar Injury Symptoms" in the title with "Leaf Area Necrosis" ("LAN") or indicate the equivalence between the two, that is, "Visible Foliar Injury Symptoms (i.e., LAN*) and H_2S. . . ."In *Table 6.22c,* the title "Inhibitors," the subject of the study, is not consistent with "Drug" of the column heading, though they represent the same materials. In *Table 6.22d,* none of the terms in the title appears to be taken from the headings.

Titles may not be accurate because they designate the subject or purpose of the research or of the table, rather than describing the data presented, or they focus on the units rather than the variables. The title of *Table 6.23a* states the purpose of the laboratory experiments, but the table actually reports the *results* of the experi-

TABLE 6.23   Titles of tables stating subject or purpose of research, not reflecting headings

a.

Table 7. Laboratory experiments testing for effects of interspecific 'aggression' by *Asterias* sp. on feeding rate of its congener

Date	Experiment[a]	Treatment[b]	Total # M.e.[c] eaten before 'aggression'	Total # M.e.[c] eaten after 'aggression'	X^2	p
6/23 to 6/27 1975	*A. forbesi* 'aggressing'	Control	122	77		
	A. vulgaris	Experimental	93	77	1.61	$0.5 > p > 0.1$
6/28 to 7/1 1975	*A. vulgaris* 'aggressing'	Control	70	56		
	A. forbesi	Experimental	75	65	0.07	$0.9 > p > 0.5$

[Oecol. 41, 1979, 263, with permission]

b.

TABLE 1

Portion of Matrix to Determine High Activity and Low Activity Learners

Manager	Supervisor Rating (0–3)	Trainers' Ratings (0–3)	Volume of Learning (0–3)	Postcourse Project (0–3)	Behavior Change (0–3)	Level of Learning (0–6)	Program Feedback (0–1)	Total

ments. A more accurate and comprehensive title would be *"Mytilus edulis* Eaten Before and After Aggression by *Asterias forbesi* and *A. vulgaris*" [or "by *Asterias* spp."] Similarly, the title *Table 6.23b* states the purpose rather than describing the types of data presented.

Titles may characterize the data rather than describing the variables (headings). Titles with categorizing words like "summary," for example, "comparison" or "analysis," may not be adequately descriptive of the data presented. The title of *Table 6.24a* allows "summary," to cover without mentioning them, five consecutive bases, and the number of occurrences and bases (not shown) for A, T, G, C residues. Titles may state abstract concepts that are related to the data in the table but that do not describe them:

Ex. 6.10

TABLE 2

PARTIAL ELASTICITIES OF MARKET VALUE[a]

Variable	Minnesota		Kansas	
	Nonlinear	Logarithmic	Nonlinear	Logarithmic

TABLE 6.24   Titles characterizing table or data, not reflecting headings

a.

TABLE I

Summary of template bases as targets for minus-one-base frameshifts

This summary includes 129 bases of *lacZα*-coding sequence and 21 bases of the regulatory sequences at which addition and deletion events have been observed. As new regulatory sequences are found to be sensitive to length variation, the numbers shown below will increase somewhat.

Number of consecutive bases	A		T		G		C	
	No. of occurrences	No. of cases	No. of occurrences	No. of bases	No. of occurrences	No. of bases	No. of occurrences	No. of bases

b.

TABLE 1

Analysis of Minority Shareholdings in the UK, 1984

Sector of target company	*Number of companies in which there is a minority holding*	*Percentage of companies in which there is a minority holding*

c.

TABLE 4.

Results on 300'' random points

		ACCELERATION					
# of Tools	Heurist.	6.667	3.333	2.000	1.333	0.667	0.444

Here the title ignores all the headings and subheadings. Titles beginning with units of measure rather than the variables shift the emphasis from the subject to measurement:

Ex. 6.11

   a. *Number* of 1-h Average O_3 Concentrations $\geq$ 0.10 and 0.12 ppm for All Sites
   b. *Values* of Solids Decay Factor F_1 as Function of Process Water Temperature and Solids Retention Time
   c. *Mean Percent* Discrimination Responses for Groups in Larger and Smaller Discriminations
   d. *Percentage* of Satisfaction Change for Specific Job Aspects After Migration

The emphasis can be corrected by shifting the units to the end of the title (*Ex. 6.4c–e*) or to a footnote.

6.5 STRATEGIES FOR ECONOMICAL AND EFFECTIVE TABLES

A well-designed table is clear, concise, and well ordered. Indeed, the mark of an effective table is regularity—in the arrangement of the elements and in the consistency of its form and content. When a table is wordy or repetitious, or when large areas of blank space separate elements, the regularity or pattern of the elements is disturbed, and the table is difficult to read. Such repetition, irregularity, and excessive blank space usually signal defects in the construction of the table.

The objective in designing a table then is to make it economical and consistent with scientific objectives. These are not mutually exclusive objectives, because economy can lead to conciseness. The major responsibility for designing an effective table devolves upon the writer, since editors and reviewers cannot be expected to redesign tables and do not edit them for form as closely as they do the written text.

6.5.1 Condensation

Economy can be achieved by condensation, which in a table consists largely of elimination—of unnecessary elements, repetition, and areas of blank space. A table is not a place for sentences, which require words just to signal syntactic functions. Sentences can be reduced to phrases or to words, where possible (see revision in *Table 6.48*). Articles can be omitted, and abbreviations can be used freely in restricted spaces. Familiar abbreviations can be used without comment; temporary abbreviations can be defined in footnotes. Data can be combined to avoid separate columns. For example, if space is limited, statistical data, such as standard deviation, and ranges can accompany the associated means rather than be set off in separate columns. Level of significance (*p*) can be indicated by bullets next to the values and explained in a footnote, rather than set in a separate column. Enumeration should be avoided except for economy, as in cross-correlation tables. Whenever the elements of a table consist of words, the writer must make a special effort to condense entries (*Table 6.47, 6.48*).

The table itself may not be worth including in the paper, because the data are so few that they can be as easily and more economically presented as part of the written text as in *Table 6.25*. Here the tables have a heavy superstructure, yet only four and eight values. *Table 6.26a*, which lists five diseases associated with a percentage, takes up the 5.5-in. width of the page in the journal. *Tables 6.26b,c* consist of six and eight columns, respectively, with only one value each, although this treatment is more economical than the vertical listing in *Table 6.26a*.

6.5.2 Ineconomy of Repetition

When elements in a table are repeated, they not only waste space, and add to typesetting costs, but they also clutter the table with unnecessary elements. Tables themselves may overlap or be redundant (*Table 6.13*) and so can often be combined. *Table 6.27a* is one of two tables 7 in. wide, one at the top and the other at the bottom of the same page in the journal. The tables have identical titles and headings except that one refers to "soft disk" and the other to "hard disk." Their separation hinders a comparison of the corresponding sets of data, yet they can easily be combined (1) by placing the two sets of values in one table one below the other, with "soft disk" and "hard disk" as row headings, each with the set of four frequencies

TABLE 6.25   Tables with too little data to warrant cost of table; data overshadowed by head note, headings, and rules.

a.

TABLE II

Carboxymethylation of intact phospholamban and ribonuclease

Phospholamban and ribonuclease were carboxymethylated with iodo[2-^{3}H] acetic acid in the presence of 2% hexadecylpyridinium chloride as described under "Experimental Procedures." Incorporated radioactivity was quantitated by cutting labeled protein bands from polyacrylamide gels and subjecting them to liquid scintillation counting. Quench correction was by the external standard method.

Protein Reacted	^{3}H label incorporated	
	− Dithiothreitol	+ Dithiothreitol
	dpm/mg protein × 10^{-6}	
Ribnoculease	1.4	117.6
Phospholamban	136.4	118.5

[J. Biol. Chem. 261, 1986, 13338, with permission]

b.

TABLE 3

SIGNIFICANT GRADE INTERACTION FOR OFF-LINE EVALUATION MEASURES

	PRIMARY PROBE QUESTION		SECONDARY PROBE QUESTION	
	Congruence		Congruence	
GRADE	Congruent	Incongruent	Congruent	Incongruent
3............................	.85	.35	1.00	.53
6............................	.70	.55	.89	.96

[Child Dev. 57, 1986, 1411, with permission]

as subheadings; (2) by intercalating the hard-disk values between the soft-disk values, one above each other, with a footnote to identify the two values; or (3) by setting the soft-disk and hard-disk values side by side with an abbreviated form of "soft disk" and "hard disk" as subheadings under each column heading. This last alternative would be the most economical of space.

Table 6.27b is one of two tables in which the title differs only in the condition of the test run. The tables, 5.5 in. wide, have excessive blank space, between the columns, which would allow the two tables to be combined by setting the values labeled R1 and R2 side by side and explaining the run in a footnote:

Ex. 6.12

R1 = Run 1: As delivered, 0.13 NM^3/min-275 sefh[a]

R2 = Run 2: 0.35 NM^3/min-740 sefh[a]

TABLE 6.26   Tables with too little data to warrant cost of table, consist of list or rows of elements, each heading with one value

a.

Table 12. *Mean absolute deviation (percentage) from trend in cause-specific burial series in London Bills 1675–1825*

Consumption	7.2
Fevers	16.0
Infancy	5.8
Smallpox	32.5
Other causes	8.7

b.

TABLE 4

An Analysis of Corporate Turnover in the Canadian Liner Shipping Industry, 1977–82

Number of Liner Companies Operating in 1977	Number of Liner Companies Operating in 1982	Number of firms entering	Number of firms leaving	Hit-and-Run Entries	
				1 year or less	2 years or less
75	86[a,b]	60[c]	40[d]	13	6

c.

Table 3. Chemical composition (wt%) of the commercial-grade magnesia-stabilized zirconia electrolyte used by Iwase et al. [12]

SiO_2	TlO_2	Al_2O_3	Fe_2O_3	CaO	Na_2O	K_2O	MgO
0.2	0.14	0.4	0.05	0.4	0.015	0.05	3.1

[**a.** J. Landers. *Mortality and Metropolis: the case of London 1675 to 1825* Pop. Stud. 41, 1987, 73; **b.** J. Trans. Econ. Pol. XX, 1986, 309; **c.** J. Appl. Electrochem. 16, 1986, 717 (Chapman and Hall, pub.); all with permission]

Moreover, the two CNC columns can easily be eliminated and a footnote added to indicate the four values (in table not shown) that were not zero. *Table 6.28* is one of two similar tables on the same page in the original publication. It has enough space for the two columns from the other table, thus saving the space of a table, title, and row headings, and gaining a more readable alignment of values for comparison.

Within tables, repetition takes various forms. Often within a row or column, one or a set of words or values is repeated throughout, as illustrated in *Table 6.29*. Except where the research scientist's intent is to emphasize the unvarying elements by the visual impact of the repetition, for example, in *Table 6.27b*, that the CNC count was always zero, such columns can be omitted, and a footnote used to supply the value. In the last row of *Table 6.29*, "TX" is repeated at the beginning of each four-letter acronym in each column heading, obscuring the informative part of the

TABLE 6.27　Tables suitable for combining with companion tables (not shown) because of repetition and excessive intercolumn space

a.

TABLE II. Calculated complex pressures for an acoustically soft disk of 35-cm radius. Source location: $z = 100$ cm. Observation point locations: $z = \pm 10$ cm.

Frequency (kHz)	Infinite plane $z = + 10$ cm	Finite disk $z = + 10$ cm	Finite disk $z = - 10$ cm	Diffraction-reduced complex pressure
1	$0.185 + 0.822i$	$0.171 + 0.717i$	$-0.014 - 0.105i$	$0.185 + 0.822i$
2	$0.135 + 1.50i$	$0.031 + 1.54i$	$-0.104 + 0.039i$	$0.135 + 1.50i$
5	$-0.101 + 1.75i$	$-0.195 + 1.69i$	$-0.094 + 0.061i$	$-0.101 + 1.75i$
10	$-0.101 + 1.75i$	$-0.052 - 1.65i$	$0.049 + 0.101i$	$-0.101 - 1.75i$

[J. Acoust. Soc. Am. 80, 1986, 24, with permission]

b.

TABLE 4. 25-cm Membrane Filter Cartridge Test, Run 1: As Delivered, 0.13 Nm³/min-275 scfh[a]

	CNC		LAS-250X		
				Counts/hr	
Hours	Counts/hr	Counts/scft	0.2-0.3 μm	>0.3 μm	Total/scft
1	0	0	15	1	2.5
2	0	0	15	1	2.5
3	0	0	13	0	2.1
4	0	0	17	0	2.7
5	0	0	15	0	2.4
6	0	0	14	1	2.4
7	0	0	8	1	1.4
8	0	0	20	0	3.2
9	0	0	19	0	3.0
10	0	0	14	1	2.4
11	0	0	16	0	2.5
12	0	0	16	0	2.5
13	0	0	15	0	2.4
14	0	0	17	0	2.7
15	0	0	18	0	2.9

[Kapper and Wen. Aerosol. Sci. Tech. 5, 1986, 179, © 1986 Elsevier Science Pub. Co., Inc., with permission]

acronym. Usually where sets are repeated, as in *calculated* and *found,* a footnote to the first set of values can explain the sets.

Repetition may take the form of two banks or sets of data, as in *Table 6.30,* where the subheadings under the two row headings are identical, so that the sets can be combined, with a juxtaposing of comparable values that is likely to be more effective as well as more economical. In *Table 6.30a* the columns of data for females can be placed beside the corresponding columns for male data, and ''M'' and ''F''

TABLE 6.28   Table suitable for combining with companion table (not shown) because of similar headings and ample space between and outside columns

TABLE 2.—P-M Wave Velocity Spectrum ARMA Coefficients

Coefficients (1)	ARMA (7,7) (2)
b_0	$1.3405 \; 10^{-1}$
b_1	$-5.4189 \; 10^{-1}$
b_2	$6.7955 \; 10^{-1}$
b_3	$-1.0322 \; 10^{-2}$
b_4	$-5.9175 \; 10^{-1}$
b_5	$3.5353 \; 10^{-1}$
b_6	$2.4591 \; 10^{-2}$
b_7	$-4.7741 \; 10^{-2}$
a_1	-3.3289
a_2	4.1066
a_3	-1.3543
a_4	-1.8138
a_5	2.2574
a_6	-1.0283
a_7	$1.8782 \; 10^{-1}$

[J. Engr. Mech. 112, 1986, 753, with permission]

included as column subheadings. In *Table 6.30b,* putting the columns of summer data alongside the corresponding columns of spring data would allow ready comparison and would halve the space taken by the table. If the desired comparison is within columns, then the summer values can be inserted below the spring values and ranges moved to separate columns. The % sign before each row heading should be removed to avoid repetition and the obscuring of the variables.

Repetition may actually impede communication. In row headings, the repetition of variables with qualifying conditions obscures the different variants *Table 6.31a,b,* see also *6.12*). Extracting the repeating variable as a heading and treating

TABLE 6.29   Columns and rows in tables illustrating uneconomical repetition of elements

calcd	heart	one-stage	4.1	highest
found	red	two-stage	4.1	nat'l avg.
			4.1	lowest
calcd	heart	one-stage	4.1	
found	red	two-stage	4.1	highest
			4.1	nat'l avg.
calcd	heart	one-stage	4.1	lowest
found	red	two-stage	4.1	
[for 12 compounds]	[for 7 pairs]	[for 12 variables]	[for 14 polymers]	[for 7 countries]

g   g   g   g   g   g
TXMM   TXGL   TXOG   TXPF   TXCP   TXIP   TXRC   TXTR   [for 14 variables]

a.

TABLE 3-3. Paroxysmal EEGs in Male and Female Symptom and Control Groups

Study groups	n	% with paroxysmal EEGs	Comparison with controls (Fisher's exact test)
Males			
Control group	24	29.2	—
Suicide ideation only	16	68.8	$p = .016$
Suicide ideation and attempts	6	100.0	$p = .003$
Assaultive—destructive (no suicide)	18	66.7	$p = .017$
Females			
Control group	20	15.0	—
Suicide ideation only	16	37.5	$p = .004$
Suicide ideation and attempts	15	60.0	$p = .008$
Assaultive-destructive (no suicide)	13	53.8	$p = .024$

[Suic. Life-threat Behav. 16, 1986, 150, with permission]

b.

Table IV. Percentage of total proximity time devoted to feeding, play, grooming and other activities in sibling, strong and weak peer relationships

	Sibling		Strong	Weak
Spring				
N	9		13	12
% Feeding	27		35 *	21
	(17–48)		(16–57)	(0–44)
% Play	13		23	31
	(2–23)		(3–45)	(5–87)
% Grooming + % Other	53 *		40	48
	(47–76)		(22–74)	(4–76)
% Grooming	4 †		0	0
	(0–15)		(0–22)	(0–40)
% Other	51 †		39	47
	(34–68)		(17–74)	(0–76)
Summer				
N	11		17	12
% Feeding	14 *		28	14
	(4–53)		(10–54)	(0–50)
% Play	9		6	16
	(1–22)		(0–28)	(0–78)
% Grooming + % Other	77 *		52	57
	(30–92)		(17–87)	(12–98)
% Grooming	7		10	0
	(0–26)		(0–75)	(0)
% Other	73 **		51	57
	(62–84)		(4–78)	(12–98)

[Anim. Behav. 33, 1985, 966, with permission]

TABLE 6.31   Row headings of tables (a, b) with repetition interfering with identification of variables, (c, d) revised to differentiate headings from subheadings

a. Managerial employees, administration
Managerial employees, production
Managerial employees, sales
Managerial employees, transportation
Managerial employees, building

Skilled workers, production
Skilled workers, maintenance
Skilled workers, transportation
Skilled workers, building

Unskilled workers, production
Unskilled workers, maintenance
Unskilled workers, building

c. *Managerial employees*
Administration
Production
Sales
Transportation
Building

Skilled workers
Production
Maintenance
Transportation
Building

Unskilled workers
Production
Maintenance
Building

b. TABLE I—CORRELATION BETWEEN URBANIZATION AND OCCUPATIONAL/DEMOGRAPHIC VARIABLE BY DISTRICT 1981

INDEPENDENT VARIABLE*

Literates as % of pop.
Male literates as % of male pop.
Female literates as % of female pop.

Cultivators as % of working pop.
Male cultivators as % of male-working pop.
Female cultivators as % of female-working pop.

Agr. laborers as % of working pop.
Male agr. laborers as % of male-working pop.
Female agr. laborers as % of female-working pop.

Household laborers as % of working pop.
Male h.l. as % of male-working pop.
Female h.l. as % of female-working pop.

Other workers as % of working pop.
Male other workers as % of male-working pop.
Female other workers as % of female-working pop.

Marginal laborers as % of working pop.
Male m.l. as % of male-working pop.
Female m.l. as % of female-working pop.

Nonworkers as % of pop.
Male nonworkers as % of male pop.
Female nonworkers as % of female pop.

[Geog. Rev. 76(2), 1986, 175, with permission]

d. *Cultivators*
Population
Male
Female

Agricultural laborers
Population
Male
Female

Household laborers
Population
Male
Female

the rest as subheadings (*Table 6.31c,d, 6.23a*) is more economical and highlights the elements.

6.5.3 Ineconomy of Excess Space

Patches of blank space or the disproportionate distribution of space signal the likelihood of an ineffective arrangement. In a table, even indentation may waste space, unless it is necessary for clarity. Space may be wasted (1) within the table, (2) in the layout of the page, and (3) in the overlapping of similar tables (*Table 6.27, 6.28*).

Tables that have only two or three row headings, each associated with columns of several to numerous values, waste space between the row headings (*Table 6.32–6.34*). In *Table 6.32a,* a wide column is wasted on two row headings. Since there is also excess space between columns, the two sets of data can be set side by side, with the two ligands, "Polymer" (abbreviated) and "DMAP" set as headings over the

TABLE 6.32   Table showing uneconomical and ineffective disposition of elements due to large areas of blank space

TABLE 2

Specificity for PPO production as a function of the ligand/Cu ratio. $(OH/Cu)_0$, [Cu(II)] and the nature of the DMAP-based Cu(II) catalyst

Nature of ligand	Ligand/Cu	$(OH/Cu)_0$	[Cu(II)] $(mmol\ dm^{-3})$	% PPO
Polymer (2): $\alpha = 0.226$	1	1	3.32	80
	2	1	3.32	90
	3	1	3.32	93
	4	1	3.32	95
	5	1	3.32	96
	6	1	3.32	96
	4	0	3.32	93
	7	0	3.32	97
	10	0	3.32	97
	4	1	1.66	89
	2	0	0.83	56
	3	0	0.83	72
	4	0	0.83	81
	4	1	0.83	88
	10	0	0.83	91
	10	1	0.83	92
DMAP	2	0	3.32	80
	2	1	3.32	91
	4	0	3.32	94
	4	1	3.32	95
	8	0	3.32	96
	8	1	3.32	96

[React. Polym. 4, 1986, 299, with permission]

set of column headings. The table would consist simply of two banks of the data across the table. Actually the table is wide enough for three banks of data. Also the values of DMAP can be set alongside the polymer values for comparison. In *Table 6.33a* the columns can be redisposed for economy and conciseness. In the revision (*Table 6.33b*), (1) the first column is omitted because "mitochondrial phospholipid" is included in the title, and ACTH and control are made column subheadings; (2) the column "phospholipids added" becomes the stub; (3) the repetitious μ M column is relegated to a footnote; and (4) the amount and percentage of pregnenolone for ACTH and control group become paired columns. This results in a shorter, more

TABLE 6.33 Table showing (a) uneconomical and ineffective disposition of elements due to wide separation of row headings and repetition of elements in columns, (b) revision of column headings to eliminate repetition and wasted space

a.

TABLE I

Effects of rat adrenal mitochondrial phospholipid on cholesterol side chain cleavage activity in unstimulated mitochondria

Unstimulated mitochondria from cycloheximide/ACTH-treated rat adrenals were incubated at 20 min, 37°C with or without mitochondrial phospholipid prepared from adrenals of $ACTH_{1-24}$-treated or cycloheximide/ACTH-treated rats. Pregnenolone formed from endogenous cholesterol was determined by radioimmunoassay. . . .

Experiment	Phospholipid added	μM	Pregnenolone formation	%
			ng/mg protein/20 min	
A. Mitochondrial	None	0	160 ± 14	100
phospholipid	Mixture	200	246 ± 52	154
from ACTH-	PI	200	142 ± 24	89
treated rats	PC	200	144 ± 24	90
	PS	200	252 ± 8	156
	CL	200	274 ± 28	171
	PE	200	302 ± 36	189
B. Mitochondrial	None	0	192 ± 21	100
phospholipid	Mixture	200	211 ± 20	110
from control rats	PI	200	129 ± 18	67
	PC	200	116 ± 10	60
	PS	200	239 ± 28	124
	CL	200	414 ± 40	216
	PE	200	244 ± 22	127

[J. Biol. Chem. 261, 1986, 14120, with permission]

b.

Phospholipid added	Pregnenolone formation ng/mg protein/20 min			
	ACTH	%	Control	%

coherent table, without the distraction of the repeated elements. Similarly, in *Table 6.34* the rank and mode columns can be combined, releasing some intercolumn space to the last column, which has the most text. It will then have the most space, and so fewer lines of text will run over. The repetitious headings *"Form," "Intent,"* and *"Role dims"* can be abbreviated to *F, I,* and *R* or *RD* for added space.

Tables with a column of row headings and one (or two) columns of corresponding values (*Tables 6.35, 6.36*) usually occupy the full width of the page with a wide expanse of blank space within or outside the table (*Table 6.26a*), or both. When the listing consists of numerous items (*Table 6.35a*), dividing the table into two columns saves space. *Table 6.35a*, which takes up the 5-in. page width in the journal and has over 2 in. of blank space in the middle, can easily be divided before the DeMay winery. The table would then break between the wineries with capacities

TABLE 6.34 Table showing uneconomical and ineffective disposition of elements due to wide separation of row heading, excess intercolumn space, and repetition of elements

Appendix I.
Familiarity Ranks and Verbal Response Mode Forms, Intents and Role Dimensions

Rank	Mode	Form/Intent/Role dimensions
8	Advisement	*Form:* Imperative, or second person ("you"); verbs of permission, prohibition, or obligation. *Intent:* Attempts to guide behavior: suggestions, commands, permission, prohibition. *Role dims:* Presumptuous, directive, informative.
7	Interpretation	*Form:* Second person ("you"): verb implies an attribute or ability of the audience; terms of evaluation. *Intent:* Explains or labels the audience: judgments or evaluations of audience's experience or behavior. *Role dims:* Presumptuous, directive, attentive.
6	Confirmation	*Form:* First person plural ("we") where referent includes audience. *Intent:* Compares speaker's experience with audience's; agreement, disagreement, shared experience or belief. *Role dims:* Presumptuous, acquiescent, informative.
5	Reflection	*Form:* Second person: verb implies internal experience or volitional action. *Intent:* Puts audience's experience into words: repetitions, restatements, clarifications. *Role dims:* Presumptuous, acquiescent, attentive.
4	Disclosure	*Form:* Declarative: first person singular ("I"). . . .

[Adapted from Soc. Psych. Q. 49, 1986, 78, with permission]

TABLE 6.35   Tables consisting of row headings and one or two columns of corresponding values, with excessive blank space within and outside columns

a.

TABLE III—WINE-PRODUCING CAPACITY 1985

WINERY	CAPACITY (gals.)
Taylor-Great Western-Gold Seal	30,000,000[a]
Canandaigua Wine	12,000,000[a]
Widmer's Wine Cellars	3,000,000
Penn Yan Cellars	300,000
Bully Hill	250,000
Wagner Vineyards	40,000
Glenora Wine Cellars	35,000
Eagle Crest Vineyards	30,000
Heron Hill	25,000
Vinifera Wine Cellars	20,000
Wickham Vineyards	20,000
Casa Larga	15,000
Plane's Cayuga Vineyard	15,000
Herman J. Wiemer	15,000
DeMay Wine Cellars	10,000
Finger Lakes Wine Cellars	10,000
Lucas Winery	10,000
McGregor Vineyard	10,000
Poplar Ridge Vineyards	10,000
Chateau Esperanza	9,900
Four Chimneys	6,000
Ambrosia Farms	5,000
Giasi Winery	5,000
Rolling Vineyards	5,000
Hazlitt 1852 Vineyards	4,500
Knapp Farms	4,000
Americana	3,200
J. LeBeck	3,200
Frontenac Point	3,000
Lakeshore Winery	3,000

[Geogr. Rev. 76, 1986, 315, with permission]

b.

TABLE 2.   Physical Constants

Quantity	Value	Units
r_e	0.6378×10^7	m
r_a	0.6498×10^7	m
H	0.1200×10^6	m
μ	0.3986×10^{15}	$m^3 s^{-2}$
k	0.5415×10^2	—

[J. Astron. Sci. 34, 1986, 12, with permission]

of 15,000 gal or more and those with 10,000 gal or less. *Table 6.35b* has two columns associated with the row headings and the columns of space are too wide. It is one of several similar tables occupying the full page width (5 in.). Two could be combined side by side into a double table. Actually the units column of one of the tables (not shown) has only one unit listed for its 11 quantities, and three of the five units of *Table 6.35b* are the same. The four units could easily be supplied in a footnote. So combined, the table would have ample space for including the definitions of the quantities, which are presented separately in the text. Such a revision replaces two tables with one, yet includes additional information.

Table 6.36 wastes space between the rows as well as the columns because of the two values in the estimate column. This, together with the additional space required between the pairs, causes the table in the journal to take up almost the full page. Placing the two values in separate columns would reduce the table by one-third and make it more readable. Doubling the table so that the left half is for the variables and the right half is for the parameters could then halve the table.

TABLE 6.36   Table considered to be uneconomical and ineffective due to large areas of blank space between rows and columns, and crowding of double values in one column

Table 1.   Dual Output Homothetic Production Function Parameter Estimates[a] (Dependent Variable Is Ln of Gallons of Potable Water Produced)

Variable[b]	Estimate
Constant	11.38 (4.50)[c]
Ln(miles of pipeline)(λ)	0.25 (3.71)
Ln(maximum treatment capacity)	0.59 (13.31)
Ln(Annual man-hours)	0.30 (7.45)
Ln(percentage of potable water delivered to nonresidential customers)	0.15 (3.32)
Parameter[d]	
δ	0.89 (19.99)
θ_1	0.37E–10 (2.94)
θ_2	0.17E–9 (2.58)
α_1	0.66 (24.61)
α_2	0.34 (12.47)

[S. Econ. J. 53, 1986, 470, with permission]

    Table 6.37a appears to be similar to *Tables 6.35a* and *6.36,* with two sets of row headings in relation to five column headings. Actually it is analogous to five tables like *Table 6.35a,* set one below the other with their values in different columns across the page. The form is therefore misleading, and very space consuming. The table can be accurately laid out by dividing the row headings into three columns (*Table 6.37b*): one "attribution item," beginning with the row heading "Drinking

TABLE 6.37   Table (a) with five sets of row headings each with corresponding values in one column across and down table, resulting in excessive blank space and misleading display; (b) with revision of column headings to reflect list character.

a.

Table 2. Attribution and Evaluation Factors in the Unemployment Interview

Attribution Items	Behavior	Character	Societal Biases	Luck	Evaluation
Drinking on the job	.97				
Fighting on the job	.95				
Rude to supervisor	.88				
Did not do job well	.85				
Frequent tardiness	.83				
Laziness	.70				
Carelessness on job	.68				
General irresponsibility	.64				
Way he was raised		.89			
Family background		.79			
General level of intelligence		.69			
Actor's destiny		.48			
Biases against people like him			.86		
Race			.80		
Biases of those in power			.48		
Bad luck				.95	
Fate				.91	
Chance				.89	

Evaluation Items					
Responsible/irresponsible					.77
Motivated/not motivated					.74
Impulsive/in control					.72
Driven to get ahead/lazy					.65
Stable/unstable					.60
Very moral/low morals					.59
Honest/dishonest					.58
Sociable/not sociable					.44

[Soc. Psych. Q. 49, 1986, 160, with permission]

b.

Attribution item	Behavior	Attribution item	Character, Biases, Luck	Evaluation item	Evaluation

on the job," a second attribution item beginning with "Way he was raised," and the third with the evaluation item. The corresponding columns of values are placed beside these. This revision reduces the table to one-third and makes it a more accurate reflection of the data. A more irregular distribution of blank space, which makes the table difficult to read, is seen in *Table 6.38*. Here the use of leaders or dashes in empty cells would make the values easier to associate with the row and column headings.

Tables may show both excessive crowding and excessive blank space. In *Table 6.39,* which occupies about 5.5 in. of the 7-in. page width of the journal, all the data in the table are in percentages; therefore the sign can be removed and a note

TABLE 6.38 Table with irregular distribution of areas of blank space among values, making table difficult to read

Table 1. The mass spectra of perfluoroalkylether substituted silanes

m/z	Me_3SiR_f	$Me_2Si(R_f)_2$	$MeSi(R_f)_3$	$HSi(R_f)_3$	Me_2SiR_fOH
			Relative intensity		
59	23				
69		14	100	100	18
73	92				
75					100
77	46	100			
79					20
81		6	7		
83				47	
85		7	19	7	
147		8	10		
165	65				
181				11	
183				7	
193	10	4			
195					0.3
197		4	3		
233				2	
235				10	
243	100	98			
247	7	67	4		
249				61	
266	22	4		18	10
297			10	6	
333				45	
347		5	28	28	
433				10	
513			26	15	
532			2	1	

[Saba and Eisentraut Organ. Mass Spectrom. 21, 1986, 807, © 1986 John Wiley & Sons, Ltd., with permission]

TABLE 6.39   Table visually ineffective because of large areas of blank space, crowding, and repetition

Table III. Metal and Ligand Contributions to the LUMOs of the Complexes[a]

	metal d	NO	C-containing ligand
$[(\eta\text{-}C_5H_5)Re(NO)(CO)(PPh_3)]^+$		$\pi_x{}^*$: 10.05% $\pi_y{}^*$: 34.21%	$\pi_x{}^*$: 16.95%
$[(\eta\text{-}C_5H_5)Re(NO)(CHO)(PPh_3)]$	$d_{x^2-y^2}$: 8.51% d_{xy}: 10.52% d_{xz}: 10.95% tot: 33.83%	$\pi_x{}^*$: 21.34% $\pi_y{}^*$: 21.19%	
$[(\eta\text{-}C_5H_5)Re(NO)(CHOH)(PPh_3)]^+$	tot: 25.40	$\pi_x{}^*$: 20.19% $\pi_y{}^*$: 16.47%	$\pi_y{}^*$: 16.85%
$[(\eta\text{-}C_5H_5)Re(NO)(CH_2OH)(PPh_3)]$	$d_{x_2-y_2}$: 8.91% d_{xy}: 11.52% d_{xz}: 10.35% tot: 35.89%	$\pi_x{}^*$: 21.66% $\pi_y{}^*$: 19.64%	
$[(\eta\text{-}C_5H_5)Re(NO)(CH_2OH_2)(PPh_3)]^+$	$d_{x^2-y^2}$: 9.14% d_{xy}: 11.37% d_{xz}: 9.53% tot: 33.09%	$\pi_x{}^*$: 22.77% $\pi_y{}^*$: 19.04%	
$[(\eta\text{-}C_5H_5)Re(NO)(CH_2)(PPh_3)]^+$	d_{xz}: 31.48% tot: 35.46%		Cp_x: 51.97%
$[(\eta\text{-}C_5H_5)Re(NO)(CH_3)(PPh_3)]$	$d_{x^2-y^2}$: 8.60% d_{xy}: 10.88% d_{xz}: 10.82% tot: 36.21%	$\pi_x{}^*$: 21.08% $\pi_y{}^*$: 20.61%	

[Organometal. 5, 1986, 2319, © 1986 Amer. Chem. Soc., with permission]

added in the title or a footnote. The last column has only three values. If they apply only to the corresponding complexes, they might be removed and footnotes substituted for them. In the "metal d" column, the totals are not actual totals of the three values, and so they can be set off in a separate column. Also, because the values are repeated in sets of three (with two exceptions), the designations might be omitted, and a footnote added to explain the ordering and the single d_{yz} value. The π_x and π_y designation in the NO column might similarly be relegated to a footnote. Indeed if there were some way of emphasizing the group in the "complexes" that seems to differentiate them, the table would be far more effective. The revised table would have three columns of data: metal d, total, and NO, and would be more economical and concise. It would be narrow enough to set the data in two ranks.

Space may be wasted because the shape of the table does not allow it to fit economically on the page. Tables should therefore be planned with the printed page in mind. *Table 6.40* is a long, narrow table (2.75 in.) more than two pages long from a journal with a 5.25-in. printed page width. It is therefore bordered by a wide band of blank space on each side and also has excessive blank space between the columns. With the border space and intercolumn space available, the table can be

TABLE 6.40 Table is long and narrow, uneconomical and ineffective because of excessive blank space between and outside columns

Table 1. The values of $c(N, B)$ and $s(N, B)$ for the triangular lattice

N	B	$c(N, B)/6$	$s(N, B)/6$
1	1	1	1
2	2	3	10
2	3	2	2
3	3	9	59
3	4	8	28
3	5	6	10
4	4	27	280
4	5	32	218
4	6	24	110
4	7	20	46
5	5	79	1179
5	6	122	1282
5	7	108	874
5	8	76	440
5	9	70	202
6	6	233	4 614
6	7	422	6 416
6	8	470	5 472
6	9	366	3 512
6	10	264	1 778
6	11	216	798
6	12	20	34
7	7	679	17 145
7	8	1 458	29 148
7	9	1 766	29 356
⋮	⋮	⋮	⋮

[J. Phys. A19, 1986, 3290, with permission]

set double across the page and take up half as much space. It can be made even more condensed by eliminating the repetition in the N column, and using a line between the ranks of numbers. *Table 6.41,* which is also several pages long, can also be doubled up by (1) adding the *quarter* as a superscript to the year and (2) placing the number of the force level directly before the description, thus associating the two closely.

Table 6.42, a table 2.5 in. wide and 6.25 in. long on its journal page 5.25 in. wide, is unnecessarily narrow, because the hypotheses are treated as notes to the table. The table can be extended to occupy the full width of the page, and the hy-

TABLE 6.41 Table long and narrow, uneconomical and ineffective because of excessive blank space and ineffective disposition of elements

Appendix: Nuclear-Capable or Major Uses of Force, 1949–1976

Event Number[a]	Force Level[b]	Quarter[c]	Year[c]	Description[d]
29	1	3	1950	Korean War: Security of Europe
30	3	3	1950	Political developments in Lebanon
31	3	1	1951	Security of Yugoslavia
38	2	3	1953	End of war in Korea
39	1	3	1953	Security of Japan/Korea
40	3	1	1954	France–Viet Minh War: Dienbienphu
41	1	2	1954	Guatemala accepts U.S.S.R. aid
42	3	3	1954	France–Viet Minh War: Dienbienphu
43	3	3	1954	British airliner shot down by China
44	1	3	1954	China-Taiwan: Tachen Islands
⋮	⋮	⋮	⋮	⋮
136	3	1	1964	Security of Panama Canal
138	2	1	1964	Cyprus-Greece-Turkey crisis
139	3	1	1964	Coup in South Vietnam
143	2	2	1964	Civil war: Laos
149	3	3	1964	Cyprus-Greece-Turkey crisis
151	2	3	1964	North Vietnam fires on U.S. ship in Tonkin Bay
155	3	4	1964	Viet Cong attack Bien Hoa barracks
157	3	1	1965	Viet Cong attack Pieiku
158	3	1	1965	Viet Cong attack Qui Nhon
159	2	2	1965	Civil war: Dominican Republic
163	3	3	1965	War in Vietnam: Withdraw troops from Europe
164	3	3	1965	Political developments: Cyprus
166	3	3	1967	Civil war: Dominican Republic
174	3	2	1967	Arab-Israeli War
178	1	1	1968	Pueblo seized by North Korea
⋮	⋮	⋮	⋮	⋮

[Am. Pol. Sci. Rev. 80, 1986, 562, with permission]

potheses can be described *in* the table, in a separate column beside with the corresponding equations. With the added width, most of the hypotheses would take up two lines or less, resulting in a table less than one-half as long as the original. The table can be made even shorter if the explanations are made more concise, and if the repeated ''Nested in hypothesis'' is explained in a footnote.

Very broad tables may have to be printed along the length of the page, that

TABLE 6.42 Table which is long and narrow, uneconomical and ineffective because of
excessive blank space outside table and ineffective disposition of elements

TABLE B-2. Testing Equality of "Present" and "Missing" Samples

	Chi-square	df	Diff. chi-square	df
H1: $\Sigma^1 = \Sigma^2$	1299.93	36		
H2: $\chi = \Lambda \chi \xi + \delta$	210.83	34	1089.10	2
H3: $\Phi^1 = \Phi^2$	292.17	40	81.34	6
H4: $\Phi^1_{3\,3} = \Phi^2_{3\,3}$	284.68	39	7.49	1
H5: $\Phi^1_{3\,2} = \Phi^2_{3\,2}$	249.65	38	35.03	1
H6: $\Phi^1_{3\,1} = \Phi^2_{3\,1}$	235.99	37	13.66	1
H7: $\Phi^1_{2\,1} = \Phi^2_{2\,1}$	234.48	36	1.51	1
H8: $\Phi^1_{2\,2} = \Phi^2_{2\,2}$	212.80	35	21.68	1
H9: $\Phi^1_{1\,1} = \Phi^2_{1\,1}$	210.83	34	1.97	1

Hypothesis 1. Assumes an equal covariance matrix for both groups.
Hypothesis 2. Assumes a common factor structure for both groups.
Hypothesis 3. Assumes a common factor structure for both groups
and that the factor variances and covariances are equal.
Hypothesis 4. Nested in Hypothesis 3, this test relaxes the assumption
that the Deviance factor has equal variances in both
groups.

Hypothesis 8. Nested in Hypothesis 7, this test relaxes the assumption
that the Disposition to Deviance factor has equal variances in both groups.
Hypothesis 9. Nested in Hypothesis 8, this test is a repeat of H2, and
relaxes the assumption that the Self-Rejection factor
has equal variances in both groups.

[Soc. Psych. Q. 69, 1986, 127, with permission]

is, "broadside." If the table is short as well as wide, about half the page may then
be left blank. When such tables have excessive blank space or the elements are not
effectively disposed, they can often be rearranged to fit across the width of the page.
Table 6.43a, occupies only 2 in. of the 5-in. width of the printed page, yet the space
between columns is so wide that simply reducing this space would allow the table
to fit across the width of the page, especially if the repetitious "yellow" column
were replaced by a footnote. In *Table 6.43b* the "direction" column is very wide,
but the two elements can be reduced to symbols, for example, $N>P$, $P>N$. Then
by reducing the space between columns, omitting indentation in the stub, and substituting abbreviations for "present illness" and "past history"—or even making it a
column with headings PI and PH—the table can fit across the page.
    Table 6.44, on clinical nutritional training, extends over numerous pages (8 in.

TABLE 6.43 Tables short and wide, set broadside on journal page with excessive space outside and inside table; capable of being revised to fit width of page by eliminating repetition, condensing entries, and reducing intercolumn space

a.

TABLE II

Synthesis and Characterization of Polyazoxyarylethers

Polyazoxyarylether[a]	Reaction conditions					Properties			
	Solvent DMSO:PhCl (v/v)	Temperature (°C)	Time (h)	Yield (%)	Color	η^{b}_{inh} (dL/g)	df	Nitrogen content (%)	
								Calcd.	Found
3	3:1	140	4	93	Yellow	0.27	1,276	6.64	6.58
4	3:1	145	5	90	Yellow	0.20	1,278	13.21	13.11
5	3:1	145	5	95	Yellow	0.21	1,277	6.31	6.29
6	3:0	140	3	91	Yellow	0.25	1,279	6.54	6.47

b.

TABLE 3-8. Qualitative Aspects of Suicide: Paroxysmal EEG versus Nonparoxysmal EEG

Group and scale item	Direction	F	df	p
Patients with suicide ideation (paroxysmal, $n = 70$; nonparoxysmal, $n = 211$)				
1. Degree of plan formulation (present illness)	Nonparoxysmal > paroxysmal	3.14	1,276	.08
2. Reactivity of ideation (present illness)	Paroxysmal > nonparoxysmal	6.35	1,278	.01
3. Frequency of ideation (past history)	Paroxysmal > nonparoxysmal	3.77	1,277	.05
4. Intensity of ideation (past history)	Paroxysmal > nonparoxysmal	4.62	1,279	.03
Patients with suicide attempts (paroxysmal, $n = 42$; nonparoxysmal, $n = 130$)				
1. Degree of plan formulation (present illness)	Nonparoxysmal > paroxysmal	4.25	1,167	.04
2. Medical seriousness of attempt (present illness)	Nonparoxysmal > paroxysmal	3.45	1,170	.06
3. Past history of ideation	Nonparoxysmal > paroxysmal	4.22	1,170	.04
4. Frequency of ideation (past history)	Paroxysmal > nonparoxysmal	5.38	1,169	.02
5. Intensity of ideation (past history)	Paroxysmal > nonparoxysmal	7.00	1,170	.009
6. Reactivity of ideation (past history)	Paroxysmal > nonparoxysmal	3.59	1,170	.06

[a. Mandal and Maiti. J. Poly. Sci. 24, 1986, 2452, © 1986 John Wiley & Sons, Ltd.; b. Suic. Life Threat. Behav. 16, 1986, 14; with permission]

TABLE 6.44 Table set broadside and occupying full journal page, with crowding and excessive blank space; capable of being revised to fit width of page, by condensing columns, rows, and entries, eliminating irrelevant entries, and reducing intercolumn space

TABLE 1 (*Continued*)

State	Institution/Director/Address	Admin dept/dir	Yearly start date	Duration *yrs*	No. of applicants 1983	No. of positions/ yr	Total current trainees	Admission criteria
CT	Norwalk Hospital Martin Floch, MD Yale University School of Medicine Norwalk, CT 06856 (203) 852-2368	Int Med/ Gastr	The program focuses mainly on Gastroenterology.					
GA	Emory University Steven B. Heymsfield, MD 1364 Clifton Rd NE Atlanta, GA 30322 (404) 329-7215	Int Med	Flexible	1	1	1	1	PG training in Med or Surg
			The main focus of this program is clinical research.					
GA	Medical College of Ga Elaine Feldman, MD Rm BD 101 Augusta, GA 30912 (404) 828-4605	Int Med	7/1	1-2	4	1	1	3 PG yrs in Med
IL	Ravenswood Hospital Med Ctr Mitchell Kaminski Jr, MD 6000 W Touhy Ave Chicago, IL 60648 (312) 774-7717	Surg	7/1	1	4	1	1	2 yr PG Med or Surg and Nutr Support Background

[Extra art, Amer. J. Clin. Nutr. 42, 1985, 154, © Am. J. Clin. Nutr., with permission]

TABLE 6.45 Tables short and wide, set broadside on journal page, but with excessive space (a,b) outside and (b) inside the table; capable of being reorganized to fit width of page

a.

TABLE 7. *The effect of Pisum sativum stocks differing at the Sn and Dne loci on lateral outgrowth and flowering in LDH Lathyrus odoratus scions*

Graft combination			Total lateral length (mm)					No. of expanded leaves on main stem				NFI			Aerial laterals on day 45*
			Day 21			Day 35		Day 21		Day 35					
Scion (genotype)	Stock	(genotype)	Mean	s.e.	n	Mean	s.e.	Mean	s.e.	Mean	s.e.	Mean	s.e.	n	
LO5 (Dnʰ)	I₃	(sn Dne)	4·83	0·49	7	135·00	29·91	4·14	0·14	9·50	0·50	14·00	1·00	2	0
LO5 (Dnʰ)	K218	(Sn dne)	10·89	1·64	18	255·56	17·93	4·53	0·12	9·39	0·12	12·89	0·11	18	11
LO5 (Dnʰ)	Torsdag	(Sn Dne)	50·32	4·96	19	390·16	21·75	4·47	0·12	9·26	0·10	>41†	—	—	0

b.

TABLE 3

CLUSTER CONFIGURATIONS

Cluster 1:
WTUL
Shirtwaist Strike organizing
Consumer's League
Socialist party
Congressional Union for Woman's Suffrage
Woman Suffrage party
College Settlement
Child Labor Committee
Columbia University
Henry Street Settlement
Progressive party
Social Reform Club
Suffrage Campaign
Woman's party . . .

Cluster 2:
Sorosis
AAW
General Federation of Women's Clubs
Anti-Slavery Advocacy
Woman's Parliament
DAR
Republican party (after 1910)
Cluster 3:
Garrisonian abolitionism
Married Woman's Property Act Campaign
National Woman's Loyal League
Temperance Advocacy
Anti-Slavery Society
Dress Reform Advocacy . . .

Cluster 4:
Charities Aid Association
U.S. Sanitary Commission
Cluster 5:
Spiritualist Practitioner
Woman's Rights Advocacy

[a. Ann. Bot. 36, 1985, 853; b. Amer. J. Soc. 90, 1039; with permission]

long) in the journal. Yet it can be modified to fit across the width of the page. The director's name and address can be removed from the table, which is designed to describe the program, to an appendix, reducing the five to six rows to one; and the state names can be made italic headings, over the various medical institutions. The columns can be brought closer together, thus releasing additional column width for the last column, which has the second longest lines of text. Efforts can then be made to reduce its two- to three-line entries to one or two lines to match the revised first column. This revision will make the table fit across the width of the page, reducing its length by more than half and making it easier to read.

Tables set broadside may not have excess blank space; still it may be possible to rearrange them to fit across the width of the page. *Table 6.45a* is space consuming because of its narrowness, the repetition, and the separated elements that could be combined. The scion and genotype are the same for all values, and so can be included in the title or as a footnote. The single value for the "Aerial laterals" can be placed in a footnote. The stock and genotype and the mean and standard error can be combined in two columns. Often tables set broadside can fit across the width of the page if turned 90 degrees. Then the row headings become the column headings, and the column headings become the row headings (*Table 6.46*). In *Table 6.45b*, the clusters constitute five lists of items, which do not necessitate three columns; therefore the clusters can be set in two ranks across the page as effectively and much more economically.

Tables in which the field has written words require careful planning to make them economical. The objective is to allot the column-width space in proportion to the average amount of text in the entries of each column, so that the corresponding elements across a row take up about the same number of lines of type. This means that the columns with the most text are allotted more column width. Where the discrepancy between the columns with long texts and short texts is still too great, it may be desirable to combine short columns to allow more column width for long

TABLE 6.46 Table 6.45a with column and row headings interchanged to fit across page

Measure of growth	Stock and genotype*		
	I_3 (*sn Dne*)	K218 (*Sn dne*)	Torsdag (*Sn Dne*)
Total lateral length (mm)			
Day 21			
Day 35			
No. expanded leaves on			
main stem			
Day 21			
Day 35			
NFI			
Aerial laterals			
on day 45			

*Scion LO5, Dn^h genotype

TABLE 6.47 Part of word table showing uneconomical and ineffective disposition of elements with excessive blank space around short entries and crowded long entries

TABLE 5. Brief Summaries of Studies Extracting Past Temperature Trends From Temperature Profiles in Boreholes in Dry Media

Reference	Area and Period Examined	Temperature Rise, °C and Remarks
Mellor [1960]	MacRobertson Land, Antarctica, to 1958	Ice temperature profiles to about 30 m at 12 points along a 600-km glacier traverse south of Mawson all showed negative gradients (i.e., cooling) below 15 m, the depth reflecting the annual mean surface temperature. The magnitudes of the gradients were considered to exceed that due to ice motion and therefore to reflect recent climatic warming.
Paterson [1968]	Meighen Island, Canada, ~1880 to 1965	A borehole through the virtually stagnant Meighen Island ice cap revealed temperatures increasing rapidly with depth to the base at 121.2 m. This was deconvoluted to indicate a surface warming of 3.5°C in the period 1880–1940 and a cooling of 1.5°C in the period 1940–1965.
Cermak [1971]	NE Ontario, Canada, ~1000 to 15 B.P.	At two of three boreholes to depths of 600 m just south of the permafrost boundary, temperature profiles were deconvoluted to obtain estimates of surface temperature departures from the reference temperature since 1000 B.P. Best fits in the two holes were as follows: for the Little Ice Age, 1500–1800 A.D. at -1.01 ± 0.23 and -0.97 ± 0.21°C; for the medieval little optimum, 900–1300 A.D. at 1.55 ± 0.42°C and 1.09 ± 0.38. For recent trends, 3.31 ± 0.16°C for period 1830 to 1970 and 2.21 ± 0.6°C for period 1770 to 1970. (Note this implies circa 1970 surface temperatures 1.31 and 1.93°C warmer than during the medieval little optimum at these two sites.)
Gold and Lachenbruch [1973]	Northern Alaska, ~1800 to 1960	Boreholes at Cape Thompson, Barrow, Cape Simpson, and Prudhoe Bay on north coast of Alaska show surface temperature rises of 2° −4°C since circa 1850.
Lachenbruch et al. [1982]	Prudhoe Bay, Alaska ~1880 to 1980	Nine boreholes through permafrost gave temperature profiles with departures from a straight line in upper 160 m which was calculated to indicate an average surface temperature warming of 1.8°C over the past 67–117 years.

[Rev. Geophys. 24, 1986, 773, with permission]

215

TABLE 6.48   Revision of *Table 6.47* to telescopic form illustrating (a) combination of 3 entries in first 2 columns, (b) condensation of 2 reports on temperature profiles.

a. 1960: *Mellor*   1973: *Gold & Lachenbruch*   1982: *Lachenbruch et al.*
      MacRobertson Land   Northern Alaska   Prudhoe Bay, Alaska
      Antarctica, to 1958   ≈ 1800 to 1960   ≈ 1880 to 1980

b. *Temperature Rise °C and Remarks*

1. Borehole through stagnant ice cap; temp. increasing rapidly with depth to base 121.2 m. Deconvoluted, indicating surface warming of 3.5 °C 1880–1940, cooling of 1.5 °C 1940–1965. [*Paterson*, 1968; 44 words to 24]
2. Boreholes (9) through permafrost; profiles departing from straight line in upper 160 m. Calculated to indicate avg. surface temp. warming of 18°C over past 67–117 yr. [*Lachenbruch et al.*, 1982; 33 words to 24]

texts. *Table 6.47* illustrates a type of table that is frequently used in dissertations and sometimes in journals to summarize the literature. The first two columns have more space than is needed, and the last column is so crowded as to be almost unreadable; yet there is much blank space. The entries in the first column can be combined, as headings, with those of the second column as subheadings, as in *Table 6.48a*, thus making three-line units for each row heading. This revision leaves the rest of the width of the table for the last column.

The long text of such tables can also be shortened by using various strategies for condensation. The sentence form consumes space unnecessarily and can be replaced by elliptical clauses and phrases. Abbreviations can be used, for example, "temp." (*Table 6.48b*) and self-evident words like "temperature" with "profiles," can be omitted. Also, in some word tables, notes or repetitious elements can be deleted. Such strategies have been applied to entries in *Table 6.47* and have shortened them by almost one-half, as shown in *Table 6.48b*.

ILLUSTRATIONS

6.6 OVERVIEW

Illustrations, much more than tables, have a strong visual impact on readers. The immediacy of an illustration cannot be matched by a table or by the clearest, most concise prose. Form communicates meaning more directly than words. Writing is sequential and linear, and to visualize a structure or understand a concept from written text, readers must read the description or explanation word by word, gradually developing a visual and cognitive structure of it. To derive such a cognitive structure from an illustration, they need only look at it; they can grasp the image or concept illustrated immediately and as a whole without close examination of the separate parts. One can form an immediate image of a cardinal from an illustration, yet not have a clear image of it even after reading the detailed scientific description of the species (*Ex. 3.22*). Compare looking at *Fig. 6.6* or *6.20* with reading a full written description of them. The structure that is so readily apparent from the illustrations would be extremely difficult to conceptualize visually from a written description in the text. Because of the immediacy of illustrations, descriptions are rarely written in full detail in the scientific paper, except when a formal scientific description is required (*Ex. 7.4*). Usually illustrations supplement descriptions. They thus save words as well as being vivid.

One of the chief uses of illustrations is to present configurations and details—both physical and conceptual—which are difficult to communicate in words. For actual physical entities, illustrations help readers to develop a visual concept of the structure; for concepts, they help to clarify the written interpretations, models, or interrelations by giving them concrete form.

6.6.1 Planning and Preparation

Writers should anticipate the illustrations that they may need and plan for them. One does not wish to discover when writing, that one should have taken a particular photograph at the site, which is now thousands of miles away. Illustrations require planning because (1) the text is dependent on the illustrations, (2) illustrations are not reviewed as thoroughly as the written text, and (3) writers are dependent on others to prepare them. Since illustrations and the text must be consistent with each other, it is desirable to have at least final sketches available for writing the review draft. Having them on hand while writing allows one to include only those details that are needed for the development in the text.

The writer bears much more of the responsibility for the quality of illustrations than for the quality of the writing. The written text is examined word for word by reviewers, editors, and printers, all of whom may contribute to meeting high standards in writing and printing. Illustrations receive no such cumulative review or incremental improvement. The author provides camera-ready illustrations. If they are very inadequate for their scientific purpose, reviewers or editors may suggest modifications or redrawing; if they are very poorly prepared technically for reproduction, the editor or printer may ask to have the defects corrected. Such changes are usually one-time changes, because changes are not easy to describe and are too difficult, costly, and time-consuming, to allow repeated revisions. Consequently, illustrations are usually forwarded to the photoengraver, without much revision or improvement. Although the photoengraver and compositor can ensure the quality of the *reproduction,* they cannot make up for serious deficiencies in the illustration submitted to them. Therefore the visual quality of an illustration must be built into it before it reaches the photoengraver.

Research scientists must therefore be aware of the visual capabilities and limitations of the different types of illustrations, the different methods of reproducing them and their cost, and attributes of an illustration that make it visually effective. They can then select the type of illustrations appropriate to their purpose, plan them for effectiveness, and ensure that they are of high quality and visually effective. Although research scientists can rely on illustrators and photographers to relieve them of the actual drawing, photographing, and designing, they are dependent on their work schedules. A writer cannot expect them to match his or her schedule and work into the night to meet his or her deadlines.

6.6.2 Design and Execution

Because illustrations are costly to reproduce, the writer must use them only when they are necessary, plan them so that they are economical in size, use the less costly methods of reproduction when possible, and work for economy of means.

Criteria for illustrations.   The criteria for illustrations must address their scientific content and visual form, and their professional quality and tone. Although form is the essential criterion for quality, any design or artistic objectives are relevant only as they advance the clarity of the message. The purpose of illustrations must be clear. They must present enough information that is sufficiently important and directly relevant to the research to warrant the cost of graphic reproduction, and they must transmit their message accurately, concisely, and meaningfully; otherwise, they should be eliminated. For example, the reciprocal relationship shown in *Fig. 6.3a* between sales agent and purchaser can easily be expressed in words, and the four values in *Fig. 6.3c* would take up much less space in the text than in the graph, and be as effective. *Figure 6.3b* would be more effectively presented as a table or summarized in the text.

Illustrations should match the professional quality of the printed text and the research orientation of the journal. In *Fig. 6.4,** the freehand drawing and lettering, though clear and neat, are not in keeping with the printed page, and the difference between the printed letters for the axis labels and cursive letters for the points is not distinctive enough or readily apparent. The graph does not merit the space that it takes up in the journal ($5\frac{3}{4}$" × 7") and the upper half is wasted space. In *Fig. 6.5* the hand-drawn arrows are conspicuously large and not in keeping with the diagrammatic character of the illustration. The other arrowheads are also drawn freehand although the small size and fine lines keep them from being obtrusive. The typed labels are too light and crowded and lack the professional character of printed letters.

Each illustration should have a central focus, toward which the research scientist directs his or her efforts for visual effectiveness. Separate illustrations may be more effective than crowding extraneous, superfluous, or excessive details into one figure. The illustrations for the pump limiter device (*Fig. 6.6*) illustrate the division of labor among three illustrations, with *Fig. 6.6b,c* having a strong documentary function and *Fig. 6.6a* having little detail and a strong communication function.

---

*In this and subsequent figures, the first part of the legend addresses the particular objective of the figure. Comments after "Note" refer to other features.

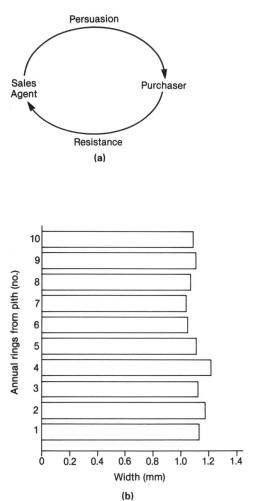

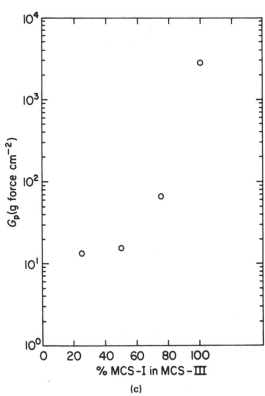

Figure 6.3. Illustrations not informative enough to warrant graphic reproduction and not required for visual impact of message. (*a*) Easily described in text; (*b*) as effectively presented as table of values; (*c*) large graph not warranted for only four points. [(*c*) Polymer. 27, 1986, 1728 by J.M. Pochan et al. and by permission of the publishers, Butterworth & Co. (Publishers) Ltd. ©

Figure 6.6c shows one of the two limiters in great detail; *Fig. 6.6b* shows the two pump limiters, greatly reduced and simplified, as part of the pump, to show their position and connections to the pump, and *Fig. 6.6a* shows the schematic diagram of the experimental apparatus, in which the pump limiters are included but are too subordinate to the overall apparatus, to even be illustrated in the figure.

To focus the illustration, the most important part of it should be positioned for optimum effectiveness. It should be centered and in the foreground, and on as large a scale as is consistent with other constraints. Figures are "read" from left to right and from the top down, and the center predominates over the periphery, and the foreground over the background. In a photograph, emphasis can be focused by cropping.

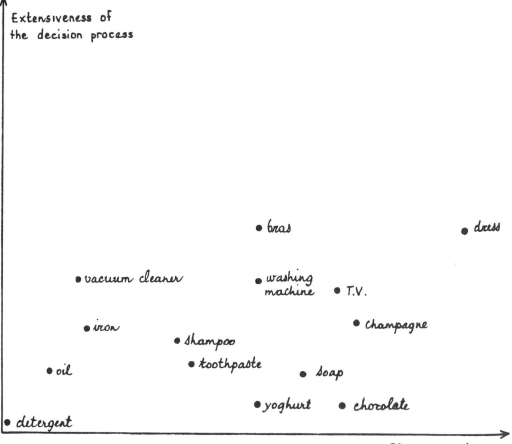

Figure 6.4. Graph unprofessionally drawn. Points and arrowheads drawn freehand; labels handwritten, partly cursive, partly printed. *Note.* Upper half of graph wasted space; no scale marks to indicate position of points on scale; vertical axis labeled across the top and inside the graph. [Reprinted from G. Laurent & J.N. Kapferer J. Market. Res. 22, 1985, 48, published by and with permission of the Amer. Mktg. Assoc.]

The tone of the graphics should be scientific and professional. Visual effects that draw attention to themselves distract readers from the meaning of the illustration. The informal character of the hand lettering in *Fig. 6.4* is not consistent with the formal character of the printed page or the research. Frames or circles around labels only add to the lines that readers must interpret. Labeling devices should not be noticeable for their decorative character. Arrows with vivacious heads and sweeping or lightning curves distract attention from the message of the illustration. And humorous or cartoon effects have no place in illustrating a scientific paper.

Illustrations should be labeled to identify the parts for readers. Labels may be

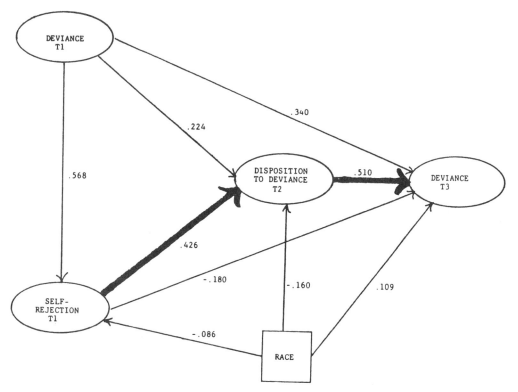

Figure 6.5. Path diagram with unprofessional drafting. Arrowheads and bold arrows drawn freehand, these arrows dominating all other elements; line width of other elements undifferentiated; labels and numbers typed. *Note.* Bold arrows not explained in legend. [Soc. Psych. Q. 49, 1986, 121, with permission]

written in full, if space is adequate and the labels are not too long. They should be orderly arranged, not radiating out in all directions. When space is limited, labels can be abbreviated, with standard abbreviations; any temporary abbreviations can be explained in the legend. When illustrations do not have space for labels, numbers or letters may be used, and then explained in the legend (*Figs. 6.21b, 6.22a*). Terms, abbreviations, symbols, and numbers used in the illustrations should be identical with those in the legend and in the text. The positions, size, and weight of the numbers can often be used to differentiate different sets of numbers. If lines to be labeled are too crowded, a key may be used, especially in graphs, to remove the explanation to a more spacious part of the drawing (*Fig. 6.61*). When a figure consists of two or more parts or illustrations, each part is labeled. The numbers or letters used to designate the parts should be distinct in size and form from those used in labeling the illustrations, and in good proportion to them. The legibility of labels after reduction is determined not only by the size of the lettering, but also by the width of the lines, the spacing of the letters, and the style of type.

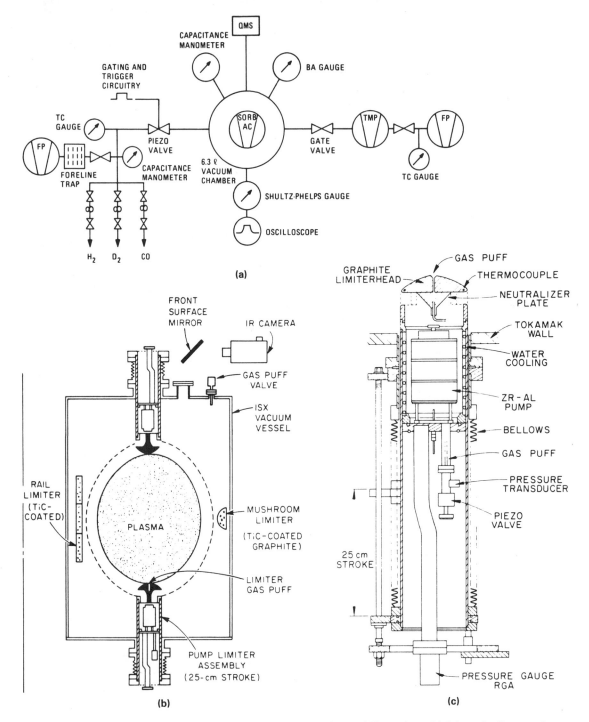

Figure 6.6. Diagrams illustrating different focus of illustrations. (*a*) Schematic diagram of experimental apparatus with (*b*) and (*c*) not visible; (*b*) component with two pump limiters shown in outline; (*c*) detailed structure of pump limiter. [J. Vac. Sci. Tech. A2, 1984, 1583–4, with permission]

222

Related illustrations may be combined into one figure. The illustrations may be related parts as in *Fig. 6.6,* or they may be different but related illustrations. In arranging the parts of a figure, the research scientist must consider the purpose of the figure, visual effectiveness, and the effective use of the printed page. Two illustrations that are very similar or closely related, but too small for the page width can be effectively and economically combined. The two illustrations in *Fig. 6.7a* were each separate, centered on a printed page 5¼ in. wide in the journal. The three parts of *Fig. 6.7b* were part of the same figure arranged vertically and occupying only 2¾ in. of the 5½-in. page width. Placing them side by side saves one-half the

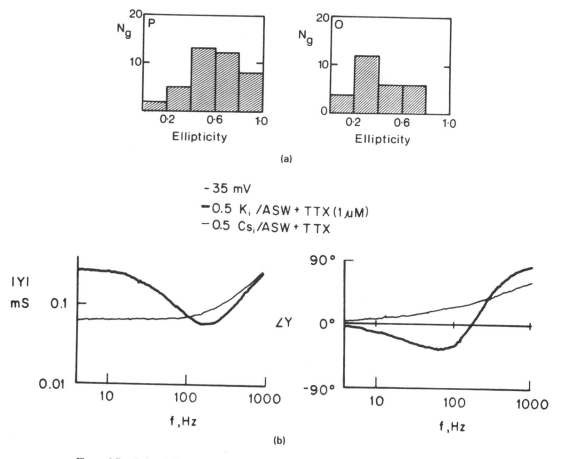

Figure 6.7.   Related illustrations rearranged to save space. (*a*) Published on separate pages (5 1/4 in. wide); (*b*) published one above the other (on page 5 1/2 in. wide). *Note.* (*a*) Hatching of bars producing optical effect. (*b*) Scale marks on outer axes too long; lines in key too short, not designating curves clearly. [(*a*) Mon. Not. Roy. Astron. Soc. 222, 1986, 679–80; (*b*) H. M. Fishman, Prog. Biophys. Mol. Biol. 46, 1985, 148, © 1985 Pergamon Journals, Ltd.; with permission]

length of the page that they occupied. In *Fig. 6.8,* the two graphs were set one above the other on the journal page (8 in. × 5 in.). They can be placed side by side, as here, or even replaced by one graph, because the only difference between the two is the difference in two pairs of points (open circles), which could be differentiated.

Size of illustrations.   Illustrations are prepared larger than their published size. The large size makes it easier to draw and letter them, and any small irregularities in the edges, width, or direction of lines are reduced in reproduction and so become unnoticeable and the drawing more finished. A photograph may be used in its original size, if it fits the dimensions of the page and the importance of the detail requires that reduction be minimal.

To choose dimensions for a figure that will fit the dimensions of the printed page or column when reduced, the illustrator or author can use the dimensions of the printed page to derive figure sizes proportional to them (*Fig. 6.9*). The dimensions of the printed page, *ABCD* in *Fig. 6.9a,* or column can be outlined on a sheet of paper larger than the page of the journal or on wide-gauge graph paper. A diagonal is drawn from the lower left corner, *A,* to the upper right corner, *C,* and extended beyond it, *ACR.* The sides *AB* and *AD* are also extended. Then any point

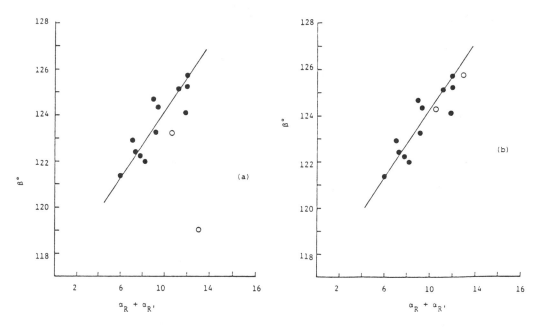

Figure 6.8.   Illustrations reoriented to more economical horizontal arrangement from vertical arrangement occupying almost full page (8 in. × 5 in.) in journal. Reduction to single graph possible by replacing one pair of open points with a different symbol. *Note.* Lines not differentiated; curve originating too far from axes; origin not labeled; points disproportionately large; lettering typed, too small. [Phosph. Sulf. 28, 1986, 109, by R.A. Shaw, reprinted with permission © 1986 Gordon and Breach Science Publishers S.A.]

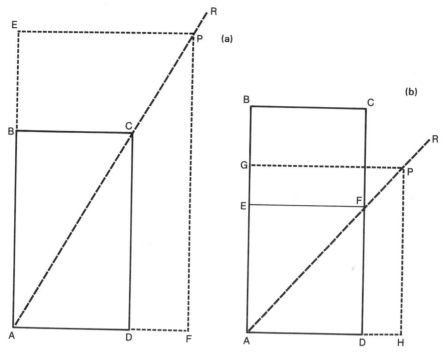

Figure 6.9. Diagram illustrating method of determining dimensions of figure proportional to page or column. (*a*) For page or column size; (*b*) for part of page or column; (*a, b*) *ABCD* is size of column or page. (*a*) *AEPF* is larger than *ABCD*; (*b*) *AGPH* is proportional to *AEFD* and smaller than *ABCD*.

on the extended sides *AB* or *AD* or on the diagonal *ACR* can be used to define a rectangle (*AEPF*) larger than the original *ABCD* rectangle and proportional to it. The dimensions should be chosen to allow the size of lettering and width of lines to be drawn large enough for reduction. If the illustration is to be less than the length of the column or page width (*Fig. 6.9b*), the dimensions of the full column or page width (*ABCD*) are again used. On this rectangle the desired size of the illustration is outlined, *AEFD,* and a diagonal line is drawn from *A* to *F* and extended beyond to *R*. A horizontal and vertical line drawn from any point on the diagonal will give a rectangle, *AGPH,* that is proportional to the rectangle of desired size, *AEFD.*

An illustration is often designed so that its two dimensions are in a ratio of about 1:1.5. The actual dimensions or outside limits of the illustration are then established, usually about twice the dimensions of the page or column width or length, whichever is the limiting dimension. If it is a very long illustration, it must be adapted to fit the length of the page; if very wide, it must fit the width of the page, or it may have to be set broadside on the page. Except for long, narrow figures, the width is usually the limiting dimension. When a figure takes up less than

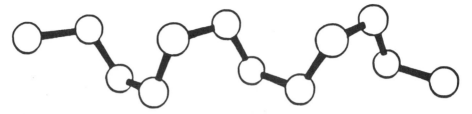

Figure 6.10.   Diagram oriented horizontally on page to save space; oriented vertically in journal, occupying column 5 in. × 3 1/2 in. Large, simple enough, and lines heavy enough to reduce to fit horizontally in column. *Note.* Bonds disproportionately heavy (unless representative of size). [J. Chem. Phys. 85, 1986, 6462, with permission]

the width of a page or column, the extra space across the page is wasted, whereas extra space above or below a figure is regularly set in type. For example, *Fig. 6.10,* a simple drawing of a molecule, is oriented vertically in the journal, where it takes up almost 5 in. of the column length and little more than 1 in. of the 3.5-in. column width, and so is very space consuming. Yet the structure is simple, the elements large, and the lines wide, so that it could be easily reduced by one-half, with little loss of detail. Furthermore, unless the vertical orientation of the helix is conventional or necessary, the drawing can be oriented more economically with the helix horizontal (as here) and reduced to fit to the column width.

Authors may specify the reduction on the illustration, either by a fraction, for example, "reduce ½" or better by the final width or length, for example, "reduce width to 5 in." Comparable illustrations should be planned for the same reduction to allow comparison. The reduction of the elements of the illustration must also be taken into account in preparing the illustration. When the dimensions of a figure are reduced by one-half, all the elements are correspondingly reduced. The graph in *Fig. 6.11* shows elements of different sizes and line widths, which are reduced one-half in *Fig. 6.12.*

6.6.3 Titles and Legends

Much of what has been said about titles for tables applies to titles for figures as well; however a figure may require more explanation than can be provided by the title and labels to be understood without reference to the text. Such explanation is usually supplied in the legend, which combines the functions of the headnotes and footnotes in a table. Legends may become extremely long (*Fig. 6.18*), and may take up as much space as the figure. Figures should therefore have titles to describe their content and to serve as a focus for the explanation in the legend. *Fig. 6.13* lacks a title; the text below the figure is not a legend, since it makes no reference to the various parts of the figure.

Titles should be concise and may be more like labels or headings when the

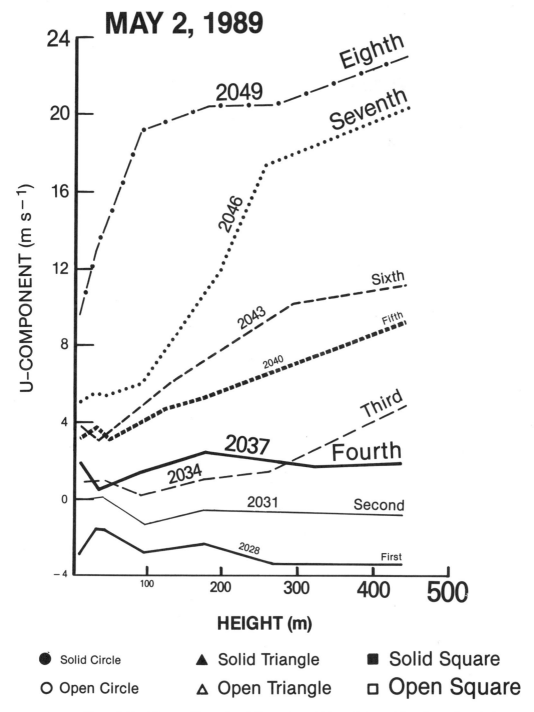

Figure 6.11. Diagram illustrating different line widths and sizes of lettering and symbols before reduction.

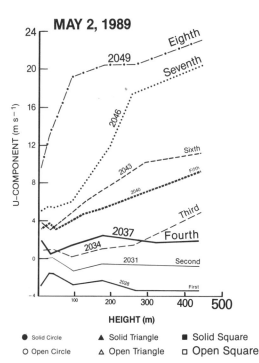

Figure 6.12. Fig. 6.11 reduced by one-half to show effects of reduction.

legends are extensive. However, unless there is an adequately explanatory legend, general, broad titles like the following are not informative enough:

Ex. 6.13

Basic Model	Schematic Diagram	Snowcover Chronology
Electronic Circuits	Dispersed Benefits	Summary of Model

For a graph, the title may simply relate the abscissa to the ordinate, for example, "Carrier Temperatures vs. Time Delay at 10 K." In a figure, the title should label the subject, describe it briefly, or indicate the main parts of the figure, leaving further explanation for the legend.

The title of a figure is not part of the figure and should not be lettered in by the illustrator. It is set in type, and its form is determined by the style of the journal (*Appendix 6B*). The figure designation may be abbreviated or written in full; capitalized, all in capitals, or in capitals and small capitals; and it may be in roman, italic, or boldface type. The number may be followed by a period or colon, with or without a dash, or by no punctuation. The number and figure designation may be separated from the title, centered, or displayed at the left side. They may be followed immediately by the title, beginning at the left, either flush or indented. The title may be in the same font as the figure designation and number or different, and it may be in

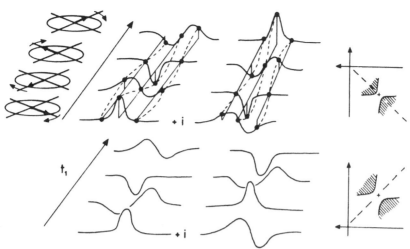

FIG. 39. An absorption mode line is intense in the centre and weak in the wings whereas the dispersion mode has considerable intensity in the wings and is zero at the centre. The variation in intensity at given frequencies (dashed lines) in a phase modulated signal therefore appears to have its phase inverted on passing through the centre of the resonance. This produces the 'phase twisted' lineshape of Fig. 37. If the sign of the frequency is the same during evolution and detection. the contours of zero intensity cut across the main diagonal. A distribution of frequencies caused by an inhomogeneous magnetic field then produces destructive interference between the peaks associated with different volume elements which spread along the diagonal. producing a broad. flat peak. By contrast. if the frequency changes sign between evolution and detection as in a refocussed signal. there is constructive interference between different volume elements which leads to a sharp ridge along the diagonal.

Figure 6.13.   Figure lacking title. "Legend" not descriptive of figure, apparently part of text related to figure. [Prog. Nucl. Magn. Reson. Spectr. 17, 1985, 343, by D.L. Turner. © 1985 Pergamon Journal, Ltd., with permission]

a different font from the legend. The subsequent lines of the title or legend may be flush with the figure number, indented below it, or flush with the first word of the title. Sometimes the last line of the legend is centered, even though the other lines are flush with the margin. The writer should consult the journal of choice for the style used (see *Appendix 6B*).

The legend follows the title on the same line or on the next line, with the same or different margin and typeface. Its function is to help make the figure understandable without reference to the text. It amplifies on the title and explains the labels that are not explanatory in themselves, describes methods used for data obtained, comments on particular parts of the illustration, qualifies data, and so on. Legends should be as condensed as possible. Symbols may be explained in the legend, but it is more effective to have keys in the illustration. In some journals the different kinds of points or lines in graphs are set in type in the legend to identify them (see *Appendix 6B*), but these are not as effective as labels and keys *in* the graph. The symbols may even be described in the title or legend,

Ex. 6.14

Na and K channel densities before (solid symbols) and after (open symbols) . . . photodestruction.

but this is a very ineffective way to identify them.

The labels in the legend may be set off visually from the text of the legend, for example, italics versus roman, boldface versus roman, or capital letters. When the figure consists of illustrations labeled with a letter or number, the title is followed by the series of letters or numbers, in consecutive order (*Appendix 6B*). Following each letter or number is a description of the illustration. The labels or labeled structures of each part of an illustration are then explained, before the next illustration is described. Sometimes, when the explanation is extensive, but it is desirable to identify the illustrations first, the illustrations are first titled (labeled) in the legend, for example, "(A) Lateral view, (B) Dorsal view of head," and then the labeled parts are described or explained.

6.7 PHOTOGRAPHS

Photographs are the closest that a research scientist can come to showing readers what he or she actually observed. One therefore elects to use photographs because they present precise details and subtle differences better than drawings. Photographs have two main uses in reporting scientific research: (1) the direct representation of usually gross, areal, or surface features, for example, topography, macroscopic structures and (2) the indirect representation of minute or invisible entities as in chromatographs and roentgenograms.

The reproduction of photographs is a highly developed technical process. Photographs cannot be reproduced directly because of the continuous tonal gradations, so they must be reproduced by halftone reproduction. They are photographed through a halftone screen or grid, with 150–300 lines per inch. The lines break up the light from the photograph into dots of different size and distribution, in proportion to the light and dark areas of the original. The screened photograph can then be used to make a plate that has printable dots.

Because halftone reproduction breaks the photographic image into dots, the printed reproduction is never as sharp or as precise as the original photograph. The photograph must therefore be sharply focused, and glossy to start with, and it should have adequate contrast and a wide range of tones. *Fig. 6.14* reproduces two photographs that were not of high quality They were not sharply focused and had a narrow range of tones.

The photograph should not require much reduction. It is preferable to crop the photograph so that focal area is emphasized. Cropping consists of eliminating anything in the background or at the periphery that would detract from the part to be emphasized—unnecessary or unimportant or distracting objects, equipment, people, or blank space. This allows the focal part of the photograph to be reproduced on as large a scale as possible. The area that should be cropped can be deter-

(b)

(a)

Figure 6.14. Photographs illustrating poor quality for reproduction and gray area resulting from spaced mounting. Not sharply focused; low contrast; narrow range of tones; (*a*) in need of cropping. [(*a*) Principes, 32, 1988, 31, with permission; (*b*) courtesy of N. W. Uhl]

mined by masking the edges of the photograph with two L-shaped bands of stiff paper to determine the boundaries for an effective illustration. For a solid background, a light background allows the subject to dominate. If a scale is needed, it is best included in the photograph, so that the magnification stated is proportional to any reduction of the original.

When several photographs are to be reproduced, they can be combined for economy, if they are related and do not vary widely in contrast and range of tones. They should be cropped and trimmed, and mounted on white cardboard. If they are mounted with space between them, the space between them reproduces as a light gray area in reproduction, as in *Fig. 6.14*. The more professional method of mounting photographs is to leave no space between them, as shown in *Fig. 6.15*. This is done by cropping and trimming the photographs sharply and squarely so that they fit closely together, then mounting them with the edges abutting as closely as possible, so that a sharp, neat separation results (*Fig. 6.15*). Halftone reproduction may be combined with line engraving, as in wash drawings that have labels extending outside the drawing (*Fig. 6.16*). This is a complex process, requiring absolutely exact superposition and so an experienced illustrator.

Figure 6.15. Photographs of quality suitable for reproduction and well mounted. Abutted in mounting; in sharp focus; with high contrast and wide range of tones; effectively cropped; [(*a*) *Genera Palmarum*, p. 134, with permission; (*b*) (*c*) courtesy of N. W. Uhl]

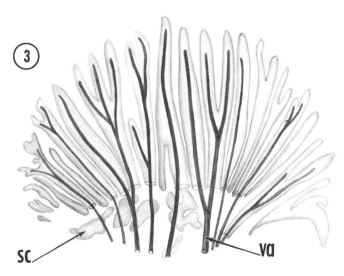

Figure 6.16. Wash drawing illustrating combined use of half-tone and line-engraving reproduction. [Am. J. Bot. 63, 1976, 98, with permission]

Because halftone reproduction can show fine gradations of shading, the surface of photographs must be carefully protected. Any bending, indentation, scratches, cracking, pressure marks, or other blemishes are likely to mar the surface and may be reproduced. Therefore, any instructions, lettering, labeling, or arrows must be shown on a sketch or marked with a felt-tipped pen on a transparent film, laid over the photograph. Cropping lines can also be indicated on the overlay. Instructions and the designation ''Top'' may be put on the overlay or on the margin of the photograph with a soft pencil.

Lettering photographs is difficult because of the glossy, gray and black surface, which must not be disturbed. Any labels should therefore be lettered on the photograph after it is cropped. They should be as brief as possible, preferably very short abbreviations or letters or numbers, because of the difficulty of drawing on the glossy surface. The lettering should be simple and easy to read, and the size chosen should be large enough for the lettering and numbers to remain legible after any reduction. If the photograph is labeled directly, india ink should be used on light areas and opaque white ink on dark areas. However, lettering on the photograph itself is best avoided, since a single mistake may make the photograph unusable.

It is easier to use preprinted numbers, letters, lines, and arrows on transparent adhesive transfers. Sheets of these are available in black and white, and the illustrator simply affixes them to the photograph. One must select black or white labels appropriately, to ensure enough contrast between the shades of gray around the label, because the letters and symbols affixed are also reproduced by the halftone process and so are reduced in contrast. A white letter on a light gray may become difficult to distinguish, and a black arrow may become invisible across a dark area. Labeling outside the photograph will require a separate line engraving; therefore, it is preferable to label the photograph with numbers or letters and explain these in the legend. Whether the lettering is done directly or by transfers, it requires great skill to align them correctly, and so it is best delegated to the illustrator. Photographs are usually mounted and provided with a protective overlay (see *Chapter 9*). In certain types of photographs, for example, X rays, the film is preferred to a print. For colored photographs, the editor must be consulted, because the author may have to underwrite part or all of the cost.

6.8 LINE DRAWINGS

Line drawings differ from continuous-tone illustrations in that their color is discontinuous and discrete; black lines and black areas break up the white background into areas of black and white.

Where photographs are representational, line drawings are abstractive and show the entity as the meaningful configuration that the writer (or illustrator) has abstracted from the entity observed. Therefore, where a photograph gives a complete view—a synthesis of all the subtle and minute details—a line drawing is analytical and so may be used to supplement a photograph. Photographs approach the actual observation; freehand drawings (*Fig. 6.17, 6.18*) abstract the characteristic structure, and the mechanical drawings (*Fig. 6.20–6.23*) show the essential or analytical structure. In fact, the analytical character of mechanical drawings makes them useful for representing structure that is not physical, such as relationships, processes, or concepts, in path diagrams and graphs. Line drawings are therefore very versatile in the variety of information that they can convey.

A line drawing must be able to transmit this variety of information with only lines, dots, and areas of black and white. Consequently, every line, every area of black or white, every letter, symbol, or label in a drawing, as well as its size, thickness, density, and form, contribute to making a statement. Lines and dots can be used to achieve a fineness of shading that approaches a photograph in showing texture and defining shape and form three-dimensionally (*Fig. 6.17a*). Such a drawing can be more informative than a photograph.

Line drawings are drawn on a material that allows erasing or redrawing: freehand drawings are drawn on good-quality drawing paper; mechanical drawings on tracing paper or plastic film. Jet black ink must be used, usually india ink, because pencil does not make the solid black lines or areas required for reproduction. Drawings may be drawn the same size as the object, or larger or smaller, depending on its size. Several drawings may be combined into a figure, but they can be drawn separately and then trimmed and mounted on a board, because the cut edges will not show in the reproduction. Photocopies can be used to find an effective and economical arrangement.

Line drawings are photographed directly, without a screen. In the resulting

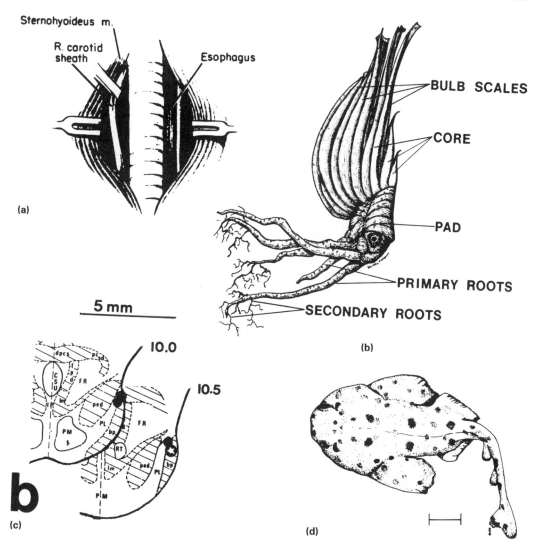

Figure 6.17.   Various types of freehand drawings. (*a*) Finely shaded three-dimensional drawing. (*b*) Drawing shaded with lines and some stippling; lettering too large and heavy for drawing. (*c*) Drawing of section, with hatched areas; letter "b" too large and bold, lettering in drawing too small, lettering of values to large. (*d*) Habit sketch largely stippled. [(*a*) courtesy of L.C. Hudson Ph.D. thesis, Cornell Univ., 1983; (*b*) Econ Bot. 41, 1987, 271 by H. Gentry et al.; © 1987 The New York Botanical Garden; (*c*) Beh. Neurosc. 100, 1986, 883 © 1986 Amer. Psych. Assoc.; (*d*) Copeia No. 4, 1986, 990; all with permission]

line engravings, the black areas are raised and pick up ink, and the white areas are depressed and do not. Usually the original drawings are submitted for publication. If photographs of drawings are to be submitted, they should be photographed by a

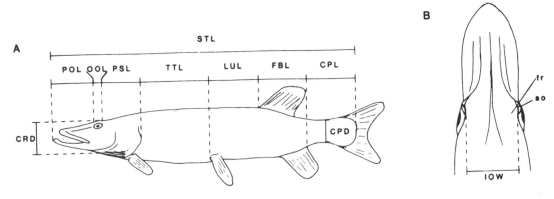

Figure 2. Morphometric variables measured. (A) Lateral view. (B) Dorsal view of head. Abbreviations: *fr.* frontal bone; *so,* supraorbital bone. The variables measured are defined as follows: POL (preorbital length), anterior of orbit to tip of lower jaw with mouth closed; OOL (orbital length), length of orbit between margins of hard tissue; PSL (postorbital length), posterior of orbit to most posterior projection of bony operculum; CRD (cranium depth), skull roof to bottom of gular region, passing through the centre of the pupil; IOW (interorbital width), shortest distance across the frontals in the orbital region (immediately posterior to the supraorbital bones); TTL (trunk lenth), distance from the posterior edge of the bony operculum to the origin of the pelvic fins; LUL (lumbar length), origin of pelvic fins to origin of dorsal fin; FBL (fin base length), origin of dorsal fin to insertion of anal fin; CPL (caudal peduncle length), insertion of anal fin to the end of the fleshy part of the caudal peduncle; CPD (caudal peduncle depth), least depth of the peduncle; STL (standard size measure), top of lower jaw to end of caudal peduncle.

Figure 6.18. Unprofessional drawing; lines irregular, shading ineffective; long legend. [Can. J. Zool. 65, 1987, 1226, with permission]

professional photographer with high-precision photographic equipment, to ensure sharp lines for reproduction.

6.8.1 Freehand Drawings

Freehand drawings allow the illustrator the greatest freedom in drawing, within the bounds of pen, black ink, and sometimes a dry brush. The chief objective is accurate representation of the object. No regularities need be observed in lines or black areas or their orientation or distribution, except as required for the drawing. Freehand drawings are therefore particularly adapted for habit or diagnostic sketches of plants and animals, or their parts, as seen with the naked eye or with magnification (*Fig. 6.17, 6.18*). The criterion for freehand drawing is accuracy and verisimilitude, and balance in the weight of lines and the weight and size of lettering.

Hand lettering is sometimes used on freehand drawings, but lettering with lettering guides looks more finished, although it must be in balance with the lines of the drawing. In *Fig. 6.18,* all the lettering has been done mechanically, although the drawing is largely freehand. In *Fig. 6.17c,* the letter "b" is so disproportionately large that it overshadows the illustration; the smallest lettering is disproportionately too small, and the numbers too large. The lettering in *Fig. 6.17b* is disproportionately large and heavy for the drawing. *Fig. 6.17* and *6.18* also show the variation in drawing techniques from the outline drawing in *Fig. 6.18* to the stippling (*Fig. 6.17d*) and hatching (*Fig. 6.17c*) and subtle shading in *Fig. 6.17a.*

6.8.2 Mechanical Drawings

Mechanical drawings have smooth, even lines, often drawn on the square and in a diagrammatic form; and the inked areas have sharp, regular edges. They are prepared by using drafting tools—T-squares, curves, stencils, letter guides, and special pens that produce a line of uniform width. These tools help the illustrator to control mechanically the thickness of lines, the shape of the curves, and the size and weight of letters. *Fig. 6.19* illustrates the unevenness of lines or dots drawn freehand. Mechanical drawing is particularly adapted to the geometric diagrams required for graphic display of length, area, volume, and for the points, curves, and bars in graphs.

The thickness of lines, the size of lettering, and the space between lines and dots must be chosen to allow for reduction. Lines that are too thin or not solid black will be broken in reproduction. Separate structures can be differentiated from one another by different hatching or stippling patterns (*Figs. 6.20, 6.21*). Open shading with light lines or dots well spaced is less obtrusive than close, heavy shading. If the lines or dots are too close, they may become confluent, and the pattern of the hatching or stippling may then be broken. Dots that are too small may become invisible. Lines and dots that are too heavy may be too bold for the printed page.

When uniform shaded or hatched areas are needed, the illustrator can use transparent, adhesive, preprinted pattern sheets. These sheets are available with stippling of varying weights and densities, hatched lines of numerous weights, densities, patterns, and so on. A suitable pattern of stippling, lines, or cross hatching is cut slightly larger than the space to be covered. The transfer is set in place and a sharp knife used to cut away the excess; it is then affixed to the paper or burnished onto it. Such patterns save the illustrator much time and are usually more regular than can be achieved if drawn by hand. The graphs in *Fig. 6.19a, c* show the uneven hatching and stippling that may result when drawn freehand and *Fig. 6.19b, d* the uneven lines. *Figure 6.19c* also shows arrowheads large and blurred by unskilled drafting, as well as disproportionately large letters and small numbers.

Although numbers and letters can be applied from preprinted pattern sheets, full labels are more effectively lettered with a lettering guide. Lowercase letters are less prominent and a block face is easier to read than italic or cursive (*Fig. 6.68a*). Arrows or lines from the label to the structure must be clear but light, and arrowheads unobtrusive. Labels are aligned horizontally, not following the direction of the arrow.

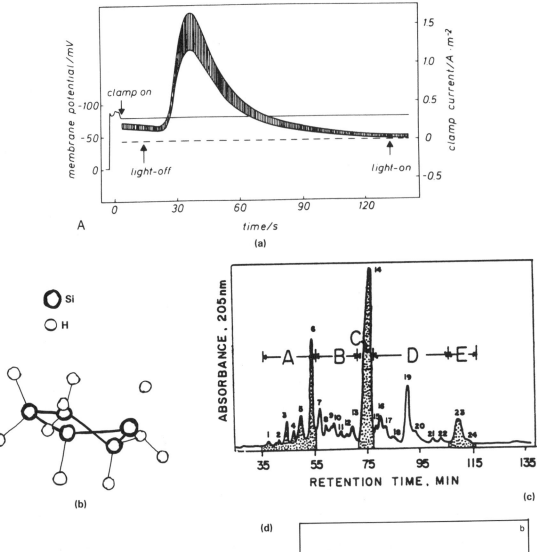

(a)

Si

H

(b)

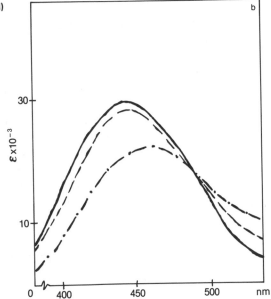

(c)

(d)

Figure 6.19. Diagrams illustrating irregularities resulting from freehand drawing. (*a*) Hatching uneven; (*b*) lines uneven, especially heavy circles; (*c*) stippling irregular; (*d*) lines uneven. *Note.* (*a*) Axis labels all lowercase, disproportionately smaller than scale labels; relation of curves to axes not clear. (*b*) Labels too small. (*c*) Letters and numbers in graph disproportionate in size; arrowheads too large, blurred. (*d*) Axis labels too small or lacking. [(*a*) Planta 167, 1986, 70; (*b*) Phosph. Sulf. 28, 1986, 57 by E. Hengge, © 1986 Gordon and Breach Science Publishers S.A.; (*c*) J. Biol. Chem. 261, 1986, 14122; with permission]

238

Diagrams are used for spatial configurations, for example for actual structures and instruments, and parts of these, or for the disposition of structures in space, such as schematic diagrams of apparatus. Such diagrams are an abstracted form of concrete objects or structures. Diagrams may also be used to represent abstract concepts, however, such as interrelations in a theoretical model, or explanation of a process or interpretation of a concept (*Fig. 6.25*).

In diagrams of equipment, the uniformity and regularity of lines and dots are important devices to delimit parts of the structure. The disposition of the lines in hatching and dots in stippling can differentiate the parts. In *Fig. 6.20a,* the different types of hatching are not readily distinguished. In *Fig. 6.20b,* the shading and edges of the cut-away edges cause some confusion in visualizing the construction. In *Fig. 6.21a,* the reduction has brought the lines and dots of the shaded and hatched areas so close together that the different shading patterns, and so the parts, are not easily distinguished. The lettering, though quite legible, is as small as permissible for readability. The arrows to the parts end in a dot, which is not always visible in the narrow bands of hatching and stippling. *Fig. 6.21b* is so detailed that the parts are simply numbered; the lines are too fine and too close together to make the structure clear. It is questionable how informative the illustration is.

In *Fig. 6.22a,* the parts of the equipment are schematic, and they are numbered rather than labeled to save space and avoid clutter. The numbers are circled, however, and the circles add clutter, because they have no meaning. The labeling lines and arrows are of the same weight as other lines and so make it difficult to distinguish structure from labels. The circling in the legend is superfluous, since the numbers obviously refer to the labels.

In the abstract representation of structure, such as molecules, the schematic elements may be reduced to geometric forms that conform to conventional representations. In such drawings, the distribution of the elements, the size of units, and the balance in weight of lines, the size and weight of letter and the uniformity of corresponding elements must be carefully planned to make the diagram visually effective. In the diagram of the molecule in *Fig. 6.22b,* the labels within the circles and ovals are too crowded, yet too small, especially the superscripts and subscripts, as are the numbers on the lines. Moreover, the abbreviations GLY, SER, and so on, despite their larger size, do not effectively designate the part of the molecule that they are intended to identify, and the roman numeral is inordinately large.

Path diagrams are designed to show sequential relationships between parts. The parts may be actual physical components, as in block diagrams, or conceptual entities, as in organizational charts. Path diagrams can be used effectively to structure and clarify interactions and relations in a concept or theory, or the steps in a process. Since these drawings consist only of words associated with lines or arrows, the shape and size of the enclosures, the weight, disposition, and type of lines, and the weight of the letters are important in making the relationships clear (see *Fig. 6.5*). Path diagrams may be entirely diagrammatic or semidiagrammatic. In *Fig. 6.23a,* the parts of the system have been reduced to blocks except for two lenslike forms. The labels and lettering in the blocks are too large and crowded, and not

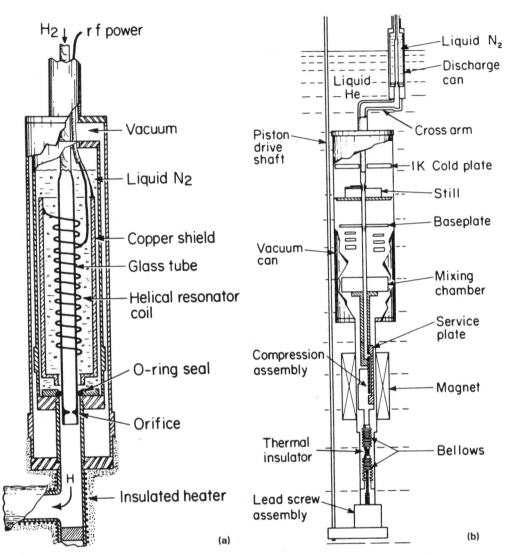

rigure 6.20. Diagrams illustrating use of lines and dots to show construction of instruments. *Note. (a, b)* Patterns of hatching not readily distinguished; *(a)* lettering somewhat large, *(b)* cutout sections to show three-dimensional structure con.using, [Phys. Rev. Cond. Matt. B 34, 1986, 7675, with permission]

cleanly drawn. In *Fig. 6.23b* some of the parts represent the structure of the component if only diagrammatically or symbolically, so that labels are placed outside the components. There is little differentiation between the weight of lines for components, blocks, and connections or the lettering. This undifferentiated character does not highlight differences between components and path. In *Fig. 6.24* the stages rep-

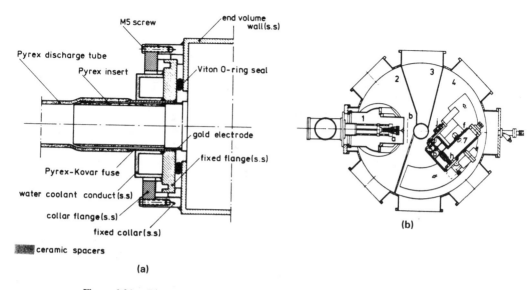

Figure 6.21.   Diagrams drawn on too small a scale for structure to remain clear after reduction. (*a*) Patterns of hatching difficult to identify; lettering too small; dots at ends of guide lines intrusive in narrow hatched areas. (*b*) Parts with detail too reduced to show structure. [(*a*) Physica B, C 142, 1986, 250; (*b*) J. Chem. Phys. 85, 1986, 6399; with permission]

resented are temporal as well as physical. The solid black for the dye, the heavy black lettering, and the bold title above the figure (which belongs in the printed title) are all too heavy for the printed page.

   The path diagrams in *Fig. 6.25* illustrate the reduction of a path diagram to words joined only by arrows or connecting lines. In *Fig. 6.25a* the arrows dominate the typewritten verbal elements of the model, which are too small and not in keeping with the printed page; in *Fig. 6.25b,* the verbal elements are so small that they have become illegible, and the lines are so fine as to risk breaking in reproduction. The path diagram in *Fig. 6.26* shows little differentiation in width of lines. The boxes are not set off from the arrows and lettering, and even the difference between the italic and roman lettering is not noticeable.

6.9 GRAPHS

Graphs are the commonest type of drawing used in scientific papers. They represent graphically the distribution of values relative to two or more variables, and their quantitative character makes them well suited to the abstract representation of mechanical drawing.

   In preparing graphs, the writer must remember that however excellent his or her selection of information, design of the graph, and its construction, the graph

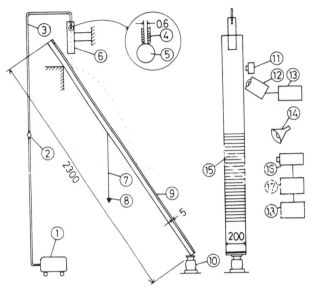

Fig. 4   Experimental equipment ①: Vacuum cleaner, ②: Glass valve, ③: Vinyl tube, ④: Metal tube, ⑤: Sphere, ⑥: Acrylic pipe, ⑦: Thread, ⑧: Weight, ⑨: Acrylic plate, ⑩: Jack, ⑪: Camera, ⑫: Stroboscope ⑬: Oscillator, ⑭: Light, ⑮: Marked line, ⑯: Video camera, ⑰: Video tape recorder, ⑱: Monitor TV

(a)

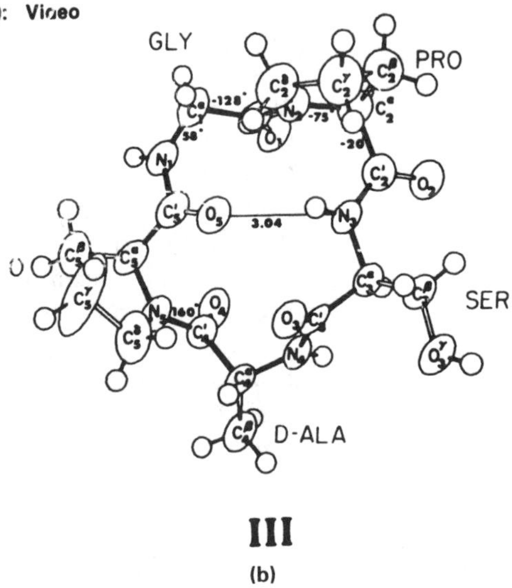

(b)

Figure 6.22.   Diagrams illustrating ineffective labeling. (*a*) Circles around enumerating numbers obtrusive, not needed in legend; line widths of circles, guidelines, and structure not adequately differentiated. (*b*) Capitalized abbreviations not dominant enough to highlight section encompassed; other lettering too small and too crowded; differences between connecting bonds not readily distinguished; roman numeral too bold. [(*a*) Tsuji et al. J. Flu. Engr. 107, 1985, 485; (*b*) Int. J. Pept. Prot. Res. 28, 1986, 421, © 1986, Munksgaard International Publishers Ltd., Copenhagen, Denmark; with permission]

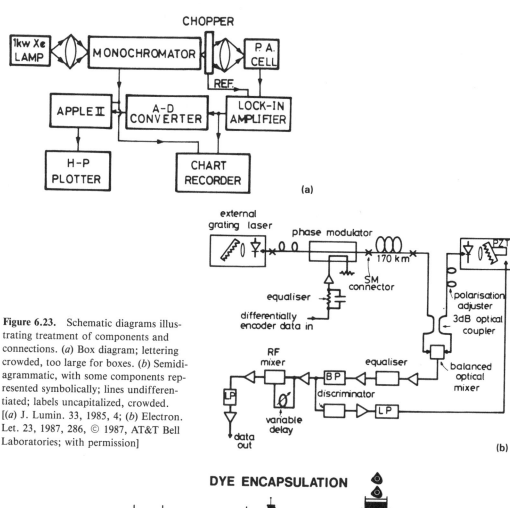

(a)

(b)

Figure 6.23. Schematic diagrams illustrating treatment of components and connections. (*a*) Box diagram; lettering crowded, too large for boxes. (*b*) Semidiagrammatic, with some components represented symbolically; lines undifferentiated; labels uncapitalized, crowded. [(*a*) J. Lumin. 33, 1985, 4; (*b*) Electron. Let. 23, 1987, 286, © 1987, AT&T Bell Laboratories; with permission]

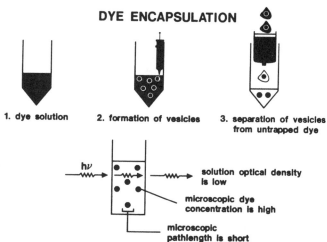

DYE ENCAPSULATION

1. dye solution   2. formation of vesicles   3. separation of vesicles from untrapped dye

solution optical density is low

microscopic dye concentration is high

microscopic pathlength is short

Figure 6.24. Diagram of process, too bold for printed page. *Note.* Title part of figure; relation between lower diagram and upper diagrams not explained. [Photochem. Photobiol. 44, 1986, 455 by A. Plant, © 1986 Pergamon Journals, Ltd., with permission]

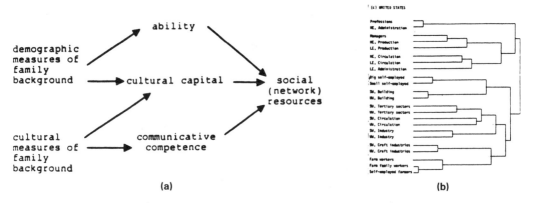

Figure 6.25. Ineffective path diagrams. (*a*) Dominated by arrows; labels typed, labels type-written. (*b*) Lettering small, illegible; lines too thin. [(*a*) Amer. J. Sociol. 90, 1985, 1256; (*b*) Amer. Soc. Rev. 50, 1985, 591; with permission]

will not be visually effective unless readers can perceive and interpret it. The writer must attend to the reader's two tasks in reading a graph: (1) the visual task of identifying and perceiving the elements in the graph and (2) the cognitive task of interpreting them. The first task includes the perception of position, length, slope, angle, area, density, and so on. The second includes estimates of position (distance), length, slope, and so on. Estimates of position along a scale are more accurate than estimates of length, angle, or slope. Estimates of length are more accurate than estimates of area. Estimates of position along a common scale are more accurate than estimates of position along identical but separated scales. Therefore one would

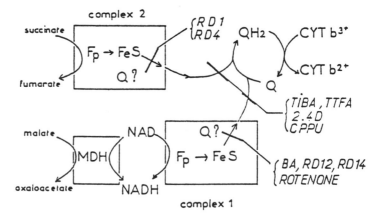

Figure 6.26. Path diagram illustrating inadequate differentiation of lines (boxes, arrows, guidelines and lettering). *Note.* Lettering disproportionately large or small. [J. Pl. Physiol. 123, 1986, 63, with permission]

elect graphs requiring position estimates in preference to those requiring length (divided-bar graph), angle (circular graph), or area estimates.

Graphs allow one to display graphically a large amount of data as a whole in summary form; therefore clarity and conciseness are as necessary as in writing the text. They must be designed for high visual discrimination. The main pitfall in preparing graphs is their construction, either in their planning and design or in the execution of the design. Graphs often have an excessive amount of variety of data, or they may not highlight the data. Graphs may not be drawn to make the data visually prominent, and they often have inaccuracies and mistakes. As in writing, an effective graph cannot be produced without revising, reviewing, and editing. Graphs may also be planned or drawn so that they are not reduced or reproduced effectively. Other common pitfalls are inadequate labeling or explanation of elements and more substantive deficiencies, such as ineffective choice of type of graph or kind of data for the message to be conveyed.

Two main types of graphs are used in scientific papers: line graphs and areal graphs, such as bar graphs and circular (pie) graphs. Line and bar graphs usually have a vertical axis and a horizontal axis that intersects it at right angles. They may have an additional vertical and horizontal axis, forming a rectangle that frames the data. The two additional axes permit a more accurate estimate of the position of points as they become more distant from the point of origin. For scalar variables, scale values progress away from the point of origin continuously in regular scale units or scale intervals along the axis.

6.9.1 Bar Graphs

Function and form.   Bar graphs may display the relationship between a scalar variable and another continuous variable, but more usually one or more discontinuous variables or several categories. Therefore both axes may be scalar and continuous (*Fig. 6.27*), but in most bar graphs only one axis is. Even though the discontinuous variables are graduated, the graduations may not have a continuous relation with the continuous variable. In *Fig. 6.29a*, one cannot expect that the percentage in 1975 would fall halfway between that of 1970 (14.0) and 1980 (10.3); the decrease cannot be expected to be regular during the decade.

In most bar graphs, the length of the bars is quantitative and represents the observed or calculated value, but the width is not. It depends on (1) the number of bars or groups of bars in the graph, (2) the space available on the page for the bars, (3) the length of the labels for the bars, and so on. Most bar graphs have vertical bars, as the tendency is to conceptualize quantity vertically. If the labels of the bars are long and would be crowded below the vertical bars, then aligning the bars horizontally allows the labels to be written at the side of the graph, where they are not limited by the width of the bar (*Fig. 6.28*). This is usually preferable to labeling the bars vertically, as in *Fig. 6.30a*, or using letters or numbers that must be explained in the legend (*Fig. 6.30b, 6.33a–c*). A graph with short bars (*Fig. 6.28*) may be more economically drawn with horizontal bars to fit within a single column.

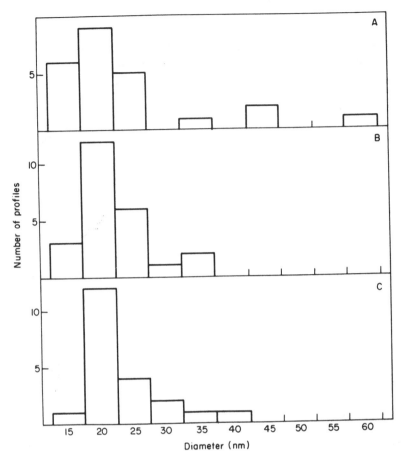

Figure 6.27. Bar graph with both axes continuous, scalar. *Note.* (*a*) Lettering of axis labels smaller than scale labels, origin not labeled; some bars extending beyond scale marks. [Ann. Bot. 56. 1985, 460, with permission].

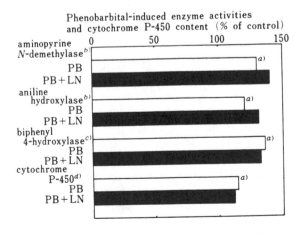

Figure 6.28. Graph illustrating advantage of horizontal bars for long bar labels. *Note.* Axis labels crowded, lettering of bar labels too large, other lettering too small; label of horizontal axis not clear. [Chem. Pharm. Bull. 34, 1986, 1310, Pharmaceutical Soc. of Japan, with permission]

The difference in axes is reflected in their differential treatment. The axis parallel with the bars has scale marks at regular intervals. These are usually inside the axis (*Fig. 6.27, 6.28, 6.30, 6.31b*), but they may be set outside (*Fig. 6.29b, 6.31a, 6.32*), or across it (*Fig. 6.33b*). The scale marks are labeled with numbers outside the axis, and the axis is labeled with a descriptive label (*Fig. 6.27–6.33*). The axis to which the bars are attached is usually not scalar. If it is scalar, or the bars contiguous, scale marks may be drawn between bars inside (*Fig. 6.27*) or outside (*Fig. 6.29b, 6.31a, 6.32*) the axis, and then invite placing scale marks outside the axis on the scalar axis for consistency. Each bar or set of bars is labeled with a number (*Fig. 6.29b, 6.33a*), letter (*Fig. 6.30b, 6.33c*), abbreviation (*Fig. 6.28*) or preferably descriptive words (*Fig. 6.28, 6.30a, 6.33c*), which do not need to be explained in the legend.

The ends of the bars may have error bars and letters indicating significant differences (*Fig. 6.30b, 6.31, 6.32, 6.33a,c*). The error bars may extend both above and below the end of the bar (*Fig. 6.30b, 6.31a, 6.33c*), or more usually, only beyond the end (*Fig. 6.31b, 6.32, 6.33a*). Even when there is no shading, the extension of the error bar into the bar may add an intrusive additional line to the graph (*Fig. 6.31a, 6.33c*). The error bars may be simply lines (*Fig. 6.30b, 6.31b*) or be terminated by crossbars (*Fig. 6.31a, 6.32, 6.33a,c*), but they should not be conspicuous (*Fig. 6.32, 6.33a,c*). Error bars may be variously interpreted, therefore they require explanation.

Pitfalls in preparing bar graphs.   The outline, spacing, and differentiation of the bars influence the effectiveness of bar graphs. Bars represent the data and so should dominate the space; they should be wider than the spaces between them. In *Fig. 6.30a,* and to some extent *Fig. 6.30b,* the equal width of the bars, spaces, and

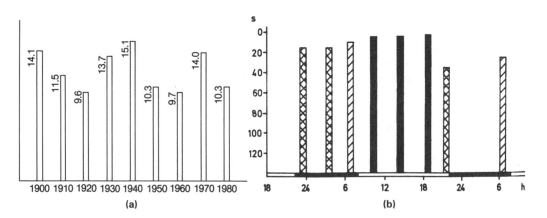

Figure 6.29.   Bar graphs showing narrow bars, dominated by space between them. Bars too narrow (*a*) for labeling year or values, (*b*) for effective cross-hatching. *Note.* (*a*) Vertical axis lacking scale units. (*b*) Axis labels too small; vertical scale confounding comparison of bars. [(*b*) *Planta* 167, 1986, 340, with permission]

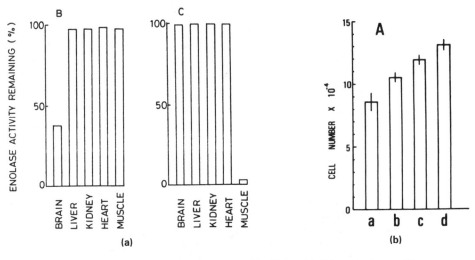

Figure 6.30. Bar graphs with bars not readily distinguished from background because of equal width of bars and interbar spaces. *Note.* (*a*) Bars labeled vertically. (*b*) Axis labels and figure label too bold; lettering of vertical axis crowded; labels of horizontal axis uninformative. [(*a*) J. Biochem. 98, 1985, 1532; (*b*) J. Exp. Zool. 239, 1986, 136; with permission]

lines together with the lack of stippling or hatching and the even length of most of the bars, makes it difficult to distinguish the bars from the background space. In *Fig. 6.29*, the bars are so narrow that they are dominated by the space between them. In *Fig. 6.29a*, they are too narrow to fit well over the dates or accommodate the percentages across the top of the bars in a readable orientation. The bars in *Fig. 6.29b* are too narrow for the cross-hatching to be effective. In *Fig. 6.31a*, the bars are so narrow, that the numbers included in some of them are almost unreadable. The graph also has at least three sets of bars, one set constituting part of a graph within the larger graph. *Fig. 6.31b* is also crowded with details—(1) roman numerals above each bar, made unnecessarily conspicuous by their size and hand lettering, (2) numbers crowded in the base of the narrow bars, (3) scale lines, lines outlining bars, and error bars inadequately differentiated, and (4) three types of hatching in short narrow bars.

Different patterns of dots (stippling) and lines (hatching) can be used to differentiate sets of related bars. The weight and density of the dots and lines determine the visual effect. Dots and lines that are too heavy or too close will make for a dark bar that becomes darker with reduction (*Fig. 6.32, 6.33a*). Lines or dots that are too widely spaced may seem structural. The hatching may be too wide or too heavy for the bars (*Fig. 6.29b, 6.31b, 6.33a*), and diagonal lines are likely to have disturbing vibrating optical movement (*Fig. 6.29b, 6.32*). In *Fig. 6.33a*, the weight of one of the diagonal patterns and their juxtaposition emphasize the vibratory effect. In *Fig. 6.33c*, the black bars have a marked optical effect due to complex outlining of

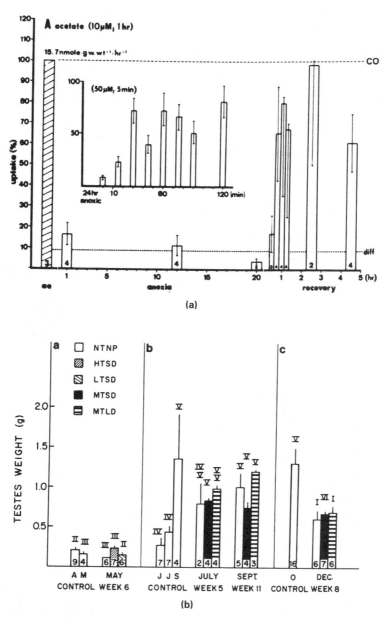

Figure 6.31.   Bar graphs with too many elements, ineffectively displayed. (*a, b*) Elements not readily differentiated. (*a*) Interbar space dominant over bars, lettering disproportionately small; scale marks on horizontal axis almost imperceptible; numbers in narrow bars illegible. (*b*) Hatching patterns not effectively distributed, two similar hatchings restricted to single short bars; lines little differentiated; roman numerals crowded, drawn freehand, irregular; numbers in bars crowded. [J. Exp. Zool. (*a*) 240, 1986, 302; (*b*) 239, 1986, 126; with permission]

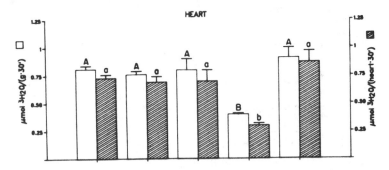

Figure 6.32. Bar graph with close, fine, diagonal hatching of bars showing optical effect. *Note.* Lettering of vertical axes too small, crowded, almost illegible; axis labeled only with unit. [J. Nutr. 116, 1986, 738, © 1986 J. Nutr, American Institute of Nutrition, with permission]

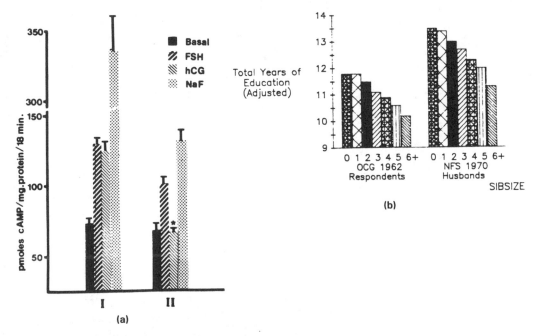

(a)

(b)

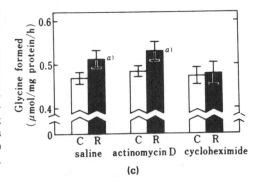

(c)

Figure 6.33. Bar graphs with differentiated sets of bars showing marked optical effects. (*a*) Bars too dark, with optical vibration; (*b*) bars with confusing patterns, some not visually effective; (*c*) bars with strong lines and masses; black bars with conspicuous outlining of error bars in white; breaks in axis and bars obtrusive. *Note.* (*a*) Error bar not included within axis; lettering of vertical axis smaller than scale labels, of horizontal axis too bold; break in vertical axis and bars imperceptible; axis labeled only with units. (*b*) Label of vertical axis lettered horizontally, lettering smaller than scale numbers. (*c*) Lettering disproportionately large or small; cross bars on error bars too dominant. [(*a*) J. Biol. Chem. 261, 1986, 13110; (*b*) Soc. Biol. 33, 1986, 9; (*c*) Chem Pharm. Bull. 34, 1986, 2941, Pharmaceutical Soc. of Japan; with permission]

the error bars in white within the black bar, the wide cross bars, and the conspicuous wavelike break in the bars. Such a strong depiction of elements in a graph (especially the small graph shown) makes form intrusive. The break in *Fig. 6.33a,* on the other hand, is barely noticeable in the axis and looks like a tear in the preprinted pattern in the bar. One must also exercise restraint in selecting patterns to differentiate bars to avoid a calico quilt effect like that in *Fig. 6.33b.* Hatching should not be used in areas too small to differentiate the bars; *Figure 6.31b* has three types of hatching in very narrow bars; the two diagonal types are in bars so small as to make the bars almost indistinguishable from each other.

In some bar graphs, the bars are divided into stippled or cross-hatched segments that represent percentages of the length of the bar (*Fig. 6.34*), with the bars representing 100 percent (*Fig. 6.34b*) or varying percentages (*Fig. 6.34a*). In differentiating the segments of the bars in such divided-bar graphs one should avoid obtrusive patterning (*Fig. 6.34a*). When the total percentages vary, there may be space

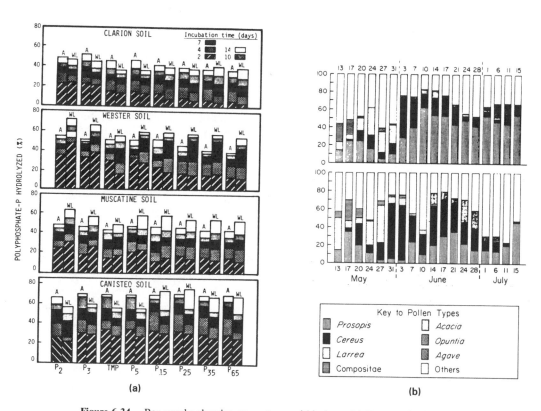

Figure 6.34. Bar graphs showing percentages within bars. (*a*) Bars varying in total percentage; key in graph. (*b*) Bars all 100 percent; key outside graph. *Note.* (*a, b*) Hatching of bars too obtrusive; lettering of axes and bars disproportionately small. (*b*) Patterns of crosshatching and stippling not readily distinguished. [(*a*) Soil Sci. 142, 1986, 135 by R.P. Dick and M.A. Tabatabai, © by Williams and Wilkins 1986; (*b*) Oecol. 69, 1986, 496; with permission]

in the graph for a key to identify the different segments (*Fig. 6.34a*); otherwise, the key may be set outside the graph (*Fig. 6.34b*). Divided-bar graphs are difficult to decode accurately. The reader can compare the basal segments by position relative to the shared axis, but the other segments have no common baseline for comparison and must be compared on length, a less accurate estimate.

Circular and other types of graphs.   Circular graphs, also known as pie charts or pie diagrams, are analogous to 100-percent divided-bar graphs (*Fig. 6.34b*), so that the wedges represent the percentages of the whole. However, angle judgments are less accurate than judgments of position; therefore converting a circular graph to a bar graph makes for greater accuracy in perceiving and decoding the data. The percentages in a circular graph may be shown inside or outside each wedge, or different types of shadings can be used to differentiate the wedges (*Fig. 6.35*). Circular graphs are limited by the size and number of wedges, and the variety of stippling and cross-hatching required. The illustrator must select shadings that can abut without producing strong optical effects, or that are not too strong in themselves (*Fig. 6.35c*). When wedges are too numerous or too narrow (*Fig. 6.35a, b*), they may be difficult to label or differentiate. Then at least the narrow wedges must be labeled outside the graph (*Fig. 6.35a*), but even the labels may become too crowded to relate to the wedges. If the wedges can be differentiated by shading but are too narrow to label, a key may be placed outside the graph (*Fig. 6.35b*). In general, circular graphs are most effective visually for a limited number of wedges that can be distinguished easily and that are not too narrow to permit clear labeling, as in *Fig. 6.35c*. This is unnecessarily large (3.5 in. diam. in the journal) and the labeling is inconsistent.

Three-dimensional graphs, which can vary widely from three-dimensional towers rising up from a grid to very complex topographical line graphs (*Fig. 7.1a*), are easily drawn with computers. The ease with which they are constructed should not lead one to forget that the objective is to display data so that readers can perceive the elements and then decode them.

It has been suggested that for nonscalar data or variables, point or dot graphs are more appropriate than bar graphs, and less misleading (Cleveland, 1984, 1985). Bars in bar graphs represent magnitudes by their area configuration, and using them only to designate position by their ends ignores the statement that their area makes. A point or dot graph is a more accurate representation of the data as well as a more economical use of space. In *Fig. 6.36,* the data for each journal consist of three points, for number of figures, illustrations, and graphs, each terminating a dotted line. In *Fig. 6.37,* the three points have been positioned along one dotted line.

6.9.2 Line Graphs

Line graphs show the relationship between a scalar variable along one axis that varies correspondingly with a scalar variable along an axis at right angle to it. The relationship is shown by points and by lines, so that one has only lines, points, and

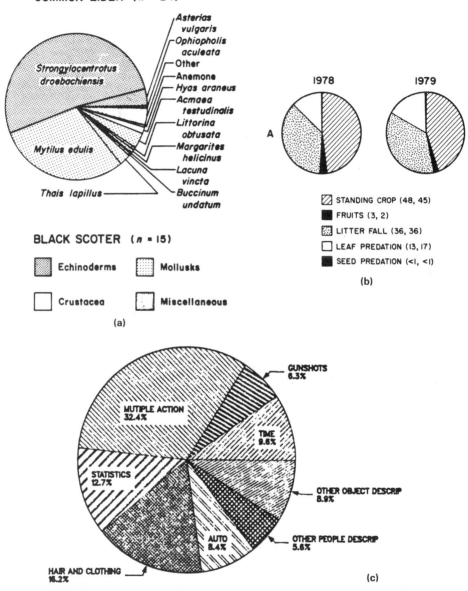

Figure 6.35. Circular graphs showing limitations. (*a*) Segments too small to be readily distinguished, resulting in crowded labeling; (*b, c*) segments large enough to be differentiated by cross-hatching. (*b*) Segments not all large enough to allow labeling, key outside graph. (*c*) Segments large enough to include all labels, location of labels inconsistent. *Note.* (*c*) Patterns of hatching too obtrusive. [(*a, b*) Ecol. 67, 1986, (*a*) 1478, (*b*) 1346; (*c*) J. Appl. Psych. 71, 1986, 297 by Yuille and Cutshall, © 1986 Amer. Psychological Assoc.; with permission]

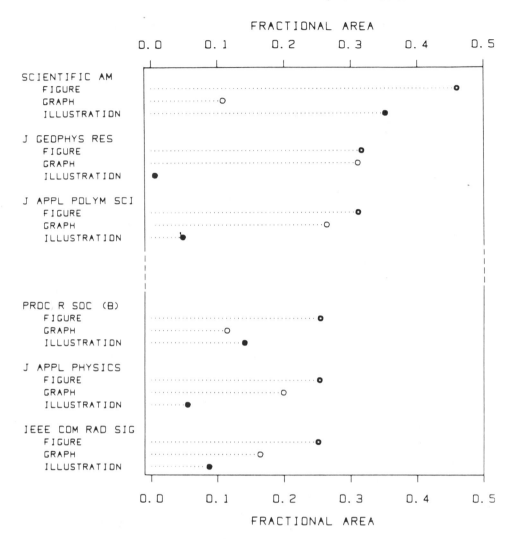

Figure 6.36   Dot graph substituting points at the end of scalar dotted lines for bars. [W.S. Cleveland, Amer. Stat. 38, 1984, 275, with permission]

space to represent quantitative relationships between variables. Therefore differences in the size and form of points, the width and form of lines, and the spatial relationships become critical in differentiating the elements of the graph. Because of these limited means, uniformity and consistency become important objectives for meaning. The uniformity in the width and length of lines, in the size and weight of lettering, in the size and shape of points, and the spacing of points or lines make possible the consistency of form that allows the graph to make its statement.

To be visually effective, that is to permit detection of elements, the elements

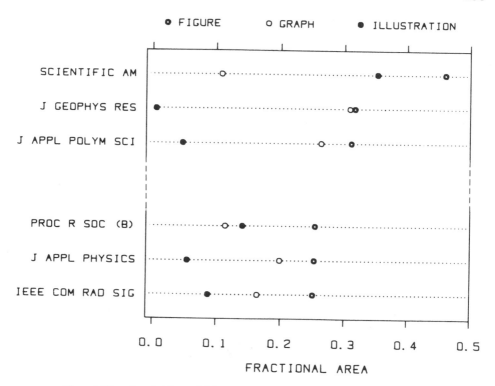

Figure 6.37.   Data in *Figure 6.36* rearranged so that the three values for each journal drawn as three points on single scalar dotted line. [W.S. Cleveland, Amer. Stat. 38, 1984, 276, with permission]

of a graph, though uniform themselves, must be differentiated from other elements; they should not all be uniform, as in *Fig. 6.70, 6.71.* In *Fig. 6.38* the lines and lettering are uniformly too light for the printed page or even to differentiate easily. The graphs in *Fig. 6.39* and *6.54d* are uniformly too heavy for the printed page, and are better adapted for slide projection. Uniformity in the weight of elements is particularly undesirable in graphs with numerous elements (*Fig. 6.70, 6.71*).

Planning and preparation.   Graphs should be as simple and as concise as possible. A graph should not be crowded with excessive data or too many elements or sets of data; however, neither must a great deal of space be given to a simple curve. A graph with excessive sets of data tends to lose its visual impact and become a documentary or study graph. *Fig. 6.40,* has six sets of three curves, with six sets of points, and six labels in parentheses, with three arrows to the curves. It also has a heavily ruled boxed key, with three values for each curve as well as its key symbol, and labels. Similarly the small graph in *Fig. 6.41a* includes five curves, of two types, three diagrams with arrows to the curves, three sets of two values, a key, and the value for AR, as well as prominent scale marks on all four axes.

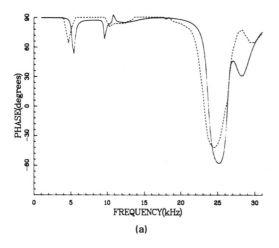

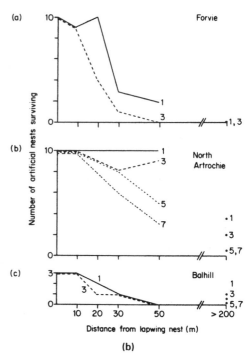

Figure 6.38.   Graphs with lines too fine for effective reproduction. *Note.* (*a*) Axis labels disproportionately bold; letters crowded; scale numbers of vertical axis lettered vertically. (*b*) Lettering too small. [(*a*) J. Acoust. Soc. Amer. 80, 1986, 18; (*b*) Anim. Beh. 33, 1985, 313; with permission]

The graph in *Fig. 6.41b* is so crowded with data that it is almost impossible to distinguish the various elements. The symbols are numerous, large, overlapping, and extremely varied; the curves are difficult to distinguish from the arrows, the arrowheads are difficult to detect. The underlining of the labels and parentheses is superfluous, and the conspicuous scale marks and labels fill most of the space around the points and curves. Graphs with multiple sets of data can be separated into two or three juxtaposed graphs to allow readers to perceive and decode more of the data. However, although juxtaposed graphs make the data easier to detect, they make comparison of data more difficult; therefore when possible, superposing data is preferable to juxtaposing them. Where data are excessive or crowded, the graph should be redesigned to highlight the important elements. The scale units can be changed to allow more separation of elements, some elements can be placed outside the graph, or a logarithmic scale can be used, especially if the data are skewed. Crowded or multipurpose graphs may contain much important data, but readers must be able to perceive them fully to decode the graph.

Graphs should, however, be as economical as is consistent with making their statement effectively. The disposition of elements within graphs should be planned to prevent the waste of space seen in *Fig. 6.42,* where the curve falls within the middle 6 of the 19 scale units. Graphs may take up too much space for the amount of information they present, as in the simple curve in *Fig. 6.43.* The simple graphs

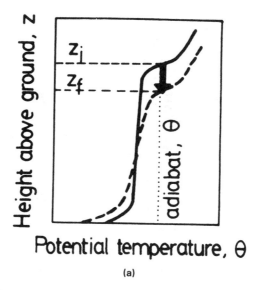

(a)

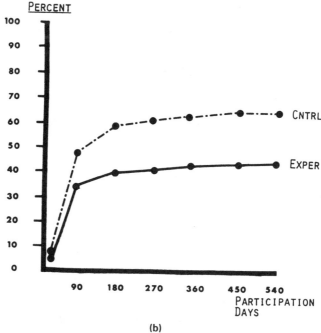

(b)

Figure 6.39.   Graphs with lines or lettering disproportionately large or bold. (*a*) Lettering disproportionately large and bold; (*b*) axes, scale marks, numbers bold. *Note.* (*b*) Scale marks too long, outside axis, lacking on horizontal axis despite scale values; vertical axis labeled across top; labels extending beyond axes; lettering disproportionately fine-lined. [(*a*) J. Clim. Appl. Meteor. 28, 1986, 1093; (*b*) Gerontol. 26, 1986, 154; with permission]

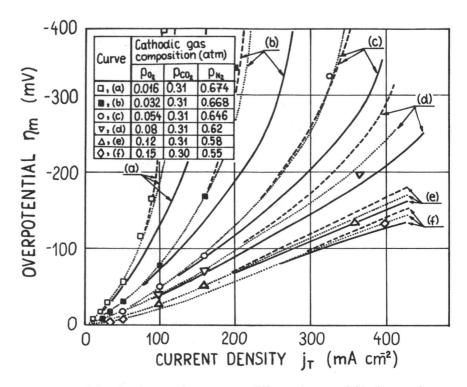

Figure 6.40. Graph presenting too many different elements: Grid; six sets of curves, three to each set; labels for sets in parentheses, with arrows to individual curves; boxed key with vertical and horizontal rules; with symbols and parenthetical labels for sets, and three values for each curve. *Note.* Line widths of elements not effectively differentiated; lettering of axes and scale labels disproportionately large; lettering irregularly drawn. [J. Appl. Electrochem. 16, 1986, 652, by J. Jewalski, (Chapman and Hall, pub.), with permission]

in *Fig. 6.44* and *6.45* each take up a full page in the journal. In *Fig. 6.44* (originally 6¼ in. × 4 in. on a page 7 in. × 5 in.), the graph can be easily reduced, and the scale intervals can be reduced without crowding the points. The graph can also be turned 90 degrees, with Co as the vertical axis. Then with slightly shorter scale intervals of the Ni axis, the graph could fit across the width of the page and take up only about one-half of the page. The graph in *Fig. 6.45* (originally 3¼ in. × 4¼ in.) is set broadside on a page 7¾ in. × 5 in. The typing made further reduction impossible—a clear disadvantage of using typescript for lettering. Yet the graph could fit across the width of the page and save more than half a page.

When a figure consists of a set of juxtaposed graphs, the objective is to work for economical disposition on the page and effective display. The graphs in *Fig.*

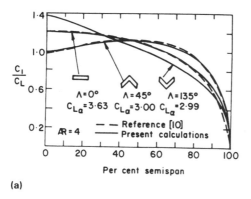

(a)

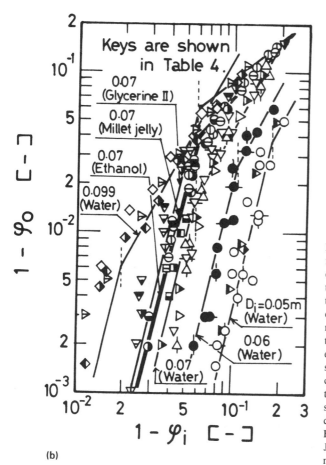

(b)

Figure 6.41. Graphs crowded with (*a*) numerous different elements, (*b*) with excessive elements. (*a*) Graph too small; two types of curves; three diagrams with three pairs of values, and three arrows to curves; key to curves; conspicuous scale marks on all four axes. (*b*) Numerous types of symbols; symbols and lines not easily differentiated; labels in parentheses, with values and underlining arrows; conspicuous scale marks. *Note.* (*a*) Lettering of horizontal axis smaller than in scale labels. (*b*) Lettering and numbers disproportionately large. [(*a*) Katz, J. Flu. Engr. 107, 1985, 441 (ASME); (*b*) Can. J. Chem. Engr. 64, 1986, 721; with permission]

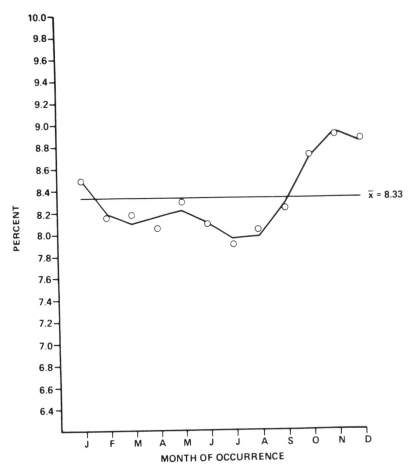

Figure 6.42.   Graph illustrating uneconomical use of space within graph. Curve in center of graph, with excessive space above and below curve. *Note.* Vertical axis not broken above zero, labeled only with units; scale marks outside axes, scale labels crowded; lines not adequately differentiated; label in graph extends outside graph. [Soc. Biol. 33, 1986, 123, with permission]

6.46 are two of a set of three graphs placed broadside in two rows in the journal occupying a full page (8 in. × 5½ in.). They can be made to fit the width of the page by terminating the horizontal axis at day 80 or 90 and save about one-half the page. The graphs in *Fig. 6.47,* which have a single curve, take up a full page (8½ in. × 6¾ in.) in the journal; yet the figure could have been reduced to page width, or less, and take up about one-half the page.

Graphs in a series should be consistently drawn (*Fig. 6.47*) to allow ready comparison. In *Fig. 6.48a,* all the vertical axes have the same scale, and so could be

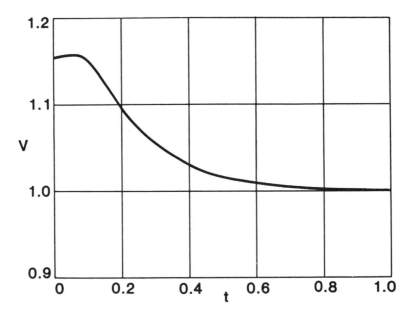

Figure 6.43. Graph illustrating uneconomical use of space for simple curve. *Note.* Scale numbers disproportionately large; grid superfluous. [J. Astron. Sci. 34, 1986, 14, with permission]

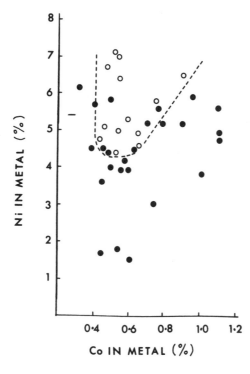

Figure 6.44. Graph illustrating uneconomical use of space on page and within graph. Occupies most of page (7 1/2 in. × 5 in.) in journal; lettering and points large, heavy; scale intervals unnecessarily long (all improved in reduction shown here); origin not labeled. [Meteorit. 21, 1986, 89, with permission]

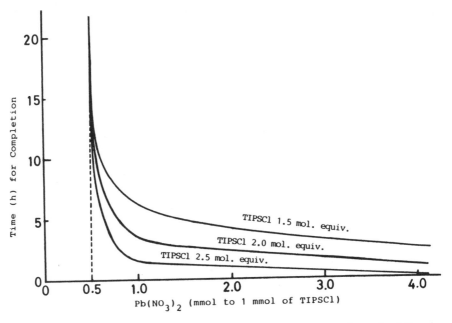

Figure 6.45. Graph illustrating uneconomical use of space on page. Occupying full page (7 3/4 in. × 5 in.) broadside in journal. *Note.* Curves extending beyond scale marks on vertical axis; labels typewritten, too small; TIPSC label repetitious; scale numbers too large, [Nishino et al. J. Carbohyd. Chem. 5, 1986, 204, courtesy of Marcel Dekker, Inc. N.Y.; with permission]

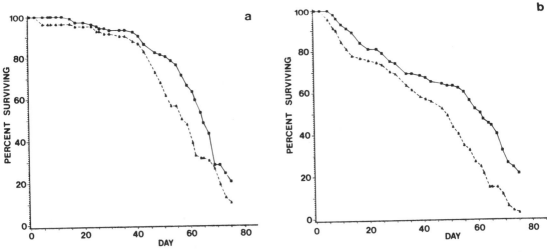

Figure 6.46. Rearrangement of graphs for more economical use of page. Two of three graphs broadside on page and occupying full page (8 1/2 in. × 7 in.) in journal. *Note.* Points too small, too close for dashed lines to be differentiated easily; lines too narrow, not adequately differentiated easily; unit preceding variable; scale marks imperceptible. [J. Exp. Zool. 239, 1986, 89, with permission]

262

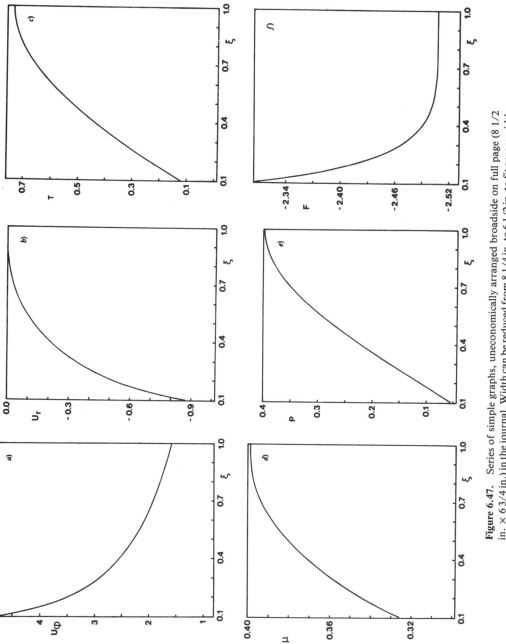

Figure 6.47. Series of simple graphs, uneconomically arranged broadside on full page (8 1/2 in. × 6 3/4 in.) in the journal. Width can be reduced from 8 1/4 in. to 6 1/2 in. to fit across width of journal page. *Note.* Lines inadequately differentiated; lettering too small. [Astrophys. J. 310, 1984, 201, with permission]

263

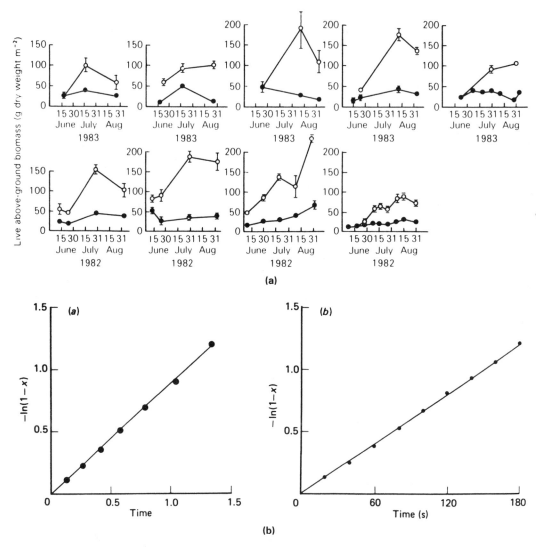

Figure 6.48. Sets of similar graphs not consistently drawn. (*a*) Vertical scales terminating at 150 or 200; (*b*) points varying in size; units of horizontal axis not consistent. *Note.* (*a, b*) Lettering of axis labels small relative to scale labels. (*a*) Labeling of axes repetitious and crowded. (*b*) Scale marks too long; serial letters of graphs too small, in parentheses. [(*a*) J. Ecol. 7, 1986, 699; (*b*) Biochem. J. 239, 1986, 233, © 1986 The Biochemical Society; with permission]

extended to 200 for uniformity. It would also be clearer and more economical to label only the vertical axes in the leftmost graph of each row and the lowermost horizontal axes, and to omit the scale number, 15. In *Fig. 6.48b,* the two graphs have the same axes, but the units of time are not consistent and the lines and points are not similar in weight.

When graphs are grouped, they should be arranged in some meaningful order for the reader. The progression can be designated by position, with letters (*Fig. 6.38b, 6.46, 6.47, 6.52a*), numbers, labels (*Fig. 6.60c*) or abbreviations. Whenever possible, letters used to label the individual graphs should be placed in the same position on each graph to identify them as enumerating elements, in an area that is free of graphic elements. Usually lowercase letters are used, but they should be dominant in size, weight, or both, over other lettering in the graph. They may be unnecessarily enclosed by parentheses (*Fig. 6.38b, 6.48b*) or followed by a period (*Fig. 6.73a*).

Axes.   In line graphs, the axes usually represent the Cartesian coordinates. Conventionally the horizontal axis is the abscissa or x-axis, and represents the independent variable; the vertical axis is the ordinate or y-axis, and represents the dependent variable. Usually both axes are positive, but either may extend in the negative direction. The graph in *Fig. 6.49* shows the four quadrants defined by the x and y coordinates and the curves extending through all four quadrants. The scale marks are drawn within each quadrant, as appropriate, those across the central axes representing the scale marks for the adjacent quadrants. In *Fig. 6.50a,* the scale marks on the horizontal axis should, strictly, extend across the axis. In graphs with positive and negative values or with points falling at zero, the axis may not be set at zero (*Fig. 6.50a, 6.57b,c*); one axis extends beyond zero, and the intersecting axis is drawn to include the curves, not at zero.

Both axes of line graphs are scalar, and the axis is divided into scale units or

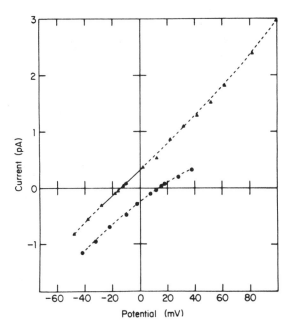

Figure 6.49.   Graph illustrating four quadrants defined by x and y axes, with negative and positive values along both axes. *Note.* Lines inadequately differentiated; lettering of axis labels disproportionately small; points too small, not well differentiated, some distorted. [J. Gen. Physiol. 89, 1987, 855, © 1987 Rockefeller Univ. Press, with permission]

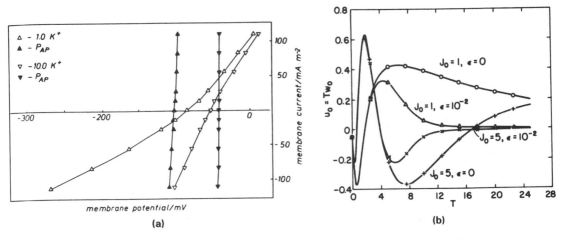

Figure 6.50. Graphs illustating axes with negative values. (*a*) Horizontal axis drawn at zero; negative, vertical axis labeled at right. (*b*) Horizontal axis drawn below zero. *Note.* Some points distorted by effect of curve on points. [(*a*) Planta 167, 1986, 71; (*b*) Phys. Flu. 29, 1986, 2058; with permission]

intervals marked by scale marks, or "ticks," which are labeled. When the scale units are subdivided, the divisions are not usually labeled, except for some in logarithmic scales. The point of origin of the axes is often zero; however, when this reduces the resolution or leaves a large area of the field of the graph open, the point of origin can be set above zero. Excluding the zero, unless it misleads readers, may increase the information displayed by the graph. If both axes start at zero, they may be so labeled (*Fig. 6.45, 6.46, 6.51a, 6.52a, 6.64b*), or a single zero may be set at the intersection of the axes (*Fig. 6.48b, 6.51b, 6.58a, 6.60b, 6.73b*), but the point of origin should not be left unlabeled (*Fig. 6.44, 6.51c, 6.62b, 6.64a, 6.70*), even though it can be inferred. When the point of origin is not zero, it should be appropriately labeled (*Fig. 6.41b, 6.51d, 6.56d, 6.63a*), or it should be set at zero and the axis broken (*Fig. 6.52a, 6.74b*).

A graph should be planned so that the points all fall within the data region and are effectively distributed in it. They should start close to the axes unless this misrepresents the data. If the points all fall in the upper part of the graph (*Fig. 6.42, 6.56a*), one should consider shortening or breaking the vertical axis to bring the points close to the horizontal axis (*Fig. 6.57c*). Similarly, if the points fall far from the vertical axis, one should consider shortening or breaking the horizontal axis (*Fig. 6.52a, 6.74b*). The break between the point of origin and the first scale mark makes it clear that the first scale mark encompasses more than one scale unit. But a break can be made at any point along the axis to shorten the graph (*Fig. 6.52b, c*). When the axis is broken, the ends of the axes at the break are terminated by short lines, diagonal (*Fig. 6.38b, 6.52a, 6.57a,c*), or perpendicular to the axis (*Fig. 6.52c*), or by a zigzag or z-shaped line (*Fig. 6.52b*). The break may be left

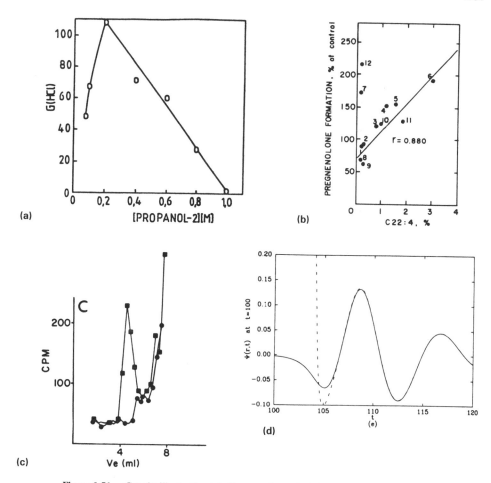

Figure 6.51.   Graphs illustrating labeling at points of origin of axes. (*a*) Zero labeled for both axes; (*b*) single zero at intersection of axes; (*c*) origin not labeled; (*d*) points of origin not zero; both labeled. *Note.* (*a–d*) Line width of curves not adequately differentiated from axes. (*a*) Letters crowded. (*b*) Graph too small. (*c*) Curve extending beyond axis; scale marks outside axis, almost imperceptible; serial label too large. (*d*) Lettering of axis labels disproportionately small; serial letter crowded against axis label, too small, in parentheses; scale marks very small. [(*a*) Mills & Henglein Radiat. Phys. Chem. 26, 1985, 396, © Pergamon Press, PLC.; (*b*) J. Biol. Chem. 261, 1986, 14122; (*c*) J. Biol. Chem. 261, 1986, 13823; (*d*) Phys. Rev. D 34, 1986, 396; with permission]

unmarked, as a gap (*Fig. 6.54d*), but this may be so unobtrusive as to go unnoticed. When the break in the axis includes curves, the curves must show the corresponding break (*Fig. 6.52b, c, 6.54d, 6.57a*). When the break occurs in a series of graphs, the break must be repeated in each graph (*Fig. 6.38b, 6.52a*).

Because such breaks are unobtrusive, readers tend to read the curves across

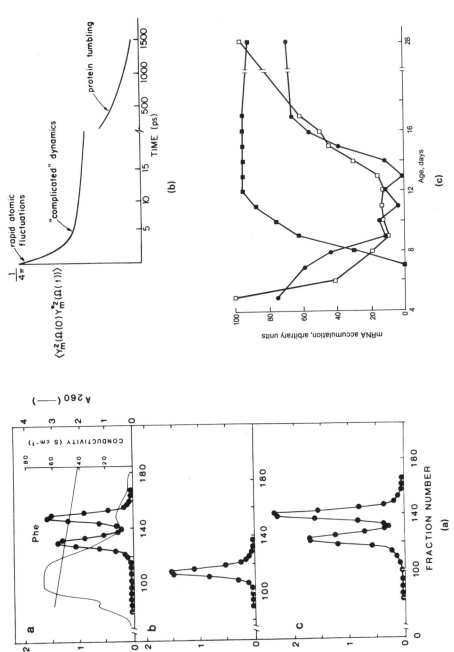

Figure 6.52. Graphs illustrating breaks in axes and curves. (*a*) Between zero and first scale mark, terminated by diagonal lines; (*b*, *c*) within curves; (*b*) shown by *z*-shaped fold in axis, break in curve; (*c*) terminated by vertical lines in axis and curves. *Note.* (*a*) No commentary. (*b*) Lines little differentiated; mathematical term lettered horizontally, space-consuming. (*c*) Lettering of axis labels disproportionately small. [(*a*) J. Chromat. 368, 1986, 119; (*b*) J. Amer. Chem. Soc. 103, 1981, 995, © 1981 Amer. Chem. Soc.; with permission]

the breaks without introducing the break in their interpretation of the graph. Full-scale breaks are more accurate; in such breaks, the entire graph is divided into two panels (*Fig. 6.53*). If the points fall so close to the axis that the curves fall on the axis or are too close to be readily distinguished, the axis may be shifted away from the zero point of the other axis so that the two axes do not intersect (*Fig. 6.54a, b*). Where there are points on each side of the zero point, the axis may be shifted away from zero (*Fig. 6.50, 6.57b,c*). A lightly dashed reference line may be used to designate zero or other important divisions of the graph (*Fig. 6.68b*).

The axes should be long enough to enclose all the points in the field; otherwise, readers must estimate the value of points outside the axis (*Fig. 6.45, 6.48a, 6.51c, 6.57c*). However axes should not extend more than one scale unit beyond the most distant point along the axis (*Fig. 6.42, 6.60c*), unless the extension is essential to include a key or labels. Labels should not fall outside the axes; they should be

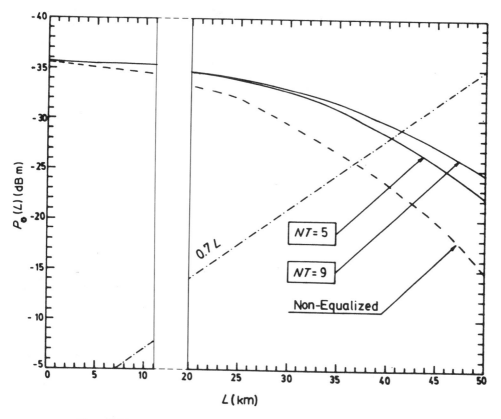

Figure 6.53.   Graph divided into two panels, illustrating full-scale break. *Note.* Lines scarcely differentiated; boxed labels conspicuous; scale marks on all four sides, making axes dominate the graph, length not adequately differentiated. [Adapted, Opt. Quant. Electro. 18, 1986, 196 (Chapman and Hall, pub.), with permission]

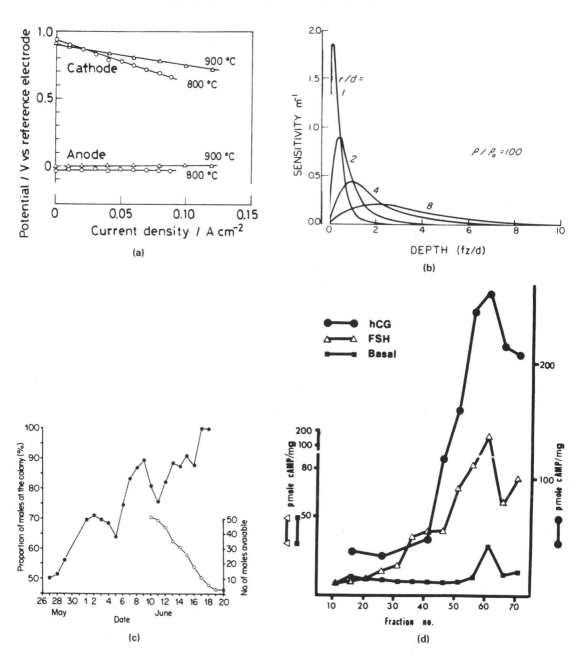

Figure 6.54. (*a, b*) Graphs showing axis shifted from zero to highlight curve along axis: (*a*) horizontal axis lowered; (*b*) vertical axis shifted to left. (*c, d*) Graphs with vertical axes having different scales, one axis longer than the other, distinguished by difference in points. Points (*c*) identified in figure legend; (*d*) shown in key on axes. *Note.* (*a, c, d*) Lines not adequately differentiated. (*a*) Triangles distorted; scale numbers and labels on curves disproportionately large. (*c,d*) Axis labels too small; scale marks outside axes. (*c*) Lines too light; numbers on horizontal axis too crowded. (*d*) Lines too heavy; break in axis and curve imperceptible; open traingles and solid squares distorted by heavy curve. [(*a*) Iwahara et al. J. Appl. Electrochem. 16, 1986, 666 (Chapman and Hall, pub.); (*b*) Geophys. 49, 1984, 574; (*c*) Anim. Beh. 86, 1985, 613; (*d*) J. Biol. Chem. 261, 1986, 13106; with permission]

part of the graph (*Fig. 6.42*). In *Fig. 6.61c,* the boxed, gridded key which extends beyond the axis, should either be included entirely within the axis (*Fig. 6.40*) or set entirely outside (*Fig. 6.60c, 6.62a*).

Some graphs show the relationship between a variable on one axis and two other variables on the two intersecting axes (*Fig. 6.52a, 6.54c,d, 6.55*). In such graphs one of the pair of axes may be shorter than the other to match the range of the pertinent values, and may thus help to distinguish the two axes (*Fig. 6.54c,d*). Such graphs are more effectively enclosed in a rectangle of axes that frame the data (*Fig. 6.52a, 6.55*). In such graphs, the different curves must be clearly associated with their axes. They may be identified by points, usually displayed with a segment of the curve, along the corresponding axis (*Fig. 6.52a, 6.54d*), in the legend (see *Appendix 6B*), or in a key (*Fig. 6.55a*). The curves may simply be labeled with a label keyed to the axis label, as in *Fig. 6.55b,* which actually has four vertical axes. In such graphs, the labels of the right-hand vertical axes should have the same orientation as the left-hand ones, not a reversed orientation (*Fig. 6.55b*).

The scale marks on the axis should be unobtrusive and uniformly spaced. They should not too small (*Fig. 6.46, 6.50a, 6.51c,d, 6.62b, 6.72*); neither should they be too long or too heavy relative to other elements (*Fig. 6.41, 6.49*). Strictly, the scale marks belong on the inside of the field that represents the quadrant defined by the two coordinates (*Fig. 6.49*). They are preferably not drawn outside the axes (*Fig. 6.42, 6.54c,d, 6.57a,b, 6.68*), unless the curves are too close to the axes. The scale marks should not extend across the axes (*Fig. 6.67a*), except for adjacent quadrants (*Fig 6.62a*). Scale marks are sometimes omitted (*Fig. 6.39a, 6.73a*) when the values are not pertinent or when the graph has a grid (*Fig. 6.43*), but they should not be omitted, especially when points must be associated with values on the axis (*Fig. 6.65b*). When a series of juxtaposed graphs are drawn as one graph, the scale marks may be labeled on the shared axes (*Fig. 6.52a*) or only on the outermost axis (*Fig. 6.63b, 6.68a*).

The scale unit should be chosen so that it is as economical of space as the effective display of data allows. The intervals should not be so close that the points are too crowded to be distinguishable (*Fig. 6.56d*), unless the objective is to show the close distribution or confluence of points. They should not be so widely spaced as to waste space (*Fig. 6.3c*). The scale unit chosen should allow space for horizontal lettering of scale labels on the horizontal axis (*Fig. 6.56d*). If the scale intervals are so close that the scale labels are crowded (*Fig. 6.48a, 6.56a, 6.57a, 6.60b, 6.68a*), either the interval can be made longer or the labels can be restricted to alternating scale marks.

The scale intervals are not usually divided, but may be divided when the interval is long enough that several points fall within the interval, when the scale is logarithmic, or to facilitate estimating values of points in the intervals. When scale intervals are divided, the subdivisions should not be too close (*Fig. 6.56a*). The main interval is labeled and the subdivisions are not. The main scale marks should be clearly distinguished from the intervening scale marks by their longer length (*Fig.

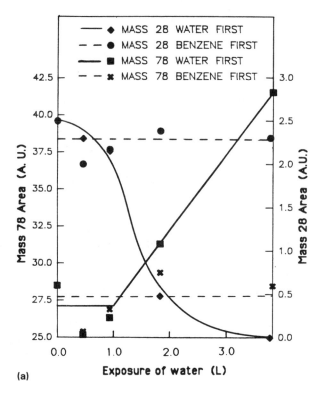

Figure 6.55. Graphs with axes framing graph, vertical axes with different scales. Distinguished by (*a*) difference in points, identified by key, (*b*) labels on curves corresponding to axis label; right axis with two scales. *Note.* (*a*) Points disproportionately large; Lettering of axes too bold, of key too large, emphasized by full capitalization. (*b*) Lettering too large, that of right axis reversed; T and R curves not differentiated. [(*a*) J. Amer. Chem. Soc. 108, 1986, 7565, © 1986 Amer. Chemical Soc.; (*b*) Astrophys. J. 310, 1986, 226; with permission]

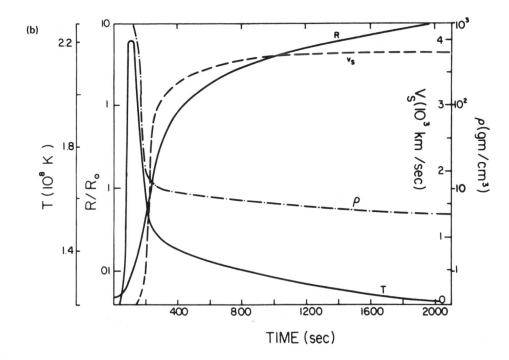

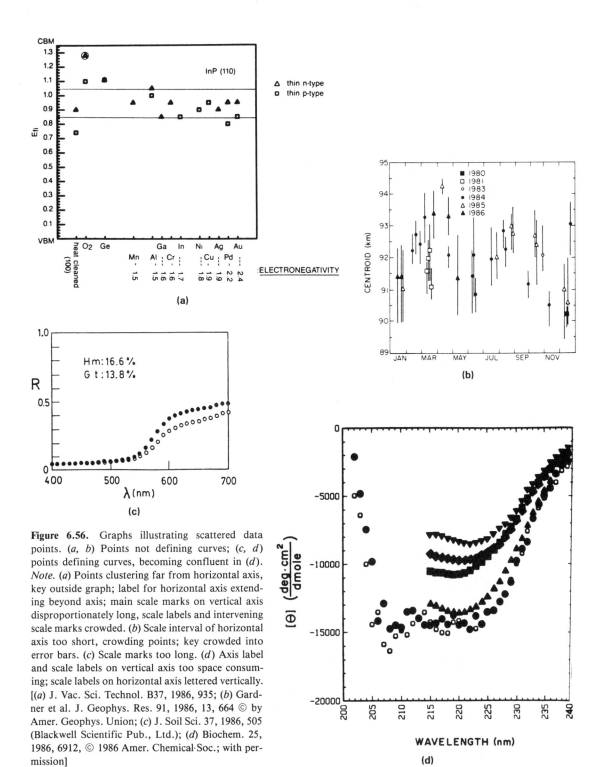

Figure 6.56. Graphs illustrating scattered data points. (*a, b*) Points not defining curves; (*c, d*) points defining curves, becoming confluent in (*d*). *Note.* (*a*) Points clustering far from horizontal axis, key outside graph; label for horizontal axis extending beyond axis; main scale marks on vertical axis disproportionately long, scale labels and intervening scale marks crowded. (*b*) Scale interval of horizontal axis too short, crowding points; key crowded into error bars. (*c*) Scale marks too long. (*d*) Axis label and scale labels on vertical axis too space consuming; scale labels on horizontal axis lettered vertically. [(*a*) J. Vac. Sci. Technol. B37, 1986, 935; (*b*) Gardner et al. J. Geophys. Res. 91, 1986, 13, 664 © by Amer. Geophys. Union; (*c*) J. Soil Sci. 37, 1986, 505 (Blackwell Scientific Pub., Ltd.); (*d*) Biochem. 25, 1986, 6912, © 1986 Amer. Chemical·Soc.; with permission]

6.56a, 6.71b); the difference is not distinct enough in *Fig. 6.53*. If a grid is used it should be wide-gauge and fine-lined.

Points.   Between the axes, data points are plotted to mark the observed, measured, or calculated values, and the curves that they determine. These are the focal elements in the graph and so should be visually prominent. Some graphs, scatter plots, display only the points (*Fig. 6.56, 6.65a*); many display only the curves (*Fig. 6.47, 6.58, 6.59, 6.62b, 6.71a*), but most graphs in scientific papers display both points and the curves they define. The scattered points may define a distinct curve (*Fig. 6.56c,d*) and may be so close together as actually to become confluent and form a curve (*Fig. 6.56d*), but they may not show a continuous relation (*Figs. 6.56a,b, 6.65a*).

The points on a graph are designated by symbols, and each set of data points is represented by a different symbol, which may be outlined ("open") or filled ("solid," "closed"). The commonest symbols are circles, triangles, and squares (see *Fig. 6.11*). These common areal, or geometric, symbols allow the writer six different sets of points—as many as is desirable for most graphs. Beyond this, the writer may use diamonds, inverted triangles, or ovals. These various geometrical symbols may be variously bisected, cross-barred, or divided into white and black areas, but when so many different types of points are used, the points and the graph may lose their visual effectiveness (*Fig. 6.41b, 6.65a*). Writers also use linear symbols, such as crosses (*Fig. 6.63a,b, 6.71b*), perpendicular signs, plus signs (*Figs. 6.63a, 6.70*), and asterisks (*Fig. 6.64b*).

Points may have error bars (*Fig. 6.57, 6.64a, 6.67a*) that extend vertically through the center of each point an equal length above and below it. When curves are crowded, only one-half of the bar may be shown (*Fig. 6.57b,c*). They may be terminated with a short crossbar (*Fig. 6.48a, 6.57a,c, 6.64a*). Error bars represent subsidiary data and so should be one of the finest lines in the graph.

Curves.   When a graph has numerous data points, each set of points is frequently joined by a curve to highlight the trend of the data. The points may even be omitted when the trend is the focus of interest rather than the values. The curves in a graph may be all the same type of line, as in families of curves, and are then usually unbroken lines (*Fig. 6.58a*). They may be labeled with a descriptive term or value if space permits (*Fig. 6.45, 6.72*), otherwise with only a letter or number (*Fig. 6.58a*), which must be explained. The curves may be differentiated, and then consist of varying combinations of lines, dots, and dashes (*Fig. 6.40, 6.58b, 6.59, 6.62b*). They may be labeled (*Fig. 6.58b*), if they do not obscure the data, or identified in a key or the legend (*Fig. 6.55a, 6.59b, 6.60b,c*). The preferred order for lines is an unbroken line, heavy or light in weight; a dashed line, of either definitely long or short dashes; a dotted line; or some distinctive combination of these. Unless the curves are well spaced, it is preferable to have no more than four to five different types of curves. Unless the different lines are well chosen, the curves may not be readily distinguishable, (*Fig. 6.38b, 6.58b, 6.59, 6.60c, 6.62*).

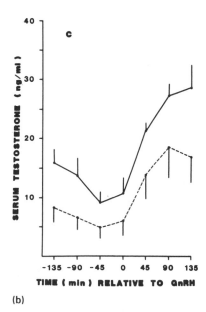

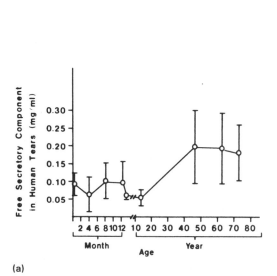

(a)

(b)

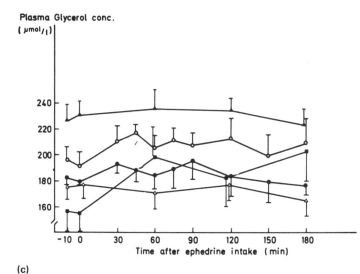

(c)

Figure 6.57. Graphs showing treatment of error bars. (*a–c*) Bars as prominent as curves; (*a*) extending above and below points; (*b, c*) extending only above or below points; (*a, c*) bars terminated by cross bar, (*b*) bars lacking cross bar. *Note.* (*a–c*) Origin of axes not labeled. (*a, c*) Vertical axis prolonged too far beyond curves, not scaled; (*a*) lettering of axes too small, scale labels crowded. (*b*) Points barely discernible; axis labels too heavy, dominating all other elements; scale marks outside axes. (*c*) Vertical axis lettered horizontally across top; error bars extending beyond scale marks; lettering of scale labels larger than that of axis labels; triangles and open points too small for width of line, distorted. [(*a, c*) Amer. J. Clin. Nutr. 42, 1985: (*a*) 283, (*c*) 90, © 1985 Am. J. Clin. Nutr, Amer. Soc. for Clinical Nutr.; (*b*) J. Anim. Sci. 63, 1986, 362; with permission]

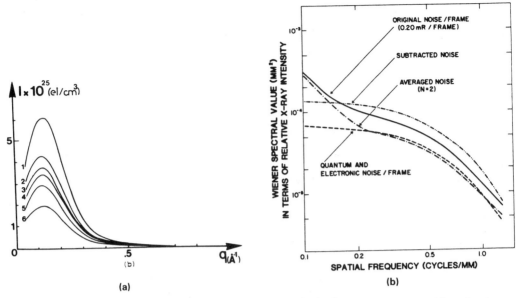

(a)

(b)

Figure 6.58. Graphs showing curves without points. (*a*) Family of curves undifferentiated, labeled by number at origin; (*b*) curves differentiated, identified by labels with arrows to curves. *Note.* (*a*) Serial letter too small, in parentheses; curves extending beyond scale labels; horizontal axis prolonged too far beyond curves; axes terminated by obtrusive arrow; axis labels too large, label of vertical axis across top, inside axis. (*b*) Dashed curves not easily distinguished; axis lettering bold. [(*a*) J. Appl. Crystal. 19, 1986, 21; (*b*) Med. Phys. 13, 1986 133; with permission]

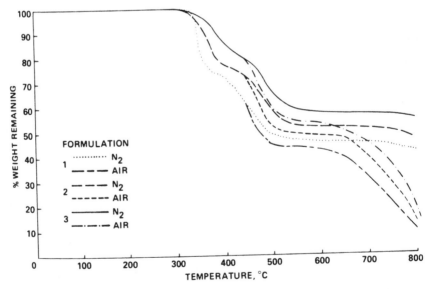

Figure 6.59. Graph showing curves without points, differentiated, identified in a key. *Note.* Long-dash curves not readily differentiated. [Kumar et al. J. Polym. Sci. A 24, 1986, 2421, © 1986 John Wiley & Sons, Inc., with permission]

276

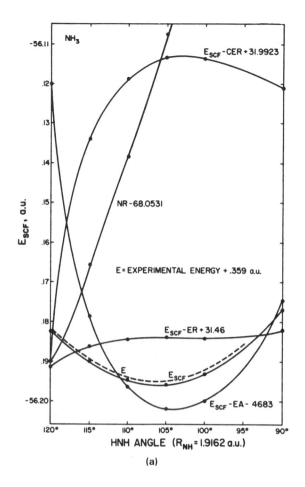

(a)

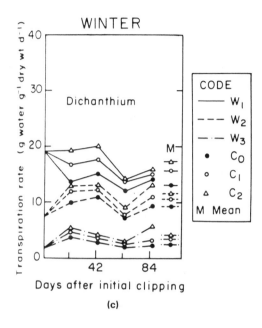

(c)

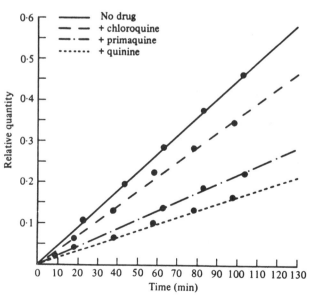

(b)

Figure 6.60 Graphs showing: (*a*) points and curves undifferentiated, curves labeled; (*b*) points and curves differentiated, identified in key; (*c*) only curves differentiated, labeled in key. *Note.* (*a–c*) Lines little differentiated. (*a*) Points, lettering of curves and scale marks too small. (*b, c*) Dash-dot curves not readily differentiated. (*b*) Lettering of axis labels too small; scale labels on horizontal axis too crowded. (*c*) Curves confined to lower half of graph; lettering of season and scale marks too large; key in graph redundant. [(*a*) J. Chem Soc. 108, 1986, 3594 © 1986 Amer. Chem. Soc.; (*b*) Q. Rev. Biophys, 18, 1985, 82; (*c*) Pl. & Soil 87, 1985, 175; with permission]

When graphs with several sets of data show both points and lines, the points and lines may all be the same, with only labels to identify the sets of data (*Fig. 6.60a*). This is sufficient in graphs with only a few data sets that are well spaced. In most graphs one of the elements is varied, the lines (*Fig. 6.38b, 6.60b*), but most commonly the points. Less commonly both the sets of points and the sets of curves associated with them may be differentiated (*Fig. 6.40, 6.60c*).

When different types of points or lines are shown in a graph, a key to the types may be used to identify the curves. This should be placed in an open part of the graph, and is usually free (*Fig. 6.59, 6.60b, 6.61a,d, 6.74b*), but sometimes boxed (*Fig. 6.40, 6.61c*) or outside the graph (*Fig. 6.60c, 6.62a*). It may be in the upper part of the graph (*Fig. 6.55a, 6.56b, 6.71b*) or in the lower part (*Fig. 6.59, 6.61c,d*), usually wherever there is ample open space to set it apart. A key is more accessible in the graph than in the legend; however, if the figure has no space for a key, or if the journal style is to use printed symbols, the symbols are explained in the legend (see *Appendix 6B*). The key in the graph shows the symbols for the points or the types of lines or both, followed by a descriptive label or quantity (*Fig. 6.55a*). Keys should be as condensed and economical as possible. The key in *Fig. 6.61a* can be condensed to highlight the points and save space, as shown in *Fig. 6.61b*. Unless there are numerous elements, keys should not be set in a ruled table (*Fig. 6.61c*), because they tend to clutter the graph (*Fig. 6.40*).

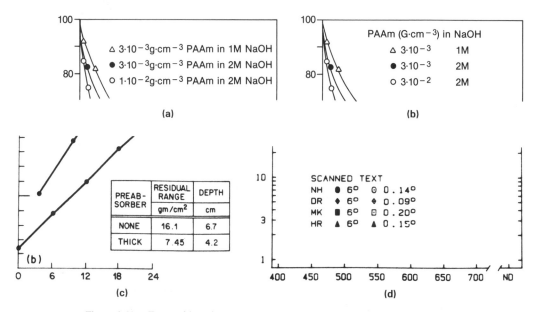

Figure 6.61. Types of keys in graphs. (*a, b*) In upper part of graph; (*c, d*) in lower part; (*c*) boxed, ruled. *Note.* (*b*) revision of (*a*) to increase visual discrimination. (*c*) With excessive rules, extending beyond axis. [(*a*) Reprinted from Kulicke, Kniewsko, & Klein Prog. Polym. Sci. 8, 1982, 381, © Pergamon Press, PLC; (*c*) Med. Phys. 13, 1986, 739; (*d*) Legge & Rubin J. Opt. Soc. Amer. 3, 1986, 45, with permission]

Choosing effective points and lines.   The choice of points and lines and their disposition in the graph should be directed toward clarity, and the effects of reduction must be kept in mind. Curves clearly differentiated by an interesting series of lines may become almost undistinguishable (*Fig. 6.62b*), or even undecipherable (*Fig. 6.62a*) upon reduction. In *Fig. 6.62a*, the curves are now difficult to differentiate even in the key. In *Fig. 6.62b* and *6.58b*, the pattern of the dashes is not distinguishable without close examination. The writer must therefore visualize the reduction in the length and width of the lines, the spaces between lines and points, and the shape and area of symbols (see *Fig. 6.11, 6.12*) in planning graphs.

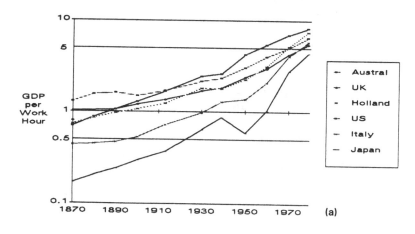

(a)

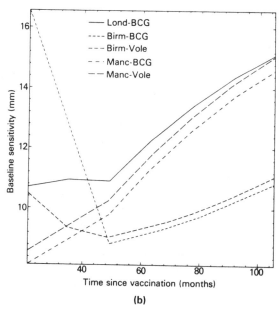

(b)

Figure 6.62.   Graphs illustrating ineffective lines for curves: line widths inadequately differentiated. (*a*) Curves and points difficult to identify. (*b*) Dashed lines not adequately differentiated. *Note.* (*a*) Label of vertical axis lettered horizontally; scale marks lacking on outer axes. (*b*) Lettering of axis labels too small; scale marks almost imperceptible; origin of axes not labeled. [(*a*) Amer. Econ. Rev. 76, 1986, 1075; (*b*) Statist. 35, 1986, 320; with permission]

Points, because they take up so little area, are most easily distorted on reduction; therefore they should be chosen for their distinctiveness. The most easily distinguished geometric symbols after reduction are solid circles and triangles, followed by outlined circles and triangles. Solid and outlined squares and diamonds follow, and inverted triangles and ovals are last choices because of similarity to corresponding forms. When symbols are crowded or very reduced, similar symbols may not be easily distinguished, for example, ovals from circles and squarish diamonds from squares. Open symbols may appear solid if they become partly filled with ink. Outlined geometric symbols with a dot to mark the actual point (*Fig. 6.64b*) and linear symbols, are most subject to distortion.

Because the kinds of elements used in line graphs are so few, writers must be constantly alert to the distorting effects of combining them. If the line joining the points is extended through the points, they are likely to be distorted (*Fig. 6.63, 6.64b*); if the line is not extended through them, there may still be distortion (*Fig. 6.40, 6.48a, 6.50, 6.71b*). If the line is drawn with a hairline space between line and symbol, the points and curve will be clear and undistorted (*Fig. 6.68a, 6.70, 6.74b*). Large points (*Fig. 6.52a*) are less likely to be distorted by the curve than small points (*Fig. 6.49, 6.50*); however, they should not be disproportionately large (*Fig. 6.64a*). If large points become crowded they may become confluent (*Fig. 6.41b, 6.56d*). When lines are heavy or points crowded and varied, solid points are less likely to be distorted than outlined or open points (*Fig. 6.50a, 6.57c, 6.64b*), although solid symbols are subject to distortion if they are small, or if there is little difference between the width of the line and the width of the symbol (*Fig. 6.49, 6.50, 6.54a, 6.57c*). In a dashed curve, the points may be distorted if the points are not prominent enough to be distinguished from the intervening dashes (*Fig. 6.46, 6.65b*).

Linear symbols are most subject to distortion because of the interaction of the lines of the symbol with the lines of the curve (*Fig. 6.63*). The lines are likely to fill in at the angles, and the juxtaposition of the lines of the curve and the lines of the symbol tend to blur the points further, especially when they are small. Therefore if linear symbols are used, it is important to break the curve for such points (*Fig. 6.50b*). Also however, since points should be symbols that have no meaning in themselves, geometrical shapes are preferable to plus signs, perpendicular signs, or letters. The objective in choosing symbols, therefore, is to choose those that maintain the distinctiveness of the symbols, yet do not disturb the smooth line of the curve.

Outlined geometric symbols with a dot to mark the actual point (*Figs. 6.61d, 6.64b*) are sometimes used for accuracy, but the dot becomes indistinct or lost in reduction, especially when the curve passes through the point. Therefore such points are of dubious value. Half-filled symbols are also difficult to distinguish, especially if numerous types are used in a graph (*Fig. 6.41b*). In *Fig. 6.65a*, 15 points are identified by 15 different symbols, many half-filled, which are explained in the legend (where the symbols are more distinctive than in the graph). These could more easily have been marked with the same symbol, with numbers or letters for identification.

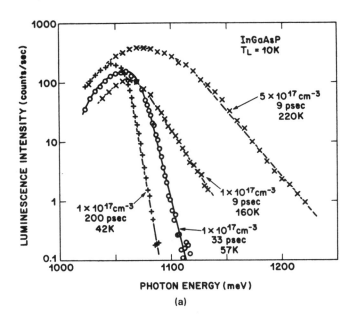

(a)

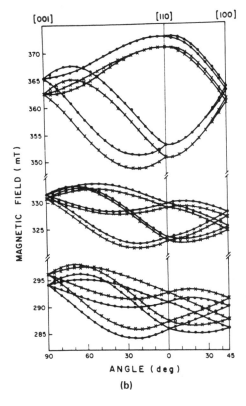

(b)

Figure 6.63. Graphs illustrating distortion of points and curves. (*a*) Linear points not readily differentiated, too close to determine whether curves are broken for linear points or dashed. (*b*) Points too small, distorted by curve, different types not readily differentiated. *Note. (a)* Labels of curves crowded with data, not readily compared. (*b*) Lettering bold. [(*a*) Phys. Rev. B. 33, 1986, 8762; (*b*) Watterich et al. J. Phys. Chem. Sol. 47, 1986, 990, © 1986 Pergamon Journal, Ltd., with permission]

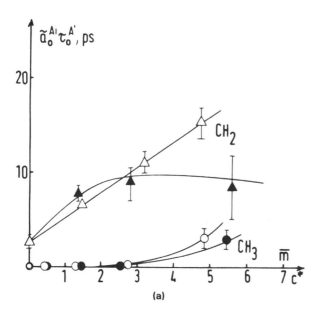

(a)

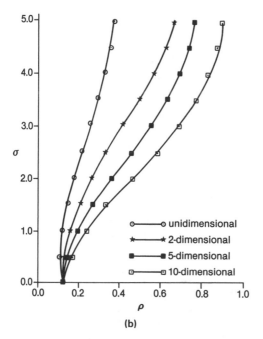

(b)

Figure 6.64. Graphs illustrating ineffective points. (*a*) Points dominating all other elements. (*b*) Points linear, open or with internal points distorted by line passing through them. *Note.* (*a*) Lines undifferentiated; axes labeled inside graph; vertical axis labeled at top; lettering of labels of curves and scale marks too large. (*b*) Lettering too small; scale marks outside axes. [(*a*) J. Sol. Chem. 15, 1986, 937, © 1986 Plenum Pub. Corp., with permission]

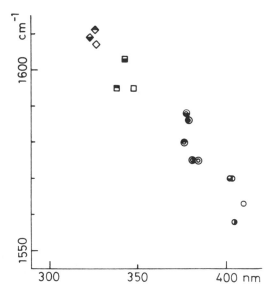

Figure 7. Correlation between the wavelength of the lowest electronic transition and the C=C stretching Raman frequency for each isomer of retinal homologues. C_{22} aldehyde: all-*trans* (○); 9-*cis* (◖); 11-*cis* (◗); 13-*cis* (◖). C_{20} aldehyde: all-*trans* (◎); 7-*cis* (◉); 9-*cis* (◉); 11-*cis* (◉); 13-*cis* (◉). C_{17} aldehyde: all-*trans* (□); 7-*cis* (◧); 9-*cis* (▣). C_{15} aldehyde: all-*trans* (◇); 7-*cis* (◆); 9-*cis* (◈).

(a)

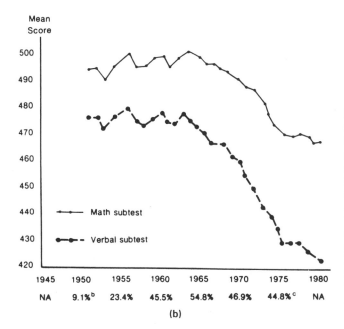

(b)

Figure 6.65. Graph illustrating ineffective differentiation of points and curves. (*a*) Points 15, each of different type, difficult to identify. (*b*) Points not adequately differentiated from curve, especially in upper fine-lined curve; lower curve with points too close to make dashed line stand out. *Note.* (*a*) Axes labeled only with units; scale labels on vertical axis lettered vertically. (*b*) Vertical axis labeled at top of axis, scale marks lacking. [(*a*) Mukai et al. J. Raman Spectrosc. 17, 1986, 393, © 1986 John Wiley & Sons, Ltd.; (*b*) Gaddy, Publ. Opin. Q. 50, 1986, 342, Univ. of Chicago Press; with permission]

The clarity of points is particularly important when there are different curves. If the points are very close, different types of lines cannot be easily distinguished between the points (*Fig. 6.41b, 6.60c, 6.65b*); if not close, the additional differentiation is probably not needed (*Fig. 6.60b*). One may, however, use the same symbol (usually closed) and same type of line (usually unbroken) if the curves are well spaced and easily distinguished (*Fig. 6.60a,* although the points in this figure are too small for the line of the curve), but then the curves must be labeled.

When different types of lines are used for curves, one can use two unbroken lines, one wide and heavy, the other narrower and lighter. Dashed lines are to be preferred to dotted lines, unless computer drawn. Lines with long or short dashes can be distinguished from unbroken lines easily, and from one another if there is enough difference between the length of the dashes. Lines consisting of dashes combined with smaller dashes or dots can be readily distinguished (1) if there is enough differentiation between the patterns, (2) if there are few curves or the curves have no points, or (3) if the points and curves are spaced far enough apart to allow the pattern of dashes and dots to be distinguished easily. In *Fig. 6.65b,* because the points on the heavy line are so close, it is difficult to identify the pattern of dashes, and the line is disconcertingly sometimes solid and sometimes variously dashed (see also *Fig. 6.46*). Also, the points are not adequately differentiated from the width of the line to make the curves visually effective. In a family of curves differentiation of curves may add unnecessary differentiation of elements, and the differences become indistinguishable where the curves become confluent. When a graph has few, well-spaced curves, they need only be labeled (*Fig 6.58b, 6.60a, 6.72*). The framed labels in *Fig. 6.53* are unnecessarily emphatic, and the arrows to the curves add three conspicuous and unimportant lines to the graph. The curves could have been unobtrusively labeled.

Labels, lettering, and lines.   Graphs should be accurately and completely labeled. Labels should be brief, preferably on the graph rather than in the legend. If the label is long, a brief descriptive word, phrase, or abbreviation can be used and then explained in the legend. Labels should not obscure the configuration of points or curves and should not clutter the graph; if they do, they should be de-emphasized or attached to a key.

Each axis should be labeled with the name of the variable measured and the unit of measure; it is not enough to have the one without the other (*Fig. 6.42, 6.43, 6.65a, 6.72*). The unit of measure should follow the axis label, not precede it (*Fig. 6.41a, 6.59*). It may be in parentheses (*Fig. 6.44, 6.55, 6.63*), or set off by a comma (*Fig. 6.52c, 6.59*). If there is a second horizontal or vertical axis, it too is labeled, and the associated curves identified on the axis or in a key in the graph (*Fig. 6.52a, 6.54d, 6.55*). The axes are labeled outside the graph, along their length, with the label centered. They are read from left to right on the horizontal axes and from bottom to top on the vertical axes. The vertical axis should not be lettered to be read down from the top (*Fig. 6.55b*), or to the left of the axis, centered (*Fig. 6.62a, 6.71b*), or at the top (*Fig. 6.66a, 6.73a*), or across the top (*Fig. 6.57c, 6.65b*), because it then

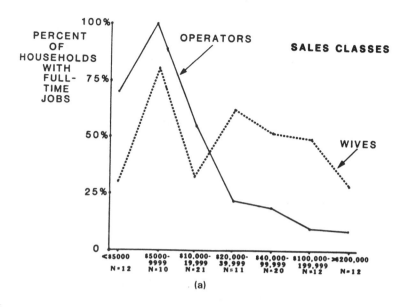

(a)

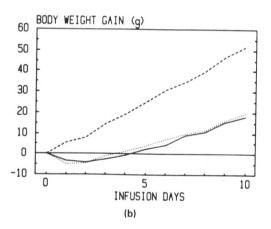

(b)

Figure 6.66. Graphs showing horizontal orientation of labels of vertical axes: (*a*) beside axis, (*b*) along upper horizontal axis. *Note.* Line widths not adequately differentiated. (*a*) Scale marks inside and outside axes; points and scale marks almost imperceptible; percentage symbol superfluous; lettering in graph too prominent, lettering on horizontal axis too small, crowded, full range of values and repetition of *N* not necessary. (*b*) Scale labels disproportionately large, those of vertical axes crowded; scale marks on horizontal axis imperceptible. [(*a*) Rur. Soc. 51, 1986, 295; (*b*) Amer. J. Clin. Nutr. 42, 1985, 63, © 1985 Am. J. Clin. Nutr., Amer. Soc. for Clin. Nutr.; with permission]

takes up excessive space. It should not be placed inside the vertical axis at the top (*Fig. 6.58a, 6.64a*), because it is not part of the field of the graph. When it is placed along the horizontal axis at the top of the graph, as in *Fig. 6.66b,* it may be mistaken for the label of the axis. Here, setting the scale of the horizontal axis from 0 to 10 days would allow space for the label along the vertical axis. The vertical axis may be labeled horizontally if the label is a symbol or brief mathematical expression (*Fig. 6.40, 6.64b, 6.71*), and may be placed at the end of the axis (*Fig. 6.47, 6.70*).

The main scale marks should be labeled, *horizontally,* on both vertical and horizontal axes. If the scale marks are very close together (*Fig. 6.42, 6.54c, 6.56a, 6.60b*), or the numbers large (*Fig. 6.56d*), only alternate scale marks should be labeled (*Fig. 6.52a, 6.55b, 6.63a*), or the scale interval may be made longer. When the scale is logarithmic, the scale interval is divided to show this, and the chief interscale marks may be labeled (*Fig. 6.41b, 6.68b*). If the numbers of the scale occupy too much space, such as numbers in the thousands (*Fig. 6.56d, 6.66a*), they can be reduced to one or two digits by incorporating "in thousands" into the unit of measure or by using an exponent. Scale marks should not be lettered vertically along the y-axis (*Fig. 6.65a*) or below the x-axis (*Fig. 6.56d*), when labeling alternate scale marks would permit horizontal lettering without crowding. When similar graphs are grouped, the same scale should be used; then they can be made contiguous and one label used for the same axis of the set (*Fig. 6.63b, 6.68a*).

Curves must be identified. If there are few curves and ample space, and the labels are short, they can be lettered along the curves (*Figs. 6.45, 6.60a, 6.71a, 6.72a*) or in the field of the graph (*Fig. 6.50b, 6.70*), sometimes with guidelines or arrows leading to the curves (*Fig. 6.52b, 6.58b, 6.63a, 6.66a*). If the curves are differentiated by different points or different lines, the points or lines can be labeled in a key (*Fig. 6.40, 6.59, 6.62*), less effectively in the legend (*Appendix 6B*).

Lettering should not be conspicuous in its size or form; its function is to draw attention to the part labelled, not to itself. Lowercase letters are more easily read than are capital letters. The first word of a label is usually capitalized; in axis labels major words may also be capitalized. All the letters may be capitalized, but this is unnecessarily emphatic (*Figs. 6.38a, 6.40, 6.66*). Well-spaced letters are more easily read than crowded ones (*Fig. 6.38a*), especially crowded, bold letters (*Fig. 6.67a*). Simple block letters are easier to read than italic letters (*Figs. 6.50a, 6.68a*) or decorative typefaces. Lettering should not be so small that it becomes almost illegible when reduced (*Fig. 6.67b, 6.68b, 6.72*), and letter-size typescript is to be sedulously avoided (*Fig. 6.67b, 6.45*).

In a figure with several related graphs, the letters (or numbers) that identify the individual graphs should dominate the other lettering on the graphs. They may be larger, bolder, or both, or emphasized by ample surrounding space. The letters are placed in a part of the graph that has no points or curves, usually in one of the corners (*Fig. 6.46, 6.47, 6.51c, 6.52a, 6.73a*). When the graphs are separate, with each axis separately labeled outside the graph, the graphs may be labeled outside the graph—above (*Fig. 6.69*), below (*Fig. 6.51d*), or at the side (*Fig. 6.38b*). These letters must be made larger or bolder than the other lettering; in *Fig. 6.38b, 6.47* and

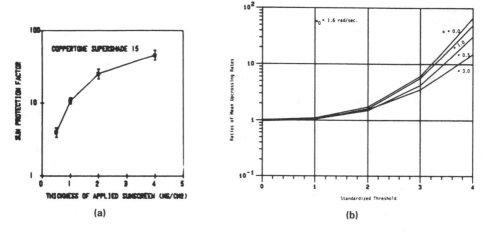

(a)   **(b)**

Figure 6.67. Graphs illustrating illegible lettering: (*a*) heavy, crowded; (*b*) type-written, too small. *Note. (a, b)* Lines not adequately differentiated. (*a*) Scale marks across axes; points and curve distorted by cross bars. (*b*) Grid too prominent; scale marks outside axes. [(*a*) Brown & Diffey, Photochem. Photobiol. 44, 1986, 511, © 1986 Pergamon Press, PLC.; (*b*) J. Engr. Mech. 112, 1986, 738, ASCE, 1988; with permission]

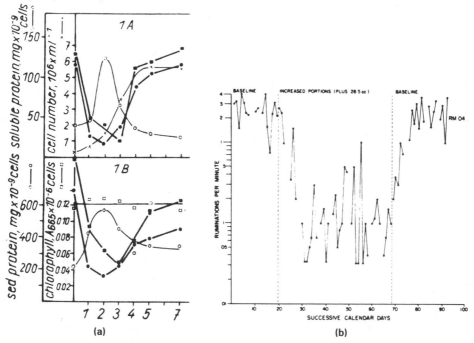

(a)   **(b)**

Figure 6.68. Graphs illustrating ineffective lettering, difficult to read: (*a*) italic, crowded; (*b*) very small. *Note.* (*a*) Points not easily identified; some numbers disproportionately large. (*b*) Points too small, curves too fine; scale marks outside axes, those on horizontal axis too small. [(*a*) Planta 167, 1986, 484; (*b*) Amer. J. Clin. Nutr. 42, 1985, 99, ©1985 Am. J. Clin. Nutr., Amer Soc. for Clin Nutr.; with permission]

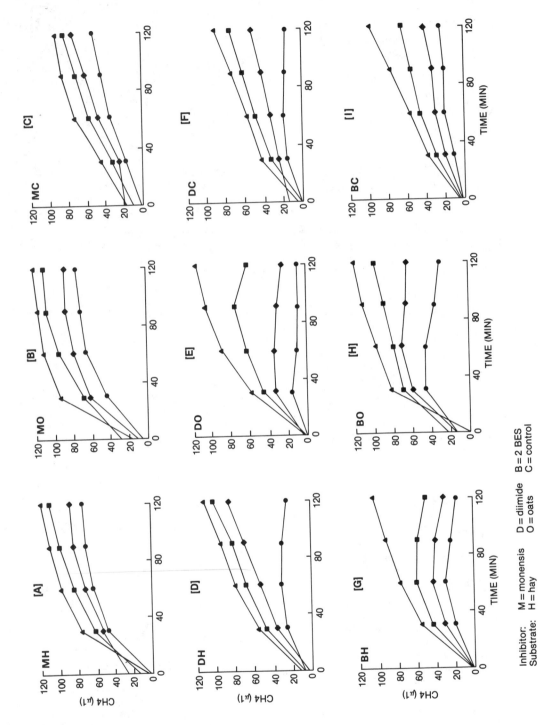

Figure 6.69. Series of graphs illustrating uninformativeness of letter labels (in brackets), replaced by mnemonic abbreviations for inhibitors and substrates. [S. Martin, 1983. MS thesis,

Inhibitor: M = monensis   D = diimide   B = 2 BES
Substrate: H = hay   O = oats   C = control

288

6.51d they are too small and fine-lined for the emphasis required, and in *Fig. 6.51d* the letter is too close to the graph.

In a figure with similar related graphs, labels may be more informative than letters (*Fig. 6.38b, 6.60c*). In *Fig. 6.69,* the graphs in the original figure were labeled only alphabetically (without brackets); actually they show the relationship between three inhibitor drugs, and three substrates. The graphs in the first row show the results of experiments with monensin and the three substrates: hay in the first column, oats in the second column, and the control in the third column. Similarly, the second row shows experiments with diimide, and the third with BES. The graphs can be labeled with a two-letter abbreviation that reflects the variables,

Ex. 6.15

M = monensin	H = hay	MH	MO	MC
D = diimide	O = oats	DH	DO	DC
B = BES	C = control	BH	BO	BC

which readers have been reading about throughout the paper.

Balance.   The objective in planning a graph is to design one that is simple, clear, coherent and highlights its message, and that avoids clutter, confusion (the visual equivalent of verbiage and circumlocution), and distortion. But lines and points are minimal visual features, which are very limiting. Therefore, balance, which in general visual design is an aesthetic objective, becomes a semantic objective in the design of a line graph.

The lines of the graph must reflect their relative importance. From a hierarchical standpoint, the axes should be dominant over the curves, but from the standpoint of communicating the message of the graph, the curves are the dominant elements. Therefore for visual effectiveness, the objective is to make the curves the dominant lines on the graph, and the axes the background frame for them. Exceptions to this are curves generated by an automatic instrument (*Fig. 6.68b, 6.74a, 7.1*), which are often too fine but must not be modified because they constitute the data. Scale marks must be less prominent than the axes, and grid lines should be among the least conspicuous. Error bars, reference lines, and arrows should be among the least conspicuous lines, as should any lines drawn to frame labels or a key (*Fig. 6.53*), which should be omitted, if possible. Points should be distinctly larger than the width of the curve to avoid distortion and maintain their identity, and their form must be selected to avoid distortion and confusion.

Generally, the lettering of labels should reflect the relative importance of the labels. Actually, the size of the letters and thickness or weight of their lines can be variously combined with the space around them and the form of the letters to reflect the hierarchical level of the different sets of labels on the graph. For independent graphs the lettering of axis labels is the dominant lettering on the graph. The units of measurement maybe in lower case letters (*Fig. 6.54b, 6.56b–d*). The scale numbers are subordinate to the axis label and should be less prominent (*Fig. 6.38a, 6.63b, 6.74b*). In the field of the graph, the lettering of groups of curves is dominant over the lettering of individual curves. The lettering of the key is subordinate to the lettering of labels of curves, but no lettering should be illegible (*Fig. 6.67a*) or too small or light to read (*Fig. 6.67b, 6.68b, 6.72*), nor should lettering be so large or bold as to dwarf other elements or the element that it is labeling (*Fig. 6.39a, 6.57b, 6.72*).

The dominance of the lettering of axes over that of curves may seem to be

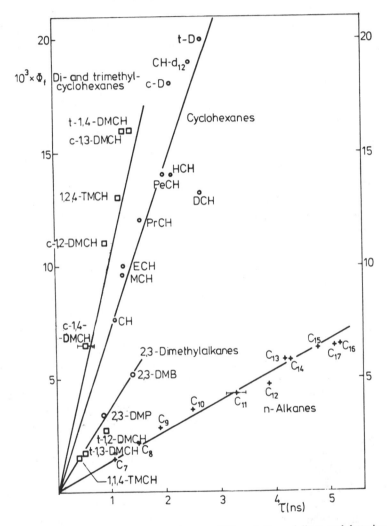

Figure 6.70. Graph showing inadequate differentiation of lines and lettering. Curves and axes not adequately differentiated; lettering of uniform size and weight; labels of groups of compounds (curves) inconspicuous, not differentiated from labels of individual compounds (points). *Note.* Points open or linear, not prominent enough to define curves amid labels. [J. Lumin. 33, 1985, 76, with permission]

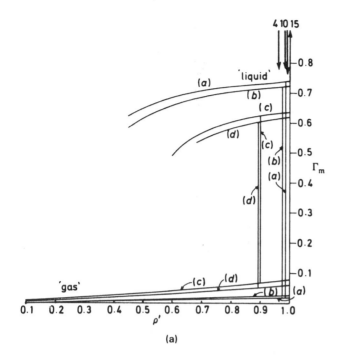

(a)

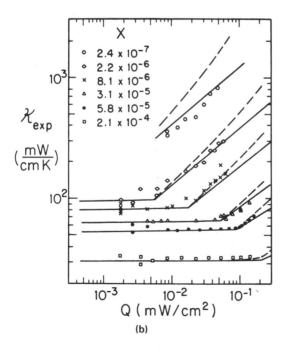

(b)

Figure 6.71. Graphs showing inadequate differentiation of lines, lettering, and points. *Note.* (*a*) Lettering of labels very small, axis labels almost undetectable; curves crowded close to axes. (*b*) Labels of axes, scale marks, and key successively smaller, but all disproportionately large; scale marks too prominent; points not readily distinguished, some distorted. [(*a*) Farad. Trans. II J. Chem. Soc. 82, 1986, 1772; (*b*) J. Low Temp. Phys. 65, 1986, 229, by M. Dingus et al., Plenum Pub. Corp.; with permission].

inconsistent with the dominance of the line width of curves over the line width of axes. The reason for this is that the labeling of the curve is subordinate to the curve, and since the space bordering the curve helps to give the curve its visual impact, the labeling of the curve should not be dominant enough to intrude on that space and so weaken the force of the curve. Furthermore, the display of several curves makes it necessary to use smaller letters, especially if the curves limit the space available for labels. The lettering in the field of the graph is therefore determined mostly by the other lines and lettering in the field. Because of these varying constraints, balance is more important than particular dimensions.

Many graphs do not show these subtle, interrelated, hierarchical relations of lines and lettering accurately. Some graphs do not show adequate differentiation among elements (*Fig. 6.38, 6.39, 6.55b, 6.71*). In *Fig. 6.70* the numerous elements

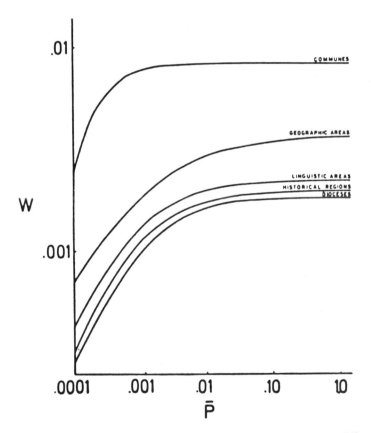

Figure 6.72 Graph with imbalance in differentiation of elements. Lines not differentiated; scale marks almost imperceptible; lettering disproportionately large, or small and illegible. *Note.* Scale intervals not consistent, those on vertical axis too long. [Ann. Hum. Genet. 50, 1986, 174, with permission].

are so little distinguished, that one does not see a hierarchy of elements. Actually, the graph includes the labels of the four curves, each with numerous labeled points, but the sets of labels are not noticeably differentiated. In *Fig. 6.71a*, the lines representing the axes and the curves are indistinguishable, as is the size of most of the lettering. Many graphs show inappropriate differentiation. Lettering may be too large or too small relative to other letters or lines (*Fig. 6.71b, 6.72*). In *Fig. 6.73*, all the lettering is overshadowed by the inordinately heavy axes or curves. In *Fig. 6.74a*, the fine lines of the curves scarcely merit the bold lettering, and in *Fig. 6.74b* the lettering of the axis labels dominates all other lettering in the graph by its capitalization, size, and boldness.

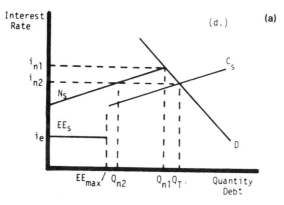

Figure 1. Effects of Economic Emergency Loans on Farm Credit Markets

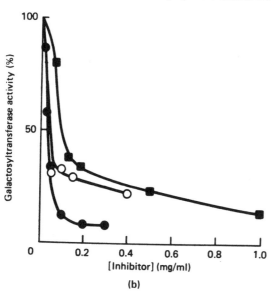

Figure 6.73. Graphs with imbalance in differentiation of elements. (*a, b*) Labels disproportionately small. (*a*) Axes too bold. (*b*) Curves too bold; points too large. *Note.* (*a*) Vertical axis labeled at top; serial label with period, in parentheses; symbols not explained in legend. (*b*) Scale marks too long. [(*a*) S. J. Agr. Econ. 17, 1985, 23; (*b*) Biochem. J. 239, 1986, 429 © 1986 Biochem. Soc.; with permission].

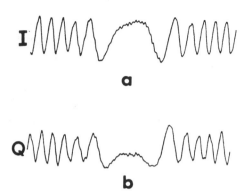

(a)

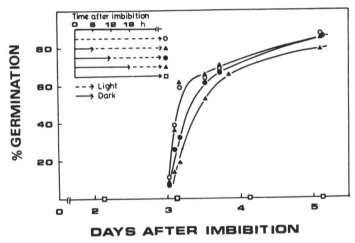

(b)

Figure 6.74. Graphs with disproportionately heavy lettering of (*a*) labels, (*b*) axis labels. *Note.* (*b*) Lettering in graph too small. [(*a*) A. M. Tai, Appl. Opt. 25, 1986, 3189; (*b*) J. Pl. Physiol. 123, 1986, 489; with permission].

Frequently curves are not as heavy as the other lines, and so may not emerge as the dominant lines on the graph (*Fig. 6.38, 6.51, 6.64a, 6.71*). In *Fig. 6.64a* the curves, axes, and error bars are barely distinguishable in width of line, and the points are disproportionately large, as is the lettering.

6.10 EQUATIONS

Mathematical equations are most common in theoretical papers or papers developing a theoretical model. They are like tables and figures in being visually independent units that must be integrated into the text. Unlike tables and figures, however, they are sentential and so must be made part of the syntactic structure of the sentence. The writer must (1) structure the sentence so that the equation forms part of its grammatical structure, (2) decide whether the equation can fit into the line of text or must be displayed, and (3) plan the spatial disposition of symbols and terms in the equation for accuracy, economy, and clarity.

6.10.1 Form and Position of Symbols

Mathematical equations consist of symbols that are alphabetic or graphic. Graphic symbols are used to describe mathematical relations or operations (*Appendix 6C*) and may function as verbs, connectives, adjectives, prepositions, or predicates. Most of them are standard symbols that are widely used ($=$, $+$, $>$) or widely understood ($\int$, $\neq$, $\simeq$, Σ). Some of them are less familiar, such as $\forall$ (for all), $\exists$ (there exists). Nevertheless, graphic symbols are usually standard and unvarying; writers do not invent operators at will. Alphabetic symbols are more variable. They are usually used to represent entities or quantities to be manipulated mathematically, and so often function as nouns. Some alphabetic symbols have become standard representations of particular quantities:

Ex. 6.16

a.	l	length	π	pi	T	temperature
b.	λ	wavelength	Q	radiant energy	C	capacitance
c.	Q	Quaternary	$\dot{Q}$	blood flow	F	filial generation

and may be widely used or understood (*Ex. 6.16a*); some are common to broad disciplines (*Ex. 6.16b*); and some are restricted (*Ex. 6.17c*) to particular disciplines. Alphabetic symbols may be adopted for a particular research problem and be restricted to the particular paper. The set of symbols used in a paper is referred to as the "notation" or "nomenclature" (*Ex. 6.41*, see also *Table 8.11b*).

Presumably one can choose any symbol and define it to represent whatever quantity one wishes; in actual practice, the writer is constrained by symbols already in use. Since there are many more quantities than there are available symbols, there is already duplication of symbols in different disciplines and sometimes in the same discipline. For example, R represents

Ex. 6.17

respiratory exchange ratio (respiratory physiology)   roentgen (radiation)
molar mass constant (chemistry)   resistance (electricity)

To avoid duplication of symbols, standard symbols should be used whenever they are available. If one requires a new symbol, it should be chosen so that it does not duplicate common symbols in the discipline and related disciplines, and is not so esoteric that typesetting is difficult. The writer has available all the letters of the Latin alphabet, upper- and lowercase, and in roman, italic, and boldface type. Letters from the Greek alphabet (*Appendix Table D.1*) are very commonly used. Subscripts, P_s, and superscripts, P^2, which may be numerical or alphabetical, can also be used with letters in forming a symbol and if necessary, letters from the German, script, and phonetic alphabets. Diacritical marks can also be used in association with the letters, either above them, below them, or beside them (*Appendix 6C*) but one should consider that letters with marks under them are difficult to set in type, as are bars over letters. Before beginning to write, therefore, a writer who expects to use many symbols should procure a list of available symbols.

In an equation, alphabetic symbols are set in italic type except for the boldface symbols used for vectors and tensors. Graphic symbols such as operational signs, and the abbreviations for trigonometric functions and other mathematical terms are set in roman:

Ex. 6.18

log	tan	lim	av	coeff	max
ln	cos	obs	calc	approx	exp (but *e*)

Numbers used in association with symbols are also set in roman. It is a convention not to begin a sentence with a symbol or abbreviation; therefore when a symbol, expression, or equation begins a sentence, for example, "V_p was assumed to be . . . ," it should be revised either to change the subject or to begin the sentence with the word or phrase that it represents: "The plasma chamber voltage, V_p. . . ." Equations may be included in the line of type (*Ex. 6.19*) or set off from the text and displayed without text (*Appendix 6E*). Equations that are important, long, or complex, are displayed, and those that are important enough or that might be referred to are numbered.

Equations are numbered in consecutive order, usually in parentheses at the right margin, but occasionally at the left, especially in the social sciences (*Ex. 6.33a, 6.37a*). In a long or complex paper or a dissertation, they may be numbered in decimal form with the first number representing the number of the chapter or section (*Appendix 6E*). When some equations are subordinate to others, letters may be used to show the relationship, both in whole: (12), (12a), (12b), and decimal: (4.1), (4.2a), (4.2b) numbers.

6.10.2 Adapting Notation for Typesetting

Equations are costly to set in type; therefore, every effort should be made to simplify them for typesetting, for example, to make them more linear and to reduce the number of levels. A mathematical expression or equation that extends above or below the line of type in the text requires extra spacing between lines (*Appendix (6F.2)*) and disturbs the visual unity of the verbal text; therefore, it should be rearranged so that it can fit into the line or displayed.

An expression or equation that is short or simple and can fit into the line of type is simply included in the text:

Ex. 6.19

The limiting current caused by power balance, $I = I_p$, can be found from the maximum of the curve I vs γ_f, i.e., from the condition $\partial I/\partial \gamma_f = 0$. We note that by setting $f_e = 0, f_m = 0$, one obtains a space-charge current limit that is greater than I_L in (1). Likewise, for the case $f_e = 1$, $f_m = 0$, one obtains $I_A = I_0 \beta \gamma_i$. [Phys. Fluids 27, 1984, p. 1898)]

(Appendix 6F). If the equation is simple but important or is likely to be referred to by the writer or readers, it is displayed, even though it could fit into the line of type:

Ex. 6.20

$$T \simeq 1/(1+\text{sd}) \qquad\qquad (13) \qquad p(s) \times 1 - m \le 0 \qquad\qquad (2.1)$$
$$1_w^s = 1 \qquad\qquad\qquad\quad (3) \qquad F = q(E + w \times B + mg) \qquad (7)$$

An equation with notation or complex fractions that require the lines of type to be spaced apart is usually displayed *(Ex. 6.21a)*:

Ex. 6.21

a.

$$1 = \int_0^T |H(t)|^\beta dt. \tag{3}$$

b.

$$\langle (v_l)^n \rangle \simeq \langle (v_m)^n \rangle + n \langle (v_m)^{n+1} \rangle / V_d. \tag{25}$$

Long equations are commonly displayed *(Ex. 6.21b)* and complex ones are necessarily displayed:

Ex. 6.22a

a.

$$\theta_3 = \theta_1 = 90° + \sin^{-1} \left(\frac{d_{4,6} - d_{1,3}}{2d_{1,6}} \right) + \cos^{-1} \left(\frac{d_{1,3}}{2d_{1,2}} \right)$$

b.

$$\gamma_d = \frac{c^2 \hbar^2 n_a \sqrt{Q}}{\sqrt{3} n^2 E_c^2} \frac{\sqrt{\Delta}}{\Delta_a^*} \int_{-\infty}^{\infty} \frac{\exp \left[-\dfrac{(E_\alpha - E_a')^2}{\Delta_a^{*2}} \right]}{\sqrt{(E_a - E_a')^2 + \Delta^2}} dE_\alpha'$$

Displayed equations may be centered or indented if short and begin at the left margin if long.

Short or relatively simple expressions with built-up fractions can be revised to fit in the regular spacing of the lines. For example, a simple built-up fraction, $\dfrac{a+b}{c-d}$, will disturb the line of type, as can be seen here; however, the slash or solidus can be used to give a form that fits easily into the line of type:

Ex. 6.23

a.

$$\frac{a+b}{c-d} \qquad\qquad a+b/c-d \qquad\qquad (a+b)/(c-d)$$

b. The path difference is $\Delta l = 2n_2 d \cos \phi_2$, and hence the phase change is $\delta = k\Delta l$. Here $k = 2\pi/\lambda$ is the propagation number. Using Snell's law, $\delta = (4\pi d/\lambda)(n_2^2 - n_1^2 \sin^2 \phi_1)^{1/2}$.

When such a term or equation is adapted for typesetting, the simplified form may require grouping of related terms in parentheses to maintain the relationships. Using the solidus, as in the middle term (*Ex. 6.23*), does not accurately represent the built-up fraction. To maintain the original relationships necessitates enclosing the numerator and denominator in parentheses. A similar confusion is seen in *6.24a*:

Ex. 6.24

a. $E(V(k)/I_1, \ldots I_n) = -E(V(k)/1 - I_1, \ldots, 1 - I_n)$

b. $E = \underset{x_\theta}{\Delta x} = [(x - x_0)/(x_0 - x)] - 1$

where parentheses enclosing the denominators of the two fractions, $I_1, \ldots I_n$ and $1 - I_1, \ldots, 1 - I_n$, would have made the relation absolutely clear; then the outermost parentheses on each side of the equal sign, following standard usage, would be replaced by brackets. In *Ex. 6.24b*, this confusion has been avoided.

It is a convention not to enclose parentheses within parentheses, but in brackets:

Ex. 6.25

$$i_{n,x} = i_{0,x} [1 - b_{ss} - (b_0 - b_{ss}) e^{-n\lambda}] \tag{37}$$

When terms in brackets must be enclosed, then the brackets are enclosed in braces { }:

Ex. 6.26

a.

$$p = -\gamma \frac{1}{r} \frac{\partial}{\partial r} \left\{ r \frac{\partial h}{\partial r} \left[1 + \left(\frac{\partial h}{\partial r}\right)^2 \right]^{-1/2} \right\} \tag{3}$$

b.

$$\left\{ \left[\left(\left\{ \left[(\ ) \right] \right\} \right) \right] \right\}$$

The signs of aggregation are used in the same order for subsequent enclosures, as in *Ex. 6.26b*.

Certain types of notation are difficult to set in type, for example, fractional exponents, symbols with overbars, superscripts, and subscripts, and arrows over vectors. Superscripts and subscripts are difficult to set in the elegant form in which both are aligned vertically, as in *Ex. 6.27a (6.22b)*:

Ex. 6.27

a.

$$1 - z_1 - K_2 C_{z1}^4 = 0 \tag{25a}$$

b.

$$M_{C1}(s) = I_{C1}/I_a^e \tag{7}$$

c.

$$r_g = \sigma_{s_x s_y} / (\sigma_{s_x}^2 \cdot \sigma_{s_y}^2)^{1/2}$$

d.

$$\theta(\zeta) = \left[\theta_c^{b+3} - \frac{E_T}{\rho\lambda} \frac{b+3}{\hat{B}} \zeta \right]^{1/(b+3)} \tag{A6}$$

e.

$$T_g = T_{ga} - \frac{2N\theta_A \rho_A}{\alpha_A M_{n_A}} \left(\frac{1}{X + (1 - X) \frac{(\alpha_B)}{(\alpha_A)}} \right) \tag{10}$$

therefore they are often offset, with subscript preceding the superscript (*Ex. 6.27b,c*). When a subscript or superscript includes a fraction, the solidus is used (*Ex. 6.27c,d*). Every effort should be made to avoid a subscript to a superscript or subscript (*Ex. 6.27e*). A complex exponential expression should not be positioned as a superscript (*Ex. 6.28b*), but reduced to a simpler form typographically (*Ex. 6.28c*):

Ex. 6.28

a.

$$y(t) = \int\int m_0 (\chi, y) e^{-i(\gamma G/\Omega) (x\cos\phi + y\sin\phi)\sin\Omega i} e^{-i/T_2} dx\, dy. \tag{14}$$

b.

$$T = \exp(-Ld) + q [\exp(-Md) - \exp(-Ld)], \tag{15}$$

In complex equations, the equal sign (or other operational sign) becomes the point of reference for all terms in the equation and for both sides of the equation. The main symbols and numbers that are not part of a fraction, the lines of division of the first-order fractions, and the other operational signs of the main terms are aligned horizontally with the equal sign in an imaginary line that forms the main axis of the equation:

Ex. 6.29

a.

$$M_1 = \frac{2K_1 C}{1 + 2K_1 C} \left[1 - \frac{(1 - \delta_1)}{Z(1 + 2K_1 C)} \frac{c_2}{c_1} - 2M_2 \right] \tag{31a}$$

b.

$$K_{1/2} = \frac{\left[i\sqrt{\rho} \left(1 - \dfrac{\rho\chi^2}{4}\right)^{-1} \Sigma^{\mu\nu} \left(\chi_\rho \dfrac{\hat{c}}{\partial\chi^\nu} - \chi_\nu \dfrac{\hat{c}}{\partial\chi^\mu}\right) + \gamma^\nu\Pi_\nu + m - \dfrac{\sqrt{\rho}}{2} \right]}{2\,(m - \sqrt{\rho})} K_0 \qquad (3.1)$$

c.

$$\frac{dP}{dt} = \frac{\dfrac{q}{c} - \left(\chi_0 + \dfrac{v_f}{v_{fg}}\right)\dfrac{h_{fg}GA}{cM}}{\dfrac{Tv_{fg}}{h_{fg}}(1 - \Gamma)} \qquad (10)$$

If the numerator or denominator includes fractions, the line of division of these second-order fractions form a second line of reference, or a secondary axis, as in *Ex. 6.29b,c.* Here the numerator of the main fraction, the lines of division, pluses, and minuses and the terms not in fractions are all centered on the secondary horizontal axis. A further fractional order would set up a tertiary axis for the associated terms and symbols.

The equal sign is also the point of reference for the disposition of the equation. If an equation is too long for the width of the line or column, and the left-hand part of the equation is very short, the terms on the right-hand side of the equation are continued on subsequent lines to the right of the equal sign with breaks coming at the operational signs:

Ex. 6.30.

$$\begin{aligned} S = &- [(C - U)/h]^2/D - 4\,(C - U)\,U'\cos\theta/hD \\ &- [U'/(C - U) + (C - U)^2\cos\theta/hD]Y'/Y \\ &+ 2(C - U)^3U'/hD^2. \end{aligned} \qquad (15)$$

If the left-hand term takes up most of the width of the page or column, then the right-hand part of the equation may begin at the left-hand margin, usually indented to set it apart from the left-hand part and beginning with the equal sign or other operational sign:

Ex. 6.31

a.

$$\mathrm{Re}\left\{ \int_0^T y(t) \sum_{k=0}^{L} \alpha_{2k}\, e^{i\,\zeta k\,\Omega t} dt \right\}$$

$$= \int\int m_0\,(x, y)s(\chi\cos\phi \div y\sin\phi)\,dx\,dy. \qquad (15)$$

b.

$$\mathrm{Re}\left\{ \int_0^T y(t) \sum_{k=0}^{L} \alpha_{2k}\, e^{i\,\zeta k\,\Omega t} dt \right\} = \int\int\int m_0\,(x, y)s(R)$$

$$\times \delta\,(\chi\cos\phi + y\sin\phi - R)\,dR\,dx\,dy, \qquad (16)$$

Long equations may continue for several lines; then the left-hand margins of the successive lines are aligned (*Ex. 6.30*). Equations should be divided at logical breaks in the equation, such as operational signs; one should not divide terms subtended by a radical sign. When there are a series of transformations, in which the left-hand term remains the same, the left-hand term is not repeated, and the equal signs are aligned, with the right-hand terms following them:

Ex. 6.32

$$
\begin{aligned}
(C_n^k)^{-1} &= (n+1)\,\frac{\Gamma(k+1)\,\Gamma(n-k+1)}{\Gamma(n+2)} \\
&= (n+1)\,B(k,n-k) \\
&= (n+1)\int_0^1 \tau^k\,(1-\tau)^{n-k}\,d\tau
\end{aligned}
$$

6.10.3 Integration into Text

Text versus mathematics.   In a mathematical proof, except for labels, such as *theorem, lemma, proposition,* and *corollary,* and descriptive terms, such as "substituting" or "assuming," and occasional verbs, the equations constitute the text; they are the words and the sentences. This staccato form is not appropriate for a scientific paper. Even if a paper depends largely on mathematics, it addresses a scientific problem; it cannot be simply a collection of mathematical proofs and derivations. The mathematics is only the means, although it may be the chief means, of solving the problem and reporting the method of solving the problem with commentary and explanations. In setting the mathematics within a scientific framework, the writer must explain the purpose, importance, and consequences of the mathematical analysis or development, and direct the analysis to the scientific objectives. Therefore, whereas in mathematics, the words are subordinate to the mathematical operations, in the scientific paper, the verbal development is dominant and, the mathematical operations must be made part of the discourse.

Structural relation of equations to text.   An equation is actually a statement in highly symbolic form. The symbols and mathematical operators represent words or phrases. The equation $E = I \times R$ can be translated into the sentence, "Voltage (E) is equal to current (I) multiplied by resistance (R)." Because of this sentential structure, equations are part of the discourse and must be made an integral part of the text. They must be inserted at exactly the point in the text at which they can make their statement and be so ordered, punctuated, and syntactically incorporated as to be part of a sentence (*Appendix 6D–F*). Individual equations, as statements, can be integrated into the text as clauses, usually noun clauses (*Appendix 6D*). They commonly serve as the complement of a linking verb or object of a verb or preposition, or stand in apposition to a noun (*Appendix 6D*).

Although sentential, an equation in a scientific paper cannot function alone as a sentence, but must be part of the structure of the sentence. In *Ex. 6.33,*

Ex. 6.33

   a. The most general procedure would be to allow each submarket to have its own parameters, as shown in equation 6.

     a.

$$(6) \qquad P_{it} = \sum_{k=1}^{K} b_{ik} X_{ikt} + e_{it,}$$

where X_{ikt} is the value of the kth explanatory variable in the ith submarket during time t, b_{ik} is the impact of the kth variable on the ith submarket, and e_{it} is a random disturbance. [J. Market. Res. 22, 1985, 419]

   b. Moacanin and Simha [21] have shown that for large numbers of segments, the functions ø and λ are described by the following equations.

     b.

$$\phi \left(\frac{\upsilon}{kT} \right) = \ln \frac{V_O}{S_y} + \frac{V_O}{(1 - V_O)} \ln \frac{V_O}{S_y^2} \qquad (27)$$

where V_o is the volume fraction of holes in the structure, and S_y and f are given by the following relationships. [J. Macromol. Sci. B26, 1900, 119]

the first sentence of each paragraph ends with a period before the equation; it does not include the displayed equation, which is followed by the definitions, as though the equation was the main clause. The equation is in apposition to "equation (6)," or "by equations"; therefore, it should not be separated from it by a period, which can be replaced by a comma or colon. Similarly, equations should not be joined simply by connectives (as they are in mathematics):

Ex. 6.34

   a. With this approximation,

$$N_2 / N_1 = (\alpha' / 4\alpha'' \alpha_{tot}) \, [-\beta + (\beta^2 + 4\alpha B_{01} N_0 I_{01})^{1/2}]. \qquad (5)$$

   b. Also

$$v_i = \frac{1}{N_i} \int\limits_{-\infty}^{+\infty}\!\!\int\!\!\int wf \, dw_x \, dw_y \, dw_z \qquad (9)$$

To derive the linearized equation of motion, we. . . .

   c. From Eq. 15

$$\langle E_g^2 \rangle_z = [n_{air}^4 / (n_g^2 + k_g^2)^2] \, \langle E_{air}^2 \rangle_z \qquad (16)$$

The mean square field decreases by a factor of thirty million as we pass from air to gold.

In *Ex. 6.34,* as in *Ex. 6.33,* the equations are made to function as independent statements. Also, although operational signs ($=$, $<$, $\cong$, $\neq$) represent verbs, such as *equals, is equal to, is less than,* and so on, they cannot be used as though they were auxiliary verbs in a finite verb to mean "must equal," "may equal," "would equal," "should equal," and so on. To express such meaning, the sentence must be revised so that the meaning is in the sentence outside the equations.

Equations are usually referred to in the text by the label "equation," followed by the number. The label may be written out but it is usually abbreviated and may even be omitted, and it may be capitalized or not:

Ex. 6.35

a. The simplest way to obtain a dip in sensitivity at longer durations is . . . to limit the integral in Eq. (2) to the appropriate time.

b. It should be noted that Eq. [1] must be an elementary reaction if Eq. [2] and [3] are to hold rigorously in the context of chemical kinetics.

c. From Eq. 24, we have:. . . . Using these relations we can deduce from Eq. 25 the relations among the semi-invariants. . . .

d. A straightforward extension of eq. 2 gives. . . .

e. . . . with *r* being a reflection coefficient computed from the interface conditions given in (5d) and (5e). . . .

The numbers are usually in the form in which they appear with the displayed equation, usually in parentheses, rarely in brackets, but sometimes not.

Dangling participles.   Dangling participles are a common pitfall in incorporating text with equations because participles are a convenient way of bringing them into the text. Frequently, in mathematical material, sentences begin with participles like "using," "substituting," and "replacing." These relate to the writer, but as structured in the sentence, they modify the term on the left-hand side of the equation:

Ex. 6.36

a. Using Eq. 53 together with Eq. 58

$$\text{Var}\,[u(x)] = \frac{2\,F^2}{(E_0 A)^2}\,[R_{FF}(0) - R_{FF}(x)] \quad\dots\dots\dots\dots\dots\dots\dots\dots\dots\dots\dots\dots\dots\dots (65)$$

b. Substituting for impedance

$$\rho_\alpha = \frac{T}{2\pi\mu_O} \cdot \frac{1}{(\sigma_1 h_1)^2}$$

Such dangling participles can be corrected by converting the participle to a gerund and making it the subject of the sentence:

Ex. 6.37

a. Substituting equation 9 into 8 yields

$$(10) \qquad\qquad b_{ik} = B_{i0} + B_{i1}D_1 + B_{i2}D_2 + \dots + B_{iG-1}D_{G-1}$$

b. Using the definition (5.13), we obtain

$$f_1(t) = \lim_{\tau \to x} A(\tau)\eta$$
$$= \beta v\eta\,(\lambda + \mu + \alpha)\,[(\beta + v)D(0)] \qquad\qquad (5.16)$$

or providing the agent (*we*) to which the participle can apply (*Ex. 6.37b*).

Definitions.   For each equation, the writer must define symbols appearing for the first time. If they are few and brief, the definitions may be included in the line of text following the displayed equation:

Ex. 6.38

   a. where s is cell speed, τ is directional persistence time (step time), σ is step size (equal to $s\tau$), R is outer radius of space around target, and A is contact radius.

   b. where s = cell speed, τ = directional persistence time (step time), δ = step size (= $s\tau$), R = outer radius of space around target, and A = contact radius.

Then a linking verb is used (*Ex. 6.38a*), not an equal sign, as in *Ex. 6.38b*. When a quantity is provided for the symbol, however, the equal sign or other operational symbol is used:

Ex. 6.39

   a. where $\lambda \to 0$ is the linearization parameter,. . . .

   b. where S = $s(1/D)^{1/2}$ is the normalized interface recombination velocity, ...

   c. where $r = \pm 1, 2$ and $\epsilon > 0$, K > 0 do not depend on *n, c, d,* or *s.*

If the definitions are numerous or long, they may be listed and displayed; then the equal sign may replace the linking verb:

Ex. 6.40

> where
>
> i = 1 for the SiO_2 layer
> i = 2 for the $LiNbO_3$
> ϵ_1 = the permittivity tensor in the *i*th region
> ρ_1 = the charge density in the *i*th region and it is assumed to be zero.

When a paper introduces many symbols, they may be collected and defined at the beginning of the paper or in an appendix, often under a heading, such as *Notation, Nomenclature, Symbols:*

Ex. 6.41

	Nomenclature		
A	= amplitude of interface profile	*Superscript*	
B	= thickness of analysis section	( )*	= nondimensional
H	= length of analysis section		
k	= conductivity	*Subscripts*	
n	= normal to interface	i	= interface
R	= resistance	TⳄ	= total
δR	= incremental resistance, = $R_{\text{1-D}} - R_{\text{TOT}}$	1-D	= one-dimensional
T	= temperature	1,2	= material identi-
W	= width of analysis section		fiers
x,y	= Cartesian coordinates		

[J. Spacecr. 23, 1988, 225]

These may be arranged alphabetically, by order of appearance in the text, or in related groups.

Punctuation.   The punctuation of equations and other mathematical material is particularly important because it helps to structure them into the sentence. However, it is often inconsistent or follows a rote consistency that ignores syntax. The inconsistency is greatest with displayed equations. Their punctuation varies from no punctuation, to close punctuation of both equations and text (*Appendix 6E*). If one recalls that an equation is sentential and functions as a clause, and that symbols and terms function as nouns or phrases, it becomes clear that one can punctuate sentences with mathematical material just as one does any sentence. The question that arises is how to interpret the blank space before and after the displayed equation. One can consider that the blank space effects the separation or break that some marks of punctuation provide and so makes such punctuation unnecessary (*Appendix 6E.1*). On the other hand, some writers apparently consider the break so great that it calls for punctuation; consequently the break may be unnecessarily or heavily punctuated (*Appendix 6E.2*).

One can take the middle path on two grounds: (1) the effect of the blank space around displayed equations is to set off the equation and so it does not require punctuation, and (2) where blank space can be used to signify separation, the omission of punctuation eliminates one more type of symbol that readers must interpret. On this basis punctuation can be omitted after displayed equations, except for the period, to mark the end of a sentence, which is not readily recognized without a period.

For punctuation before displayed equations one must consider the structure of the sentence. As the object of a verb or preposition, the equation functions as a noun clause and should, of course, not be separated from the verb or preposition by any punctuation (*Appendix 6D.a*). When an equation is in apposition to the subject, object, or any noun, but is included in the running text, the customary comma separates the equation from the associated noun (*Ex. 6.42a,b*),

Ex. 6.42

    a. Here γ_2 is the phase relaxation constant, $\gamma_2\ \gamma_2 + \gamma_s$, and. . . .

    b. Two values for α, $\alpha = 10^{-9}$ cm^2 and $\alpha = 10^{-10}$ cm^2, were used for the calculation.

    c. The simplest Beltrami flow in the family (3) has only two Fourier components with the wave vectors $\pm\ \mathbf{Q}_0 = (\pm 1,0,0)$ and amplitudes $A(\pm\ \mathbf{Q}_0) = c/2$; $v^l = c\ (0,\cos x/l,\ -\ \sin x/l)$. A hyperscale perturbation with the wave vector $(0,k,0)$ then obeys the equation of motion

but the equation may be treated restrictively and then is not separated from its noun by punctuation (*Ex. 6.42c*). When the equation in apposition is displayed, the comma preceding the equation is often omitted, the space functioning as the separating comma.

The colon may be used instead of a comma to separate the displayed equation from the noun in apposition, but it is usually too heavy a mark of punctuation for this purpose. It is appropriately used as an anticipatory or annunciative mark of punctuation and therefore before equations preceded by phraseology that leads readers to expect an equation:

Ex. 6.43

   a. To do this, the following transformation is required:

   b. In this way, other useful derivatives may be derived:

The colon (or comma) should not be used regularly at the end of the text preceding a displayed equation, especially if it separates a verb (*Ex. 6.44a,b*) or preposition (*Ex. 6.44c,d*) from its object (or complement); indeed no punctuation should separate verbs and prepositions from their objects. In the following sentences introducing displayed equations, the terminal punctuation is more appropriately omitted:

Ex. 6.44

   a. Given the inelastic labor supply assumption, national labor supply is,

   b. The total peak variance in system E then becomes:

   c. The variances of chromatographic bands can be expressed as:

   d. During the period when pressure varies, the mole fractions before and after the period are related by,

When displayed equations follow conjunctive adverbs, such as, *namely*, or introductory words or phrases such as, *that is,* they are preceded by a comma:

Ex. 6.45

   a. When the point A coincides with the center of mass, Eq. (4) and (5) reduce to the well-known relationship between torque about the center of mass and torque about 0, namely,

   b. Since II is the momentum operator, it is expected to be the generator of translations, i.e.,

Symbols or terms are treated like words in punctuating a sentence. They function as nouns and so may serve as subject, object, or appositives. Most of the symbols used in the examples in *Appendix 6F* are in apposition, but included are symbols that function as subject, object of verb or preposition, and even as an adjective, as in "the b_{ij} values". Whenever symbols or terms are defined following "where," they serve as the subject of the defining statement. When symbols or simple terms stand in apposition, they are often treated restrictively, with no punctuation (*Appendix 6F*). Commas are seen as too heavy for simple symbols, which already stand out from the text because of the italic type. In a series in which intermediate elements are omitted, ellipsis dots are used, and a comma (or operational sign) follows the last element before the dots:

Ex. 6.46

$$t_1, t_2, t_3, \cdots t_n \qquad t_1 + t_2 + t_3 + \cdots t_n$$

6.10.4 Preparing Equations for Publication

Equations are always costly to set in type. They are much less familiar to typists, copy editors, compositors, and proofreaders than words. If in addition the terms or equations are written carelessly, errors are inevitable and likely to be serious. The displacement of a symbol or the misinterpretation of an illegible symbol may change the meaning radically. Therefore, they must be prepared with finical care. The disposition of every letter, number, symbol, and line must be unequivocally clear.

First, equations should be typed; if symbols are not available for typing, they should be carefully printed. Displayed equations are typed with additional space above and below them. All elements should be very carefully aligned with respect to other elements. Superscripts and subscripts should be as simple as possible, systematic, and unambiguous. Though equations are set in italics, the writer does not underline them. The editor will mark them or instruct the compositor about them. If a symbol is *not* to be in italics, the writer can mark them—by circling them and noting the typeface in the margin. Special care must be used with symbols that can be confused, such as, zero and the lowercase o, lowercase x and Greek chi (*see Table 10.2*). If two symbols are similar enough that they may be confused, they must be examined carefully whenever they appear, and marked if necessary.

6.10.5 Verbal and Chemical Equations

Other types of equations may be used for particular purposes and in particular disciplines. In the biological and social sciences, simplified word equations or formulas may be used (*Appendix 6G.a*). In chemical studies, the equations consist largely of symbols of the elements and numbers. The symbols are one- or two-letter symbols, which are internationally used, as are the conventions for indicating the number of atoms and molecules, molecular structure, ionic form, valence, types of reactions, and so on in the equation.

A chemical equation shows the elements and compounds reacting with one another and the products of the reaction (*Appendix 6G.b*). The elements and compounds are shown in chemical symbols with subscripts indicating the number of atoms of each element in a compound. If more than one molecule of a compound takes part in the reaction, the number of molecules is placed before the compound. A plus sign is used between the reacting and reacted elements and compounds, and arrows indicate the directions of the reactions. The symbols of the elements in a compound, and bonding lines are typed without spacing, but the reaction arrows and plus signs are set off with a space on each side. Arrows may have notes above them about the reaction, such as catalyst or a temperature, and others. For complex organic compounds and their reactions, the structure of the molecule may be drawn in diagrammatic form with lines to show the most important chemical bonds (*Appendix 6G.b*). The diagram does not necessarily display the complete structure or the complete reaction, but only such parts as are necessary to make the pertinent structure and reaction clear to readers. Biochemical equations are more verbal, the names of the compounds being used rather than their chemical composition or structure (*Appendix 6G.a*)

Appendix 6A

Types of Formats for Footnotes to Tables

1. _____

SOURCE: Adapted from Meyer et al., 1983.

2. _____

Note: Data are for 1980 sales and pertain only to nonassociated gas; they do not include take-or-pays for associated (casinghead) gas.
Source: EIA (1983, *ix*).

3. **Total**   27.8   6.8   34.5

*The countries classified as small debtors are those for which debt represented less than 20% of GDP in 1982. They are: Barbados, El Salvador, Guatemala, Guyana, Haiti, Honduras, Jamaica, Paraguay, and the Dominican Republic.
Sources: Bank debt calculated on the basis of Bank for International Settlements data; Gross Domestic Product taken from Inter-American Development Bank data.

4. _____

*p $\leq$ .05.
**p $\leq$ .01.
***p $\leq$ .001.

5. _____

*,**Significant at $P \leq 0.05$ and 0.01, respectively.
†Days entered in this analysis include 0, 2, and 4 d after the second application.
‡Days entered in this analysis include 8 and 32 d after the second application.

6. _____

1. *Beijing Review,* Vol. 23, No. 16, April 21, 1980, p. 27.
2. Ullman (1961).
3. Yao et al. (1981).
4. Includes 2,000 county towns with population from 10,000–50,000.
5. Including agricultural population residing in suburbs.

7. _____

[1]Values are means $\pm$ SEM; ($n = 12$). [2]In each column means not sharing a common superscript letter are significantly different ($P > 0.05$). [3]For composition of the diets see table 1. [4]Cytosol (20000 $\times$ g, 30 min, supernatant fraction of rat liver homogenate). [5] Resuspended mitochondria (8500 $\times$ g, 15 min, pellet from the 500 $\times$ g, 10 min, supernatant fraction).

8. _____

[a]Average (standard deviation) image contrast for 21 different acquisitions at this count level for given size sphere.
[b]Cutoff frequency of Shepp-Logan filter as multiple of Nyquist frequency for 4-mm pixel size.
[c]Metz-1 is a Metz filter reported previously.[3]
[d]Metz-2 is Metz optimized for acquisition parameters of phantom.
[e]Significantly different ($p < 0.05$) from ramp filter.

Appendix 6B

Titles and Legends of Figures

TYPES OF TITLES AND LEGENDS

a. *Single-column legends*

Figure 3. Length Frequency for Combined Species (brook and brown trout), Tenmile Creek, Preconstruction and Postconstruction.

Figure 4. Temperature dependence of the g-tensor components of Ni-(pc) Br.

FIGURE 2

Conformation of cyclic(LPro-Gly-LPro-LSer-DAla) *III* with the inverse chiral sequence LDLLD (4). The β-bend is Type II.

Fig. 5 Effect of Actinomycin D or Cycloheximide on the Rhatannin-Stimulated Increase in Glycine Formation.

Rats fasted for 24 h were administered actinomycin D, cycloheximide, or saline 30 min prior to rhatannin treatment, and the glycine formation rate was estimated 4 h after the rhatannin treatment.

Data are expressed as means ± S.D. of 5 rats. *a)* $p < 0.005$, student's *t* test.

☐ C, control rats; ■ R, rhatannin-treated rats.

b. *Two-column legends*

Fig. 2. Plasma corticosterone and plasma testosterone in male alligators from Louisiana (LA 1 and LA 2). North Carolina (NC), South Carolina (SC), and south Florida (FLA). The solid bars represent the means = SEM of testosterone concentrations in each of the group and the open bars represent the corticosterone concentrations. Numbers in parentheses under each group indicates sample size.

Fig. 14. Experimental system for the demonstration of passive synthetic aperture imaging concept.

Fig. 6—Plot of number of beads (N) in the annulus as a function of segregaton time (t) 0.2 m ID cylinder and 2.4:1 particle size ratio.

Fig. 1. Body (○), liver (△) and gastrocnemius (☐) weights compared with AH-130 tumour growth (●) in rats

Body and tissue wet weights in tumour bearers are expressed as percentages of control values (see Table 1). Vertical bars denote S.D. ($n = 6$).

FIG. 2 Body weight changes in rats given sequential or continuous TPN for 10 days. _____, Sequential A TPA (SA), ·····, Sequential B TPN (SB); -----, Continuous TPN (C).

FIG. 1. Experimental set-up for automated TELISA. (1) Autosampler, (2) timer with valve (A, injection timer; B, cycle timer), (3) peristaltic pump, (4) PBS reservoir,(5) waste, (6) enzyme thermistor. (7) wheatstone bridge, (8) recorder.

c. *Page-width legends*

Figure 1.   Size and sales of farms by full-time off-farm work of operators and wives (retired and disabled farmers excluded). N = 98

Figure 5. Temperature Reponse Functions for Georgia; (a) $\lambda = 0$ (unrestricted ordinary least squares); (b) $\lambda = 1,000$; (c) $\lambda = 10,000$; (d) $\lambda = 126,936$ (optimum).

Figure 2
Relapse Rates in Controlled Comparisons of Inpatient Versus Outpatient Settings

Fig. 1: Fertilization in caged plants of Winter CMS manually pollinated in 1983, according to ovular position; 1 = stigmatic end of the ovary.

FIG. 3. Composite portrayal of grazing and survival for *Bromus tectorum* populations exposed to six grazing treatments (a-f) during 1980–81. Details in Fig. 2.

FIG. 1.—Mean heart period at 5½ years for children classified as inhibited and not inhibited at 21 months.

LEGENDS DESCRIBING PARTS OF FIGURES

1.   FIG. 2. (a) Γ-centered hole octahedron; (b) *N*-centered ellipsoids and the JG surface [after Mattheis (Ref. 17)].

2.   *Fig. 3. Schematic drawing of method of measurement. EL = electrical latency, ML = mechanical latency and CT = contraction time. Excitation-contraction latency is calculated as difference between ML and EL.*

3.   FIG. 10. Deconvolution of the excess heat capacity functions of plasminogen fragment K1-3 in solutions with different pH values: (a) pH 7.4, (b) pH 5.4, (c) pH 4.0, and (d) pH 3.4. Crosses indicate the experimentally determined function. . . .

4.   Fig. 8.   Effects of hyperpolarizing current application on H-CV in 5 mM (A) and 10mM (B) KCl solutions. Records from *E. nucleofilum.* In experiments shown in this and the next figure, current was applied through a second electrode. . . .

5.   FIG. 1. Schematic drawings showing the modification of a nocturnal stable layer growth by subsidence. The initial potential temperature sounding [solid line in (a), dotted line in (b) and (c)] grows by turbulent, radiative, and possibly advective processes in the absence of subsidence [solid line in (b), curved dashed line in (c)]. Subsidence, however, causes the growths to be slower, as indicated by the solid line in (c). The vertical dashed lines indicating the initial mixed-layer potential temperature at the start of the evening. The vertical line labeled v_0 indicates the reference sounding as modified from the earlier mixed layer sounding by uniform, background radiational cooling. The temperature jump at the surface, Δv_s, is also shown.

Appendix 6C

Graphic Symbols and Diacritical Marks Used in Equations

Mathematical operators

+	plus	·	centered dot	∂	partial derivative
−	minus	/	slash solidus	√	radical
×	multiplication sign	∫	integral	Π	product symbol
±	plus or minus	Σ	summation	∇	del, vector operator

Mathematical signs of relations

=	equals	≧	greater than or less than	::	as
	double bond	∝	proportional to	→	approaches, tends to
≠	not equal to	~	approximating, on the	⇄	in both directions
≡	identically equal to		order of magnitude of	⇌	right arrow
≃	approximately equal to	⊂	included in		predominating
>	greater than	∈	element of	↑	giving off gas
≪	much less than	⊥	perpendicular to	↓	precipitating
≦	less than or equal to	∥	parallel with	⇔	if and only if
≮	not less than			⇐	is implied by

Mathematical symbols of inclusion

( )	parentheses	⦃	barred braces	⟨ ⟩	angular brackets
[ ]	brackets		vertical bars	( )	bold parentheses
{ }	braces	‖	double vertical	[ ]	bold brackets

Diacritical Marks

Above letter

´	acute accent	^	circumflex	‾	macron
`	grave accent	ˇ	inverted circumflex	.	dot
..	umlaut, diaeresis	~	tilde	˘	breve

Below letter

₃	cedilla
₅	inverted cedilla

Above or below letter

°	circle

Beside letter

′	prime
″	double prime

Appendix 6D

Syntactic Functions of Displayed Equations in Sentences

a. *Object of verb*

1. Thus (14) and (15) both yield

$$dK_{0S}/dx \cong -36 \text{ GPa} \qquad (16)$$

for wüstite, which agrees well with the observed dependency (4).

2. For a waveguide we may write

$$\omega^2/c^2 = k^2 + k_\perp^2, \qquad (1)$$

where $k_\perp$ is the transverse wave number.

b. *Complement of linking verb*

1. Thus the disturbance-velocity component in the x_3 direction is

$$\mathbf{x}_3 \cdot \mathbf{V}_1 = f_1 = f(x_3)\exp[i(\omega t - kx_1)]. \qquad (6)$$

2. . . . the n_2 medium. These are

$$t_{12}t_{21} + r_{12}^2 = 1, \qquad (5)$$
$$r_{11} = -r_{21}. \qquad (6)$$

c. *Object of preposition*

1. *B. Time Evolution and L Eigendistributions.*
The time evolution in Hilbert space is formally
given by

$$L|\rho(t)) = i\,\partial|\rho(t))/\partial t. \qquad (4)$$

2. We next note that if the coefficient of exp $(s_2 t)$
is zero for some choice of λ, μ and χ, then (5.26)
reduces to

$$h_{sty}(t = (h_1(x))^2[1 + \exp(s_1 t)]. \qquad (5.27)$$

d. *In apposition to noun*

1. If we make the following substitution,

$$\mu = s^2\,\tau/4, \qquad (4)$$

the effects of cell speed and persistence
time. . . .

2. . . . these carcass components were modeled by
the Gompertz equation (Ricklefs 1968) of the following form:

$$W(t) = A\cdot\exp(-b\cdot\exp(-kt)),$$

where A is the asymptotic weight, b is the *ln*. . . .

e. *Part of noun clause*

1. Use of the boundary conditions reveals that

$$D = -2(i - \Delta)/[(i - \Delta)\Omega]^{1/2}K_0([(i - \Delta)\Omega]^{1/2}) \qquad (47)$$

2. For large N, it therefore follows from Eq. (22) that

$$M_1 \equiv (m_1)/N = -(f_1/z_1)(\partial z_1/\partial f_1) \qquad (24a)$$

f. *Part of adverbial clause*

Thus, lasing at 2.8 μm occurs when

$$a'aB_{01}N_0I_{01}/(2a''a_{tot}\beta) > 1 \quad \text{if} \quad \beta^2 > 4aB_{01}N_0I_{01} \qquad (6)$$

Appendix 6E

Punctuation of Displayed Equations

1. *Equations not punctuated*

A. Definitions of the Basic Quantities

The algebraic properties of the golden mean trajectory are rooted in the properties of
Fibonacci numbers and of the golden mean itself.[5] Just as the Fibonacci numbers obey the
recursion relation (1.4) so the golden mean, $-g$, obeys

$$g^{n-} = g^n + g^{n-1} \qquad (2.1)$$

and the two are related by

$$gF_n = -F_{n-1} + g^n \tag{2.2}$$

High iterates of the original mapping function (1.1) may be defined recursively by giving a quantity involving F_n iterations by

$$f_{n-1} = f_n f_{n-1} \tag{2.3}$$

with the initial conditions

$$f_0(0) = 0 - 1 \tag{2.4}$$
$$f_1(0) = f(0)$$

The periodicity condition $f(0 - 1) = f(0) - 1$, then ensures that the entire set of f_1 forms a set of functions which commutes under the recursion operation

$$f_n f_m = f_m f_n \tag{2.5}$$

The basic quantity $u_{n,j}$ is then given by

$$u_{n,j} = f_{n\,i\,i}(0_i) - 0_i \tag{2.6}$$

The value of the bare winding number Ω, is chosen to get the actual winding number to be the golden mean and thereby ensure that $u_{n,j}$ goes to zero as $n \to \infty$.

[J. Statist. Phys. 43, 1986, 397; with permission]

2. *Equations closely punctuated*

The development of temperature and electric field perturbations in a superconductor is described by a system of electromagnetic (Maxwell) and heat equations. In a linear approximation, this system has the form,

$$v \frac{\partial}{\partial t}(\Delta T) = \kappa\nabla^2(\Delta T) + j_c E,$$

$$\text{curlcurl}E = -\frac{4\pi}{c^2}\frac{\partial j}{\partial t}, \tag{4.1}$$

$$j = j_c + \sigma E,$$

where v, κ, j_c, σ represent either the values of respective quantities in a hard superconductor or the averaged values of these parameters in a composite. We now write ΔT and E as,

$$T = T_c v(x/b)\exp(\lambda t/t_k), \tag{4.2}$$
$$E = \frac{\kappa T}{j_c b^2}\,\epsilon(x/b)\exp(\lambda t/t_k),$$

where λ is the eigenvalue of the problem to be found. One can easily derive from the original equations,

$$\lambda v = \nabla^2 v + \epsilon, \tag{4.3}$$
$$\text{curlcurl}\epsilon = \lambda\beta ev - \lambda\tau\epsilon,$$

where e is a unit vector in the direction of ϵ.

Boundary conditions should be imposed on Eqs. (4.3). The thermal boundary conditions in a linear approximation have the form,

$$W_0 v - \frac{\kappa}{b}\,(\mathrm{n}\cdot\nabla v)=0,\tag{4.4}$$

where n is normal to the sample surface.

[Rev. Mod. Phys. 53, 1981, 560; with permission]

Appendix 6F

Syntactic Treatment and Punctuation of Mathematical Elements Within Text

1. The cluster galaxy density then falls to zero at a radius R_h given by $F_{\mathrm{isot}}(R_h, \beta) = C/6.06$. Bahcall (1973a) also defines a modified central surface density parameter $\alpha \equiv \sigma/(6.06-C)$; then, for small $C \ll 2\pi$, the central volume density is just $n_0 \approx \sigma_0[1-(C/2\pi)^2]/(6.06\ \beta) \approx \alpha/\beta$. [Rev. Mod. Phys. 58. 1986, 13]

2. These two effects are represented quantitatively by the passive change $(\lambda_i^!)$ and the active change $(\sum_{n=0}^{1-c^c} \Theta_i^!\ (1-n))$. The passive change equals the sum of the products of unit passive responses, b_{ij}, and the drawdowns at the beginning of the period, $s_j^!$. (see Equation 3). The active change equals the sum of the products of the unit active responses $(a_{ij}(1-n))$ and the applied stresses $(x_j^!(n))$ (see Equation 2). Both the $a_{ij}(1-n)$ and the b_{ij} values are extracted from the original set of. . . . The elements of matrices $A(\tau)$ for $\tau \leq 1$ are included as the unit active responses, $a_{ij}(1-n)$. . . . [Wat. Res. Bull. 22, 1986, 419]

3. Being interested also in situations far from thermal equilibrium, we consider the (nonlinear) response to finite changes Δe of an external parameter $e(t)$. Assuming that the system is in equilibrium for $t'0$ (where $e(t) = e + \Delta e e^{\mathrm{i}\,\bar{\mathrm{q}}\cdot\bar{\mathrm{x}}}]$, we describe the relaxation for $t>0$ [where $e(t) =. e$] by a nonequilibrium relaxation function[29] [Phys. Rev. B 12, 1975, 5263]

4. The only activity of the government in this economy is to supply money, injected as lump-sum transfers, and the money growth factor in any period t is a fixed function $g(s)$ of the shock.[4] Therefore, if $\hat{m}_{t-1}$ is per capita money in circulation in $t-1$, an agent who carries overnight balances plus invoices of m_{t-1} will have post-transfer balances in t of $m_t = m_{t-1} + [g(s_t) - 1]\hat{m}_{t-1}$. Throughout the paper, we will normalize per capita money balances to be unity: $\hat{m}_{t-1} = 1$. [Econometr. 55, 1987, 494]

Appendix 6G

Verbal and Chemical Formulas and Equations

a. Verbal Formulas

1.

Calculations. The portal appearance and net hepatic release of blood metabolites were calculated as described by Bergman (4) by using the equations reproduced below:

$$\text{Net portal appearance} = F_P(P - A) \tag{1}$$

$$\text{Net hepatic release} = F_P(H - P) + F_A(H - A) \tag{2}$$

$$F_A = F_H - F_P \tag{3}$$

2.

$$GI = \Sigma \frac{(\text{no. seed newly germinated})(i\text{th half day})}{\text{total seeds germinated by } i\text{th half day}}$$

The germination index was calculated for the first 10 seeds. . . .

3.

$$\text{UDP-glucose} + \text{D-fructose 6-phosphate} \overset{\text{SPS}}{\rightleftharpoons} \text{sucrose-p} + \text{UDP}$$

$$\text{UDP-glucose} + \text{D-fructose} \overset{\text{SS}}{\rightleftharpoons} \text{sucrose} + \text{UDP}$$

4.

The lock-in variable is defined as

$$\text{lock-in:} = \frac{\text{face value}^1 - \text{market value}^1}{\text{initial principal amount}^1 x \; \dfrac{P_1}{P_0}}$$

b. Chemical Equations

1.
$$4FeSO_4 + 6H_2O + O_2 \rightarrow 4FeO(OH) + 8H^- + 4SO_4^{2-}$$

2.
$$Cp^*_2ScCH_3 + HC \equiv CCH_3 \xrightarrow[\text{min}]{<0°C} Cp^*_2ScC \equiv CCH_3$$
$$\quad\quad \textbf{2} \quad\quad\quad\quad\quad\quad\quad\quad\quad\quad\quad\quad \textbf{16}$$

$$\tag{30}$$

3.

4.

7

Results and Discussion

7.1 RESULTS

7.1.1 Nature of Results

The results of research are the substance of science; they are the real objective of scientific research. It is in the results that the painstaking efforts to design rigorous methods and to achieve accuracy and precision in the procedures, observations, and measurements come to fruition.

The new findings are reported in the results section of the paper. This is usually called "Results," but the different types of results may be presented under separate headings:

Ex. 7.1

Experimental Results	Solutions	Results of Laboratory Studies
Computer Results	Experimental	Results of Analysis
Numerical Results	Observations	

In experimental papers, the results section may be relatively short compared to the methods and discussion section (see also *Ex. 7.5*):

Ex. 7.2

RESULTS

In the first experiment, the highest acidity (pH 2.6) significantly reduced the mass of hypocotyls, but there were no significant effects of acidity on shoots (Table 2). There also were no significant effects of anions on the mass of either shoots or hypocotyls, nor was there an interaction between acidity and anions. Both linear and quadratic terms in the dose-response functions for effects of acidity on hypocotyls were significant.

In the second experiment, simulated rain at pH 3.0 and 3.4 reduced the mass of hypocotyls compared to pH 5.0, but there were no significant effects of acidity on shoots (Table 3). Anion composition of simulated rain did not significantly affect mass of shoots or hypocotyls nor was there a significant interaction between acidity and anions. The same results were found in harvests 1 and 2 although the effects of acidity on mass of shoots was close to being significant at $a = 0.05$ in harvest 1.

Generally, hypocotyls were more susceptible to effects of acidity than shoots (Tables 2 and 3) and effects tended to become less pronounced when plants were given a recovery period after treatment (harvest 2, Table 3). Daily exposures of simulated acidic rain during the period of rapid hypocotyl expansion appeared to have a slightly greater effect than three exposures per week beginning with the seedling stage (Table 2 vs. Table 3). Anions had no effect on dry mass, alone or in combination with acidity, either in the first or second experiment. [J. Envir. Qual. 15, 1986, 303].

Ex. 7.2 is a complete results section. In descriptive papers, however, the section on results or observations may be very long.

Writers may include methods in the results section. They may describe the materials and particular procedures in the methods section, then the experimental methods or protocol in the results section (*Table 7.1*). In some results sections, the methods are described incidental to the results and may fall into a narrative form:

Ex. 7.3

RESULTS

In a survey of the ability of various cell-free and broken cell preparations of pseudoplexaurid–derived zooxanthellae to form squalene, first indications of success came on trial of a breaking medium with maleate as a buffer and the use of a Hughes press for cell breakage, see Table 1. The source of zooxanthellae in this case was *Pseudoplexaura wagenaari*. The other pseudoplexaurids, *P. flagellosa* and *P. porosa* also provided active preparations when their zooxanthellae were similarly treated. Since *P. porosa* typically yields more zooxanthellae and is common and easily identified in the field, its zooxanthellae were routinely employed in subsequent studies. Regardless of the source of zooxanthellae, maximally active preparations required both NaCL and KCI in the breaking medium. Breaking media employing buffers other than maleate gave less active or inactive preparations. This is not solely a pH effect since both medium III (MES buffer) and medium IV (phosphate) are at the pH of the maleate medium. Reduced activity is in part the consequence of inadequate breakage or of an inhibition of the squalene synthetic activity since transfer from breaking medium V (tris) to a maleate incubation medium results in moderate activity. The activity survived freezing whether frozen as intact zooxanthellae or as the broken maleate preparation, but with substantial loss over the period of a week. Consequently, only fresh preparations were employed. [Comp. Biochem. Physiol. 81B, 1985, 425]

Such a mixed treatment may also include some discussion and result in a combination of methods, results, and discussion, which are inextricably interwoven (*Ex. 7.3,*

TABLE 7.1   Results sections that include methods and discussion

a. *With Methods*

RESULTS

Overview

Accurately recalled (macro)propositions within
each experimental condition were pooled item-
by-item, and actual reproduction frequencies
calculated and then compared with macro- and
micropropositional models, and (over a sample
of the text with various other strategic models.
[*methods*]

Comparisons between the various models of recall and actual recall, and estimates of the two
statistical parameters; and $\hat{m}$ were made using
the GLIM program with stepwise procedures
(Baker & Nelder, 1978), which for a given set
of experimental frequencies (a) seeks to minimize the χ^2 values between predicted and actual recall frequencies for each parameter
value, and (b) calculates the correlation (r) between actual and predicted recall. [*methods*]

Other tests of the relationship between actual
and predicted recall were (a) comparisons between the first and second halves of the text,
and (b) comparisons over six categories of information. [*methods*]

*Full Model: Micro- and
Macropropositional
Processing Strategies*

Actual reproduction frequencies were compared
with both micro- and macropropositional
models. [*methods*] Analysis revealed that the
micromodel accounted for between 10 and
17% of actual recall variance, whereas the
macromodel accounted for. . . . [J. Mem.
Lang. 25, 1986, 302]

b. *With Methods and Discussion*

RESULTS

A series of experiments was performed to determine the resistance of brain cells to oxygen
deprivation. [*methods*] Incubation of confluent
neuroblastoma cells in a 95% nitrogen/5%
CO_2 atmosphere or in the presence of antimy-
cin (1 μg/mg protein) up to 24 h did not result
in degeneration, at least at the morphological
level as inspected by light microscopy.

However, an adequate supply of glucose proved
very important during conditions of limited
oxygen supply, since otherwise neuroblastoma
cells showed significant morphological alterations: they lost their neurites and became detached from the substrate and released lactate
dehydrogenase and protein into the medium
(Fig. 1).

Anoxia experiments were also set up with 7-day
in vitro cultured neurons. [*methods*] In contrast with neuroblastoma cells, neurons were
very sensitive to a lack of oxygen when incubated in DMEM supplemented with 10% fetal
calf serum. . . . Subsequently, the experiments
were repeated in the absence of fetal calf
serum. [*methods*] In contrast with the above
results, neurons did not show any morphological sign of degeneration indicating a toxic effect of fetal calf serum during anoxia. This
experiment also rules out a toxic effect of anti-
mycin other than inhibition of respiration. Because anoxic cells depend entirely on glycolysis
for ATP production, a difference in Pasteur
effect could be the explanation for rapid cell
degeneration. [*discussion*] Indeed. . . .

. . . .

Glycolytic activity was also measured in each cell
type under various glucose and oxygen concentrations. This was studied as follows: cells
after being washed twice were incubated in
serum-free DMEM . . . [*methods*] Under nor-
moxic conditions, glycolytic activity was much
higher in confluent astrocytes than in confluent neuroblastoma cells or differentiated
neurons (Table 1). In each cell type,. . . .

. . . .

DISCUSSION

Cultured neuroblastoma cells, cerebellar neurons, and cerebral astrocytes are characterized
by an active production of lactate; the astro-
cyte cultures especially showed a high rate of
lactate production (1.2 μmol/mg protein/h).
[*results*]. . . . [Pauwels et al. J. Neurochem. 44,
1985, 143–45, with permission]

Table 7.1b). When it is important to keep methods close to results, the protocol can be described separately and placed at the end of the material and methods section or as a last resort, at the beginning of the results section. When the protocol calls for alternating procedures and results, these can be treated like a series of experiments (*Table 3.7*). Such a section should be given a heading that clearly designates the inclusion of both methods and results.

The results consist of the observations and measurements recorded while conducting the procedures described in the methods section. The measurements can be as precisely quantitative as the development of theory, methods, and instrumentation in the discipline permits. The measurements may be direct observations and measurements; more commonly they are made through the medium of an instrument and so are indirect. The results are often in numerical form and may be converted to values of interest by various computations, statistical analyses, and other mathematical procedures. The numerical results are often more effectively presented graphically in tables and graphs, than in the written text.

Observations may not be quantifiable; then the results are described verbally, as in the description of structures (*Ex. 7.16*), species (*Ex. 7.4, 3.22*), minerals, and so on. In such descriptions special efforts are made to approach an objective, quantitative form, and words assume a quasi-quantitative role. For example, *pubescent, hirsute, hispid, hairy, puberulent,* and *hoary* describe types of hairiness on plant structures that are scientifically distinct characteristics, even though not readily measured. Furthermore, such detailed scientific descriptions are made rigorous by the use of a conventional format (*Ex. 3.22*):

Ex. 7.4

Pseudocercospora urariicola sp.nov. (Figs 4,5)

. . . *Mycelium* internal and external. Primary mycelium internal: hyphae colourless or very dilute olivaceous, 2–4 µm wide. . . . Secondary mycelium hypophyllous, external, copious: hyphae emerging through a stoma in small fascicles of up to 6, assurgent, smooth, septate, repeatedly branched, up to 130 µm long, 2·5–3 µm wide and pale olivaceous, becoming somewhat darker and wider and prolonged into the conidiophores. *Conidiophores* assurgent, moderate olivaceous, smooth, septate, repeatedly branched and intertangled, up to 325 µm long or more, 4 µm (here and there 5–6·5 µm) wide. . . . *Conidia* moderate or pale olivaceous, the shorter ones usually deeper in colour than the longer ones, cylindrical or obclavate-cylindrical, straight or slightly curved or sigmoid, smooth, 1–8 septate, not constricted, 25–90 × 4–5 µm. [Trans. Br. mycol. Soc. 88, 1987, 371]

The part is named and the various characteristics of the part follow, so that there is a sequence of nouns each followed by one or more attributive adjectives. Such descriptions may consequently be greatly contracted, and reduced almost to an itemization of characteristics. They illustrate the declarative type of description appropriate for scientific description (see Chapter 3).

7.1.2 Reporting of Results

Accuracy.   In reporting the results, the writer's overriding objective is accuracy, because the results are the primary and permanent source of scientific knowledge. The centrality of accuracy emphasizes the importance of reporting results separately from methods or discussion.

Selection and structuring of results.   The objectives in reporting results are: (1) to include all the findings in full and (2) to present them in a systematic order. The writer selects for tables and figures those data that cannot be presented as effectively in written form. Some data can only be presented in graphic form, for example, a large set of measurements or values, or records from automatic recording devices or a computer, which cannot be adequately described in the text, or even adequately drawn (*Fig 7.1*).

However, the results section should not be a collection of tables and figures connected by a thread of text. The tables and figures are important as records of the data, but they cannot provide an exposition of the data they contain. Their function is to illustrate the written text. Although they constitute the substance of the presentation, they cannot substitute for it. Therefore, the data must be presented as part of the development of the argument.

Data are then selected from the graphics for inclusion in the written text. The text should not simply repeat much of the data presented in graphics. After all, one

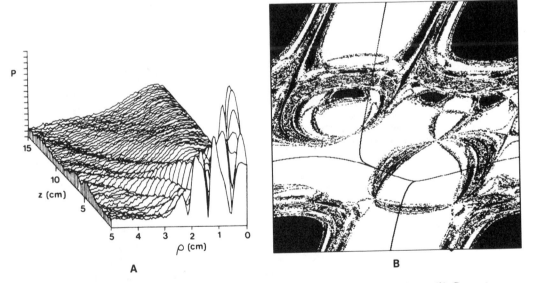

Figure 7.1   Data requiring direct reproduction. (*a*) Graph drawn by plotter (*b*) Computer graphic of fractal. [(*a*) J. Acoust. Soc. Am. 80, 1986, 8; (*b*) J. S. Thorp and S. A. Nagavi Proc. IEEE Conf. Dec. Cont., 1989, 1824 © 1989 IEEE; with permission]

purpose of the graphics is to avoid paragraphs glutted with numerals or descriptive details. However, it is not enough simply to refer readers to the graphics:

Ex. 7.5

Results

The iron concentrations and hydrographic data for the six stations are given in Table 1. Profiles for other data including temperature, salinity, oxygen, pH, nitrate, and nitrate at each station are given by Hong (1984).

Discussion

Hydrography. Previous studies have identified four water masses in the Peruvian coastal region:. . . . [Limn. Ocean. 31, 1986, 515]

and leave it to them to interpret them, for it is the writer's task to report on the data in the graphics that are important.

The most important results to include are those that answer the research question. Data that can be used to make particular points or to outline important trends are also highlighted. Also, any results that are to be discussed in the discussion section are at least highlighted or summarized; readers should not be suddenly faced with discussion of data that have not even been mentioned in the results section. The results that confirm or disconfirm hypotheses are explicitly presented, as well as the results from which the conclusions are inferred. The results of statistical analyses are included, either as the results themselves or in conjunction with the measurements analyzed. Finally, results that are related to those obtained by other investigators are highlighted, especially if the results disconfirm them or are controversial; however any *discussion* of such results is deferred to the discussion section.

When the results are negative, that is, when they are not the results expected or predicted in the hypotheses, the writer must state them clearly. Such negative results are not explained until the discussion section, unless the explanation is brief, or simple. When only some of the results are negative, it is essential that the negative as well as the positive results be reported. Similarly, when a series of hypotheses is tested, those that are not supported are of course so reported.

Results that do not pertain directly to the research question or that are much more voluminous than are required for the paper should be omitted. Data from peripheral experiments should be omitted. Such peripheral studies would not be retrievable, since neither the title nor the abstract would reveal their inclusion in the paper. Extensive data that are of general use in a discipline may be stored in data banks; then their availability can be noted in the paper and readers directed to the source.

7.1.3 Writing the Results Section

Structure and contents.   The results section must ultimately address the questions raised in the introduction and any hypotheses formulated there. Because it is a direct consequence of the methods, it is most logically organized to correspond

TABLE 7.2   Headings showing correspondence between methods section and results sections in the topics presented

a.	**Example A**
Experimental	*Results*
Steady state measurements	Steady state experiments
Laser photolysis measurements	Laser photolysis measurements
b.	**Example B**
Materials and Methods	*Results*
Plant material and growth conditions	Seed size variability among populations
Effects of environmental conditions on seed size	Environmental effects on seed size
Effects of nutrient supply	Effects of nutrient supply
Position effects	Position effects
Effects of position of seed within a fruit on seedling survivorship	Effect of position of seed within fruit on seedling survivorship
Statistical analyses	

to the methods section, as illustrated in *Table 7.2*. When the research consists of a series of methods or experiments, the results are reported in the same order that the experiments were described in the methods section (see *Table 3.7*).

The results section should be restricted to the presentation of results. Therefore, it should not begin with background discussion:

Ex. 7.6

> *Results*
>
> The gross features of the conformational energy surface of a ring molecule such as DMIS, carrying several bulky substituents, should result from the balance between strain energies (ring closure, bending, and torsional terms, including also the anomeric effect) and steric interactions. Other energy terms, such as electrostatic interactions and solvent effects, although very important in a polar molecule, are expected to show a smoother behavior, so that inclusion of these terms in the calculation may alter the relative stability of the energy minima, even quite significantly, but it can hardly yield stable conformers in regions severely forbidden by the former type of forces. Therefore, we set out to explore the energy surface of DMIS by omitting the electrostatic term from the force field previously described. [J. Computat. Chem. 7, 1986, 107]

It should not begin with a description of the methods:

Ex. 7.7

> a. *Results*
>
> In these computations, a Poisson sequence of stimulus pulses was applied to the excitable cable model. Each stimulus pulse had a 0.2 ms duration and a fixed amplitude. A sample of the response of the model axon to this type of Poisson stimulation is shown in *Fig. 2*. [Biophys. J. 50, 1986, 31]

> b. *Results*
>
> The new tetrapeptides were obtained by conventional methods in solution, using active esters for mixed anhydrides, through stepwise addition of each amino acid starting from C-terminal glycine derivatives (Schemes I–II). [J. Med. Chem. 29, 1986, 891]

Not only is this repetitious, since the introduction and methods have just been presented, but it also fails to maintain the separation established by the "Results" heading. Neither should the results be given and then followed by the methods:

Ex. 7.8

> *Results*
>
>   *Phoenicopterus chilensis* was present in the region in large numbers on each visit, but the data are not adequate to identify any pattern of seasonal change. Although 27 lakes were visited at least once, only 20 were used in this analysis; some were omitted because we either collected no zooplankton sample from them or were unable to census the flamingos accurately due to poor viewing conditions. [Limn. Ocean. 31, 1986, 461]

Methods should not be restated even in summary to recall them to the reader. When it is necessary to refer to the method, reference can be made to the method indirectly. The method can be indicated implicitly in the statement of the result:

Ex. 7.9

   a. Close examination of salivary gland squashes probed with affinity-purified anti-75 antibodies showed that the most intense immunofluorescence staining coincided with puffed regions of the chromosomes.

   b. Congo red staining of autopsy material revealed extensive amyloid deposition in the cardiac tissue.

Integration and interpretation.   In writing the results section, the writer faces two pitfalls: (1) failing to integrate the graphic results into the text and (2) interpreting the results rather than reporting them.

Integration.   Tables and figures are not textual and so must be integrated into the text. The integration should consist of more than announcing a table or figure (*Ex. 7.10a,b*) or restating its title (*Ex. 7.10c,d*):

Ex. 7.10

   a. Table 1 shows the data for the three treatments.

   b. Tw_0 is displayed in Fig. 2.

   c. The concentration and distribution of calcium, magnesium, zinc, copper and iron in whole milk, skim milk, supernatants, precipitates and sediments are shown in Table 5.

   d. Fig. 3 shows the absorption isobars (760 mm Hg) of various gases on Linde molecular sieves.

   e. The data from the three treatments (Table 1) show a gradation in the concentration of. . . .

A sentence such as that in *Ex. 7.10a,b* wastes a sentence simply to refer to a graphic, which can be referred to parenthetically (*Ex. 7.10e, 7.12*) in a statement that contributes to the development of the section.

   Also, even though the results section may be based on the tables and figures, it should not be reduced to a serial presentation of them:

Ex. 7.11

> *Table 1* is a list of the observational results. Positions are given both. . . . *Figure 2* shows the spectra of the five points. . . . *Figure 3* is a contour map of the peak temperatures for the strongest

emission line at each observed point. . . . *Figure 4* is a contour map of Td. . . . *Figure 6* is a contour plot of the radial velocity of the first CO emission line for each observed point. . . .

Such a treatment is a presentation of a series of tables and figures, not of results. The statements should be made about the subject matter, not about the tables or figures:

Ex. 7.12

a. The size and shape of the asci, ascospores, and conidia from all specimens examined (*Table I*) were essentially identical.

b. The rate of glucose utilization decreased throughout the biotransformation, as shown in Fig. 1b, and adjustments to the glucose feed were necessary.

c. The calculated 37 GH2 (horizontal polarization) brightness temperatures for dry snow over frozen ground (Fig. 8) shows the effect of changing the thickness of the depth-hoar layer.

The text should be written directly from the tables and figures in their final form so that the report and references to numbers, titles, headings, and labels in the text will be consistent with the tables and figures.

Besides the data in graphic form, the writer may have other data, which must be integrated with the data extracted from the graphics into a coherent presentation of the results.

Reporting versus interpreting.   The results section is the factual part of the paper; the results consist of bare, dry, unembellished observations and measurements. The presentation of the results must therefore be factual and informative, not interpretative or explanatory:

Ex. 7.13

a.

RESULTS

Population variation

Composite samples of seeds from many plants in 1980 and 1981 showed slightly different frequency distributions of seed weight (Fig. 1). The variation in seed size was continuous and unimodal. The 1980 sample ($n = 706$) had a mean seed size of 0.93 mg. was significantly leptokurtic with a pronounced peak (kurtosis $= 3.47 \pm 0.18$; $t = 2.55$; $P < 0.01$), and was skewed to the right (skew $= 0.32 \pm 0.09$; $P < 0.01$). Seeds in the larger sample ($n = 2416$ seeds) in 1981 averaged 1.14 mg. The frequency distribution was significantly skewed right (0.28 ± 0.05; $P < 0.01$) and also significantly leptokurtic (3.51 ± 0.32; $P < 0.01$). [J. Ecol. 74, 1986, 363]

b. **Phase Diagram and Lattice Parameter Results**

It was found that all of the single phase samples gave X-ray photographs which appeared to show the zinc blende structure typical of CdTe. . . . At high z values, lines of the nickel arsenide structure of MnTe were observed in addition to the apparent zinc blende lines. . . . The slope discontinuities in the graphs of a vs z at constant x:y ratio gave the limits of single phase behavior shown in Fig. 4. The lines of a vs z in the single phase range appeared to be linear within the limits of experimental error except for the case of the $x = 0$ line at low z, i.e. close to CuInTe$_2$. The graphs of a vs y are linear in the range $0.2 < y < 0.75$ but may show some curvature for $0 < y < 0.2$ while for $0.75 < y < 1.0$ there is appreciable curvature. These effects will be discussed below. [J. Sol. State Chem. 63, 1986, 112]

From a scientific point of view, it is important to make the distinction between reporting and interpreting, because the results—and the methods—will stand as reported or described and become part of the permanent store of scientific knowledge, but any interpretation of them may change overnight. Therefore it is desirable to keep the results separate from the less durable discussion.

The two can be distinguished by their different rhetorical structure. Rhetorically, the sentences in the results section are fundamentally declarative statements— factual, informative statements:

Ex. 7.14

1. No differences were found among six independent isolates classified within serotype A1; however some variation was evident among A2 isolates, especially in the case of VCS 1133 (Fig. 3).

2. The large anaerobic column also exhibited significant biotransformation of PCE.

3. The figure shows data for the visible (0.63 µm) reflectively as well as for 3.7 and 11 µm radiances.

4. All mice given 4-ipomeanol showed increased heart and respiratory rates within 1 hr. of administration.

5. All six facial measurements correlated highly with one another (p < 0.0001).

The sentences in the discussion section are largely statements of relationship, explanation, interpretation, commentary, inference, and so on (*Table 7.8*). Since discussion, however slight or subtle or unintentional, may distort the results, every effort must be made to present results as informative declarative statements.

Discussion of a result may be included in the results section provided that it is clearly differentiated from the results, it is brief, and it is restricted to the immediate result and does not digress into a general discussion. If there is discussion of several of the results in the results section, the writer may have to consider a combined "Results and Discussion" section.

Tense.   As in the methods section, the choice of tense is derived from the position of the research in the rhetorical setting: the focus on the past results requires the past tense. It is inaccurate to say, on the basis of *one* research study:

Ex. 7.15

a. The findings show that structural variables, such as position of employees in the division of labor, along with external market and technological conditions, *influence* job elimination, while unionization *affects* the degree of universalism in the method used to lay off employees.

To use the present tense to report results is to forget that, scientifically, one cannot make generalizations on the basis of very specific and limited experiments or studies. It is not only inaccurate, but misleading.

In qualitative, descriptive studies, for example the description of new species, minerals, and macroscopic natural structures, the present tense is used for describing the organism or object observed, following a convention in which the writer is writing the description with specimen at hand. The present tense is also a convention followed in descriptions of microscopic structure, although such studies may be part of an experimental study:

Ex. 7.16

a. Network SR is heavily concentrated over the A bands as a collection of tubular elements closely arrayed in largely parallel orientation (Figs. 1,4,6). Over the I-band portions of the sarcomere, the SR becomes less concentrated, forming a loose meshwork whose elements are continuous across the Z-line levels of the sarcomeres, regardless of the presence (or not) of transverse tubules at those levels (Figs. 1,6). Commonly there forms over the middle of the sarcomere (the pseudo-H zone) a coalescence of network SR in which circular perforations appear (Fig. 1); thus this segment resembles the "fenestrated collar" of skeletal muscle cell SR. Other regions of perimyofibrillar network SR form dilatations, not restricted to any particular sarcomere level, and Z *tubules*, which surround the myofibrils at their Z-line levels. . . . [Ultrastruct. Res. 93, 1985, 7]

b. . . . In areas of the caudate's head that lie outside of the SP cell clusters, the vividness of immunostaining is somewhat less than within the cell clusters due to a slight reduction in the density of immunoreactive processes and the presence of but a few, widely scattered SP-positive cell bodies. The SP-immunoreactive cell bodies in both the caudate and putamen are generally round or ovoid (maximum somal diameters range from 10 to 18 μm) possess 0–5 visible major dendrites, and have large nuclei relative to the perinuclear cytoplasm (Fig. 5A) when compared to the larger cholinergic striatal cells (Fig. 5B). . . . [Neurosci. 20, 1987, 558]

The sentence in *Ex. 7.16a*, "Commonly there forms over the middle . . ." illustrates how the present tense leads the writer to confer generality on what is actually a finding of the research.

The present tense is commonly used when writers are presenting data from a table or describing an illustration:

Ex. 7.17

a. *Results*

The amplitude and phase of the diurnal harmonic, determined by the method described in the previous section, is presented in Fig. 1 for each of four three-month seasons. The structure of the amplitude of the diurnal harmonic is shown in the form of contours at 2, 5, 10 and 20 W m^{-2}. The lengths of the arrows are constant for amplitudes greater than 5 W m^{-2} and decrease linearly for amplitudes less than 5 W m^{-2}. Arrows are not shown for amplitudes less than 1 W m^{-2}. The contour value is thus easily determined by beginning at the 2 W m^{-2} contour bordering a region with missing arrows and counting contours in the direction of increasing arrow length. It is immediately apparent that the amplitude is largest over the land, especially over desert areas. Significant amplitudes also appear over the oceans far from land, however. [J. Clim. Appl. Meteor. 25, 1986, 801]

b. *Results*

Figure 8 presents the incremental resistance for the saw-tooth interfacial configuration. It is seen from this figure that the influence of increasing both the interfacial amplitude and the conductivity ratio is to increase the incremental resistance and, therefore, to decrease the total resistance. The trends in the figure indicate that small interface amplitudes lead to relatively small decreases in total resistance. As the interface amplitude increases, however, the decrease in resistance grows rapidly. . . . [J. Spacecr. Rock. 23, 1986, 228]

The writer is figuratively pointing out the various details of the tables or figures that the reader has in hand, in a rhetorical "present" that the writer and reader share, not as results obtained in past research. He or she can thus present results in the present tense, without making a statement about the universality of the data. The pitfall here is that the writer is likely to continue in the present tense.

7.2 DISCUSSION

7.2.1 Objectives and Structure

Although the results are the substance of science, the discussion allows for the play of ideas that advance science. The discussion section is the part of the paper in which the writer explains or interprets the results and the research as a whole, in themselves and in relation to earlier research. The generative nature of the discussion gives it an open-ended character that allows the research scientist freedom to speculate about the research while integrating it into the scientific framework of past research.

The objective in the discussion section is to give the research, especially the results, meaning. The writer's task is to integrate—the results, the method, the related literature, the theoretical context—anything that is related to the research and that might illuminate it. Specifically, the objective is to examine the results, determine whether they solve the research question, compare them within themselves and to other results, explain and interpret them, and then draw conclusions or derive generalizations, and make recommendations for applying the new results or for further research.

The discussion section is the logical outgrowth of the results section. The interpretation of the results is the focus for the discussion and allows a synthesis across three frameworks: (1) that of the immediate research, (2) that of related research, and (3) that of research and theory in the discipline. The three frameworks are not independent of each other, nor are they all equally important in all research studies, but the writer must be prepared to address them as the research requires.

7.2.2 Discussion of Immediate Results

The discussion of the results within the framework of the immediate research may relate to the introduction, the methods, and the results themselves. The discussion must make clear the extent to which the results address the research problem or question, or the hypotheses; that is, whether they are a direct answer, whether they do not answer the question or support hypotheses, or whether they provide only a partial or dubious answer.

The methods are evaluated critically on the basis of the results obtained, and on any pertinent literature. The writer may consider whether the results are consistent with the methods or throw new light on them, or whether they raise questions about them, and whether modifications or different methods would have been more appropriate, and so on. The methods are always discussed, often in detail, when they are the subject of the research or when they are an important element in the research.

The discussion logically begins with the results in themselves, that is, with a discussion of the similarities or differences in the observations or data, high or low values, or trends among the findings that merit discussion in themselves or in the light of the research question or hypotheses. Such findings would have been emphasized in the results section in preparation for their discussion in the discussion section. In this way, the discussion continues the thread of development begun in the introduction and continued through the methods and results section.

7.2.3 Discussion of Results and Literature

The results are the point of departure for discussing the research as a whole and in relation to other research and scientific concepts. Writers analyze and interpret the results in relation to underlying relationships, their consequences, and any theoretical implications. This discussion of the results gets below the surface of the immediate findings and brings out their meaning. Writers may also identify and emphasize results that are new or important additions to the discipline and explain how they add to scientific knowledge. They must also point out the limitations of the findings and how these limit the generality or applicability of the research. This broader discussion necessitates drawing upon the literature, which the research scientist must analyze critically for its bearing on the research. The research is thus integrated with the results, methods, interpretations, and conclusions of other studies to make them part of the larger framework of the discipline. The broader discussion may even be speculative, but the writer should not extend the discussion to the results or literature on a tangential problem or question.

The research may touch on disciplinewide concepts or even scientific principles. This broader discussion is usually theoretical and may even be speculative, but the speculation must come as a terminal outline or aside; it should not be an elaborate construction erected on the narrow base of the particular research. What is wanted is an indication of the broader ramifications of the research not a rambling, desultory armchair theorizing on the subject.

Negative, discrepant, anomalous results.   The results may be inadequate, negative, or not consistent with earlier studies. Such limitations must be identified and the nature and extent of the limitations discussed as appropriate (*Ex. 7.26*). The explanation of a deficiency should be brief and pertinent to the research. Rather than try to hedge, justify, defend, or minimize the deficiency, the writer must focus on it as it affects the research.

When the results do not answer the research question or support the hypotheses proposed, or are not consistent with those obtained by other research scientists, or with established principles or theoretical concepts, the writer must try to account for them. If the method used in the other studies is somewhat different from that used in the study being reported, the discussion explains whether the difference in method might account for the difference in results. When the results are different despite similar methods, the explanation of the differences may be more complex. The discussion may lead to qualifying the earlier findings or the present ones, or to an interpretation or explanation that reconciles them. The writer may not be able, however, to fit the contradictory results into the existing framework.

Such unexplained results are part of the driving force in scientific research. They may remain unexplained until later research throws light on them or until repeated confirmation of the anomalous results necessitates a modification of the existing theoretical framework to accommodate them. The present results may actually explain earlier anomalous findings or permit a better understanding of such findings, and the discussion then explains the advance.

The interpretation of discrepant, or anomalous, results may be a source of controversy. A writer presenting controversial results must make certain that the methods and hypotheses have strong theoretical foundations and that the evidence for interpretations and explanations presented in the discussion are clear, logical, and substantial. The objective in any discussion of the controversy must be to consider and account for the difference in interpretation, not simply to reject the different interpretation or refute it out of hand. By taking a position that marshals the strongest evidence, the writer establishes a firm basis for fruitful discussion.

7.2.4 Discussion, Conclusions, and Summary

Conclusions.   Conclusions are the end product of the discussion; they state the ultimate meaning of the research. They are inferences derived from the discussion and may be drawn directly from the evidence of the research reported and indirectly from the further evidence provided by the literature. The conclusions are therefore an essential part of the discussion and an important part of the paper. They are so important that they are included in the abstract, even though the discussion, out of which they arise, is not included. Conclusions are important because they provide closure for readers.

In drawing conclusions, the writer draws different types of conclusions at different points in the development of the argument, and these have different degrees of validity or generality. Early in the discussion, the writer may begin with the direct conclusion. For example, if different dosages of a drug are administered experimentally to rats to determine the safe dosage, and there is a high percentage of mortality even at the lowest doses, the immediate conclusion is that the drug used is toxic even at low dosages. Such initial conclusions may be termed *direct* or *immediate* conclusions. They arise directly out of the results and apply directly to the specific conditions of the research. They are often generalized statements of the results. Immediate conclusions are preliminary conclusions. The results and the immediate conclusions are then examined further, both in themselves and in reference to the literature.

Further conclusions, termed *intermediate* or *indirect* conclusions, may then be derived. In the drug example, the literature may include reports of the drug being toxic only at high dosages. The immediate conclusion cannot be maintained, therefore, without a critical consideration of earlier studies. The writer may find evidence that the difference is due to the difference in the chemical structure of the drug, or in the method of administration. Therefore, the immediate conclusion must be modified, so that the intermediate conclusion, based on the review of the literature, is logically consistent with the pertinent literature. In the drug example, the indirect conclusion may be that in the chemical form used in the present experiment, the drug was highly toxic. And indeed from further review of the literature, it may be possible to infer from various studies that the toxicity is related to a particular radical, possibly to its position on the molecule. The *final* or *ultimate* conclusion derived from all the available evidence therefore may be that the toxicity of the drug is related to its chemical structure, particularly to a specific radical, and possibly to its relative position on the molecule.

For a conclusion to be supported, the writer must present adequate and sound evidence. The validity and strength of a conclusion depend upon (1) the substantiveness of the evidence, (2) the sufficiency of the evidence, and (3) the strength of the logical character of the relations between the evidence and the conclusion. The substantive character of the evidence, that is, the results, depends upon the research design and methods. If the evidence consists of accurate and precise measurements and observations based on a method in which the variables studied and the background conditions are carefully controlled, then the results in themselves make a strong contribution to the conclusions drawn from them. If, in addition, evidence can be presented from other and different types of sources, then such evidence can fulfill the sufficiency requirement. Finally there must be a direct, logical relationship between the design, the evidence, and the conclusion. This relationship should be susceptible to testing, if the conclusion is to be tenable. In the drug example, the ultimate conclusion might be substantiated by hypothesizing the relation of different positions of the radical to toxicity and performing an experiment to test the hypothesis. However, when the evidence is inadequate or there is difficulty in accounting adequately for disparate conclusions, the writer must make clear the conditions under which the conclusion does or does not hold.

Discussion and conclusions.   Both discussion and conclusions are usually included under the heading "Discussion." The conclusions should form a small part of all such discussion sections, because they are simply statements of inference derived from the discussion.

If there is extensive discussion of them, the discussion section may be called "Discussion and Conclusions." When intermediate conclusions are interspersed throughout a long or complex discussion, the conclusions may be enumerated and restated in a separate section or subsection called, "Conclusions." Such a section is very short because (1) its purpose is to identify and emphasize the conclusions, not to discuss them further, and (2) the conclusions are generally few and relatively brief statements (*Table 7.3*). Most papers do not require a separate conclusion section; yet it is common to find detailed sections labeled "Conclusions." Such sections are frequently summarizing or general discussions.

Discussion and summary.   In a paper that has several different and interrelated parts or that describes several experiments or a complicated protocol, it may be helpful to readers to summarize the structure of the research leading up to the

TABLE 7.3   Concluding sentences illustrating their rhetorical form, and their relation to specific or general results

a. *Related to Specific Results*

1. Abraded bedrock and the subsequent deposition of up to 6 mm of till indicate that some erosion of the valley occurred during the last glaciation.
2. The agreement with the observed period near 10 s indicated that the model contains the necessary elements of a realistic explanation of auroral pulsations.
3. Thus the CHS 1 gene product does not appear to be involved in the pheromone-dependent deposition of chitin.
4. Thus it seems that there is, as yet, no consistent explanation of the De Voe-Brewer data.

b. *Related to Results in General*

1. The data also indicate that learning can be measured in situations where recall does not show that learning has occurred.
2. The results from this study and others (. . .) allow us to conclude that foraging by squirrels is consistent with dispersal of some nut-bearing tree species into prairie habitat.
3. The findings from these analyses suggest that exposure to the modern health care system may influence breast-feeding practices.

discussion. This kind of summary precedes the discussion (*Ex. 7.18*) and, if extensive, the section can be called, "Summary and Discussion." In some papers, the discussion itself may be complicated and extensive; then the discussion might warrant summarizing. Such a summary is an integral part of the discussion and serves as an organizing framework for the preceding discussion, and so may be included under the heading, "Discussion and Summary." Unfortunately, however, this heading is also used when the terminal summary is a summary of the paper as a whole. The summary of the discussion together with the conclusions may be included under a separate heading, "Summary and Conclusions," though here too the "summary" may apply to a summary of the paper or the discussion. The use of the term "summary" is further confused because it is used synonymously with "abstract," for the paper. The separate heading "Summary" is used, correctly, for a summary of the paper as a whole, but this is uncommon in scientific papers:

Ex. 7.18

> Detrended correspondence analysis and percentage difference, multivariate techniques frequently used in the analysis of regional vegetational gradients, successfully identified local pattern in herbaceous communities using a limited number of discontinuous samples. Sites in dry alkaline habitats had more vegetational pattern than wetter, acid habitats. At the regional scale, moisture, grazing, and soil determined the composition of the vegetation; at the microscale, growth characteristics of the plants and termite activity were the dominant determinants. [Belsky, 1986, Unpublished]

Conclusions versus summary.   The conclusions are frequently confused with a summary (1) because they both come at the end of the paper and both provide closure, and (2) because the ultimate or general conclusions are mistakenly thought to summarize the results and discussion. Nevertheless, the conclusions are significantly different from a summary, and it is important that they be accurately distinguished and designated.

Conclusions are necessary inferences and deductions, first from the results and ultimately from all the evidence presented in the discussion. They are the logical outcome of the development of the argument in the discussion. The discussion explains, interprets, justifies, and argues; the conclusions are the logical decisions consequent upon such explanations and arguments. The conclusions are the end points of the argument; they make clear the ultimate meaning of the discussion (*Table 7.3*). They are therefore an essential part of the scientific method, and so are an inescapable part of the research and of a scientific paper. The summary, on the other hand, is primarily a rhetorical convenience. It is used to reduce the discussion (or the paper) to its main elements. As a condensation of the discussion, the summary *includes* the conclusions; therefore, it cannot be equivalent to or synonymous with them. Furthermore a summary has no scientific imperative. It is neither essential nor inevitable for the discussion or the paper. There need be no summary or summarizing, if the discussion or paper does not warrant it. A summary may be expanded or reduced to fit the objectives of the writer or editor, or the needs of the reader. Moreover, it is not readily adapted to being converted to an enumerated list. In sum, conclusions consist of relatively few, brief and discrete statements, which can be enumerated, whereas a summary is usually an integrated whole, a paragraph or more long.

Summarizing statements are general and inclusive, and encompass as many of the details of the condensed section as possible. Concluding statements are statements of inference, drawn at the end of a line of argument, and derive essential consequences. The scientific importance of the conclusions requires that they be stated clearly and explicitly, as conclusions. The form of concluding statements must indicate their function (*Table 7.3*). A concluding sentence may state directly the specific data that are the source of the inference (*Table 7.3a*), or it may refer indirectly to results in general (*Table 7.3b*).

The headings "conclusions" and "summary," are often used inaccurately. Sections labeled "conclusion" are not restricted to conclusions (*Tables 7.4a,b*), and sections labeled "summary" may include discussion and conclusions and may not include a summary (*Table 7.4c*). Most headings that include "Summary" or "Conclusions" are actually the closing of the discussion and often include a motley assortment of material.

The foregoing discussion indicates that the terms "conclusions" and "summary," if accurately used, would be used very infrequently. The term "summary" alone is best avoided because (1) it is often confused with conclusions, (2) it is often difficult for a reader to know which part of the paper is being summarized, and (3) it can be confused with the "summary" that serves as the title for the abstract of the paper (see Chapter 8). Both terms are best used in conjunction with the term "discussion."

TABLE 7.4   Examples of type of subject matter included under headings "Conclusions" and "Summary"

a. Conclusions

Regeneration of phenol-saturated carbon by acetone extraction has been experimentally investigated and mathematically analyzed. [*method*]

The rate of phenol desorption by batch mode acetone extraction is dependent upon the convective motion between the regenerant and the carbon granules. When there is no convection, the desorption rate is almost negligible. [*results*]

Mathematical models for batch and continuous modes of phenol desorption have been developed. The models simulate the desorption data reasonably well, and the desorption of the phenol by acetone extraction can be described by a linear mass transfer driving force equation. [*summary of results, conclusions*] However, the diffusion-controlling step of the desorption remains to be investigated. [*future research*] [Indust. Engr. Chem. Fund. 22, 1983, 425]

b. Conclusions

The following conclusions can be drawn from this investigation:
1 Acoustic emission energy, $(RMS_{AE})^2$, increases with wheel loading.
2 The acoustic emission energy, $(RMS_{AE})^2$, is proportional to the depth of cut, a. A linear relationship between log $(RMS_{AE})^2$ and log a exists for grinding with both dressed and worn wheels.
3 The acoustic emission signal accurately detects workwheel contact and sparkout. In both cases the sensitivity of the signal exceeds that of force measurement. [*results*]

The results of this study indicate that the feasibility for using AE sensing techniques as a process monitor for wheel loading sensing as part of process adaptive control is very good. . . . Further, the ability to sense both wheel contact and sparkout can be useful in expediting setup of grinding operations and optimization of cycle time and surface generation sequences in production. [*conclusions*] . . . [J. Engr. Indust. 106, 1984, 32]

c. Summary

The stability analysis of the two-layer configuration given herein, . . . has shown that two modes are possible: 1) the classical Kelvin–Helmholtz mode due to shear at the interface and 2) the gravitational or convective mode due to the unstable stratification in the one layer which also modifies the stable layer adjacent to it. [*conclusion*] The behavior of the Kelvin–Helmholtz mode is well known. The phase speed and growth rate are proportional, respectively, to the average wind speed and the shear across the interface. [*general discussion*]. . . . As expected, an increase in the stratification of the lower layer reduces the growth rate of the mode. The phase speed tends to be proportional to a depth-weighted average velocity of the two layers. [*results*] Hence, when the layers have equal thickness, c_r will be $U_1/2$ if the magnitudes of the stratification in each layer are equal [*conclusion*]. . . . [J. Atmos. Sci. 43, 1986, 1003]

In using the terms for headings, writers should use "conclusions" for inferences and "summary" for a condensation. When a summarizing section includes both, the heading may be "Summary and Conclusions." If the section is actually a discussion section that includes a summary or conclusions, the heading may be "Discussion and Conclusions," for a discussion section that includes conclusions; "Summary and Discussion," for a discussion section in which the first part summarizes the paper *before* the discussion; and "Discussion and Summary," for a long complex discussion section that requires summary at the end to effect closure. The heading "Discussion and Summary" can include conclusions, whereas the heading "Discussion and Conclusions" does not, strictly, include a summary. The heading "Discussion, Summary, and Conclusions" would be comprehensive, but would be too heavy for a short section. The heading "General Discussion" or "Summary Discussion" may be appropriate for the closing section of the discussion.

7.2.5 Future research

Recommendation for further study.   The discussion may carry implications for further theoretical or empirical research, or for application in the same or other disciplines—medicine, agriculture, and engineering. These implications and suggestions may be discussed as part of the discussion section (*Table 7.5*) at the end. Where application of research is important, the implications of the research for revised treatment, design, practice, and so on may be presented in detail. Where the advancement of scientific knowledge is important, the discussion is directed toward the implications for further theoretical and experimental research.

Although such recommendations are not part of the scientific method, they have come to be included in some journals because of their usefulness. Having been

TABLE 7.5   Excerpts from discussion section: (a) statements of implications of research (b) discussion of direction of future studies

a. *Implications for Research*
1. Factors unconnected with mineral nutrition are probably also at work, and more evidence is required before the disjunct and erratic distribution of this species can be explained.
2. A CIDNP study would be pertinent to see whether the radical cations proposed in the present work are involved in the photoinduced electron transfer in systems of the bicyclopentanes containing suitable electron acceptors, as shown for various cases.
3. The present study raises a number of broader theoretical questions. Can the findings be generalized to other industrial hegemonies? Can citizens' grievances in hegemonies usually be redressed only by "grieving the regime" with military force. . . .

b. *Future Research*
The apparent success of the theory in explaining some features of pulsating aurorae indicates several potentially fruitful directions for further observational research. An essential observational item is a set of precipitating electron spectra with sufficient resolution in the loss cone to test the theory including atmospheric scattering, correlated with a simulataneous search for VLF wave activity. That the backscatter of energetic electrons from the atmosphere might be important to the anisotropy of the upgoing distribution and to the control of wave growth is a concept that must be tested by electron detectors looking both up and down along the auroral field lines. Coordinated observations of the high-altitude pitch angle distributions of energetic electrons would be extremely valuable, because their anisotropy determines the minimum energy for wave growth [*Kennel and Petschek,* 1966; *Davidson,* this issue]. Any data that might provide evidence on the magnitude of the minimum resonant energy or the plasma density during a pulsating aurora would be useful in deciding the role of VLF waves. The understanding of pulsating aurorae is at a point now where a few well-designed experiments aimed at specific questions raised by the various theories are needed for further progress. [J. Geophys. Res. 91, 1986, 4435]

so closely and intimately involved in the research, the research scientist is likely to have developed insight into central theoretical or experimental problems and can point out key studies or important directions.

Such recommendations should arise logically out of the discussion and should be directly related to the research reported. They should be brief, explicit suggestions for studies, with a brief justification. The recommendations may close the discussion section, but if significant enough, they may be emphasized by being set in a separate section, for example, "Recommendations for Further Study," or "Future Directions." Such a section must be brief; it should not be taken as an opportunity to discourse generally on future research, or to outline an expansive program, or discuss the various questions that might be studied. The objective is to give colleagues the benefit of particular insights one has developed (*Table 7.5b*).

Research in progress.   Writers sometimes plan to undertake some of the research proposed and may in fact have the research in progress. Such research should not be mentioned in the paper, unless it has been completed and at least *accepted* for publication. A statement such as:

Ex. 7.19

> A study designed to determine how treatment time influences the lowering of the glass transition temperature is in progress.

may lead readers interested in the study to search for it vainly in the subsequent literature. Yet, the lack of a grant, a change in position, or preliminary research may have led to abandoning the research "in progress." If this research is included out of concern about priority of publication, it is preferable to publish a preliminary note.

7.2.6 Writing the Discussion Section

Like the introduction section, the discussion section is a difficult section of the paper to write, and for the same reasons. Both sections lack the concrete character of the methods and results sections; both have a central organizing function; both are interpretative and open-ended; and both allow great latitude in content and form. However, whereas the introduction usually brings together the threads of the background of the research, the discussion fans out from the results to the literature and beyond.

In fact the two chief pitfalls in writing the discussion section arise from its open-ended and interpretative character. The open-ended character makes it necessary to try to avoid overexpansion and digression. The interpretative character can lead to loose or ambulatory explanations and interpretations. To keep the discussion focused, the writer must hew closely to the subject of the research. To develop a rigorous and cogent argument, the writer must analyze the research critically and develop a logical line of thought that leads from the results to a conclusive explanation or interpretation of them.

The literature brought to bear on the results, and the conclusions drawn from the discussion must be directly pertinent to the research problem or question. Any interpretations, explanations, or theoretical concepts developed must not extend so far across the topic that they obscure the research question and become more a general discussion or speculation on the problem rather than an interpretation of the evidence. And any implications of the research for future application or research should not be extrapolated so far as to lose their anchor in the research. The discussion in *Appendix 7A*, which constitutes half the paper, has moved away from the specific subject of the research toward an essay or lecture on leaf miner activity and leaf abscission, in which the results are almost incidental to the general discussion of the topic.

Development and structure of section.   The discussion section most logically begins with a discussion of the results, because they have just been presented and are the most important part of the paper relative to the discussion of the research. However, the discussion may start with the conclusions, if the discussion is developed deductively (*Table 7.7*). But even then, the results should soon be brought forward as evidence. Starting with concluding or generalizing statements interrupts the logical development of the paper and may mislead some writers into opening the discussion with an introduction to the research or a statement of purpose (*Table 7.6*).

TABLE 7.6   Opening sentences of discussion sections giving purpose or introduction to subject

a. *Purpose*

1. The present study was conducted to extend our understanding of recall memory processing in young children by presenting objects and events within narrative contexts designed to be more or less similar in structure and theme to early-developing natural knowledge organizations.

2. This study addressed the problem of determining the resistance to heat flow across two-dimensional contiguous, rough interfaces.

3. Our study had three major objectives: (1) to develop a recognition measure that would not yield consistently high scores and would discriminate across stimuli, (2) To show that learning of commercial information occurs even though recall measures may be unable to tap this learning without a measure that. . . .

b. *Introduction to Subject*

1. Anderson (1972) presented an excellent three-dimensional scale model based on thin-section studies. Since his description, there have been few reports which advanced knowledge of the complete structural analysis. In the present experiment, we reexamined the basal body structure, confirmed most of his descriptions, and added some new findings.

2. Many researchers have attempted to differentiate laryngeal pathologies from normal-voiced subjects by using various methods of voice analysis in voice signals. In doing so, attention must be paid to the degree of hoarseness or the severity of disorders in the pathological group.

3. Although adrenal gland zona fascilata cell response to ACTH stimulation is known (12, 29) and various steroid intermediates and enzymes involved in corticosterone biosynthesis are established (25, 29), interaction of various subcellular structures during steroidogenesis is not well understood.

The section may begin immediately with the discussion of results if they require explanation or comment (*Table 7.7*). However, the results should not be restated or summarized in the discussion section as a basis for discussing them (*Table 7.7a*). The starting point of the discussion is the *interpretation* of the results. Results can be referred to or mentioned indirectly, as illustrated in *Table 7.7b*. If the presentation of the results is very complicated or if a series of results is to be discussed, the results may be summarized, but very briefly, as a springboard for the discussion:

Ex. 7.20

> For all dopant levels, at some temperatures and driving forces, growth fronts develop bulges, protrusions, and irregularities with widths ranging from a few to a dozen subgrains.

Writers may begin with the literature to relate the present research (*Table 7.7c*) to the earlier findings or concepts in other research. Here too they should avoid restating findings from the literature (*Ex. 7.21a*) in preparation for discussing them,

TABLE 7.7   Opening sentences of discussion sections giving results, literature, or conclusions

a. *Results Restated*

1. A plot of the effective index of refraction as a function of wavelength for TE and TM waveguide modes is shown in Fig. 2.

2. A numerical model for the diurnal mixed layer is presented, based on the one-dimensional integrated forms of the momentum equation, the turbulent kinetic energy equation, and equations for conservation of thermal energy and salt.

3. In the 9.54 μm irradiation of liquid PDMS, most of the laser energy is deposited in the first few microns of the PDMS layer. The temperature of the irradiated portion rises quickly to the predicted maximum temperatures given in Fig. 2.

b. *Results Introduced by Indirect Reference*

1. The mode in which a *Paulsenella* dinospore approaches and attacks *Streptothera* cells indicates that there are two different tactic phases: a long-distance chemotaxis phase and a short-distance reaction which allows the parasite to penetrate precisely the host hypotheca.

2. Although the standard deviations on the refined parameters presented in Table I are rather large, there is evidence to suggest that the number of $\langle 111 \rangle$-displaced oxide ions decreases with increasing yttrium concentration in the solid solution.

3. The spatial distribution of the diurnal thunderstorms shown by Figs. 3–6 indicates that there are distinct thunderstorm "regions" in the United States.

c. *Literature*

1. Information concerning the molecular biology of the reovirus multiplication cycle has been obtained primarily from studies of human reovirus replication in mouse L-cells (see references 14 and 38 for recent reviews).

2. Of the seven known anthraquinones isolated in this study, six have also been isolated from calli of *C. ledgeriana* [6], with the exception of alizarin-2-methylether.

d. *Conclusions*

1. The main conclusion of this research is that job conditions affect intellective process in older men just as much as in younger men.

2. Students in this study showed no evidence of advance or hierarchical planning.

Ex. 7.21

a. The gas-to-dust ratios have been found to be two orders of magnitude higher than the usual 100 (DeBretor, 1985). [*restatement of earlier findings*] In the present experiment, this value was not reached for *S292*. However, the effect of the dust lane crossing the surface was severe.

b. Burkhardt's exponents agree with ours to better than 1%, but his estimate for the tricritical temperature is almost 5% too low.

but should draw the literature into the discussion indirectly and integrate into the discussion. In referring to the literature, the writer should compare present results to past results, as in *Ex. 7.21a*, not past results to present, as in *Ex. 7.21b*.

The discussion should not begin with a statement or restatement of the methods:

Ex. 7.22

a. We have used physical detection of recombined DNA to expand the criteria by which meiosis- and recombination–defective mutants can be phenotypically distinguished.

b. A harmonic analysis of the starting time of thunderstorms over a period of 25 years for 450 stations in the coterminous United States was performed and spatial patterns in both amplitude and phase angles for each season were determined.

c.   **DISCUSSION**

In the present study, DNA polymerase A was purified to homogeneity from 15 to 20 kg wet weight of baker's yeast (*S. cerevisiae*), one of the eukaryotes, by a procedure which, in the early steps, involved extraction in the presence of a high concentration of PMSF (a serine proteinase inhibitor) and a rapid batchwise separation in order to minimize the degradation by proteinases. . . . [J. Biochem. 102, 1987, 751]

This confuses reporting with interpretation. Writers should not open the discussion with a background introduction (*Ex. 4.33*) and should avoid going back to the purpose (*Table 7.6a*). The writer may provide a brief introduction to the discussion section, which summarizes the preceding sections (*Ex. 7.23a*) or outlines the coming discussion section (*Ex. 7.23b*), if the sections are extensive or complex:

Ex. 7.23

a. The three areas of concentration of this investigation discussed above are the effects of wheel loading, depth of wheel engagement, and sparkout and wheel contact on acoustic emission from the grinding process. In the study of wheel loading, Fig. 6, it can be seen that the wheel loading pattern parallels that of the emission level increase, Fig. 4. Thus, at this point, it is difficult to assess. . . . [J. Engr. Indust. 106, 1984, 32]

b. In this section the relationship between the cyclogenetic sites and the upper troposphere wind field is briefly examined and interpreted in terms of the results of other investigators.

Restating or repeating the results, literature, or methods reflects confusion about the difference in rhetorical form between a sentence that makes a statement (*Ex. 7.24a*) and one that refers to a statement that has already been made (*Ex. 7.24b*, *Table 7.7b*, *Appendix 3A*):

Ex. 7.24

   a. The concentration of plasma cholesterol in the treatment groups was low relative to the concentration in the controls.

   b. The low concentration of plasma cholesterol in the treatment groups relative to the controls indicated that the. . . .

The statements in the discussion section are commentary; they are statements of relation, evaluation, comparison, subordination, chronological or serial ordering or hierarchy, and so on. Most of the sentences are therefore statements expressing relationships (*Table 7.8*) rather than reporting findings.

TABLE 7.8   Sentences from discussion sections illustrating rhetorical form and relational function

1. This wave vector is so small that its effects on the photon frequency can be neglected. [*results to inference*]
2. Ten-year-old children may be utilizing alternative means to retrieve embedded objects, and thus are able to remember equally well under strong and weak script conditions. [*explanation of results*]
3. The results regarding validity were also encouraging and consistent with other results. [*comment on results and relation to literature*]
4. The discrepancies among the various measurements of the K velocity were caused by an unfortunate choice of spectral lines. . . . [*explanation of discrepancies*]
5. Greater family cohesiveness generally found among blacks may have contributed to racial differences in subjective well-being. [*general findings as interpretation of results*]
6. The importance of health to the morale of adults in multigenerational households has important implications for the maintenance of the multigenerational family as a viable alternative to institutionalization. [*results and implications*]
7. Therefore if the intramuscular connective tissue is abnormal, the structural maintenance of the developing muscles will probably be abnormal as well. [*condition for conclusion or prediction*]
8. Although HPLC techniques for various steroids were reported (. . .), this study is the first to describe separation and assay of all five major zona fasciculata steroids. [*comparison of present study to earlier studies*]
9. These differences probably arose because of better aeration and more pronounced seasonal changes in the surface layers. [*explanation of differences in data*]

Development of conclusions.   The importance of a logical relationship between evidence and conclusion requires that conclusions be written in a form that expresses this relationship. The immediate conclusion can be stated explicitly at the beginning of the discussion; it may be implied or even stated at the end of the results section, but it is best deferred to the discussion. If the immediate conclusion can be confirmed by the literature, the argument is developed for presenting the confirming evidence. When the immediate conclusion is not decisive or requires explanation, the statement of any conclusion may be deferred until the results can be discussed in relation to the research or to the literature, or a qualified or tentative conclusion may be offered. Intermediate and final conclusions are then stated as the discussion is developed:

Ex. 7.25

    a. The main conclusion of this research is that job conditions affect intellective processes in older men just as much as in younger men.

    b. The temporal connection central to the socialization perspective is not as strong as many assume.

Conclusions are usually stated following the presentation of the evidence; however, if the development of the evidence is complicated, the conclusions may be stated first, then the evidence for them developed.

The writer may not be able to draw a conclusion from the results, but that is in itself a conclusion and should be stated and discussed. Conclusions should make clear the conditions under which they apply and the range of their validity. Both the experimental variables and the experimental conditions must be clearly associated with the conclusions. Gaps in data require that conclusions be qualified or expressed in a tentative form, and any reservations, for example about the method or some procedure, should be made explicit.

When the results or conclusions, or the methods, must be qualified:

Ex. 7.26

    a. A valid criticism of the results is that the data are a national average summation and do not consider local hydrological variations that would affect the results within a single watershed.

    b. Although estimates of capture efficiency by the laboratory technique may sometimes be higher than those in the field (Hand, 1982) for various reasons, the use of the "table" technique enabled different factors to be studied in isolation. However, it does have its limitations. For example, Dewar et al. (1982) found in the field that D-vac capture efficiency decreased as aphid density increased on mature wheat. They suggested this was a result of aphids falling to the ground when the sampler disturbed the plants. If this is the case, the laboratory method cannot be comparable to the field situation. . . . [Ann. Appl. Biol. 108, 1986, 239]

    c. . . . However, the use of simple stencil masks does restrict pattern geometries since patterns which close on themselves are not permitted. . . .

        The use of a grid support stencil mask . . . eliminates many of the restrictions of stencil masks.[1] Grid support masks, however, introduce factors which may affect linewidth control. . . .
    [12] [J. Vac. Sci. Technol. B4, 1986, 13]

the writer must decide whether to qualify them early or late in the discussion. If readers are likely to have questions or reservations as they read, then it is best to address the questions or make the qualification at the beginning of the discussion. If the reservations are minor, the question may be raised at the end of the discussion, or wherever it is most pertinent. The discussion of the reservation may be accompanied by a suggestion for a research design in which the question would not arise.

    Tense.   Because the discussion develops an argument to arrive at conclusions, and infers the relation among results, the literature, and findings, the present tense can be used more often than in the results or methods sections. However, this does not admit of using the present tense exclusively or even predominantly.

Distinctions in tense reflect distinctions in meaning and so cannot be ignored without loss of meaning.

In a discussion in the present tense, it is difficult for the reader to differentiate findings of the present research, from earlier findings, from conclusions derived from them, from general knowledge in the discipline, and from explanations of processes and so on, as in the following discussion:

Ex. 7.27

DISCUSSION

Phosfolan and mephosfolan are proinsecticides based on their activation to more potent AChE inhibitors by microsomal MFOs. Similar activation occurs on S-oxidation by peracids. Reactivities with AChE (k_i's) and 50% AChE inhibition levels (I_{50}'s) show that the S-oxides rather than the parent compounds or hydrolysis products are the potent inhibitors in the peracid system. The reactivities of phosfolan S-oxides are consistent with comparable mephosfolan S-oxides except that **6** is considerably less potent than **4** or **7**. This might result from a less favorable steric interaction for **6** at the AChE active site or a less reactive amino carbon. Detection of diethyl phosphoramidate as a NADPH-dependent MFO metabolite also implicates the S-oxides as bioactivation products because the S-oxides (obscured in the ^{31}P NMR spectrum by **1** or **2** or NADPH) are rapidly hydrolyzed to the stable phosphoramidate (*12*). Further evidence for MFO S-oxidation is provided by the similar potencies of **1** and **2** activated by MFOs or by equivalent peracid to those of the pure sulfoxides.

The deviation from linear inhibition kinetics (*13, 14*) after 10–15 min is attributable to the rapid hydrolysis of **3–7** (*12*) rather than to reactivation of the inhibited enzyme as established by the lack of recovery on dilution. AChE derivatization probably occurs at or near the ACh binding or reaction site since the substrate protects the enzyme from inhibition by **4** (*13, 15*). These findings are consistent with phosphorylation of AChE but do not rule out other addition reactions at the active site (*12*). [Biorg. Chem. 13, 1985, 349]

Such a discussion is not clear or scientifically rigorous.

In a theoretical or mathematical paper, the discussion is likely to be in the present tense, but such studies are not time-limited or time-dependent:

Ex. 7.28

Two effects dominate the static optical properties of a multiple quantum well (MQW), or superlattice (SL), near the fundamental absorption edge for the Γ symmetry point: 1) the two-dimensional (2D) exciton, and 2) the composite dielectric properties of the MQW. Both of these effects are influenced by the 2D band structure of the multiple quantum well, resulting in aniso-tropic optical properties which are different for fields polarized parallel, or perpendicular to the plane of the MQW layers. . . . [IEEE J. Quant. Electron. QE-22, 1986, 1016]

In empirical studies, the results or methods of the present or earlier studies are all past events, and discussing them in the past tense contributes to accuracy and clarity. Comments, explanations, or conclusions associated with the method or results can also be in the past tense:

Ex. 7.29

1. These differences probably *arose* because the surface layers had better aeration and more pronounced seasonal changes.

2. The monthly variation in total N concentration in soil beneath mature coastal fynbos *did* not *appear* to result from increased organic matter input or decomposition at the soil surface.

Therefore for most discussion associated directly with the research, the past tense is the least likely to cause confusion.

The present tense can be appropriately used for general knowledge or present commentary (*Appendix 3A*):

Ex. 7.30

> a. Aberrant cells *were* detected in minimal medium, especially in the late stationary phase; however the significance of the observation *is* not clear.
> b. However, the reliabilities achieved in the present studies *were* still only moderate, suggesting that multiple simulations *can* probably *be* administered to attain more stable estimates of subject skills.

It is also used to explain results referred to indirectly:

Ex. 7.31

> a. The limited data available at 1023K *necessitates* a cautious approach to the interpretation of the high-temperature results.
> b. The production of physically exchanged DNA molecules in the absence of virtually any viable haploid or diploid intragenic recombinants *can be explained* in several ways.

When tenses change from one sentence to another or within a sentence, the writer must be sure that the tenses are in logical sequence:

Ex. 7.32

> Since the variances *do* [did] not *appear* to increase with age, it *was* presumed that the higher values in the original survey *were* due to error.

The result about the lack of increase, an observation in the *past*, is in the present tense but precedes the inference in the past tense; therefore the results should have been reported in the past tense. The inference that follows is accurately stated in the past tense, but can be stated in the present tense as a present inference.

7.2.7 Discussion in Social Science Papers

In research papers in the social sciences, discussion has a more substantive function than it has in papers in the natural sciences, and it is not confined largely to the discussion section. The rationale of the research and the theoretical background are discussed in developing a theoretical framework, and the procedures and analyses are usually explained or justified. Some of these discussions may be unrelated to the discussion in the discussion section, but some are really part of the overall argument of the paper.

The pitfall in writing the discussion in a social science paper is diffuse and digressive writing. In trying to make the discussion coherent, the writer has four major hurdles: (1) determining the central line of development of the argument, (2) determining the evidence to be used and its logical relation to the argument, (3) introducing the different pieces of evidence at the appropriate points in the argument, and (4) developing each discussion parsimoniously, so that it does not become too discursive. The first objective, the line of development, is exceedingly difficult, because it is unique to the paper or section. The writer must try to find the focus that will allow a logical development of the argument, yet the focus is obscured by the various interrelationships that the various pieces of evidence highlight.

One approach to developing the line of thought is to lay out the main parts of the argument in clusters, delineating the interrelationships within the clusters. Then from these, a path diagram can be derived which shows the central line of the argument, the interrelations of the various clusters, and the relationships within the clusters.

7.3 RESULTS AND DISCUSSION SECTION

Keeping the discussion in a separate section from the results separates the actual findings from the less durable interpretation of them. Under certain circumstances, however, it may be desirable to combine the two sections into a single section, usually called, "Results and Discussion."

When the results are brief, decisive, or easily presented, then their presentation may not warrant a separate section, for example, if the results are mostly graphic, and the writer has relatively little to say in presenting them. If the discussion is brief, for example, in a paper presenting decisive results, and the introduction has presented the research question so that the confirmatory results require little discussion, then the discussion does not warrant a separate section. Finally, when the results are complex, strongly interrelated, or not sharply defined, so that the results cannot be presented effectively without discussing them, the two sections are combined.

When the combination of the results and discussion is due to the brevity of one of them, the two sections are written much as they would be if written for separate sections. When the combination is due to the close interrelationship of the two (*Table 7.9*), then the discussion may become a complicated interweaving of the various results and their interpretation, and the literature pertinent to them. This combined results and discussion section may be followed by a general discussion section.

7.4 DISSERTATION

In the dissertation, the results and discussion sections are usually separate chapters. In fact, the results section has an important archival value both for the institution and for the discipline. Often it is too costly to publish the voluminous, detailed data presented in some dissertations for the limited circulation of a journal. Yet the data may be very important for particular readers, or they may be difficult or impossible to replicate. Such dissertations can function like published monographs, but be accessed into the library, or stored in a data base, or deposited with University Microfilms International.

TABLE 7.9 Combined results and discussion section

RESULTS AND DISCUSSION

Of the five hypotheses, the first three reached significance. Concerning Hypothesis 1, consonant with Lazarus's view that anxious people regard their general environment as potentially harmful and dangerous, individuals with high trait-anxiety were more concerned over the problem of atmospheric pollution ($\chi^2 = 6.00$, $p < .05$). It may be averred that the level of air contamination, much of it visible for all to see, provided strong objective threat. The pollution impinges so strongly on one's sense of vision and smell, that denial is impossible. When combined with the tendency of respondents with high trait-anxiety to perceive threat in many situations, this finding could easily be anticipated.

Hypothesis 2, that subjects with high trait-anxiety would personally undertake more antipollution measures than those with low elevations also was significant ($\chi^2 = 7.00$, $p < .05$). In this case their concern was translated into positive action to reduce air pollution. This same activity, if replicated through the entire population would have beneficial effects on atmospheric pollution levels.

Hypothesis 3, that awareness of the pollution hazard is a positive function of internality was significant ($\chi^2 = 6.20$, $p < .05$). This finding replicates other research in which cognitive awareness has been associated with internality (Seeman, 1963; Curtis *et al.*, 1984). Internal scorers generally appraise their environment more accurately.

Hypotheses 4 and 5, namely, that obtaining information on atmospheric contamination from the media, and personal actions to reduce atmospheric pollution would be positively associated with internality, were nonsignificant. Hypothesis 4 possibly was nonsignificant because many respondents were of lower socioeconomic status and so had very limited access to the written media. It may be speculated that Hypothesis 5 was nonsignificant because the timing of the interviews corresponded with restrictions on transportation by private cars which had recently been imposed by the authorities. Internal, neutral, and external scorers alike may have believed to a similar extent that sufficient actions were being taken and that further personal actions were unnecessary.

It is apparent from this study that with regard to personal actions to reduce air pollution, trait-anxiety is a better discriminative measure than locus of control. [Navarro, Simpson-Housley, & DeMan. *Anxiety, locus of control and appraisal of air pollution.* Percept. Mot. Skills 64, 1987, 813–14, with permission of authors and publisher.]

The detailed descriptions of objects, and results of analyses, which are too detailed or numerous to be included in the main text, are usually collected in appendices. The greater detail allows the faculty an opportunity to verify the accuracy of observations, descriptions, and mathematical manipulations. The dissertation may also include results that are not directly related to the main research or that are inconclusive, since they may be suggestive for subsequent studies by other graduate students. These too may be segregated in an appendix.

The discussion is also more extensive because it is more comprehensive. Students are expected to demonstrate to the faculty their broad knowledge of the scientific literature as well as their ability to draw broadly on the literature to develop interpretations and conclusions from the research data and to set the results in a broader framework. The focus is on the research problem broadly conceived rather than on the narrow, specific research question. It extends beyond the immediate results and covers the literature more broadly than in a paper. The student may also venture farther in proposing interpretations or explanatory models, provided that he or she makes clear the extent to which the evidence is lacking to support them.

The objective, as in the research paper, is a rigorous logical development of the argument, which is more difficult to achieve, of course, because of the extensive coverage and detail.

Appendix 7A

Treatment of Discussion as General Presentation

The following discussion is written broadly in a form more appropriate for a lecture on the subject of the research rather than as a discussion of the results of the particular research.

Discussion

Abscission rates of mined and unmined leaves

Mining by *Cameraria* accelerates leaf abscission in Emory oak. Adaptive scenarios for why this should be so are easily constructed (Faeth et al. 1981a). The cost of retaining damaged leaves due to loss of water and photosynthate may outweigh any advantage of their continued photosynthetic activity. Many plants drop leaves that are diseased or damaged (Jacobs 1962; Addicott 1981). However, there are several factors that should influence this seemingly adaptive response by plants.

Leaf miner density is often low (< 5 miners/100 leaves) (Faeth and Simberloff 1981; . . . T. Bultman, unpubl. data), and perhaps too insignificant for an evolutionary response by the host. Only when leaf miner densities are chronically high might these insects exert selection pressure on their host plants. For example, at high densities, leaf miners of jarrah significantly reduce girth increase by their host plant (Mazanec 1974). In central Arizona, *Cameraria* occurs on Emory oak at relatively high densities (25 miners/100 leaves) and may influence photosynthetic rates of trees. In contrast, . . .

Size of the mine excavated by leaf miners relative to the size of host leaves may also influence the likelihood of an abscission response by the host. Consumption of large proportions of the tissue of individual leaves by leaf miners should exert stronger selection pressure on hosts to drop mined leaves prematurely relative to leaves that are only slightly damaged by leaf miners. . . . If the proportion of leaf area mined influences the likelihood of early abscission, then *Q. hemisphearica.* which has the smallest leaves, should show a stronger trend toward early abscission of mined leaves than its two congeners. This is the case (Faeth et al. 1981a).

As *Cameraria* mines nearly 80% of the surface of Emory oak leaves, mined leaves might be expected to drop prematurely. Lack of association of leaf-mining with early abscission may be a result of large leaves (relative to mines) of the study trees (Pritchard and James 1984b). . . .

A third factor that may influence magnitude of leaf miner damage to its host and, therefore, likelihood of a host response to leaf miners, is the relationship between phenologies of host plant and leaf miners. Leaves of deciduous trees are productive for a shorter period of time than are those of evergreen host plants. For this reason, leaves of deciduous trees may be more energetically costly than are leaves of evergreen trees (Orians and Solbrig 1977). Consequently, deciduous trees may not drop leaves as quickly as evergreens given the same relative amounts of damage. Studies on abscission of mined vs unmined leaves support this contention:. . . .

Another potentially important factor is the period when leaf miners are active in leaves. Insects that mine newly flushed leaves are more likely to decrease the photosynthetic productivity of leaves than insects that attack older leaves. . . .

Leaf fall and leaf miner mortality

. . . .
. . . Another factor that should influence susceptibility of leaf miners to abscission-caused mortal-ity is the temporal relationship between the phenologies of leaf miners, host plants, and parasi-toids. . . . If leaves normally fall about the time leaf miners exit leaves, then a slight variation in host phenology may substantially influence leaf miner survival. Like *Cameraria*, second genera-tion *Phyllonorycter* leaf miners on red oak often fail to complete larval development because of leaf fall (Askew and Shaw 1979). . . . If leaf miners specialize on new foliage and exit leaves well before normal leaf fall (e.g., *Caloptilia* on oaks (Faeth et al. 1981)), then risks of mortality from leaf fall should be insignificant. . . . If most parasitism occurs during pupal stages of miners then abscission may act to reduce overall mortality by allowing miners in fallen leaves to escape parasitoids (Kahn and Cornell 1983). However, if most parasitism occurs during larval stages then leaves may fall too late to allow escape of parasitoids by miners (Potter 1985).

Factors other than leaf miner damage may also influence propensity of leaves to fall. For example, ... [Oecol. 69, 1986, 118–19, with permission]

8

Abstract and Other Parts of Paper

8.1 IMPORTANCE, TYPES, AND FORM

The scientific literature is so extensive—there are over 30,000 scientific and technical journals—that it is impossible for scientists to read all the research papers that interest them, even in their discipline. Scientific papers, therefore, regularly include an abridgment in the form of an abstract.

8.1.1 Dissemination of Research

The abstract is the most important part of the paper because it is the means of disseminating the research widely. Every effort should be made, therefore, to write it as effectively as possible. Its chief function is to report the most essential *new* scientific information in the paper in a very abbreviated form. Its brevity makes it the effective vehicle for the widest dissemination of research.

Research scientists can read many more abstracts than they can read papers; they can also read abstracts as a screening device to select papers they wish to read. For papers in an unfamiliar language, the abstract in a familiar language is the only part of the paper that surmounts the language barrier, and it thus substitutes for the paper. The abstract is therefore the section most likely to be read: it is the only part of the paper that most readers read, and it may be the only part that is available to some readers.

Abstracts are collected together so that research scientists can have at least some access to papers in journals to which they or their libraries do not subscribe. They are collected in journals, as in the *Review of Plant Pathology*, but most commonly they are collected in indexes of broader disciplines, such as *Chemical Abstracts* and *Sociological Abstracts*. The importance of such indexes for the advancement of science cannot be overestimated, for besides disseminating scientific research, they make it easier to search the literature. For some scientists, the indexed abstracts may be their only access to a paper.

The abstract of a scientific paper is an accurate abridgment of the innovative parts of the paper presented without interpretation or comment. It is a self-contained unit, set off from the rest of the paper, usually at the beginning. It is treated here after the discussion because it is most accurately written after the paper is completed. There are two kinds of abstracts: (1) descriptive or indicative abstracts and (2) informative abstracts. The informative abstract is the most generally used for scientific papers, but for some types of papers, and for some parts of the paper in the informative abstract, a descriptive form may be more appropriate.

8.1.2 Abstract versus Summary

Scientific and technical journals label the abridgment at the beginning of the paper either "abstract" or "summary," (*Appendix 8A*). In some journals the abridgment is at the end of the paper labeled "summary." The term "summary" is also used for sections near the end of the paper, but such summaries have a different function (see Chapter 7).

Although the two terms are often used as synonyms, a useful distinction can be made between them. An abstract is a nonlinear reduction of the paper; a summary is a linear reduction. In an abstract, some parts of the paper may be omitted and the parts that are included are not proportionately reduced; in a summary all parts of a paper would be included and abridged more or less proportionately. The summary is intended to provide the reader with a shortened version of the original. It includes all the major parts of the original document and can vary in length depending on its objectives and the length of the original. An abstract is limited to certain parts of the paper, determined by scientific criteria, and the extent to which the parts are reduced varies, depending on scientific criteria and on the word limit set by journal editors. A summary, as here defined, is therefore often longer than an abstract. In view of the past usage of the two terms, and the confusion caused by the very varied use of the term "summary," it seems desirable to establish a distinction between them.

8.1.3 Descriptive Abstracts

The descriptive abstract is written about the *paper*; it does not transmit the information contained in it (*Table 8.1*). It is essentially a table of contents that indicates the scope and coverage of the paper. Its main function is to inform readers about the

TABLE 8.1 Descriptive abstracts

a.
This study explores the influence of demographic factors on the attitudes of individuals living in non-metropolitan, natural-resource-oriented environments toward their community. [*purpose*] Interviews were conducted in a rural area of Missouri, and data were subjected to factor analysis and analysis of variance. Strength of feeling scores were calculated. Hypotheses are offered regarding the nature of natural-resource-oriented communities and individual-community relationships. [*method*] The investigator suggests implications of his findings for adult education programs. [*reference to discussion*] [Diss. Abstr. Intnl. 32, 1971, 1822-A]

b.
Fast and slow magnetoacoustic shocks are studied in the framework of relativistic magneto-fluid dynamics with the Synge equation of state. [*objective, method*] An approximate analytical solution is presented in a particular case. [*ref. to result*] The general case is treated by numerical methods. [*ref. to methods*] [Phys. Flu. 30, 1987, 3045]

c.
The formal Hilbert space structure of classical mechanics is reviewed with emphasis on the relationship of the spectrum of the Liouville operator to regular vs. irregular motion. [*objective*] Two aspects of this approach are then described. First, eigendistributions of the Liouville operator for the harmonic oscillates and pendulum are displayed, providing insight into this alternative view of classical mechanics. Second, the dynamics of selected classical distributions in regular systems are discussed, with emphasis on the broad range of possible behavior, including periodicity, dephasing, and relaxation. [*methods*]

d.
The kinetic equations for a dilute superfluid developed previously are used to derive the Landau-Khalatnikov two-fluid hydrodynamic equations. [method, objective] *Explicit expressions for the associated transport coefficients are given for very low temperatures and for moderately low temperatures.* [ref. to results] [J. Low Temp. Phys. 58, 1985, 399]

e.
Analytical methods are reviewed with particular emphasis on liquid chromatography. Carbohydrate chemistry is also briefly considered and the possible uses of carbohydrates as drugs and diagnostic agents are outlined. (19 references). [*subjects of review*] [Anal. Abstr. Roy. Soc. Chem. 49, 1987, 431] G. C.

kind of information in the paper. It therefore gives little actual information about the research. The only actual information given is the description of the method, but this is more abbreviated and indirect than it would be in an informative abstract. The descriptive abstract is written for those interested in retrieving the article rather than in obtaining the information from it, that is, primarily librarians, bibliographers, and research scientists searching the literature. Most research scientists write descriptive abstracts only rarely, unless they write descriptive, mathematical, or theoretical papers; they write descriptive abstracts primarily for annotated bibliographies, review articles, and monographs.

Because the descriptive abstract reports on the paper, the different subjects of discussion become subjects of sentences, with the verbs related to their treatment in the paper:

Ex. 8.1

   a. The parameters affecting electron emission *are discussed*.

   b. Industrial exposure to several toxic agents *is reviewed* and case histories *are given*.

   c. The temperature dependences of the fluorescence intensity and lifetime *are described*.

   d. A stochastic model for the population regulated by logistic growth and spreading in a given region of two- or three-dimensional space *is introduced*.

This type of indicative sentence is sometimes found in the informative abstract in any reference to the paper or some part of it, for example, the discussion (*Ex.*

8.11, 8.19). Since the descriptive abstract describes the *paper*, it is usually in the present tense, except for any description of methods, which are in the past tense. The descriptive abstract should not be written in the future tense because it describes the existing paper.

8.1.4 Informative Abstract

The function of the informative abstract is to report the *research* rather than to describe the *paper*. It is written for scientists who want a brief digest of the research, that is, any new methods, results, concepts, or conclusions. Its various uses and its content influence its position in the paper as well as its form. It is usually placed at the beginning of the paper, after the title and the name(s) of the author(s), but before the body of the text (*Fig. 8.1*). It is also set off from these typographically, usually by reduced type, sometimes by italics or boldface. It is often indented from

Pathotypes of the Pinewood Nematode
Bursaphelenchus xylophilus[1]

R. I. BOLLA,[2] R. E. K. WINTER,[3] K. FITZSIMMONS,[2] AND M. J. LINIT[4]

Abstract: An isolate of *Bursaphelenchus xylophilus* from *Pinus sylvestris* in Missouri infected and reproduced in 2–3-year-old seedlings of *P. sylvestris* and to some extent in seedlings of *P. nigra*. Wilting, however, occurred only in *P. sylvestris*. *B. xylophilus* isolated from *P. strobus* in Vermont infected and reproduced only in *P. strobus* seedlings. *P. taeda* seedlings were resistant to both of these isolates. Phytotoxin production was seen only in susceptible seedling species–nematode combinations. Significant water loss occurred only in those seedlings that were wilted because of infection by a compatible nematode isolate. Our results suggest that these isolates are pathotypes of *B. xylophilus*.

Key words: nematode pathotypes, pathology, pinewood nematode, phytotoxin, *Pinus nigra, Pinus strobus, Pinus sylvestris, Pinus taeda,* resistance, susceptibility.

Bursaphelenchus xylophilus (Steiner and Buhrer, 1934; Nickle, 1970; syn. *B. lignicolus,* Mamiya and Kiyohara, 1972) is indigenous to the United States and has been reported in 27 pine and 7 nonpine species (14). Pinewilt disease has been reported from southern Canada to Texas, from the Missouri River basin to the East Coast, and from Vermont to Florida (5,8,14,20–22). Pinewood nematode is also found in California and southern Nevada. The species of pine showing highest susceptibility to infection and expression of disease pathology varies regionally (8,12,14,22). In some regions of the United States, *B. xylophilus* appears to be a primary pathogen responsible for wilting of the tree; in other areas it appears to be a secondary infection in disease stressed trees (20,21). Geographical isolation or feeding preference of transport insects may lead to evolution of specific gene pools within this nematode

Figure 8.1. Abstract showing position and typographical features relative to title, authors' names, and text. Note reduced type, indented margins, extension over two columns of text, key words. [J. Nematol. 18, 1986, 230, with permission]

one or both margins. Where the paper is set in two columns, the abstract may extent across both columns, but it may be restricted to one column yet set off typographically from the text. The heading, for example, "Abstract" or "Summary," may be in italic or boldface type or capitalized to emphasize it, but it may be lacking. The abstract may be followed or preceded by key words (*Fig. 8.1*). It may be set off by a rule above or below it or by both. The forms of physical separation emphasize that the abstract stands independent of the paper. It can therefore be extracted from the paper, without modification, for inclusion in abstracting indexes and other information retrieval systems. The abstract may include the bibliographic details, and, in some journals, is prepared *exactly* in the form for abstracting, with author(s) and title repeated in the abstract, and the surname of the author(s) first to allow for immediate alphabetizing (*Appendix 8A.6*).

The abstract usually consists of one paragraph, a format that saves space and that points up the necessity for brevity, although a few journals have abstracts written in paragraphs or list of sentences, usually enumerated (*Appendix 8.A.3a,4a,c*):

Ex. 8.2

SUMMARY

(1) Fifteen species of aquatic vascular plants were examined for the presence of alkaloids.
(2) The alkaloid content ranged from 0.13 mg g^{-1} dry weight in *Heteranthera dubia* to 0.56 mg g^{-1} dry weight in *Potamogeton crispus*. These concentrations are within a range which has been found to be pharmacologically active in previous studies, and can therefore serve as potential deterrents to herbivory.
(3) Plant species examined contained from two to nine alkaloids isolated by thin-layer chromatography. Aklaloid arrays showed little similarity among species, even among nine species of the genus *Potamogeton*. [J. Ecol. 74, 1986, 279]

TABLE 8.2 Abstract of review paper consisting of brief abstract and outline

The authors discuss the concepts on which the interacting boson model is based, in particular those of pairing and seniority, which lead to a connection with the shell model. They review some of the calculations performed so far using these concepts and briefly comment on their results.

CONTENTS

I. Introduction	339
II. The Interacting Boson Model 1	340
III. The Interacting Boson Model 2	344
A. The model	344
B. Nucleon pairs and seniority	345
C. The proton-neutron interaction and nuclear deformation	348
IV. Microscopic Calculations	350
A. Calculations in a spherical basis	350
1. The single j shell	352
2. Several degenerate shells	353
3. Several nondegenerate shells	353
B. Calculations in a deformed basis	354
V. Other Topics	356
A. F spin and Isospin	356
B. Valence nucleons versus all nucleons	358
VI. Conclusions	360
Acknowledgment	360
References	360

[Rev. Mod. Phys. 59, 1987, 339, with permission]

Journal editors often set a word limit for the abstract—about 200–250 words, though the abstract should be only as long as is needed for reporting the research. This limitation keeps the abstract within the space limitations suitable for indexing. In papers reporting comprehensive research or several unrelated experiments, it may be necessary to exceed the limit. In a review journal, an outline of the paper may largely or entirely replace the abstract (*Table 8.2*).

8.2 PARTS OF PAPER REPRESENTED IN ABSTRACT

The abstract does not represent all parts of the paper, and those included are not equally condensed. Only those parts are included that are related to the nature of scientific inquiry and the needs of research scientists; that is, the (1) research objective, (2) methods, (3) results, and (4) conclusions. The statement of the research objective is necessary to characterize the research. A description of the methods is necessary because they are the basis for validating results, which are, of course, the ultimate objective of the research, and conclusions are the final inferences.

8.2.1 Introduction

The introduction of the paper is represented in the abstract by at most one sentence. It states the subject of the paper, either as the purpose or the objective of the research or as the research problem. This statement may include a characterization of the study, such as the development of a model, especially in the social sciences.

Many an abstract gets off to a bad start by beginning with an introduction (*Table 8.3*) because the writer fails to differentiate between the function of the paper and the abstract. The introduction to the paper introduces the topic or subject of the research. It provides, in addition to the research objective, background information. However, background information has no place in the abstract, even in an abridged form, and many writers waste the first sentence or more of the abstract trying to introduce the general topic or provide some background for the research (*Table 8.3, Appendix 8A.1*). In some abstracts, the background introduction forms a major part of the abstract (*Table 8.3.3* and *4, Appendix 8A.1b–d*), although it contributes no new information.

The function of the opening sentence(s) in an abstract is not to supply readers, who are peers, with background, but to *announce* the subject or objective of the research. Readers of abstracts are interested mostly in results and conclusions, and methods as they are related to these or as they are new. Background material is not new. Moreover, the same general background can apply to closely related papers; therefore including an abbreviated version of the background would result in repetition of essentially the same introductory statement(s) in the abstracts of other papers on the same subject. Such redundancy wastes space in abstracting indexes, wastes the reader's scanning time, and reduces the scientific information disseminated. The first sentence should simply state the subject or object of the research:

TABLE 8.3 Abstracts beginning with background to topic rather than addressing research directly or devoted largely to background.

---

1. Caribbean pine (*Pinus caribaea*), an economically important tree of tropical lowlands, is at risk of SO_2 exposure in certain locales. . . .

2. Although previous research has examined the development of children's conceptions of friendship, two major limitations currently exist. First, previous investigators have relied principally on . . . the open-ended interview. Second, little is known about children's expectations of other peer relationships or how friendship expectations are distinguished from them. [*background*] These two issues were addressed in the present study [*purpose*]. . . . [Develop. Psych. 20, 1984, 925]

3. *The marama bean,* Tylosema esculentum, *is a drought-tolerant legume native to southern Africa. Its seeds are comparable to soybeans in protein content and quality, its oil content approaches that of peanuts, and the plants might be desirable as a forage legume. Although the marama bean has great potential as an arid land crop in the United States, studies of the species are extremely limited.* [background] *The current investigations have demonstrated that the plants can be grown successfully in an arid region of the United States, at least under experimental conditions, and that a healthy seed crop can be expected in about 4.5 years.* [results, conclusion] [Econ. Bot. 41, 1987, 216]

4. Modifications of material surface properties due to interactions with ambient atomic oxygen have been observed on surfaces facing the orbital direction in low earth orbits. Some effects are very damaging to surface optical properties while some are more subtle and even beneficial. Most combustible materials are heavily etched, and some coatings, such as silver and osmium, are seriously degraded or removed as volatile oxides. [*background*] The growth of oxide films on metals and semiconductors considered stable in dry air was measured. [*objective*] Material removal, surface roughness, reflectance, and optical densities are reported. Effects of temperature, contamination, and overcoatings are noted. [*ref. to results*] [Appl. Opt 25, 1986, 1290]

Ex. 8.3

   a. A study was made of the *in vivo* effects of various vegetable juices on DMBA-induced chromosome aberrations in rat bone marrow cells.

   b. The present paper reports a study of three cases presenting the pathological aspect of chronic nonevolutive MS, corresponding to a clinical picture of benign MS.

8.2.2 Literature

The abstract should have no reference to the literature; research scientists read abstracts of scientific papers for *new* developments. If the literature were included, readers might be repeatedly reading about the same earlier research. Also, citing research in one abstract that has already been reported in other abstracts takes up precious space in the index. Besides, citing one reference makes it disproportionately important. If references are used, they must be cited in full, because the abstract must stand alone. However, they then take up a disproportionate amount of the limited space. References in abstracts are therefore to be avoided.

Nevertheless abstracts do sometimes refer to earlier studies. A general reference to recent research as in:

Ex. 8.4

   a. This phenomenon has earlier been reported for ducklings by other authors, who suggested a cooperative function in attracting the parent(s).

   b. Elementary radical-molecule or radical-radical reactions have recently been investigated directly rather than by modeling of multistep mechanisms.

serves as an introduction to the topic and should be avoided. The reference may simply name the research scientist:

Ex. 8.5

    a. We consider the single-period inventory problem in which the amount received is a random variable. Silver has investigated such a problem by analyzing a general EOQ model, where he assumes a deterministic demand and develops analytic results for different situations. . . .

    b. A new probabilistic-informational concept, earlier constructed by Mugur-Schächter, is further developed.

Unless the scientist is so outstanding that the name will be recognized by contemporary and future readers, however, the writer will have failed to communicate his or her message. The abstract may have an author-date or numerical reference, in which case readers must refer to the paper for the full reference (*Ex. 8.6a*).

If a reference is absolutely *essential* for the abstract, it should be cited briefly but adequately and accurately (*Ex. 8.6b,c*):

Ex. 8.6

    a. The MANOP Site C data are well explained by using organic-matter oxidation rates in accord with the minimum O_2 fluxes suggested by *Reimers et al.* (1984).

    b. Mandel et al. (1975, *BBA* **408**:123), using *Cecropia* midget, suggested that crt. b_5 linked oxidative metabolism and active transport of potassium.

    c. DNA helix openings can be visualized, measured, and counted, during the course of *in vivo* cell differentiation within human bone marrow (*Nature* **248**, 334, 1974).

8.2.3 Methods

If the subject of the paper is stated in the title, it should not be repeated in the text of the abstract because the title always accompanies the abstract (*Table 8.6*). In such abstracts, or when the subject is obvious, the abstract may start directly with the methods (*Appendix 8A.3*), as in:

Ex. 8.7

    a. *Bacillus thuringiensis* spores and parasporal crystals were incubated in natural soil, bo' ı in the laboratory and in nature.

    b. Samples of 758 children in the United States and 220 children in Finland were interviewed and tested in an overlapping longitudinal design covering Grades 1 to 5.

    c. Exact quantum results [*results*] for collision-induced dissociation on a reactive surface [*research objective*] are presented. A modified LEPS potential-energy surface modeling the H + HD — H_2 + D system has been used [*method*].

Some abstracts begin with the methods and then return to the purpose, which is then followed by results, or the opening sentence may begin with results and then return to the methods (*Ex. 8.7c*).

The extent to which the methods are abstracted depends upon how important and innovative they are. The accent is on *new* and *significant* materials and methods: new compounds, new drugs, new equipment, new procedures, new analyses, and new modifications of old equipment, procedures, and analyses. If the method is standard or well-known to readers in the discipline, or can be referred to by name, there may be only an indirect reference to it. The abstract may go directly from the statement of purpose to the results. If the paper is primarily a report of a new method or procedure, the description of the method may make up a large part of the abstract:

Ex. 8.8

    a. A convenient method of injecting reference pulses into a gamma ray spectrum is described. The method utilizes an injection rate which is proportional to the gross gamma count rate; the circuit described automatically adjusts the pulse injection rate as the gross count rate changes. The method has been successfully tested using a wide dynamic range spectroscopy system. This injection technique has been installed in an automated scanning system designed for unattended operation. [*methods*] Good results have been obtained for measurement of hundreds of spectra having wide gross count rate variations. [*results*] [Nucl. Instr. Meth. Phys. Res. A256, 1987, 355]

    b. A simple procedure for drying and storing of polyacrylamide slab gels is described. A polyacrylamide slab gel is fixed in acetic acid plus glycerol and then sandwiched between a gel bond plastic sheet and a dialysis membrane in the presence of a minute amount of gelatin and dried on the benchtop at room temperature. [*methods*] The fixed gel can be stored indefinitely. [*discussion*] [Anal. Biochem. 163, 1987, 42]

8.2.4 Results

Results are the most important part of the abstract, because they are the new additions to scientific knowledge. In fact the results may dominate the abstract:

Ex. 8.9

    a.

Abstract—Feeding of ($\pm$)-abscisic acid to leaves of *Xanthium strumarium* resulted in formation of a new metabolite. The compound was identified as 7'-hydroxy ($-$)- R-abscisic acid by high resolution mass spectrometry of its methyl ester and monoacetate, and by optical rotary dispersion. The numbering system for abscisic acid has been extended to include the exocyclic methyl groups. Feeding racemic [2- ^{14}C] abscisic acid to *Xanthium* leaves resulted in *ca* 20% conversion of the radiolabelled compound into the new metabolite. Evidence is presented that, in *Xanthium,* only the synthetic ($-$)-R-enantiomer of abscisic acid is hydroxylated at the 7'-position. [Phytochem. 25, 1986, 1103]

    b. Studies showed that *L. willkommii* was transmitted from natural virgin forest to plantations through windborne ascocarps infecting dead buds and branches as well as wounds. Natural and planted stands seriously damaged were usually between 6 and 30 yr old and infected branches were mostly aged 3–5 yr. Apothecia were present on infected branches throughout the year. The release of ascocarps was related to precipitation and reached a peak between June and Sept. The most appropriate conditions for germination were a temp. of 15°C, 100% RH and a pH value of 5.5. [Rev. Pl. Path. 66, 1987, 221]

(see also *Appendix 8A.4*). The writer may hesitate to clutter the abstract with measurements, quantities, or values, but in reporting scientific results, the criterion for including extensive quantitative data must be their essential scientific importance, not their digestibility. Peers who are interested in the results or the research will welcome the data.

Many of the results are presented in tables, graphs, diagrams, and other graphic forms, which are too space-consuming to be included in the abstract. Besides, they have more information than is needed in an abstract. Ironically, then, the conciseness achieved in putting data in graphic form cannot be applied in writing the abstract. Referring to figures or tables in the abstract is uninformative for readers who do not have the text.

When the results include many tables or graphs, the writer may use the indicative type of statement, for example, "Histograms show the classes of truck and loads for both U.S. and European bridges." Mathematical derivations and proofs are also unsuited to the abstract and are similarly omitted. However, in some disciplines, editors accept abstracts that include equations (*Table 8.4*) and rarely even an illustration.

8.2.5 Discussion and Conclusions

Although the discussion is an important part of the paper, it is not included in the abstract, except for the conclusions. The report of the method and results will stand as an accurate report of these in the future. Both become part of the body of scientific knowledge, which research scientists in related work must include in their considerations in any subsequent research. This is much less true for the discussion; any explanation of the results may change at any time because of the many factors that influence it, for example, how many peers are studying the problem, and their results, or how well developed the theory and technology are. A recent paper, or a recently developed concept, or even one's rethinking of the research may change one's interpretation overnight, or right after it is printed. Therefore, for scientific research, the most interesting part of the paper may be the most ephemeral part. To include such ephemeral material in an abstract is to waste space on interpretations

TABLE 8.4   Abstracts with (a) equations and (b) numbers representing complex
                organic compounds

a. *With Equations*

Results from percolation theory are used to study phase transitions in one-dimensional Ising and q-state Potts models with couplings of the asymptotic form $J_{x,y} \approx \text{const}/[x-y]^2$. For translation-invariant systems with well-defined $\lim_{x \to \infty} x^2 J_x = J^+$ (possibly 0 or ∞) we establish: (1) There is no long-range order at inverse temperatures β with $\beta J^+ \leq 1$. (2) If $\beta J^+ > q$, then by sufficiently increasing J_1 the spontaneous magnetization M is made positive. (3) In models with $0 < J^+ < \infty$ the magnetization is discontinuous at the transition point (as originally predicted by Thouless), and obeys $M(\beta_c) \geq 1/(\beta_c J^+)^{1/2}$. (4) For Ising ($q = 2$) models with $J^+ < \infty$, it is noted that the correlation function decays as $\langle \sigma_x \sigma_y \rangle(\beta) \approx c(\beta)/|x - y|^2$ whenever $\beta < \beta_c$. Points 1–3 are deduced from previous percolation results by utilizing the Fortuin-Kasteleyn representation, which also yields other results of independent interest relating Potts models with different values of q. [J. Statist. Phys. 50, 1988, 1]

b. *With Compounds Represented by Numbers*

The (cycloalkylidenemethyl)triphenylphosphonium salts **2b–d** were synthesized in high yields by phenylselenylation of (cycloalkylmethylene)triphenylphosphoranes with benzeneselenenyl bromide to the [cycloalklyl(phenylseleno) methyl]triphenylphosphonium salts **1b–d** and subsequent oxidative elimination of the phenylseleno moiety, while the synthesis of the (cyclopropylidenemethyl)triphenylphosphonium salt **2a** was unsuccessful. Hydrolysis of **2b–d** in aqueous THF and methanol containing sodium hydroxide was studied. The reactions of the (cyclohexylidene-methyl) phosphonium salt **2d** and 1.1 equiv of butyllithium with aldehydes **9** gave alkenylcyclohexenes **10d–f** in 56–81% yields, whereas similar reactions using 1.5 equiv and 2 equiv of butyllithium produced allenes **11d–f** as major products together with small amounts of **10d,e**. Similar reactions of the (cyclopentylidenemethyl) phosphonium salt **2c** and butyllithium with **9** gave alkenylcyclopentenes **10a–c** in 56–80% yields, regardless of the amount of butyllithium used. The formation mechanism of **10** and **11** was discussed. [J. Org. Chem. 53, 1988, 2937]

that might soon be inapplicable or invalid. Furthermore, even if one included the discussion it would take up an inordinate amount of space:

Ex. 8.10

> Analysis of questionnaire data from 51 matched male/female pairs of engineers shows that even though these pairs are in very similar positions and share similar orientations to the role of work in their lives, the women have a more ambivalent attitude toward technical expertise. [*method*, *result*] The reasons seem not to lie in complications stemming from women's multiple roles, but relate, rather, to the singular, more "masculine" way that technical work is defined. In managerial roles, in contrast, where criteria of effective performance are more difficult to specify, women seem to be engaged in new models which are associated with less ambivalence. Such diversity, it is argued, is useful for all employees, male and female. [*discussion*] [Hum. Rel. 41, 1987, 299]

(see also *Appendix 8A.5*). Such abstracts report very little research. For these reasons, the discussion should be omitted from the abstract. However, if the discussion in the paper is extensive or important, reference can be made to it indirectly, as in a descriptive abstract (see also *Ex. 8.19, 8.20*):

Ex. 8.11

  a. Three different physical interpretations of the yield stress *are considered*.

  b. An alternative picture where the higgsino is the lightest supersymmetric particle *is* briefly *discussed*.

  c. The results *are discussed* in terms of population strategies in hazardous environments.

From the discussion, therefore, only the conclusions, that is the inferences drawn, are included in the abstract. Being inferences, conclusions are usually single statements:

Ex. 8.12

  a. It is concluded that no special theories positing status frustration are required to account for moral-reform social movement adherence.

  b. Overall, the findings reflect the strong contribution of cognitive development to the growth of gender understanding.

  c. Attempts to determine the identity of phosphate minerals from solubility product measurements were inconclusive.

  c. In broad terms, the results favor some form of an open model of the magnetotail and are inconsistent with diffusive entry of protons across a closed magnetotail field with a time constant substantially greater than a few minutes.

and cannot be greatly condensed, being essentially a restatement of the conclusions in the paper (*Ex. 8.22, Appendix 8A*).

Recommendations for further research and plans for further research should be omitted from the abstract. They do not supply scientific information. The ideal abstract therefore consists largely of methods and results, an opening sentence stating the research objective and a closing sentence stating the conclusion(s). Except when the order is prescribed, as in chemical abstracts, the order may be varied.

Some abstracts show special variations. They may contain statistical material, illustrations, or equations (*Table 8.4*).

8.2.6 Omissions, Additions, and Variations

As has been indicated, some parts of the paper are not represented in the abstract because: (1) they provide no *new* information, (2) they are space consuming, and (3) they may be ephemeral. Some elements are omitted because the abstract must be self-contained and separable from the rest of the paper, for example, references to the literature or cross-references to tables, figures, or equations. Such references are superfluous for readers who have the paper available and unavailing to readers who have access only to the abstract.

Although the abstract must be self-contained, it is derived entirely and directly from the paper. Therefore, nothing must be added to abstract that is not included in the paper. It is expected to be a condensation of the *paper*, not a report of the writer's knowledge about the research at the time of writing it. However, indexing results in some bibliographic additions—the title and author(s), journal, volume, date, and pages. Indeed in some journals, the full bibliographic citation may be incorporated in the abstract in the paper, as in:

Ex. 8.13

> Keller, A. 1987. Modeling and forecasting primary production rates using Box–Jenkins transfer function models. Can. J. Fish. Aquat. Sci. 44: 1045–1052
> Box–Jenkins transfer function models were developed for time series of integrated hourly primary production rates. A 28-mo record of 56 biweekly measurements collected from seven mesocosms during a nutrient addition experiment was analyzed. Incorporation of two input variables (phytoplankton biomass and hourly light) significantly improved the fit of the models. When compared with standard regression models, the time series models all had reduced residual variance. The forecasting ability of the final fitted model for a control system was demonstrated with independent data from the two replicate control mesocosms.

Such an abstract has associated with it, all the information needed for indexing the paper (*Appendix 8A.6*) and can be transferred directly to indexes or information retrieval systems without requiring that the editors add the bibliographic information to it.

Key words are an entirely new addition to the abstract. These are subject headings under which the paper can be indexed:

Ex. 8.14

> a. Key words: Glycoprotein; Concanavalin A; Antigen isolation; (Hamster pancreas)
> b. INDEXING KEY WORDS: lactation . milk synthesis . energy expenditure . metabolic efficiency
> c. KEY WORDS: geostatistics, kriging, Bayesian statistics

Their chief function is to help in the retrieval of the paper. They are usually placed after or before the abstract (*Appendix 8A. 3,4,6*) and can be arranged to be machine read.

8.3 WRITING THE ABSTRACT

8.3.1 Objectives

The abstract should be the most effectively written part of the paper, because it is the main source of information for most scientists in the discipline. The objective in writing the abstract is to convey as much *new* scientific information as possible to scientists in the same or related discipline in as *few words* as possible—accurately. Writing the abstract is a continual balancing of substance against words. Despite their brevity, abstracts are often not economically or effectively written. The abstract is undoubtedly seen as a sort of summary, but because the meaning of the term "summary" is variously interpreted, objectives vary widely and abstracts vary accordingly.

Accuracy and brevity.   Even more than in the paper, accuracy is a stringent requirement for the abstract. Since the contents of the abstract are almost entirely new and the only source of the information for many readers, the writer must be meticulous about every piece of information included. Numbers, symbols, and other nonword elements should agree with those in tables, figures, and the text of the paper, and new names, new terminology, new localities, and new compounds should be consistent with the text and correctly spelled.

The abstract must be short; for some papers, it may consist of only a few sentences (*Table 8.5*). It is obvious from the limited space allotted to the abstract that brevity is the overriding requirement rather than conciseness. The objective is maximum content per word. As a result, the abstract should be the most densely written part of the paper. In abstracting the paper, first the parts to be included must be drastically reduced; then wordiness and repetition must be eliminated (*Ex. 8.15*) because they take up space that could be taken up by words that carry information:

Ex. 8.15

    a. *In confirmation of previous findings*, nigrotectal neurons *which had been* identified by the retro-grade transport of. . . .

    b. Analyses *of the following characteristics* were made *while the* tomatoes *were* at the table-ripe stage of maturity:. . . .

    c. *A direct and very efficient approach* for obtaining sensitivities of two-point boundary value problems solved by Newton's method is studied. . . . *This approach is employed in*. . . .

    d. *This is because* nitrogen use declines *in response to* a reduction in energy use. *On the other hand a reduction in energy use* brings more cropland under tillage and thus increases pesticide use.

    e. Treatment of primary leaves of asparagus bean (*Vigna sesquipedalis* Fruhw.) [omitted if in title] with salicylate *had different effects on resistance to local lesion development caused by tobacco necrosis virus (TNV), depending on the concentration used for treatment.* At 1 mM salicylate . . . , while at 3 mM it. . . .

TABLE 8.5   Abstracts consisting of only a few sentences, all descriptive abstracts

a. Three aspects of "quantum chaos" are discussed: ergodicity, loss of memory of the initial state, and the distribution of eigenvalues to be expected in the spectrum of a "chaotic system." [J. Phys. Chem. 88, 1984, 4823]

b. Abstract—1. An *in vivo* microscopy technique is described for observing the microcirculation in the shell gland of the anesthetized Japanese quail.
       2. The preparation allows visualization and quantitation of changes in luminal diameters of arterioles and venules in the intact shell gland. [Comp. Biochem. Physiol. 82A, 1985, 521]

c. Monte Carlo simulation can provide a direct determination of the distribution of quarks inside hadrons. Such distributions are useful for a variety of purposes, including the study of confinement. We discuss the theoretical and practical issues involved. A detailed study of two-dimensional QCD is described. [Ann. Phys. 168, 1986, 284]

d.   ABSTRACT
    Some measurements, taken at the Näsudden wind energy test site, for small separation distances at a height of 65–77 m, are analyzed in terms of a theoretical coherence model for isotropic turbulence, and for Davenport similarity. The decay parameters for the horizontal wind components are found to be about the same for vertical and horizontal separation in the inertial subrange. [J. Clim. Appl. Meteor. 26, 1987, 1770]

Transitions and connecting phrases also have no place in an abstract (*Ex. 8.15*), because the abstract is not a coherent paragraph. It is in paragraph form for economy of space. It actually consists of a series of sentences stating: the objective, the methods, the results, and the conclusions. Therefore empty phrases like "in this case" and "in order to" are to be avoided, as are other wordy or ambulatory phrases (*Appendix 8A.7*).

Repetition is an indication of commonality, and can be avoided when revising the abstract. For example, a generalizing statement (*Ex. 8.16a*) should not be used to introduce a more concrete or specific statement that (*Ex. 8.16b*) presents the same information:

Ex. 8.16

     a. The treated and control groups showed a marked difference in enlargement. The diameter of the lumen was twice that in the treated group.

     b. The diameter of the lumen in the control group was twice the diameter of that in the treated group.

Abbreviated forms, although undesirable in a paper, can be particularly useful in the abstract, where the need to make maximum use of space allows for freer use of standard or common abbreviated forms and greater freedom in coining them. Some journals provide contributors with lists of accepted abbreviated forms. In abstracts of papers in organic chemistry, boldface numbers are used for complex compounds to save space (*Table 8.4b*), although the reader must then go to the paper to retrieve them.

8.3.2 Process of Writing

Reducing paper to paragraph. The process of writing an abstract is a repeated balancing of words against information—now deleting words or phrases or condensing sentences, now determining whether to sacrifice some information. One cannot expect therefore to dash off an abstract in final form in one draft.

The abstract is best written from a copy of the paper. First, one can examine the introduction and the discussion, and highlight the research objective and the conclusions. Then one can dismiss these two sections and read the remaining sections—the methods and results, bracketing material that is important enough to *consider* including in the abstract and circling important words for key words. This method of "compiling" the abstract has the advantage of eliminating certain sections from consideration from the start. It also helps to avoid starting the abstract by giving the background of the research. The writer can then write a preliminary draft. This will be too long; however, the extra length allows more latitude in excising verbiage and less important information.

Now begins the process of balancing information against words. First, the writer should read for obvious wordy or redundant phraseology. Next the deletion of information must be considered. The writer should compare the results with the methods, to determine an appropriate apportionment of space and then delete details accordingly. If the abstract is still clearly too long by three or more sentences, further information may have to be sacrificed. The draft is now ready for serious revision and editing—the restructuring of sentences to avoid wordiness and repetition and to retain as much information as possible. This final draft can now be edited for accuracy to ensure that figures, quantities, values, symbols, and so on, are consistent with those in the text.

Opening sentence.   The opening sentence is the main pitfall in writing the abstract. It may state the purpose of the research or paper, but if the title makes this clear, the first sentence need not repeat or paraphrase it (*Table 8.6*) because the

TABLE 8.6   Title of paper and first sentence of abstract illustrating redundancy

1. "When *Caenorhabditis elegans* (Nematoda: Rhabditidae) Bumps into a Bead"
    When *C. elegans* bumps into a bead, it. . . .

2. "New Criteria for Polynomial Stability"
    New criteria for real-polynomial stability developed during 1957–1978 are introduced.

3. "A Method for Staining Pollen Tubes in Pistil"
    A quadruple staining procedure has been developed for staining pollen tubes in pistil.

4. "Stress Intensity Factors for Notched Configurations"
    A method is presented for approximating stress intensity factors for notched configurations without cracks.

5. "A Low Cost Apparatus for the Automatic Determination of Precise Oxygen Equilibrium Curves on Red Blood Cell Suspensions"
    An appartus for the automatic recording of precise oxygen equilibrium curves on red blood cell suspensions, as well as hemoglobin solutions, is described.

6. "Flexible Manufacturing Systems: A Review of Analytical Models"
    This paper reviews recent work on the development of analytical models of Flexible Manufacturing Systems (FMSs).

title always accompanies the abstract. The abstract can then begin directly with the method.

If an introductory sentence is to be used, it should not introduce the general topic or provide background for the abstract. One way to avoid slipping into a general introduction when beginning an abstract is to start with a sentence that addresses the paper or the study: "this paper," "this study," "this research" (*Ex. 8.17a, c*). There will then be little danger of falling into introducing the *topic* to provide general background as one does for a paper. Such sentences can then be made more concise by shifting to the substance of the research, as in *Ex. 8.17b, d.*

Ex. 8.17

   a. This paper explores the potential role of family socioeconomic factors in school achievement outcomes at two separate periods in the life course—early in childhood and during late adolescence.

   b. The potential role of family socioeconomic factors in school achievement outcomes was explored for early childhood and late adolescence.

   c. This paper develops and tests hypotheses about the characteristics of organizations and their environments that favor the proliferation of detailed job titles to describe work roles.

   d. Hypotheses are developed and tested about the characteristics of organizations and their environments that favor the proliferation of detailed job titles to describe work roles.

Methods and results.   Methods and the results follow and should be presented without commentary or discussion. If the title states the purpose clearly, the abstract can begin with an abridgment of the methods (*Appendix 8A.3*). The methods may consist only of one sentence:

Ex. 8.18

   a. Seven parental tansy (*Tanacetum vulgare L.*) plants were crossed by artificial pollination in the experiments in order to develop a schematic model for the genetic control of the thujone skeleton monoterpenes: thujone, sabinene and umbellulone.

   b. Positron annihilation lifetime spectra were measured for solutions of $1,2,3,5\text{-}C_6H_2Cl_4$ in hexane, toluene, *m*-xylene, and mesitylene, CCl_4 in hexane and toluene, and C_2HCl_3 in *n*-hexane for concentrations below 1M and at various temperatures between $-30°$ and $67°C$.

In papers in which the methods are complex, they are reported in greater detail. If the new methods or techniques are the focus of the research, the methods may make up most of the abstract (*Ex. 8.8*). Then the results may be brief and may make up a small part of the abstract.

Since the results are the part of greatest interest to peers, the major part of the abstract is given to results; in fact the abstract may be largely results (*Ex. 8.9, Appendix 8A.4*). The objective is to report the key values and principal results, or to summarize the results in as few words as possible. In research that consists of a series of experiments, the results should be treated together if possible, since treating them separately is likely to be repetitive, and so space consuming.

Discussion and conclusions.   Discussion is not included in the abstract, although reference can be made to an extensive or important discussion (*Ex. 8.11*). Such a reference can be a general statement:

Ex. 8.19

> a. Interpretations for these findings are discussed.
> b. The physiological implications of the results are discussed.

but it can be more substantive or informative, and still avoid interpretation or commentary:

Ex. 8.20

> a. The results are discussed in relation to other work implicating selective elicitors in the leaf rust-wheat system and in the interaction between *Cladosporium fulvum* and tomato.
> b. The possible significance of the pattern of termination of striatonigral fibers in the substantia nigra is discussed with reference to the known dendritic arborization of nigral neurons.
> c. The role of the different compounds in relation to Dutch elm disease symptom development are discussed.

Statements giving interpretations, explanations, or commentary, such as the following, do not belong in the abstract:

Ex. 8.21

> a. Observational learning undoubtedly plays a role, but its role may be no more important than the attitude changes that TV violence produces, the justification for aggressive behavior that TV violence provides, or the cues for aggressive problem solving that it furnishes.
> b. The first result validates our experimental procedure, but the second result suggests that. . . .

The conclusion, which is the only part of the discussion that belongs in the abstract, usually closes the abstract. It may be stated as the interpretation of the findings or as a suggestion about the meaning of the results (*Ex. 8.12*):

Ex. 8.22

> a. The study suggests that the concept of cultural capital can be used fruitfully to understand social class differences in children's school experiences.
> b. The PiB having a sudden onset are probably the magnetic manifestation of the stepwise motion of the westward electrojet current wedge poleward and/or westward.

The order that has been described follows the order of the sections of the paper and so is easiest for the readers to translate back to the paper. However, the order may differ from this, sometimes because of the nature of the research or the emphasis desired and sometimes because of confusion about the function or structure of the abstract. *Appendix 8A* shows the structure of various types of abstracts and illustrates differences in their organization.

Key words.   For some journals authors are asked to provide key words for indexing the paper; it is presumed that the authors best know the various audiences that will be interested in their paper and the subject headings under which readers might search for it. Before choosing key words, the writer might consult subject headings in common abstract indexes or other information retrieval systems. In some disciplines, authors are expected to use key words derived from a standard subject index, as for *Index Medicus*. In compiling the key terms, the writer may include only the key words in the title, if the title is comprehensive. However, the paper may cover more subjects than are indicated in the title, or it may be of interest to peers outside the discipline; therefore, additional key words may be taken from the abstract. Key words also have been identified in reading the paper in preparation for the abstract. The various terms are then collected, and the most pertinent ones are included with the abstract, according to the style of the journal.

Camera-ready copy.   For some journals and for conferences, camera-ready copy is required for the abstract, and sometimes for the paper as well. This is best done by typing an advanced draft of the abstract within the boundaries provided. If the excess is small, then one should try to eliminate wordiness, empty phrases, and repetition. If the excess is greater, it may be necessary to sacrifice some information. Using a type of smaller size as a means of including more words must be considered in the light of decreased legibility after reduction.

8.3.3 Author, Person, and Audience

The informative abstract is written as by the author of the paper. Even when it is written by an abstractor, the abstractor writes in the role of the author, not by reference to the author. It is written about the paper or the research, not about the author. It should be written in the third person, even if the writer felt impelled to use the first person in the paper. Writing it in the first person is letting a fetish of style deflect one from one's scientific objectives. Consider scientists searching the literature having to read through an index in which abstract after abstract is written in the first person (italics added in *Ex. 8.23*):

Ex. 8.23

    a. *We* report on experiments performed with droplets of supercooled liquid H_2, maintained in a state of neutral buoyancy in pressurized ^{4}He fluid. *We* have measured the nucleation rate for solidification as a function of temperature, and have observed and analyzed several interesting hydrodynamic effects relating to the motion of the drops and interactions between them. *We* discuss the possibility of supercooling liquid H_2 to sufficiently low temperatures to observe the superfluid phase. [Phys. Rev. B 36, 1987, 6799]

    b. *We* calculate the site-dependent susceptibility and the neutron scattering from factor S_Q of the random-anestotropy model. . . . *We* obtain for S_Q a Lorentzian and a Lorentzian-squared term. . . .

Although abstract indexes have a much broader audience than journals, research scientists cannot write the abstract in language that any user of library services can understand. They cannot convey enough scientific information to peers by addressing such a broad audience, and with the enormous proliferation of scientific research, research scientists' need to know makes them the primary audience for abstracts. Therefore the audience to address in the abstract consists largely of scientists, mostly peers. Authors must also take cognizance of the international circulation of abstracts, which is much greater than that of journals. They must consider readers for whom English is not a first language and should avoid colloquial, popular, or humorous usage and use shared terminology, where possible.

8.3.4 Tense

Because of confusion about the form of the abstract, the use of tense varies widely.

Past tense.   The past tense is the only tense that can be used consistently throughout the abstract. It is the most precise and rigorously scientific tense for reporting research. The tense of the opening sentence is especially important for establishing the universe of discourse: it determines whether the abstract is giving background (*Table 8.3*) or reporting on research. The abstract may report the subject of the paper in the past tense and can then continue in the reportorial mode reporting methods and results in the past tense:

Ex. 8.24

   a. This study *undertook* a developmental analysis of the exploratory patterns of 112 children between 4 and 12 years of age.
   b. Dominance relations among adult female American bison (*Bison bison*) *were studied* in the National Bison Range.

The abstract will then all be in the past tense, with no shifting of tense and no confusion about how much of the abstract is reporting research.

When the title makes it superfluous to repeat the purpose or subject in the first sentence, the writer can begin by reporting the methods (*Ex. 8.25a,b*), then report the results (*Ex. 8.25c*), both in the past tense, because they were past events:

Ex. 8.25

   a. A CW magnetic resonance study of phosphorus-13 *was made* in samples of bone from humans and various animals.
   b. Glucose, glucitol, and sucrose in 0.2M-NaOH *were determined* by flow-injection analysis with pulsed amperometric detection at a gold electrode.
   c. Deaeration of the HPLC phase *improved* detection, but polishing the electrodes *resulted* in an unstable baseline.

When the sentence presenting the results is referential,

Ex. 8.26

   a. The results *provided* little evidence that immune suppression *was associated* with the improvement in the clinical condition.
   b. The results *show* that aggression frequencies *were* highest in habitats where forage *was* sparse and heterogeneous in quality.

  c. The electrophysiological data *demonstrate* that the electromotoneurones *behave* similarly *in vivo* and *in vitro* brain slices.

it is most rigorous when both the referential statement and the results are in the past tense (*Ex. 8.25a*). It is least rigorous when both are in the present tense (*Ex. 8.26c*), but at least the results should be reported in the past tense (*Ex. 8.26b*). If conclusions are included, they are most rigorously stated in the past tense, as part of the reporting of past research. In general, therefore, whenever the abstract is reportorial, as it usually is, it is best written in the past tense.

  Present tense.   The writer may take the stance of addressing the reader with paper in hand, announcing the purpose or subject of the paper; then the first sentence can be in the present tense (*Ex. 8.27a,b*):

Ex. 8.27

  a. This paper *explores* the effects of environmental variability and grain on the niche width of organizational populations.
  b. This paper *describes* a microcomputer–based event recorder for recording behavioral observations.
  c. Relative elastic differential cross sections for 10–50 eV positrons (electrons) colliding with argon *are being measured* [were measured] in our crossed beam apparatus with an angular range of 30–135°.

The writer must then shift to the reportorial mode, that is, to the past tense, in presenting the methods and results. Using the present tense for describing methods misrepresents them as ongoing procedures and so is inaccurate. In *Ex. 8.27c*, the fact that the cross sections are still being measured at present does not warrant the use of the present tense, since the paper is not reporting on ongoing research.

  From a scientific standpoint, using the present tense for stating experimental results is a more serious misrepresentation (Chapter 3). *Ex. 8.28* illustrates how the present tense states the results of a single or particular experiment as though they were a general finding or general knowledge and how the statements become more specific and restrictive—and more scientific—when stated reportorially in the past tense:

Ex. 8.28

  a. In all cases, SI-hypersensitivity sites *are located* [were located] in known or presumed regulatory regions.
  b. Aggregate time *exerts* [exerted] powerful and pervasive effects. Sociocultural variables have [*had*] nonproportional effects—that is, their effects *vary* [varied] with individual time.
  c. Excitation spectra *have* [had] a strong pressure dependence consistent with the formation of a variety of argon clusters, which rapidly *predissociate* [predissociated] upon excitation.

  When the present tense is used for reporting methods or results at the beginning of the abstract, they may be read as introductory or background information:

Ex. 8.29

    a. Raman scattering experiments are carried out [*were carried out*] in single crystals of 2H-NbS$_2$. From the results it *appears* that some crystals undergo. . . .

    b. Hand-reared bar-headed goslings (*Anser indicus*) isolated in pairs *tend* to alternate their distress calling. The phenomenon has earlier been reported by other authors, who suggested a cooperative function in attracting the parent(s). In this paper six possible functions *are considered*. . . .

    c. Immunofluorescence of sections of infected elms treated with an anti-glycopeptide serum *showed* intense fluorescence only in the cell walls of the fungus. [*result*] Apparently the glycopeptide *is* [was] abundantly present in or on fungal cell walls.

In *Ex. 8.29a*, the first sentences appear to be a generalization, and it is not until the next sentence that readers can deduce that the first sentence was *reporting* results. Similarly, in *Ex. 8.29b*, the first sentence seems to be a general observation, but the second sentence makes it appear to be a finding. Not until the third sentence does it become clear that the research is designed to try to account for the finding.

    Conclusions are in the past tense, if rigorously stated, but they are often stated in the present tense (*Ex. 8.12*), as inferences valid in the present. Writers may be led to use the present tense for the final result or immediate conclusion, when the past tense would be more effective.

    In mathematical or theoretical papers, the abstract may be largely or entirely in the present tense. In such papers, what the writer did is directly *before* the reader; it is not time-dependent. Therefore the abstract can be in the present tense:

Ex. 8.30

    THE GOVERNING field equations and boundary conditions for stress assisted diffusion are derived from basic principles of continuum mechanics and irreversible thermodynamics. Attention is confined to elastic and viscoelastic material response, with emphasis on isotropy and the case of small strains. It is shown that both the diffusion process and the saturation levels are stress dependent. In viscoelastic materials, the time-dependence of the moisture boundary condition causes the saturation level to drift with time. In addition, the diffusion process becomes non-linear in stress, in spite of the linearity of the mechanical behavior. [J. Mech. Phys. Sol. 35, 1987, 73]

(*Ex. 8.31a*). Since the abstract is too short for even an abridgment of the theoretical development, the abstracts of theoretical papers are descriptive abstracts, in the present tense (*Ex. 8.30, 8.31a*):

Ex. 8.31

    a. *Theoretical*

    The energy relaxation of electrons in inert gases is examined with a discrete ordinate method of solution of the Boltzmann equation. Realistic election moderator momentum transfer cross sections are employed. The time evolution of the election velocity distribution function is obtained as well as particular averages such as the energy and directed velocity. Each property can be expressed as a small sum of exponential terms each characterized by an eigenvalue of the Boltzmann collision operator. [J. Phys. Chem. 88, 1984, 4854]

    b. *Review*

    This is a continuation of a previous review of modern aspects of acoustics. The present article deals with the interaction of acoustic, magnetic, and internal waves in fluids. Both dispersion relations for high-frequency waves and analytical solutions in terms of special functions for waves of arbitrary frequency are used. Nonlinear phenomena, mode coupling, and dissipation are included [Rev. Mod. Phys. 59, 1987, cover]

Abstracts of review papers are similarly in the present tense (*Ex. 8.31b*).

In experimental papers, the abstract should not be in the present tense. In *Ex. 8.32a*:

Ex. 8.32

a. *Experimental*

An argon ion beam is combined with a Cl_2 gas jet to etch silicon. Auger electron spectroscopy is used to measure the surface coverage of chlorine simultaneous with the etch process. This steady state surface chlorine coverage is enhanced by an increase in the Cl_2 flux or a decrease in Ar ion flux. The coverage varies from 10% to 100% of saturation as the Ar^+ to Cl_2 flux ratio varies from 0.4 to 0.004. The Si etch rate increases with increasing Cl_2 flux giving an enhancement of as much as a factor of 3 over pure Ar^+ sputtering at 500 eV. A general model is presented which is based on the mass balance of ionic and neutral species to and away from the surface region. Assumptions are made to apply the model to the $Ar^+ + Cl_2 + Si$ system. Expressions for both the surface Cl coverage and silicon etch rate are then obtained as functions of the incident Ar^+ and Cl_2 fluxes. The model is fit to the data and accounts for all of the observed trends in surface coverage and etch rate [J. Vac. Sci. Technol. B1, 1983, 37]

b. *Descriptive*

(1) Larvae of the ant-lion *Macroleon quinquemaculatus* (Hagen) pupate in the rainy seasons which occur in the Dar es Salaam area in March–May and September–December. The size of pupating larvae varies greatly, all larvae above 120 mg and some between 90 and 120 mg pupating in a rainy season. Hunger can induce pupation. The time when a larva pupates within a rainy season is inversely related to larval weight. Larvae usually pupate within 2 days of leaving their pits but some larger larvae delay pupation for up to 4 months.

(2) All pupating larvae emerge as adults, but there are three distinct size classes of larvae: those < 140 mg become adults that are a decreasing percentage of larval weight, whereas larger larvae become adults that are a constant or an increasing percentage of larval weight. Adult females show two length–weight relationships, small adult females (from larvae <140 mg) being relatively shorter than large adults. Mean egg size and fat content of adults increases with adult weight.

(3) The above relations strongly suggest that larvae pupating when small become adults of reduced fitness. Various larval behaviours are in accord with this explanation. The minimum size for pupation appears to be determined by the quantity of fat stored for pupal consumption. [J. Anim. Ecol. 54, 1985, 573].

the ion beam is not combined with the gas jet in the paper before the reader's eyes; it is not even combined while the writer is writing the abstract or the paper. All the procedures and the results are past events. Reporting them in the present tense makes it difficult to identify the research and also misrepresents it. Such a use of the present tense carries with it an air of authority or generality that is incompatible with the method of scientific inquiry and alien to rigorous scientific research. Such writers are arrogating to the results a generality that is scientifically unfounded.

The present tense is sometimes used for descriptive studies, presumably because of the direct observation. In the abstract on the ant-lion (*Ex. 8.32b*), the writer describes the observations made, all in the present tense. The statements are made as for *all* larvae; yet one cannot make accurate scientific statements about all larvae, since there is no way of ensuring examination of all larvae. The global *all* can apply only to the particular set observed in the study. The past tense would represent this more accurately.

The present tense is of course appropriate for *present* commentary; however, such commentary or discussion is usually inappropriate in an abstract. In summary then, the present tense can be used for (1) the first sentence stating the purpose or subject of the *paper*, (2) a mathematical or theoretical, paper-in-hand, development-before-your-eyes presentation, (3) a concluding statement, and (4) a referential statement or present commentary. The past tense can be used equally well for these purposes, except present commentary, which has no place in the abstract. In other uses the present tense may have serious scientific consequences and cause confusion, and so is best replaced by the past tense.

Present perfect tense.   The perfect tense is sometimes used in the abstract. Its use may cause confusion if not skillfully used. Used in the opening sentence of the abstract, it may lead the writer to present background information:

Ex. 8.33

   a. An equation *has* often *been made*, especially but not exclusively by Marxists, between radicalism and the rational understanding of objective interests.

   b. Concrete deterioration due to the alkali-silica reaction *has been reported* from nearly all major geographic areas of Canada.

   c. Clusters of the form $NH_4 (NH_3)_n$ *have been generated* by neutralizing a fast beam of ions in the electron transfer reaction. . . .

When there is no cue as to the function of the statement, an opening sentence in the present perfect tense is likely to be confusing. In *Ex. 8.33c*, it is not clear whether the opening sentence refers to the literature, describes the method, or states the result; therefore if an opening sentence is in the present perfect, it must be "labeled" as to function, to avoid confusion.

The present perfect is sometimes used for the research objective, methods, results, or conclusions:

Ex. 8.34

   a. The detailed biochemical composition of the epidermis and cuticle from the foregut of *Carcinus maenus has been examined*.

   b. An abundance analysis of eight G and K supergiants *has been performed*, with particular emphasis on two stars Peg and 12 Pup.

   c. We *have observed* a broad, structureless, fluorescence excitation spectrum in jet-cooled isoqunoline. . . .

   d. The above conclusions *have been obtained* assuming that photinos are lighter than gluinors and live long enough to escape collider detectors.

There may be little confusion about the function of such sentences in the abstract, but they are not as clear as statements in the past tense. The use of the present perfect can lead to an inappropriate shift into the present tense:

Ex. 8.35

Qualitative and quantitative aspects of the reaction between NO_2/N_2O_4 and hexane soot *have been studied* by FT-IR. The rapid reaction near room temperature *yields* [yielded] several surface species. . . . The formation of these functionalities *has been* [were] *confirmed* through reaction with. . . .

8.3.5 Dissertation

The abstract of a dissertation may be longer than that of a paper, because the dissertation is more comprehensive than a paper, but the parts abstracted should be the same. If the dissertation includes several separate studies, they may be abstracted separately in the abstract. The faculty advisors or graduate division may, however, require a summary or a more detailed abstract. This permits one to include more parts of the paper and more details from the methods and results. For a more detailed summary, the student condenses the various parts as required.

8.4 TITLE

The title states the subject of the paper. Its function is to identify and describe the contents of the paper accurately and specifically. This is of central importance to both writer and reader in the retrieval of the paper, because it brings reader and paper together. It is the title that may be decisive in the reader's decision to read it. Therefore, it is essential that titles be as accurate, informative, and as comprehensive as possible, yet as short as possible. As in the abstract, the objective is maximum content with minimum wordage.

8.4.1 Brevity, Comprehensiveness, and Specificity

Brevity is an essential requirement; readers do not want to read mini-abstracts, such as:

Ex. 8.36

a. Sequence of *psi*, a Gene on the Symbiotic Plasmid of *Rhizobium phaseoli* Which Inhibits Exopolysaccharide Synthesis and Nodulation and Demonstration That Its Transcription Is Inhibited by *psr*, Another Gene or the Symbiotic Plasmid.

b. The Striatonigral Projection and Nigrotectal Neurons in Rat. A Correlated Light and Electron Microscopic Study Demonstrating a Monosynaptic Striatal Input to Identified Neurons Using a Combined Degeneration or Horseradish Peroxidase Procedure.

The writer must try to avoid wasting words. General introductory phrases, such as:

Ex. 8.37

a. *An Approach to* Analyzing Single Subject's Scores Obtained in a Standardized Test with Application to the Aachen Aphasia Test (AAT)

b. *An Investigation of* the Organization of Pigtail Monkey Groups Through the Use of Challenges

c. *Some Considerations on* the Yield Condition in the Theory of Plasticity

clutter the beginning of the title with unimportant or empty words. Even the initial article can usually be omitted as well as articles and other unimportant or wordy phrases within the title.

A title can be too short, however, if it does not provide enough information. If it is too general,

Ex. 8.38

Ferritin	Environmental Studies	Development Features of Explorations
Blood Coagulation	Curosity in Zoo Animals	Recrystallization of Metals and Alloys
The Lifted Beam	Strategic Implementations	Heavy Metal Tolerance in Plants
Professional Ethics	Contingent Valuation Surveys	Land Breezes and Nocturnal Thunderstorms
Pulsating Aurorae	Work, Nonwork, and Withdrawal	

it will not make the particular subject of the paper clear; besides, scientific papers are rarely reports on general topics of research. General titles can be used for review papers, though the title should indicate that they are reviews. Titles must be comprehensive enough for ready retrieval of the paper. They should include as many of the most important elements of the paper as possible.

8.4.2 Types of Titles

Titles may be indicative or informative. The informative title states the results and conclusions of the research, usually in sentence form (*Table 8.7b*); indicative titles state the subject of the paper (*Table 8.7a*). Some indicative titles are informative in indicating but not stating results:

Ex. 8.39

a. Evidence for a Pathogenetic Role of Xanthine Oxidase in the "Stunned" Myocardium.

b. Unique Spermatozeugmata in Tester of Halfbeaks of the Genus *Zenoarchopterus* (Telostei: Hemiramphidae)

c. Long-Term Variations of Calorie Isolation Resulting from the Earth's Orbital Elements

d. Distortion of Twisted Orientation Patterns in Liquid Crystals by Magnetic Fields

Informative titles can often be effectively converted into indicative titles (*Table 8.7b*). Titles may ask a question:

Ex. 8.40

a. Is the Predominant Period of Cell Arrest and Presence of Endoreduplicated Cells Coincident with Production of Secondary Vascular Tissues in Intact and Cultured Roots of *Rapharices sativus* L.?

b. Is a Zero Cosmological Constant Compatible with Observations?

Such questions usually begin with a weak verb and frequently sound artificial and sometimes pretentious or condescending.

TABLE 8.7 Indicative titles and informative titles with their indicative form

a. *Indicative*

1. Temperature Dependence of the Inhibition of Positronium by Chlorine-Substituted Hydrocarbons in Non-Polar Liquids
2. Structural Optimization and Properties of First Row Monolayers
3. Drainage-Base Characteristics of Nordaustlandt Ice Caps
4. Application of Physical Effects for the Detection of Smallest Ion Currents in a Vacuum
5. Estimation of the Survival Rate of *Anopheles arabiensis* in an Urban Area (Pikine-Senegal)

b. *Informative*

1. Lactation Increases the Efficiency of Energy Utilization in Rats.
   Increased Efficiency of Energy Utilization in Lactating Rats
2. Hepatic Arterial Pressure-Flow Autoregulation Is Adenosine Mediated.
   Mediation of Autoregulation of Hepatic Arterial Pressure Flow by Adenosine
3. Hostile Attributional Biases Among Aggressive Boys Are Exacerbated Under Conditions of Threats to the Self.
   Exacerbation of Hostile Attributional Biases Among Aggressive Boys Under Conditions of Threats to the Self

Research scientists are sometimes tempted to write a scintillating title in imitation of titles in general magazines. They may use questions, figures of speech, literary illusion, humor, or emotive language:

Ex. 8.41

    a. The Optimal Rate of Inflation Revisited
    b. The Wandering IQ: American Culture and Mental Testing
    c. The Sausage Machine: A New Two-Stage Parsing Model
    d. The Match Game: New Stratigraphic Correlation Algorithms
    e. A Tale of Forecasting 1001 Series: The Bayesian Knight Strikes Again
    f. Supersymmetry: Lost or Found
    g. Self-consistent Cosmology, The Inflationary Universe, and All That

Such titles are likely to be neither adequate nor accurate because of the rhetorical language (see Chapter 3).

8.4.3 Double Titles

If the research necessitates a long title, it may be necessary to divide it into two parts. Double titles are often used in serial papers; however, many editors do not accept papers in series, and many prefer to avoid double titles. Such titles do not have a conventional form. The first part may be the general title and the second the specific, or vice versa. The first part of the title may give the subject and the second part indicate the type of study (*Table 8.8a*), thus removing from the beginning of the title a word or phrase that merely designates a category. Consider every review

TABLE 8.8   Double titles with second title descriptive of type of study or methods, or explanatory

a. *Descriptive of Type of Study*

1. Qualitative Research Traditions: A Review
2. Source of Intrinsic Innervation of Canine Ventricles: A Functional Study
3. Agricultural Development in Three Asian Countries: A Comparative Analysis
4. Symbolic Gesturing in Language Development: A Case Study

b. *Descriptive of Method*

1. Large Divalent Cations and Electrostatic Potential Adjacent to Membranes: A Theoretical Calculation
2. How Vervet Monkeys Perceive Their Grunts: Field Playback Experiments

c. *Explanatory*

1. Solid Waste Disposal with Intermediate Transfer Stations—An Application of the Fixed-Charged Location Problem
2. Memory for Pictures: A Life-Span Study of the Role of Visual Detail
3. Psychosocial Processes of Remission in Unipolar Depression: Comparing Depressed Patients with Matched Community Controls

article beginning "A Review of . . .". Where the type of study is not important, it should be omitted. The main title may state the subject and the second title indicate the method (*Table 8.8b*) or give an explanation of the first part (*Table 8.8c*).

Both parts of the title may be substantive, with the first part specific, the other more general (*Ex. 8.42a,b*), or vice versa (*Ex. 8.42c–e*):

Ex. 8.42

   a. Late Quaternary Sedimentation in the Western North Atlantic: Stratigraphy and Paleoceanography
   b. Mesoscale Lake-Effect Snowstorms in the Vicinity of Lake Michigan: Lunar Theory and Numerical Simulations
   c. Men and Women at Work: Sex Segregation and Statistical Discrimination.
   d. Human Retinal Development: Ultrastructure of the Outer Retina.
   e. The Inactivity of Animals: Influence of Stochasticity and Prey Size.

The two parts can often be combined into one, often with a gain in brevity. In serial titles, the general subject is usually first:

Ex. 8.43

   a. Minimum Reflex Conditions. Part 1: Theory.
   b. Sound Transmission and Mode Coupling at Junctions of Thin Plates, Part I, Part II.
   c. Modeling of Chemical Reactors—XXV: Cylindrical and Spherical Reactor with Radial Flow
   d. The Biology of Cartilage II. Invertebrate Cartilages: Squid Head Cartilage.

e. Development of Antipredator Response in Snakes: I. Defensive and Open-Field Behaviors in New-borns and Adults of Three Species of Garter Snakes (*Thamnophis melanogaster*, *T. sirtolis*, *T. butleri*).

Double titles lend themselves to asking explicit questions, usually in the second title:

Ex. 8.44

a. The Striated Cholinergic Interneuron: Synaptic Target or Cholinergic Interneuron?
b. Environmental Impact Statements: Instruments for Environmental Protection or Endless Litigation?
c. Squatters or Suburbanites? The Growth of Shantytowns in Oaxaca, Mexico.

Such titles are generally not effective titles for scientific research papers, especially issue-oriented ones.

8.4.4 Developing the Title

A title may be derived from the statement of the problem or by condensing the abstract. It may be formed by collecting all the pertinent key words into a title. In formulating the title, one can begin by including all the details or key words that seem important and then shortening it. An important part of the title is often the methods, and so the title may focus on methods:

Ex. 8.45

a. Tornado Detection by Pulsed Doppler Radar
b. Studies with an Artificial Endocrine Pancreas
c. Diffraction by Band-Limited Fractal Screens
d. High-Sensitivity Helium Resonance Magnetometers
e. Phase Diversity Techniques for Coherent Optimal Receivers

The organism, participants, particular materials, or locality should be specifically named in the title:

Ex. 8.46

a. The Researcher and the Manager: A Dialectic of Implementation
b. Studies on the Transmitters of the Afferent and Efferent Pathways of the Striatum and Their Interaction in the Baboon, Cat, and Rat
c. The Diurnal Variability of Florida Rainfall
d. Social Identification of Toxic Diets by Norway Rats (*Rattus norvegicus*)
e. Late Holocene Sea Levels and Coral Reef Development in Vahitahi Atoll, Tuamotu Islands (Izmir, Turkey)

Papers with unspecific titles may not reach interested readers:

Ex. 8.47

a. The Arylsulphatases and Related Enzymes in the Livers of Some Lower Vertebrates.
b. A Comparative Study/Germination in a Local Flora.

The writer can begin abbreviating the title by eliminating duplication, super-fluous words and phrases, and by condensing clauses:

Ex. 8..48

a. Two Phosphorylated Subclasses of Polyomavirus Large Antigen *That* Differ[ing] in Their *Modes of* Association with the Cell Nucleus.

b. *The* Estimation of *the* Generalized Covariance *When It is* [from] a Linear Combination of Two Un-known Linear Covariances.

c. Spontaneous Verbal Rehearsal in Memory Tasks *as a Function of* [with] Age.

d. *How to Share* [Sharing] Memory in a Distributed System.

e. *On the Application of* [Applying] an "Entropy Maximizing Principle" in Flow Regime Predic-tions.

If the title is still too long, substantive elements may have to be eliminated.

The most accurate way to write the title is to capitalize only the first word; then words and symbols that must be capitalized or lowercase will be in their correct form when the title is edited for capitalization. If initial capitals are used, articles and prepositions are written in lowercase, except for the initial article, and preposi-tions of five or more letters:

Ex. 8.49

a. Thermospheric Dynamics *During* the March 22, 1979, Magnetic Storm, 1, Model Simulations.

b. A Model-Based Comparison of Switching Characteristics *Between* Collector-Top and Emitter-Top HBT's.

c. Finite-Element Analysis of Multiphase Immiscible Flow *Through* Soils.

Symbols are kept in capital or lowercase letters as appropriate.

8.5 AUTHORSHIP

The title is followed by the byline, that is, the line listing the name of the author(s). The author of a scientific paper is the research scientist who conducted the research and wrote the paper. This definition is based on the traditional paradigm of the in-dividual scientist working independently with little assistance, but today most sci-entists work with others in a laboratory, department, research group, or institution. Moreover, they do not always perform the experimental procedures. They may di-rect the research of postdoctoral students, research assistants and associates, and graduate students. Therefore determining the authorship of a paper is not a simple decision and may give rise to dissatisfaction.

8.5.1 Problems and Their Source

Most problems are related to the prestige of authorship. Directors of laboratories may expect to be included as coauthor in papers issued from their laboratory, though they may have provided primarily administrative support. Faculty members may expect to be named coauthors in a graduate student's paper, though the student has initiated the research and performed it independently, albeit with guidance. Colleagues may wish to be named coauthors, though their contribution may not be considered significant enough. A sponsoring group may preempt authorship or senior authorship, although the cooperating group may have conducted most of the research. Partly because of these problems, the number of coauthors of papers has tended to increase:

Ex. 8.50

    a. R. L. Arnoldy, R. Rajashekar, L. H. Cahill, Jr., M. J. Engebretson, T. J. Rosenberg, and S. B. Mende

    b. F. S. Felber, F. J. Wessel, N. C. Wild, H. U. Rahman, A. Fisher, C. M. Fowler, M. A. Liberman and A. L. Velikovich.

There are three main difficulties in determining the authorship of a paper in most research settings. First, estimating the contributions of members of a research group is inherently difficult. Second, there are no standard or explicit criteria for authorship or for ordering coauthors of a paper. Third, even if one could arrive at a fair ordering, there is no effective or accepted format for communicating that estimate in the byline. The byline of a coauthored paper may therefore be uninformative or misrepresentative of the actual contributions of the authors.

8.5.2 Determining Authorship

A few research scientists on a small project can come to an early understanding informally on the order of names on the title page. In a larger formal research project with more numerous research scientists, authorship should be explicitly and openly discussed and determined at an early conference on the research and reviewed before the writing of the paper.

    Three decisions must be made in determining authorship: (1) the senior author, (2) the coauthors, and (3) the order in which the names of coauthors are to be listed. The senior author is the participating research scientist who assumes leadership for the scientific formulation, planning, execution of the research, and the writing of the paper. He or she is to be distinguished from the research scientist who has a scientific-administrative role and supervises the planning and execution of research administratively. Coauthors can be determined on the basis of criteria for evaluating contributions.

    Whatever the decision-making mechanism of the group, collaborating research scientists should work toward definite criteria for authorship. Such criteria promote open discussion and focus the evaluation on objective rather than personal criteria. With definite criteria, even unilateral decisions can be made more accurately representative of the contributions. Establishing criteria for authorship may also counterbalance the constraints placed on a research scientist by the hierarchical position of the authors and the informal influence relations among them.

Participation in the research is a minimal criterion for authorship; authorship cannot be conferred. Participation that merits authorship includes:

Ex. 8.51

 a. Having a substantial knowledge of the literature.

 b. Having a full understanding of the scientific problem and question.

 c. Planning the research, i.e., assisting in developing hypotheses, the research design, and procedures and equipment.

 d. Performing experimental procedures, unless this is done by assistants.

 e. Analyzing the data, contributing to the discussion, and interpreting the results.

 f. Writing the paper or some part of it, or revising it substantively.

Ultimately, the scientific importance of participation must be determined for a more accurate estimate of a research scientist's contribution. This requires regularized guidelines or procedures.

8.5.3 Communicating Contribution

The linear character of the byline makes it difficult to reflect variations in contributions. It can only indicate the major (first) and least (last) contributor, and the descending order of the intervening coauthors from the first to the last. In papers with more than two authors, the ordinal character can be used to reflect equal participation and contribution by alphabetizing the names of the authors. However it is rare for all authors of a paper to have contributed equally in a uniformly descending order. In most multiauthor papers therefore, the ordinal byline is not well adapted to communicating the contribution of the various authors. There is no simple procedure for resolving the treatment of authorship, because of the varying conditions of the research, the varying contributions, and different usages. Until the problem is addressed by the broader scientific community, it will remain difficult to clarify the contributions of coauthors within the confines of present usages.

8.5.4 Writing and Authorship

One of the main tasks of research scientists on a joint project is planning the writing of the paper(s). There is an increasing tendency to separate writing from the research. Although the research scientists assigned to writing may be among the main contributors, they may excuse themselves from writing because of presumably more important commitments. Therefore, the writing of the paper tends to be delegated to a few members of the group, often junior authors or even junior colleagues or assistants who are not to be coauthors, which may result in a tacit plagiarism. Because the practice emphasizes research and administration over writing, it reinforces the subsidiary position of writing and devalues it.

   In the by-line, each author is listed with given name and initials first and surname following (*Ex. 8.50*), but if the surname is a common name, the full name may be necessary to identify the author. In the United States, the authors' names are not usually followed by academic degrees; they are usually followed by the author's institutional affiliation, or the affiliation may be supplied in a footnote. The institutional address of one of the authors, the corresponding author, is often provided in a footnote to direct correspondence. The institution listed may represent the institution with which the author was affiliated at the time of the research and which supported the research, or it may represent the author's present affiliation. A footnote is added if clarification is needed.

8.6 BIOGRAPHICAL SKETCH

A biographical sketch is commonly included in dissertations and is requested for some journals after a paper is accepted for publication. In a dissertation it may include personal information as well as professional information. It is largely chronological, but some topical separation is desirable to avoid marriage and children amid degrees and graduate assistantships. It may begin with date and place of birth and early education, and may even include siblings and parents. For a more professional biography only the birth date and birthplace are given. These are followed by undergraduate and any graduate degree, undergraduate majors and minor, entrance into graduate school, candidacy, and graduate field, graduate major, and minors. The biography may also include graduate assistantships and other academic experience, academic awards, and professional affiliations. This academic part may indicate sources of one's interest in a discipline or research problem or area, but should consist largely of concrete events. It should not be a personal narrative of one's journey of exploration in arriving at one's chosen field. Personal data, such as marriage, spouse's name, number of children, if included, are placed at the end of the biographical sketch. It can be written in the first or third person, and in one to several paragraphs depending on its length.

For a scientific journal, the biographical sketch has a narrower focus—in time, subject, and tone. Its purpose is to provide the editor and ultimately readers, with information about the author's institutional affiliations, and scientific and professional activities and interests. It is written in the third person and tends to be short. Occasionally a biographical sketch is written to honor a research scientist; then it is usually written by a colleague. It is much longer and includes a laudatory appraisal of the scientist's contributions and even personal commentary. For a paper, the journal editor usually extracts from the biographical sketch the biographical material to be printed with the paper:

Ex. 8.52

   a. Assistant Professor, Department of Forest Resources, Utah State University, Logan, Utah 84322.

   b. Human Development and Family Studies, University of Arkansas, Fayetteville, Arkansas. Received his Ph.D. in Life-span Developmental Psychology from West Virginia University in 1984. Research interests in dialectical models of adolescent development. [J. Youth Adolesc. 16, 1987, 1]

TABLE 8.9 Biographical sketches

a.

JOHN J. BETANCUR, born in Colombia, is currently a Research Associate at the Center for Urban Economic Development of the University of Illinois at Chicago. He holds a Ph.D. in public policy analysis and a master's degree in urban planning and policy from the same university, and also has degrees in sociology and philosophy from Colombian universities, where he was assistant and associate professor between 1970 and 1979. He has conducted extensive research in the areas of housing, political participation, and economic development.

[Envir. Behav. 19, 1987, 286]

b.

Paul D. Burrow was born in Oklahoma City, OK, in 1938 and received his S.B. degree from MIT and his Ph.D. from the University of California, Berkeley, both in physics. After nine years in the Engineering and Applied Science Department of Yale University, he joined the Department of Physics and Astronomy of the University of Nebraska where he is a Professor. His research interests concern low-energy electron-scattering processes, in particular scattering from excited atoms and molecules and the formation and decay channels of temporary negative ions. [Chem. Rev. 87, 1987, 557, with photograph]

The biographical data may be put at the beginning of the paper, in a footnote at the end of the paper, or in a section, "Notes on Contributors." Biographical sketches may include more extensive professional information, (*Table 8.9*), and may be accompanied by a photograph of the author.

8.7 ACKNOWLEDGMENTS

Acknowledgments have a twofold purpose: (1) to express the research scientist's appreciation publicly to persons, groups, or institutions that have provided assistance in performing and publishing the research, and (2) to make clear the scientific contributions that others have made to the research. Because of the scientific and professional importance of such assistance, writers must be sure to acknowledge assistance they have received.

8.7.1 Form and Content

A writer may acknowledge contributions of ideas or assistance with the actual performance of the research or writing of the manuscript. It is most important to acknowledge any ideas that make a significant scientific contribution to the research or the paper (*Table 8.10, Appendix 8B*).

Where human beings have been studied—children, parents, teachers, patients, employers, employees, students—their participation should be acknowledged. Grants and funding are acknowledged, briefly. For some government grants, the acknowledgment is usually required, and the position or content of the acknowledgment is often specified. The acknowledgments should include the agency or organization providing the support and the serial number of the grant. If more than one agency has funded the research, they can be ordered in some appropriate order.

TABLE 8.10   Types of assistance acknowledged

Conceptual or general	*Analysis*
Advice on research	Statistical analyses
Substantive comments	Computations or calculations
Specific comments or ideas	Programming
Stimulating discussions	Developing of computer program
Constructive comments	
Suggestions on methodology	*Editorial*
	Reading draft
Materials and methods	Improving manuscript
Participation as subjects	Making editorial comments
Performing measurements	Helpful comments on manuscript
Providing cultures, samples, etc.	Translating
Collecting data	
Animal care	*Support*
Experimental analysis	Grant or financial support
Donating or loaning material	
Use of facility	*Personal*
	Encouragement
	Interest in the research

The work of assistants who are regularly employed is not usually acknowledged in scientific papers, but may be acknowledged in dissertations. The research scientist may, however, wish to acknowledge them in a paper, when they have made exceptional contributions. Personal support is not usually acknowledged in the scientific paper; it is more common in the dissertation, but should not be effusive.

The order of acknowledgment may be formally hierarchical; that is, supervisors and peers are acknowledged first, then assistants or others. Or it may be functional, that is, those who made the greatest contributions are acknowledged first. Persons are acknowledged by name without titles, for example, M. R. DeBrion, L. K. DeCanet, rather than Prof. M. R. DeBrion and Dr. L. K. DeCanet, although in dissertations, titles may be used. If titles are used, one should avoid demeaning hierarchical distinctions. To thank an assistant as Mary Ann Green when others are thanked as Prof. W. C. DeLacor or Dr. E. R. DeBretor tarnishes the courtesy of the acknowledgment to Ms. Green. The names should, of course, be spelled correctly. Types of assistance acknowledged are listed in *Table 8.10,* which is suggestive rather than definitive, and some examples are included in *Appendix 8B.*

Acknowledgments may be included in a footnote to the title or to a reference in the text, or set off at the end of the paper before the reference section. Often grants and funding are acknowledged in a title footnote, and general acknowledgments are placed in a section at the end of the paper. Acknowledgments of materials are often made in the introduction, but they intrude on its logical development, and it is preferable to place them in a footnote. In some settings it may be necessary to make a disclaimer that the opinions expressed are those of the author and are not to be attributed to the agency or organization sponsoring or funding the research. The acknowledgments section at the end of the paper may be set in italic or reduced type, but may be distinguished only by position. It is typed on a separate page with its own heading. Acknowledgments treated as footnotes are typed together with other footnotes.

8.7.2 Dissertation

In the dissertation, the acknowledgments form a separate section before the beginning of the text. They tend to be more numerous than those in a paper. As novices in research, students may require more help than full-fledged research scientists, and so have more assistance to acknowledge than an experienced investigator. The acknowledgments also tend to be less formal and usually acknowledge personal as well as professional support. Faculty assistance is usually acknowledged first, then professional and then personal support, although the student may have both a professional and personal relationship with some coworkers. Faculty members directing a graduate student's research are usually acknowledged, by at least a formal acknowledgment of official function:

Ex. 8.53

> a. The writer would like to thank Professor E. R. DeBretor for serving on her graduate committee.
> b. I would like to thank Professors M. R. DeBrion, L. K. DeCanet, and W. C. DeLacor, for directing my graduate studies.

One can then make more specific or warmer acknowledgments for the assistance of the more helpful and supportive faculty members. The acknowledgments section, like the dissertation, is a public document; therefore, the tone of the acknowledgments should reflect this writer-audience relationship and the superlatives should be kept in rein.

8.8 APPENDICES

An appendix is a body of material that is set apart from the main text of the paper or dissertation to avoid a digression that would interrupt the line of development, yet retain the detailed information that readers may require. The material is either excessively detailed or supplemental. It is usually placed after the reference section, often set in reduced type, and may even be reproduced directly from typescript.

Editors usually discourage the use of appendices because they are space consuming. They are, therefore, not customary in scientific papers and should not be used unless they are essential. Appendices are much more common in dissertations, which may have several appendices. There they are a particularly effective way to remove from the development of the main text those parts that need to be presented in almost indigestible detail.

Appendices are used for supplementary illustrative details, supporting data, such as procedures, mathematical symbols or derivation, and for any detailed development or explanation. The appendix may include information that would be a real excrescence in the text, for example, a questionnaire used in survey research (*Table 8.11a*) or a list of initialisms and definitions (*Table 8.11b*). The material may include secondary or supportive data, such as the details of computations (*Appendix 8C.1*) and mathematical derivations (*Appendix 8C.2*) or explanations. An appendix should not be used as a reservoir for assorted information. It is limited to one subject, which may be stated in an introductory sentence (*Ex. 8.54a*) or title (*Ex. 8.54b*):

Ex. 8.54

    a. *Appendix A*
       The following equations are growth rates for $U_{ei} >> U_{en}$:

    b. *Appendix A*: A Canonical Decomposition of SO(3,1) Object-Valued Tensors Associated with a Slicing of Space Time

When a paper has more than one appendix, each appendix is designated serially by capital letters (*Ex. 8.54, Appendix 8C*), but sometimes by roman numerals (*Appendix 8C*), or even arabic numerals. Tables and figures in the appendix are designated as Table A.1, Table A.2, Fig. A1; or Fig. I-1, Table I-3, with or without the decimal point.

TABLE 8.11 Part of appendixes listing (a) questionnaire items, (b) definitions of acronyms

a. *Questionnaire*

Appendix
CATCH questionnaire[†]
1. I wouldn't worry if a handicapped child sat next to me in class.
2. I would not introduce a handicapped child to my friends.
3. Handicapped children can do lots of things for themselves.
4. I wouldn't know what to say to a handicapped child.
5. Handicapped children like to play.
6. I feel sorry for handicapped children.
   .
   .
   .
23. I would feel good doing a school project with a handicapped child.

24. Handicapped children don't have much fun.
25. I would invite a handicapped child to sleep over at my house.
26. Being near someone who is handicapped scares me.
27. Handicapped children are interested in lots of things.
   .
   .
   .
32. I would not go to a handicapped child's house to play.
33. Handicapped children can make new friends.
34. I feel upset when I see a handicapped child.
35. I would miss recess to keep a handicapped child company.
36. Handicapped children need lots of help to do things.
[Develop. Med. Child Neur. 29, 1987, 334]

b. *Definitions of Acronyms*

FGGE	First GARP Global Experiment.	SMS	Synchronous Meteorological Satellite.
GARP	Global Atmospheric Research Program.	SSM/I	special sensor microwave imager.
GMS	Geostationary Meteorological Satellite.	SST	sea surface temperature.
GOES	Geostationary Operational Environmental Satellite.	SSU	stratospheric sounder unit.
HIRS	high-resolution infrared sounder.	STAT	statistical retrieval method.
HIS	high-resolution interferometer sounder.	TOVS	TIROS operational vertical sounder.

[Rev. Geophys. 24, 1986, 730]

Reference must be made to the appendix at least at the point in the text where the appendix is pertinent to the development. Some indication may be given of the substance of the appendix, so that readers will have the information needed without stopping to turn to the appendix. It is not enough to say as in *Ex. 8.55a*:

Ex. 8.55

a. Subjects were sent information in packets (Appendix A).

b. Subjects were sent packets, which included the questionnaire and instructions for the program, and a letter explaining the research and thanking them for their participation (Appendix A).

readers need to know more about the packet (*Ex. 8.55b*).

Appendix 8A

Analysis of Abstracts

The abstract may include (1) objective or subject of research, (2) methods, (3) results, and (4) conclusion. It should not include background, literature, or discussion. The various parts are characterized by the notes in brackets. The notes apply to the preceding text between brackets.

1. Abstracts Beginning with Background or Largely Introductory*

a.

Abstract—Almost a decade has passed since Gul'elmi and Dovbnya (1974) introduced the high latitude class of ultra low frequency (ULF) pulsations known as "Serpentine Emission" (SE). [*literature*] The purpose of this paper is twofold; firstly, to present new morphological results of SE observed at Davis, Antarctica and secondly, to stimulate further experimental and theoretical interest by pulsation researchers. [*objectives*] Analyses presented include the first cited amplitude–time structure of SE along with maximum entropy power spectra and digital sonagrams. [*results*] SE polarizations are investigated and a limited diurnal morphological survey is presented. [*ref. to objective and results*] [Planet. Sp. Sci. 35, 1987, 313]

b.

Discourse can be organized in different ways; four of these ways are comparison, problem/solution, causation, and a collection of descriptions. These four discourse types correspond to schemata that vary in their organizational components; these differences were expected to result in differences in processing text. The more organized discourse types of comparison, problem/solution, and causation were predicted to yield superior recall of information than when this same information was cast as a collection of descriptions about a topic. [background, general hypotheses] *The data from two studies support the hypothesized facilitation of the more organized types of discourse and have implications for understanding memory and writing instructional materials.* [results, ref. to discussion] [Am. Educ. Res. J. 21, 1984, 121]

*In these analyses, the term "objective" is used to include research question, purpose of research, and subject of research.

c.

Since the passage of the Civil Rights Act of 1964, various forms of employment discrimination have been prohibited by law. The courts have enforced these laws vigorously, often requiring plaintiffs and defendants to present quantitative analyses of applicant and employee records as evidence. In fact, statistical analyses have become a regular feature of employment discrimination cases. Consequently, there has been an explosion of interest in the production of statistical assessments of discrimination, as more and more social scientists and statisticians have become involved as consultants and expert witnesses on both sides of discrimination cases. [background] *This paper discusses current practices in the production of statistical assessments of discrimination and suggests nonstatistical techniques that can be used to aid the interpretation of the results of statistical analysis.* [objective] *Statistical assessments (1) may contradict accumulated social scientific knowledge about decision making, in general, and discrimination, in particular, and (2) may also contradict the spirit of civil rights legislation concerning discrimination.* [discussion] *These shortcomings motivate the introduction of a method of analysis that incorporates principles of Boolean algebra, an approach that allows holistic comparison of categories of similarly situated individuals.* [method, comment on method] [Am. Soc. Rev. 49, 1984, 221]

d.

The recent impetus of the semiconductor industry toward submicrometer feature sizes on integrated circuits has generated an immediate need for measurement tools and standards suitable for these features. Optical techniques have the advantages of being nondestructive and of having high throughput, but the disadvantage of using wavelengths comparable to feature size which results in complex scattered fields and image structures that are difficult to interpret. Although submicrometer optical linewidth measurement is possible for 0.3 μm feature sizes, current instrumentation and linewidth standards, particularly for wafers, will have to radically improve in accuracy as well as in precision to meet the anticipated needs of the integrated circuit (IC) industry for submicrometer dimensional metrology. [*background*] This paper discusses the effects of inadequate precision and accuracy on process control in IC fabrication and suggests some ways of circumventing these limitations until better instrumentation and standards become available. [*objective, discussion*] [J. Res. Natl. Bur. Stand. 92, 1987, 187]

2. Abstracts Beginning with Objective

a.

Abstract—1. The detailed biochemical composition of the epidermis and cuticle from the foregut in *Carcinus maenus* has been examined. [*objective*]

2. The cuticle was composed mainly of protein and chitin with a relatively small lipid component, whilst the epidermis was rich in both protein and lipid. [*result*]

3. The sterol fractions of both tissues were composed almost entirely of cholesterol but the component phospholipids were quite distinct. [*result*]

4. The phospholipids of the epidermis were composed mainly of long chain, highly unsaturated fatty acids, whilst those of the cuticle were dominated by short-chain saturated fatty acids. [*result*]

5. These compositional differences are discussed in relation to the likely permeabilities of the tissues. [*reference to discussion*] [Comp. Biochem. Physiol. 82B, 1985, 695]

b.

The surface-dressed optical Bloch equations for a two-level atom near a metal surface are solved for the case of a strong driving field. [*objective*] An analytic form is obtained for the adatomic resonance fluorescence spectrum. [*result*] Due to the multiphoton effects of the surface-reflected field and the surface plasmon resonance, the three-peak spectrum is strongly influenced by the surface. [*discussion*] A unique surface-induced asymmetry in the side peaks is revealed. [*result*] [J. Phys. Chem. 88, 1984, 4801]

c.

ABSTRACT   The changes in adipose depot weight, cell size, cell number and body composition during pregnancy, lactation and recovery were studied in Osborne-Mendel rats fed standard or high fat diets. [*objective, materials*] Rats were killed on day 21 of pregnancy, after 21 days of lactation, and after 21 or 22 days of a postlactational recovery period. Nonpregnant control groups were killed at the beginning and at the conclusion of the experimental period. [*methods*] The high fat-fed, mated group was always fatter than similarly treated animals fed standard diets throughout pregnancy and lactation. However, by the end of the recovery period, carcass composition of the animals fed high fat or standard diets and the nonpregnant groups were not statistically different. The weight of the parametrial, retroperitoneal and subscapular depots was higher in the high fat-fed animals at the end of the recovery period, and in the latter two pads, this increase was statistically significant. [*results*] Thus, despite the extensive lipid mobilization that occurs during lactation, the high fat-fed animals appear to be predisposed to postpartum obesity. [*discussion, conclusion*] [J. Nutr. 114, 1984, 1566]

3. Abstracts Beginning with Methods

a.

Summary

A troop of Guinea baboons living in an enclosure was exposed every day and for twelve consecutive days to a new object. The new object and the object(s) of the previous day(s) were presented simultaneously in the compound. [*method*]

The troop as a whole demonstrated excellent abilities to rapidly react to the new objects: 11 out of 12 new objects were discovered within a maximum of 3 min of their first presentation and were furthermore the first to be approached. [*results*] An analysis conducted on data from age and sex subgroups showed the preponderant part played by juveniles and by some adult males in the discovery process and subsequent contacts with objects. [*method, result*]

The results are discussed within the conceptual frame of "cognitive mapping." In addition, the extent to which social factors (*e.g.,* dominance) and perceptual and cognitive factors might determine the differential role of subgroups in the exploration and manipulation of objects is examined. [*reference to discussion*] [Beh. 96. 1986, 103]

b.

Summary

Leaf mesophyll protoplasts of a streptomycin resistant nitrate reductase deficient mutant of *Nicotiana tabacum* were fused with leaf mesophyll protoplasts of a wild type variety of *N. plumbaginifolia* in a simple electrofusion apparatus. [*method*] Somatic hybrid colonies, and plants were readily recovered after fusion by selection of colonies in medium with nitrate as sole nitrogen source and also containing streptomycin. [*results, method*] The value of this simple electrofusion apparatus (Watts and King, 1984) lies in its simplicity of construction and use and also in its ability to deal with literally millions of protoplasts in any one experiment. [*literature, discussion*]

Key words: Heterokaryons, electrofusion, streptomycin resistance, nitrate reductase, cell fusion, Nicotiana, mesophyll protoplasts. [J. Pl. Physiol. 129, 1987, 111]

c.

A fluorometric method for the determination of the histidine in water samples is described. [*reference to method*] The histidine is first extracted with dichloromethane containing dibenzo-18-crown-6. The histidine is oxidized by bis(trifluoroacetoxy)iodobenzene to give a fluorescent product. [*method*] The experimental variables and interferences in this determination are studied. [*objective*] [Microchem. J. 36, 1987, 169]

d.

Abstract—Grafting acrylic acid into air-irradiated Teflon-FEP films was investigated. Pre-irradiation doses ranged from 0.5 to 10 kGy. [*objective, method*] Grafting occurred at 45 or 60 C. [*results*] Homopolymerization inhibitors, ferrous ions or methylene blue, were added to the system. [*method*] It was found that after completion of the reaction, within 40-100 min. membranes were obtained with very low electric resistivities. [*results*] The influence of added inhibitors, pre-irradiation dose and grafting temperature was studied. [*problem*] From the results it is concluded that the initiating centers in air-irradiated Teflon–FEP are, on the one hand, peroxides of structure POOP′, in which P is a polymeric radical and P′ a small fragment, and on the other hand trapped PO_2 radicals. [*conclusions*] The latter only react after losing their oxygen. [*result*] In the presence of polymerization inhibitors, initiation involves a redox process which reduces the overall activation energy. [*conclusion*] [Radiat. Phys. Chem. 32, 1988, 193]

4. Abstracts Consisting Largely of Results

a.

Abstract—1. The plasma clearance rate (PCR) of radioactivity after a single intracardial injection of ^{3}H-cortisol was elevated during the spring in yearling coho salmon, *Oncorhynchus kisutch.* [*result, method*]

2. Graphical analysis suggested a seasonal correlation between PCR and gill Na/K-ATPase activity. An explanation for this correlation is suggested. [*method, result, reference to discussion*]

3. The major metabolite of ³H-cortisol in plasma was ³H-cortisone. It appeared rapidly following injection of the original radiotracer. [*result*] [Comp. Biochem. Physiol. 82A, 1985, 531]

b.

ABSTRACT

Pachnocybe ferruginea from Douglas-fir utility poles in western Oregon occurred in two forms. Both forms consisted of dikaryotic mycelium that gave rise to brown captitate basidiocarps. However, one form had associated blastic-sympodial conidia, larger basidiocarps, chlamydospores, and a slower growth rate. Single uninucleate basidiospores of both forms produced dikaryotic mycelium with simple septal pores and holobasidia in which karogamy and meiosis occurred. [*results*] Thus, *P. ferruginea* has a primary homothallic life cycle. [*conclusion*] Spores from conidial isolates gave rise to both basidiocarps and the conidial form demonstrating that the conidial form is the anamorph of *P. ferruginea*. [*result, conclusion*] Nevertheless, results suggest that the *P. ferruginea* variants form a single pleomorphic species corresponding to published descriptions of *P. ferruginea*, but with an anamorphic stage. The species should be placed in the Chionosphaeraceae. [*conclusions*]

Key Words: *Pachnocybe ferruginea,* Heterobasidiomycetes, Atractiellales, Chionosphaeraceae. [Mycol. 78, 1986, 334]

c.

Abstract—1. Peroxisomes and mitochondria were prepared form livers of rainbow trout fed diets containing either 15% crude fish oil (CFO) or 11.5% partially hydrogenated fish oil (PHFO) plus 3.5% CFO. [*method*]

2. Peroxisomal preparations from the two dietary groups showed similar rates and substrate specificity patterns for acyl-CoA oxidation. [*results*]

3. The peroxisomal oxidation rate was highest with 12:0-CoA and decreased with increasing chain length, being negligible with 22-carbon acyl-CoA's. The *trans* isomer of 18:1(*n*-9) was oxidized at a higher rate than the *cis* isomer only by peroxisomes from the PHFO + CFO group. [*results*]

4. Mitochondria prepared from both groups of fish exhibited a broad chain-length specificity for the oxidation of acylcarnitines. Both polyunsaturated and *trans* monoenoic fatty acids were readily oxidized. [*results*]
[Comp. Biochem. Physiol. 82B, 1985, 79]

d.

ABSTRACT

Absolute Hβ nebular fluxes are presented for a total of 97 planetary nebulae (PN) in the Magellanic Clouds. [*results*] These new fluxes are compared with all previously published data. [*method*] Nebular masses are derived for 54 objects and are found to lie mainly in the range 0.01–0.35 M. A relationship between density and ionized mass ($M_{neb} \propto 1/n_e$) for a subset of the nebulae is used to show that these objects are optically thick. [*method, results*] Another relationship between Hβ flux and nebular density is examined. [*objective*] The point at which the nebulae become optically thin is seen as a change in the slope of this curve. From these relationships a nebular mass-radius relation of the form $M_{neb} \propto R^{3/2}$ is found to apply to optically thick nebulae, while optically thin nebulae evolve at constant M_{neb}. [*result*]
Subject headings: galaxies: Magellanic Clouds—nebulae: planetary [Astrophys. J. 329, 1988, 166]

5. Abstracts with Extensive Discussion

a.

Abstract—Interferometric measures of HCl, ClNO$_3$, HNO$_3$, NO$_2$ and NO obtained over the Antarctic in 1986 (Farmer *et al.*, 1987, *Nature* 329, 126) are used to constrain models for the chemistry of the atmosphere in the region of the Ozone Hole. [*method, literature cited, objective*] It is shown that the abundance of stratospheric HCl was exceptionally low in early September over McMurdo Station (78°S, 167°E) and that it recovered below 20 km at a rate consistent with the gas phase reaction of Cl with CH$_4$. [*results*] The low abundance of HCl is attributed to incorporation of HCl in polar stratospheric clouds (PSCs), and subsequent reaction of HCl with ClNO$_3$ providing a source of chlorine radicals. [*conclusion*] There is evidence for net loss of HNO$_3$ from the stratosphere. [*discussion*] This is attributed to condensation of HNO$_3$ in PSCs followed by precipitation. The abundance of odd nitrogen is suppressed also at high altitudes in the vortex, reflecting in part, it is suggested, the influence of the reaction of N with NO in the cold summer mesosphere. [*result, discussion*] Model results are consistent with observed temporal trends of the species reported by Farmer *et al.* (1987, *Nature* 329, 126) and also with trends for total column ozone reported by Stolarski *et al.* (1986, *Nature* 322, 808). Loss of O$_3$ is attributed to the catalytic influence of halogen radicals through reactions suggested by McElroy *et al.* (1986b, *Nature* 321, 759) and by Molina and Molina (1987, *J. phys. Chem.* 91, 433). [*discussion, literature*] [Planet. Sp. 1988, 73]

b.

The responses of single units in the cerebellum, the vestibular nuclear complex and adjacent regions of the brainstem and in the oculomotor nucleus were studied in decerebrate, paralysed rainbow trout (*Salmo gairdneri*). [*objective*] Natural vestibular stimulation was provided by horizontal, sinusoidal oscillation of the fish and extraocular muscle afferents of the eye ipsilateral to the recording were activated either by passive eye-movement or by electrical stimulation of the trochlear (IV) nerve in the orbit. Unit responses to vestibular and/or orbital stimuli were examined in peristimulus–time histograms interleaved in time. [*method*] The effects of the orbital signal were usually phasic but rare tonic responses also occurred. . . . Of the eight units carrying both signals, histological confirmation that the recording site lay in the column of cells forming the oculomotor/trochlear nuclei was obtained in four. The responses and interactions were similar to those found in the brainstem. [*results*] The results present two principal points of interest. i. They reinforce the accumulating body of evidence that, in species with widely different oculomotor and visual behavior, signals from extraocular muscle proprioceptors reach the vestibulo-ocular system; this, in turn, suggests that these signals may play some rather fundamental role in the oculomotor system. 2. The hypothesis that extraocular muscle afferents are involved in oculomotor control requires that an effect of these signals be apparent at the output of the system—for example, in the oculomotor nucleus—though, if the action were to alter the characteristics of the system rather than to act from moment to moment, such effects might not be detected in acute experiments. In fact, the results reported here confirm that such a signal is found in the oculomotor nucleus of a bony fish in acute experiments and that it can alter the effect of vestibular drive to units in that nucleus. [*discussion*] Thus, the evidence in this paper further supports the hypothesis that a proprioceptive signal from the receptors in the extrinsic ocular muscles plays a part in the control of eye-movement. [*conclusion*]. [Reprinted from Ashton, Milleret, & Donaldson. Neurosc. 31, 1989, 529 © 1989 Pergamon Press, PLC, with permission]

6. Abstracts Accompanied by Bibliographic Citation

a.

BERTENTHAL, BENNETT I., and CAMPOS, JOSEPH J. *New Directions in the Study of Early Experience*. CHILD DEVELOPMENT, 1987, **58**, 560–567. In this commentary, we review Greenough, Black, and Wallace's conceptual framework for understanding the effects of early experience, and illustrate the applicability of their model with recent data on the consequences for animals and human infants of the acquisition of self-produced locomotion. [*objectives*; *note* that this is a descriptive abstract]

b.

REFERENCE: Spencer, J. L., "Theory, Characteristics, and Operating Parameters of Portable Optical Emission Spectrometers for the On-Site Sorting and Identification of Steels," *Journal of Testing and Evaluation*, JTEVA, Vol. 15, No. 4, July 1987, pp. 231–238.

ABSTRACT: Portable optical emission spectrometers are evolving as important tools for the on-site sorting and identification of metals. Their analytical precision and accuracy, while not quite as good as laboratory systems, are more than adequate for sorting mixes and most grade verification requirements. [*background*] The intentions of this paper are to provide a brief review of the technology and history of emission spectrometers, and then to describe parameters, operation, capabilities and limitations of the device for plain carbon, low alloy, and stainless steels. [*objectives*]

KEY WORDS: optical emission, chemical analysis, verification, sorting, scrap segregation, excitation discharge, steel spectra

c.

ABSTRACT

The elution and sorption characteristics of selenate and selenite were measured in column studies. [*objective, methods*] A composite of sandy loam textured strip-mine overburden was adjusted to pH 2, 3, 5, 7, and 9 and packed into glass columns. Selenium was added to the surface as the Na salt of selenate (SeO_4^{2-}) or selenite (SeO_3^{2-}). Individual columns were leached with 1.5, 10, or 50 pore volumes of 0.01 M $CaCl_2$. [*methods*] Selenate was mobile at all pH values and was completely leached from columns with <3 pore volumes of solution. Selenite was rapidly sorbed at all pH values. Very slight selenite movement was observed when high pH overburden was leached with 50 pore volumes of solution. The least movement was recorded when pH values were <7. [*results*]

Additional Index Words: Se, leaching columns, strip-mine overburden.

Ahlrichs, J. S., and L. R. Hossner. 1987. Selenate and selenite mobility in overburden by saturated flow. J. Environ. Qual. 16:95–98.

7. Abstracts with Nonsubstantive or Repetitious Words and Phrases

a.

We discuss *some* factors *that appear* to dominate *the process of* pattern recognition by the human eye *with special reference to the question* of whether significant filamentary or cellular structure exists in the observed large-scale distribution of galaxies and clusters in the Universe. We illustrate *the way in which* photographic techniques can alter the perceived picture of galaxy clustering. *A number of* examples *are given which* demonstrate the role *played by* nearest-neighbor distances, orientations, visual inertia, local point densities and point sizes in biasing pattern recognition *by the eye*. We list *a number of* necessary criteria *to be satisfied by* any useful measure of filamentary structure in a point data set. Some methods *employed* to quantify pattern recognition *by the human eye* are critically examined and *some specific* improvements suggested. We display "double-poisson" point patterns *which have* a regime in which filamentary patterns are produced. [Q. J. R. Astron. Soc. 28, 1987, 109; italics added]

b.

ABSTRACT   *In this paper* we propose a new model for *the study* of microcomputer *technology* in agriculture.[2] We conceptualize diffusion/adoption of innovations as a structural process. *This process* is affected by *what we call* "access conditions"; these result from (1) research and development of technological innovations, (2) the intrinsic characteristics of the technology (*i.e., labor saving, knowledge intensive*), and (3) the distributional characteristics of the innovation (*i.e., private and public diffusion infrastructures and strategies, such as commercial franchizing and public-funded extension*). Our thesis is that potential adopters respond more to access conditions than to attitudinal variables. We conclude *by arguing* that the diffusion/adoption of microcomputers in farming will *bring about important* changes in the organization of agricultural production. [Rur. Soc. 51, 1986, 60; italics added]

Appendix 8B

*Acknowledgments**

a. Specific Technical and Professional Assistance

1. We are grateful to G. J. Sexton for his help with the 4-index transformation.
2. We thank Joan Eynon for typing the manuscript.
3. The authors (K. U. and T. S.) are indebted to Professor Koichi Itoh for his remark on the rotation of radical cations in the solid. They thank Dr. Akira Kira for his generosity to allow them to use the ESR facility at the Institute of Physical and Chemical Research, Wako, Japan.
4. The authors thank LuAnn Johnson for statistical evaluations, Nancy Driscoll for help with animal care, and James Normandin, Terry Shuler and Cheryl Stjern for technical assistance.
5. We thank Gary Radke, Steve Athey, Becky O'Donnell, Anne Rockhold, Harvard Townsend, and David Roeder for assistance in the field. We are grateful to Mark McGinley and Siobhan Sullivan for their work on the data and figures. James Higgins gave us advice on statistics. O. J. Reichman made helpful comments on the manuscript.

b. Discussion

1. The authors gratefully acknowledge helpful discussions with J. M. Simonson during the work, especially regarding the examination of hydration effects in terms of molar volumes at high temperatures.
2. I wish to thank especially Y. T. Chiu and M. Schultz for many helpful and enlightening discussions.

c. Programs: General and Specific Acknowledgment

1. I would like to thank all of the environmental professionals, local, state, and federal government officials, and individual citizens in New England who assisted Wendy Rundle, Phyllis Robinson and myself in the preparation of the case studies. Larry Susskind played an important advisory role throughout the preparation of this special issue.

*These acknowledgments are at the end of the article preceding the reference section, mostly accompanied by acknowledgments of grants and other support.

2. We gratefully acknowledge the help of our psychometrists, Joan L. Rocheleau and Nancy N. Matthews, who administered the tests described here; our outreach workers, Abigail Halperin and Janet Gillespie, who located and scheduled the children; Veronica Buffington and Seymore Silbertstein, who helped in the development of the equipment; and Nydia Fabian for technical support. We also acknowledge the cooperation of the Obstetrics and Nursing Services of the Group Health Co-operative of Puget Sound and the University of Washington Hospital, where the prenatal phase of this study was carried out, and the study mothers, whose cooperation has been indispensable.

Appendix 8C

Appendices Presenting Computations and Mathematical Derivations

1.

APPENDIX B

The depth of sediment in diffusive contact with seawater was calculated by equating the time derivative of the root mean square of the diffusion length (given by the Einstein-Smoluchowski equation) with the sedimentation rate. The depth at which they are equal is given by D'/S, where D' is the diffusion coefficient, corrected for porosity and adsorption, and S is the sedimentation rate. At greater depths diffusion from the seawater/sediment interface cannot keep up with burial.

$$D' = \frac{D}{1 + K\rho \frac{(1 - \phi)}{\phi}}$$

where:
D = porosity corrected diffusion coefficient
ρ = average density of the sedimentary material
K = adsorption coefficient
ϕ = average porosity.

In the calculations the following values were used:
$D = 4.0 \times 10^{-6}$ cm²/sec
$\rho = 2.7$ gm/cm³
$K = 3.5$
$\phi = 0.5$
$S = 1 \times 10^{-3}$ cm/year

which gives a depth of 120 meters for D'/S.
[Geochim. Cosmochim. Acta 51, 1987, 1949]

in Eqn. (2.2), the Bernoulli separation technique yields

$$\frac{R''(x)}{R(x)} + D_2 \frac{Z''(y)}{Z(y)} = 0 \qquad (A3.2)$$

which implies

$$\frac{Z''(y)}{Z(y)} = -\lambda^2 \quad \text{a constant} \qquad (A3.3)$$

The solution of this separated equation for each value of λ is

$$Z_\lambda(y) = A_\lambda \cos(\lambda y) + B_\lambda \sin(\lambda y)$$

The boundary conditions in Eqn. (3.3) require, for each λ,

$$Z_\lambda'(0) = Z_\lambda'(1) = 0$$

Thus, $B_\lambda = 0$ and $\lambda = n\pi$, $n = 0,1, \ldots$; and

$$Z_n(y) = C_n \cos(n\pi y) \qquad (A3.4)$$

The equation for $R(x)$ is therefore

$$R''(x) + \frac{1}{x} R'(x) - D_2 n^2 \pi^2 R(x) = 0. \ldots$$

[Bio Syst. 20, 1987, 173]

2.

Appendix III: A trigonometric series representation of the solution

If we assume

$$U_2(x, y) = R(x) \cdot Z(y) \qquad (A3.1)$$

9

Reviewing, Rewriting, Revising, and Editing

Papers must be critically reread and reviewed, revised, and edited before they can be submitted to a journal. These postwriting processes are directed at making changes so that the paper is a clear and rigorous presentation of the research and ready for publication. The preparation of a paper is here divided into five stages, each associated with a particular draft.

9.1 PREPARATIVE STAGE: PRELIMINARY OR ROUGH DRAFT

The objective at the preparative stage is simply to write some sort of rough draft of the paper. This stage therefore consists mostly of writing, with some rewriting. The aim is to get the research down in writing, without stopping to choose words, develop structure, correct usage, or integrate ideas. Such revision can be deferred to the next stage.

The rough draft need not be well structured or complete; indeed it may be a heterogeneous collection of drafts. Nevertheless, at the end of this stage the author has some sort of written draft of all sections of the paper. In fact, soon after the research is under way, research scientists who write the paper concurrently with the research can have a rough draft of all the following parts of the paper,

Ex. 9.1

Title	Methods	Illustrations (some)	Acknowledgments
Introduction	Tables (some)	References	Biographical sketch
Theoretical framework			

with only the results, discussion, and abstract remaining to be written.

9.2 FORMATIVE STAGE: FIRST DRAFT

During the formative stage, authors prepare a complete first draft, the first really complete, structured draft of the paper.

Collecting materials.   Authors can greatly facilitate their writing by assembling all the materials that they will need when they start to write the first draft. These include the rough draft written during the preparative stage, notes, research data, analyses, and computer printouts. These can all be categorized as to section, labeled, and ordered, so that they are readily available as needed when the different sections of the paper are written.

The original references and photocopies and notes on the literature should also be assembled, annotated, and ordered. Notes on the literature should be brought together by section or in some orderly arrangement. A copy of the notice to contributors from the chosen journal or any article issued by the journal about writing should be at hand, as well as the appropriate disciplinewide manuals, a scholarly hard-cover desk dictionary, and a general handbook of usage. Tables should be in final form except for data to be obtained from analyses, and complete sketches of the figures should be available, with the final labels. The objective at this stage is to make everything readily available so that one can resist the temptation to interrupt one's writing to go to the library or laboratory to verify a point. One should instead make notes in the text for later verification.

Writing the first draft.   During this stage, the rough draft is reviewed and largely rewritten. The main objective is a comprehensive draft of the paper, which includes all the information, data, concepts, and ideas about the research that the writer might ultimately wish to have in it. Then in the revisions that follow, the changes will be largely cutting and condensing, which are usually easier tasks than integrating new material into a written manuscript. The second objective is to group the main information and ideas logically into the sections of the paper. Authors should therefore make little attempt at this stage to attend to other than writing and rewriting. Since the first draft is planned for cutting and condensing, the draft is certain to require major revisions. At the end of this stage, therefore, the manuscript is not advanced enough to elicit useful commentary from colleagues. The first draft need not be a pristine copy. One can write notes to oneself, for example, "condense description of phase instrument," "elaborate, see Green's '83 paper," "combine tables 2 and 4?'' The first draft is now ready to be revised to a unified, coherent, closely organized paper. But by this time the author is likely to be surfeited with writing. This is a good time, therefore, to set the paper aside for a period of gestation, so that when one returns to revise it, one can see it with a fresh eye.

9.3 PROCESS OF REVISION

After the first draft is written, the writer begins a process of repeated reviewing, rewriting, revising, and editing to improve the manuscript. In the early stages of this process, authors concern themselves more with content and the structure of sections and paragraphs; in the later stages, they concern themselves more with details of structure, form, and usage in sentences.

9.3.1 Definitions

Reviewing is the critical reading of a manuscript. It is a prerequisite for rewriting, revising, and editing any draft.

Rewriting includes adding, deleting, shortening, expanding, or restructuring sentences, paragraphs, or sections. Rewriting may change the content or structure of the part rewritten so that it is noticeably different from its earlier form. *Revising* refers to changes that do not entail major changes or a great deal of additional writing, for example, changes in words, phrases, or clauses or sentences in a paragraph. Revising may also include rearranging one or several paragraphs. This necessarily requires some rewriting to reestablish the line of development. ''Revising'' is also used in the broad sense to include reviewing, rewriting, revising, and editing.

Editing is the very close rereading of the manuscript for details of form; that is, for consistency, grammatical structure, punctuation, and so on. Most editorial changes do not disturb the major structure of the paper. Editing is part of writing a finished manuscript and is the responsibility of the author. The editor's function is supplementary—to check for the author's omissions, errors, inconsistencies, and so on.

The early postwriting stages consist largely of rewriting and revising; the final stages mostly of editing, with less revising and almost no rewriting. The term ''proofreading'' is used loosely by writers for these last stages but proofreading is really part of the publication process.

9.3.2 Author as Reviewer and Editor

The author assumes the reviewer-editor role mostly to improve the manuscript, but also because some parts of the task cannot be delegated to others. Only the author can know whether the data are complete and accurate, or can supply missing details in the development of a section or paragraph, or determine whether the development clearly expresses his or her ideas.

The first step in revising a paper is to read it for unity—to ensure that the main sections and parts do indeed include the material designated. This is a verification for content. Next the writer can read for coherence, the ordering and integration of the parts. In revising the paper, the writer can facilitate the process by following some ordered pattern of revision. One pattern is a sequential one from writing to editing, with each step entailing successively fewer changes:

Ex. 9.2

WRITING →*reviewing* → REWRITING → *reviewing* → REVISING →
reviewing → EDITING → *reviewing* → EDITING

In this pattern, writers start by spending most of their time writing and rewriting the paper, then gradually spend more time revising and editing. Their attention during a particular step is concentrated on that particular process. In another pattern, the writer revises the paper in successively smaller units. At first authors read the manuscript as a whole, focusing their attention on overall content and structure. Next, they read each section to improve the central line of development of the section; they then give their attention to the structure of the paragraphs and sentences in the sections. Finally, they attend to details of the writing, such as diction, capitalization, punctuation, and so on.

The author does not necessarily start at the beginning of the paper with each rereading, but continues from a point slightly before the last stopping place. Whenever an author thinks that he or she does not have a grasp of the whole, it is better to reread from the beginning. At the last stage of revision, the process is one of great concentration of effort. During this stage a copy should be made of important drafts and kept in a separate location.

9.3.3 Gestation Periods

When authors have worked too concentratedly or too long on a draft, they tend to become satiated or inured to the manuscript. They then lack the concentration required for clear writing, critical reading, or thoughtful careful revision. They may be too full of all the details to write coherently; they may also be writing with conflicting objectives; or they may not have fully resolved their thinking about a particular part or section or interpretation. It is then more efficient to turn to other tasks to allow one's ideas to gestate; however, such periods should be used to refresh oneself rather than to procrastinate.

Gestation periods seem to serve as periods of subconscious integration, so that when one comes back to the writing, one finds that problems that had been intractable earlier can now be solved. They allow authors to develop a sharper insight and an alertness that make deficiencies leap off the page and facilitate correcting them

STAGE	GESTATION PERIOD: NONWRITING ACTIVITIES	DRAFT
Preparative		Preliminary or Rough
	Develop outline for first draft.	
	Draw sketches for illustrations.	
	Prepare tables in outline.	
	Collect materials for first draft.	
	Perform other nonwriting activities.	
Formative		First
	Draw sketches in final form.	
	Seek preliminary review of illustrations by co-authors and colleague-reviewers.	
	Prepare tables in final form.	
	Review journal requirements.	
	Research or other nonwriting activities.	
Revision		Review
	Submit review draft to coauthors and colleague-reviewers.	
	Perform other nonwriting activities.	
	Reread review draft critically in preparation for reviewer's comments.	
Editorial		Final
	Review journal requirements for submitting paper.	
	Collect parts of papers and items to send to journal editor.	
	Write letter of transmittal.	
	Submit copies to coauthors.	
Publication		Manuscript
	Mail manuscript to journal editor.	

Figure 9.1. Stages in writing a paper from rough draft to submission of manuscript, showing stages, draft produced, and activities during intervening gestation periods.

more readily and effectively than before. However, gestation periods are not likely to be productive unless one has worked intensively at revising and trying out various alternatives. Whatever mental or subconscious processes are at work seem to require material to work on. Gestation periods should be long enough to allow one to reread and review a manuscript with a fresh eye and alert mind, but not so long that one must spend too much time reorienting oneself to the manuscript because of having forgotten the structure and details of the argument. During the gestation period, the writer can attend to nonwriting tasks such as tables and figures or research, or secondary writing tasks, such as the biographical sketch, acknowledgments, or title (*Fig. 9.1*).

9.4 REVISION STAGE: REVIEW DRAFT

9.4.1 Revision

Objectives.   During this stage the draft is revised to prepare it for informal review by colleagues and coauthors. This review draft should, therefore, be approaching the final draft, so that the reviewers can give useful feedback. The objectives during this stage are (1) to eliminate parts of the first draft that are superfluous and condense parts that are overexpanded, too detailed, or too wordy; (2) to strengthen the structure at all levels, so that the paper is closely and logically ordered; (3) to ensure the scientific accuracy of the manuscript, figures and data; (4) to write the abstract; and (5) to begin seriously to edit the manuscript.

Requirements of the journal.   Before revising the first draft, authors should learn the requirements of the journal chosen. If a journal has not been chosen for publishing the paper, it should be chosen now. The author should identify the most suitable ones or may wish to consult the reviewers. He or she should examine recent issues and the notice to contributors for information about the scope of the journal, type of research published, and type of articles preferred.

The notice to contributors and any article or pamphlet that the association publishes describing the style format and requirements of the journal should then be consulted for the style followed in the journal for references, treatment of scientific terms, symbols and conventions preferred, limitations on length of paper, tables and figures, and directions for submitting the paper.

Accuracy.   Throughout the revision of the manuscript, the author should be alert to possible inaccuracies—in data, in tables, figures, references, statements, and especially in numbers. Numerals in the text, tables, and figures should be scrupulously verified for accuracy. This stage should be the checkpoint for scientific accuracy, because the author has the opportunity to solicit the advice of coauthors and colleagues on scientific questions. The author should also be alert to misstatements, contradictions, inconsistencies, misrepresentations, and fallacious reasoning.

9.4.2 Graphics

During this stage, tables and figures should be completed and prepared in final form for the typist and illustrator; otherwise authors will not have the benefit of colleagues' review of them. The graphics should be examined in themselves for economy of design and for visual effectiveness. They should be complete with number, descriptive title, any notes, and legends in final form.

At the end of this stage, tables and figures should be reviewed to ensure that (1) they can be understood without reference to the text; (2) they are referred to in the text by number; (3) they accurately illustrate the description or explanation in the text; and (4) the number, quantities, values, labels, terminology, and description in the tables and figures are consistent with those in the text. Equations should be given the same careful scrutiny and should be examined scrupulously for accuracy. During this stage the author should consider the number of figures and tables. Because of the cost of reproduction, graphics that do not have enough information or explanatory power should not be included. Decisions should be made about eliminating overlap and combining tables or illustrations. Here authors have an opportunity to consult their colleagues.

9.4.3 From First Draft to Review Draft

Content.   Since the first draft was intentionally overwritten, some parts will clearly be superfluous, supplemental, or too detailed. The key words for editing for content at this stage are *cut* and *condense*. The parts that are too detailed are condensed, and unimportant or irrelevant details are deleted. This is a good time to consider the length of the paper relative to the requirements of the journal. If the paper is too long, the writer should attempt to shorten it or query coauthors and reviewers about likely parts to condense or eliminate. Each section should be examined for superfluity; for example, the introduction and the discussion may be too extensive or diffuse.

Structure of text.   The deletions and condensations will, of course, interrupt the line of development across the deleted material, so that the paper must then be read and revised for logic and coherence. During this stage, authors should first make certain that the elements included in the various sections belong there. Then the task of organizing the paper can be reviewed, such as the logical ordering of sections, the logical development of each section and each paragraph, and the logical connections between sections, paragraphs, and sentences. Making the text read logically and smoothly is one of the most difficult of writing tasks and may take much revising. Chapter 11 addresses some common writing problems.

Each section should be examined and revised to make sure that it fulfills its function. Each paragraph should be revised so that (1) there is a progression in the line of thought through the paragraph, (2) it makes the point it is intended to make, and (3) it is logically and substantively connected to the paragraph before it and after it. Very long and very short paragraphs should be examined to determine whether their length is appropriate to their function and content. Sentences should be examined for their internal order and structure and for their logical connection to neighboring sentences. The line of development may not carry across sentences because of a confused internal order or because the sentences are not in logical order. Transpositions may necessitate revisions of the sentence, and connecting sentences, to reestablish the line of development.

Revision of nontextual parts.   The parts external to the text must also be revised. For references, the author-date system is more efficient during revision. The title is important enough to put in final form for colleagues' review. It should be descriptive, complete, but concise and specific, and it must accurately describe the content of the paper (see Chapter 8). The abstract is important enough to have it reviewed, and the paper is now advanced enough to write one for review. Headings must be revised, so that (1) they accurately, clearly, and concisely describe the material covered and (2) their form and position mark their rank.

At the end of this stage, the author begins the close editorial revision that will form a major part of the postreview stages. A decision must be made as to how close to a final draft, the review draft is to be. It is not desirable to edit it to a final draft, since changes can be expected after colleagues have reviewed it. Also authors may not be certain about how to treat a particular part and may wish to ask for reviewers' suggestions for the parts that are not carefully written. The author is not approaching an advanced stage, the writer will not have the benefit of the reviewer's suggestions for the parts that are not carefully written. The author should, therefore, edit to an advanced draft those parts that might benefit from colleagues' scientific expertise.

Review draft.   It is difficult to say at what point the first draft becomes a review draft. An author may write one section in advanced form at one sitting, yet rewrite another section five times and still not be satisfied with it. Let us say that the manuscript is a review draft when the draft (1) includes everything that the writer thinks important to include, (2) is in a well-structured, logical form, (3) closely expresses the author's ideas, and (4) is carefully although not finally edited for accuracy, consistency, and clarity. Copies of the draft can now be distributed to colleagues together with the author's queries. The author may set a tentative deadline but should allow reviewers adequate time to review the manuscript.

9.4.4 Colleague Review

Reviewing at this stage is a more informal process than the formal reviewing of the manuscript after it has been submitted for publication (see Chapter 10). The reviewers are usually colleagues in the same department, though occasionally the author may ask interested peers at another institution engaged in similar research to review the paper. The reviewer chosen should not be so favorably disposed as to be uncritical. The ideal reviewer is one who is exacting and constructively critical of deficiencies.

The main function of reviewers is to act as scientific advisors, not as editors. Their great usefulness is their ability to review the paper from a different perspective than that of the author. Authors can expect them to note overdevelopment, omissions, faulty logic; they should not be expected to read careless, opaque writing, or have to edit the manuscript—though they may edit in passing.

Toward the end of this period, authors should review the paper themselves, preparatory to reading the comments and suggestions of the reviewers. They will then discover changes that they would now make in the manuscript and will be able to compare such changes with those suggested by the reviewers. With the reviewers' comments in hand, the author can now undertake the preparation of the final draft.

9.5 EDITORIAL STAGE: FINAL DRAFT

The objectives during the editorial stage are (1) to incorporate the suggestions of coauthors and reviewers into the paper, (2) to revise the paper to include all that is needed to present the research, (3) to review the organization of the paper to make it coherent, and (4) to edit the draft in minute detail, in preparation for submission to the journal.

9.5.1 Reviewing Reviewers' Comments

Reviewers' comments are important, as a scientific response to the paper and as an early clue of the response of other peers. Authors should therefore consider the comments thoughtfully. Reviewing a colleague's paper informally is a professional courtesy; therefore it is advisable to discuss changes with reviewers, especially changes about which the reviewers have strong opinions or strong differences. Such discussions of comments are important in forming part of the ground bed for scientific research.

After consulting with reviewers, the author can come to a resolution about the problems and revise the manuscript accordingly. After the changes have been made, the manuscript must be reviewed and revised for content and coherence, as in the revision stage.

9.5.2 References

During this stage, the references are put in final form for publication (unless this has been done earlier), so that they do not have to be edited again before reading proof. This is a major editorial task. First the references themselves must be verified against the original sources to ensure that the entry is complete and *absolutely accurate*. For this task, all the publications, or photocopies of the pertinent pages, should be available, to save time and to avoid errors. Next, each entry is examined to ensure that the style follows the format required by the journal. The text is now read to make certain that every source cited in the text is represented in the list of references. References not applicable to the text are deleted from the text and bracketed in the list of references, to be removed if not referenced elsewhere in the paper. The references in the list of references that have not been cited in the text are now deleted.

If there is any change in the references in the text after the references are edited, the cross-check must be repeated. If the list of references is numbered, it must be reordered or renumbered after such changes, and the references cited in the text must be renumbered. When the references are not numbered, only the added or deleted reference need be cross-checked.

9.5.3 Editing

The final task during this stage is editing for usage and form; therefore authors must be thoroughly familiar with the requirements of the journal. This is the fine-tuning precision stage in writing the paper. The chief objective is thoroughness; each sentence, each word, and each mark of punctuation must be scrutinized. The watchwords at this stage are accuracy and consistency, both for scientific and for editorial reasons.

Every effort should be made at this stage to prepare a manuscript that is entirely accurate. This is a responsibility of the author, which cannot devolve on reviewers and editors. It is also easier to ensure accuracy at this stage, when all materials are at hand or fresh in one's mind, than months later when one receives queries with the copyedited manuscript or galley proofs. Moreover, errors that cost nothing to correct at this stage become costly if they are carried over into proofs. Accuracy is particularly important for numerals. Editors cannot be expected to realize that the value 167 in a table has become 176 in retyping or recopying. Therefore, numerals in the manuscript must be doubly checked, and whenever they are retyped, they must be verified again. Equations, because they consist entirely of numerals and symbols, should be treated in the same way.

Editing entails careful reading for many different elements: paragraph and sentence structure, conciseness and clarity, the form of tables, figures, and references, consistency in terminology and usage, redundancies, punctuation, capitalization, spelling, and so on. It is difficult, however, to read simultaneously for so many elements. Therefore, the author should prepare a checklist of all the various editorial tasks and then read through the manuscript, attending to only one or two particular tasks at a time. Brief notes to the editor, for example about confusing symbols, can be written in the margins.

At the end of this stage, the draft should require only minor checks for preparing it for the printer; therefore, copies of this edited final draft can be distributed to coauthors. This will avoid delaying the mailing of the manuscript later while coauthors read the draft.

9.6 PUBLICATION STAGE: MANUSCRIPT

This final stage centers on preparing the manuscript for publication. The objectives are (1) to meet the journal requirements for submission of a paper and (2) to prepare the manuscript for printing.

Before the manuscript is assembled for mailing, it should be given a final reading and editing. Since the manuscript has fewer changes than earlier drafts, it is easier to identify errors; however, it is also easy to miss them, because the pristine appearance of the manuscript lulls one into scanning. In this last editing the author should pay particular attention to anything that may have been changed or to errors that might not be easily detected, such as numbering of footnotes, tables, figures, equations, and cross-references. A final editing is particularly important for tables and figures, where the cost of corrections may be prohibitive.

9.6.1 Preparation of Illustrations

During this stage the author should carefully review the directions for submitting a manuscript and prepare the illustrations in the final form. Line drawings should be examined for breaks in the lines and ink spots or other flaws; photographs should be examined for sharp focus, contrast, and any marring of the surface. Titles, headings, labels, and legends should be checked in themselves, with reference to the table or figure, and with reference to any description or discussion of them in the text. For convenience, illustrations should be mounted. Mounted illustrations are less likely to be damaged and they provide margins, which can be used for directions to the photoengraver and others processing them. If possible, the mounting boards should all be of the same size to facilitate wrapping and mailing. Illustrations should be labeled ''top'' at the top, if they have no lettering on them to prevent incorrect orientation.

Because illustrations are treated separately from the text in printing, they should be carefully identified with figure number and author's name. These should be clearly written in the margin, usually at the top, outside the actual illustration. For line drawings they are written on the front of the drawing; for photographs, they are written on a tab attached to the photograph or on the back of the photograph, *in the margin,* very lightly in soft pencil so that the glossy surface of the print is not marred. Mounted illustrations can be labeled on the margin of the mount.

The author should consult the editor about the number of photocopies required for reviwers and for the printer. If the journal editor prefers photographs of the drawings rather than the original drawings, the photographs should be prepared by a professional photographer having experience in preparing photographs for publication. It is important to start with as good a copy as possible, because the best of reproductions does not improve on the original. The various parts of the manuscript can then be assembled in the order required for the journal (see *Table 9.1*).

9.6.2 Preparation of Manuscript

Typescript or printout.   The manuscript sent to the journal should be good clear copy prepared to meet the needs of the various members of the editorial and production staff, all of whom handle it and write on it. The paper used should therefore be good-quality, opaque, white, bond paper, which will resist abrasion. It

TABLE 9.1   Checklist for preparing parts of manuscript for submission to journal

1. *Cover page.* Not numbered; provides information to editorial and production staffs: title of paper, name(s) of author(s) and affiliation, mailing address and telephone number of corresponding author and corresponding editor, parts of manuscript included, with number of tables, figures, and footnotes.

2. *Title page.* Page 1, not typed on page; title; name(s) of author(s) and affiliation of each; date.

3. *Running head.* Page 2, this and subsequent pages numbered consecutively, typed on page; short title extracted from full title to help printer and readers to identify article; one-half width of page, typed at upper right or lower left of page.

4. *Abstract.* Page 3, labeled "Abstract" or "Summary" or not labeled, according to journal style.

5. *Body of text.* Page 4, typed consecutively unless journal requirements specify new page for each section; exclusive of graphics but notes their placement, e.g., "Fig. 3 about here."

6. *List of references.* Begin new page, typed according to journal style.

7. *Acknowledgments.* Begin new page, typed according to journal style.

8. *Footnotes.* Begin new page, numbered and typed in order of appearance in text, each treated as separate paragraph.

9. *Tables.* Each typed on separate page, according to journal style; numbered and paged in order of appearance in text.

10. *Figure titles and legends.* Begin new page, separated from illustrations; numbered in order of appearance in text; typed according to journal style.

11. *Illustrations.* Separate from text, not paged, except for page-size photocopies; mounted and protected; labeled with figure number and author's name.

12. *Biographical sketch.* If required, not paged.

13. *Photograph.* If required; not paged.

14. *Letter of transmittal.* Including title, guarantees, restrictions.

should be of standard size, $8\frac{1}{2} \times 11$ in., with ample margins (1 to $1\frac{1}{2}$ in.) for notes and directions.

The type should be clear, black, and large, to allow easy reading, since editors and printers spend most of their working day reading typed or printed material. Decorative or elaborate typefaces should be avoided, as should an open-dot matrix or type with vibrating optical effects. Copies should have sharp, dark, clear characters and a white background.

The manuscript should be typed double-spaced throughout, even quotations. Paragraphs should be clearly indented and marked with the paragraph symbol (¶) if there is any doubt. If pages are deleted, then the page number of the missing pages should be included in the numbering of the pages, for example, 11, 12–14, 15, 16. If pages are inserted, letters are added alphabetically to the page number, beginning with the page preceding the insertion and continuing until the last page inserted, for example, 18a (originally 18), 18b (first inserted page), 18c (second inserted page), and 19 (originally 19); page 17 carries the note "followed by 18a" and page 18c carries the note "followed by page 19."

Corrections and insertions.   Corrections and insertions can be made directly with word processors; otherwise they are inserted by hand if few, or the page is retyped. The margins of the manuscript are the channel for communication between author, editors, and compositor. Queries, directions, and notes are enclosed in balloons (circles) close to the pertinent part of the manuscript—hence the need for wide margins. Authors should write notes to identify and clarify diacritical marks, Greek letters, symbols, confusing letters, and numbers (*Table 10.2*).

9.6.3 Preparation for mailing

Assembling for mailing.   When the manuscript is ready to be submitted to the journal, the various parts of the paper are assembled for mailing in the order required for the journal (see *Table 9.1*). The appropriate number of copies are then made, the copies for co-authors being distributed to them promptly for final approval. The author of course retains at least one file copy as insurance against loss and for reference in subsequent correspondence. For oversized or mounted graphics, the author should provide photographs or photocopies for coauthors. The assembled pages and illustrations should not be stapled, pinned, or bound.

The illustrations should be shipped flat, not rolled or folded, and packaged carefully, so that they do not bend, wrinkle, or develop folds. Pins, clips, staples or other sharp or hard objects should not be allowed to touch a photograph, as they may mar the surface. Illustrations should be placed face down on a protective cover sheet of light paper and taped at the edges. The protective cover is then taped to thin cardboard. These three-ply illustrations are then interleaved between firm sheets of thin cardboard larger than the illustrations. The four edges of the packet are taped together. This packet is then taped to a double-faced corrugated cardboard of a size wider than the packet, which is taped to a corrugated cardboard of similar size at the edges. This taping keeps illustrations flat, protected, and rigidly in place. If the illustrations are large or bulky, they may be mailed separately. Manuscript and illustrations can be sent, in a padded mailing envelope, if they have been firmly packaged. If they are sent separately, each package should include a letter of transmittal.

Letter of transmittal and other attachments.   A letter of transmittal to the editor of the journal accompanies the manuscript. The purpose of the letter is (1) to identify the manuscript being sent and (2) to provide information that the editor may need in publishing the paper. If there has been previous correspondence or telephone consultation, this can be mentioned, as pertinent. Otherwise the letter starts by stating that the paper is enclosed, giving the title of the paper and the names of the coauthors if any, and briefly characterizing the paper.

The letter must inform the editor of any restrictions or limitations. The author may not be able to transfer copyright to the journal or may not be able to read proofs before disappearing into the wilds of Africa for a sabbatical. The author may offer suggestions about storing the excessive data in a data bank or about reviewers for the paper, if there are few peers conversant with the research.

The author should also provide the editor with two important guarantees—on originality and ethical practices. It is essential that observations, findings, and concepts be original. Also, journal editors require an explicit statement that no part of the research has been published before or that the paper has not been submitted to another journal. If the author has published a paper that might be considered related to the paper submitted, he or she can cite the paper, explain clearly how it is different, and provide the editor with a copy of it. The author must also provide guarantees of informed consent from persons participating as subjects in the research and of having followed accepted practices in the treatment of animal subjects. The author should also provide the editor, where appropriate, with the full name, complete address, and telephone number of the corresponding author who will be responsible for answering substantive scientific questions and of the corresponding editor who will be responsible for editorial and administrative decisions. The package should be sent to the editor and addressed as specified in the journal, usually the editorial office. It is worth repeating that the author should retain a copy of everything he or she is mailing.

9.6.4 Camera-Ready Copy

In some journals, the typed pages of the manuscript are photographed and printed directly, as submitted, without changes in text or page layout. For such journals, the author is asked to submit camera-ready copy. To maintain uniformity in the page layout, the editor usually provides authors with specifications or special paper that outlines the boundaries of the page for typing. The typist must type within the boundaries set. The pages are then reproduced photographically for publication. Because the typed page becomes the "printed" page, the author and typist must follow the directions provided *exactly*.

With this method, the paper goes from typed publication draft to "printed paper," with no intervening editorial processing or transcription into print. This results in more expeditious publication of research, but it places a heavy responsibility on the writer: (1) for the editorial processing that would otherwise have been the responsibility of journal copy editors and proofreaders and (2) for the physical appearance of the printed page. More important, there is no opportunity to make changes or corrections. Therefore the writer must treat the typescript or printout like page proofs and edit the manuscript intensively—in every minuscule detail. The paper must be perfectly typed and every effort made to give it the solidity of print. The typeface must not intrude itself on the reader's attention. The lines must not be too far apart or too crowded, and the letters, spacing, and margins must be consistently styled.

9.7 DISSERTATION

In writing a dissertation graduate students should inform themselves of the requirements of the institution for the dissertation before they begin to write, so that they can fulfill them in the course of the writing. They should also confer with their supervising faculty to ascertain the faculty requirements or expectations.

9.7.1 Parts of Dissertation

A doctoral dissertation usually consists of chapters, which often represent the sections of a scientific paper. The chapters are more extensive than the corresponding sections of a scientific paper. When the subject matter of the research is extensive or when the research consists of a series of experiments or studies, these may be reported in separate chapters. Because of its length, the dissertation may include parts not included in a scientific paper, such as appendices, a table of contents, list of tables, and so on. Also, the sections, being long, are variously divided so that there are more headings and a broader hierarchy of headings. It may not be possible to distinguish the various hierarchical levels with the limited variation in type available in typing; therefore, an alphanumerical system, such as that used in many outlines, may be used with the headings. Tables and figures are numbered as a unit for each chapter, such as *Tables 2.1, 2.2, 2.3,* and so on.

9.7.2 Stages in Writing

The stages in writing a dissertation are similar to those for writing a paper, except that in the revision stage, the dissertation is prepared for submission to the graduate faculty for review and conditional approval; in the editorial stage, the dissertation is edited in final form for final approval, and in the final stage, corresponding to the publication stage, it is prepared for submission to the graduate school faculty.

   If students have been writing concurrently with their research, they may have only the results and discussion sections to write and are almost ready to write a first draft. The stage at which students start writing strongly affects the time required to write the dissertation, and the ease of writing it. If students continue to be employed part time after completing their research, they can alternate writing their dissertation with their work; then it is not of great import whether they begin to write before completing their research. If students have only to write the dissertation, they will not be able to use all their time productively in writing, because they cannot write steadily all day effectively. Even though they may have some nonwriting activities, such as verifying references, drawing sketches for the illustrator, or "thinking" about the research, they will still have time that they cannot use effectively in writing, although they may welcome working leisurely before starting a position.

   For many students, however, the pressure to finish writing the dissertation is as great as the pressure was to complete the research. They may not have funding while writing their dissertation, or they may have a position waiting, dependent on their having the degree. Such students must be able to write the dissertation in as short a period as possible. Since they will not be able to write all day and into the night, it is important for them to have been writing concurrently with their research. If they have, they will have drafts of the earlier sections of the paper and can now write the results and discussion sections and soon have a first draft or better of all the sections of the dissertation.

After the first draft, students can proceed to the revision stage and concentrate on rewriting and revising, so that they can submit a revised draft to the faculty for review. They then revise the dissertation to meet their advisors' suggestions and edit it to produce a final draft, which is submitted to the faculty for formal approval. Then the dissertation is ready for final editing before being filed with the appropriate official body.

9.7.3 Preparing a Paper from the Dissertation

The author of a dissertation usually publishes one or more papers based on the research reported in it; indeed doctoral students are expected to publish their research. In preparing a paper from the dissertation, the author begins at the revision stage. The dissertation has a fully developed structure and contains more information and ideas and more detail than can be included in the paper, so the task is largely one of cutting and condensing at first. The objective is to identify the parts to be deleted or condensed. The deletions can be bracketed, and the condensations can be marked with a marginal line. Tables and figures can be examined to select those suitable for the paper. Notes can be made about how to convert it to a paper.

The result is a rough draft that consists of excerpts. The draft will require more rewriting than is usual at this stage to make the paper a logically ordered paper, but the author can soon have a good first draft, and from this, the succeeding stages will follow easily. The dissertation may include the substance for several papers. It may be structured so that the separate chapters or discrete parts lend themselves to separate papers. The separate parts can then be delimited and shaped into papers as has just been described.

10

Publication

10.1 ADMINISTRATIVE PROCESS

10.1.1 Overall Process

The receipt of the manuscript at the editorial office is the beginning of the publication process. This includes (1) the review process, in which the author works with the journal editor and reviewers, primarily on scientific questions; (2) the editorial process, in which the author works with editors, largely on form; and (3) the publication process, in which the author works with editors and printers, on tasks related to printing and publishing the paper. The editorial and publication processes represent the production processes of the journal. Parallel to all these processes is the administrative process, by which the manuscript is circulated among author, editors, and printers until the manuscript is printed and distributed to subscribers. In this process, the author works with administrative assistants or a managing editor.

The editor of a scientific journal, the "journal editor," is a scientist of some professional standing, who as a member of a scientific or professional association has been selected to edit its journal. Journal editors shape the journal both scientifically and editorially. The broad objectives and scope of a journal are determined by the scientific association, but they execute the objectives, often with the assistance of an editorial board. Journal editors decide on the different types of contributions that the journal will publish, for example, research papers, review articles, notes, and book reviews, and determine how regularly and how frequently to publish them; they ultimately decide which papers will be published. Journal editors establish the editorial style of the journal, usually together with the board and managing editor. They also decide on the makeup, composition, and emphasis of each issue. They establish general administrative policies, but delegate their execution to the managing editor. Ultimately therefore they are responsible for the scientific, editorial, and administrative decisions.

Managing editors are paid professionals, with editorial as well as administrative qualifications, but usually without training in the discipline. They are responsible for the administrative tasks and are at the center of the actual exchanges between author and journal editor, reviewers, copy editors, and printer. Copy editors have an entirely editorial function: to edit the manuscript for the printer. They may be on the editorial staff or be free-lance copy editors.

An outline of the path of the manuscript after it is received by the journal editor is shown in *Fig. 10.1* and *10.2*. When the manuscript is received, a record is made of it as to the date received, the title, author(s), affiliations, corresponding author, telephone numbers and so on. The manuscript is then examined for (1) the inclusion of the various parts (see *Table 9.1*); (2) the numbering of pages, tables, and figures; (3) and completeness and compliance with journal requirements. If there are deficiencies, they are addressed to the author. When the manuscript is complete and in order, the paper is given to the journal editor, who reads it (1) to determine whether to consider the paper for publication, (2) to decide on suitable reviewers, and (3) to assign it tentatively to a future issue.

10.1.2 Priority of Publication: Delays and Deadlines

Prompt publication is important because new research contributes to further research. Because original research is strongly valued in science, the date of first publication becomes a significant date scientifically. It is not easy to determine this because of the many delays. Most journals include a date relative to the submission of the manuscript to the journal. Often two dates are given (*Table 10.1*). These may include (1) the date of receipt of the manuscript at the editorial office or the date of the letter of transmission, (2) the date on which the revised version is received, or (3) the date of acceptance of the paper for publication. For taxonomic journals, the date of delivery of the journal to the post office for distribution sets the priority date. These dates are usually printed in the journal (*Table 10.1*).

It is obvious that with so many exchanges there are inherent delays in the publication process. There is delay simply because of the exchange itself. Journal editors, being the gatekeepers, have the most exchanges, and so contribute to the delay; yet the editing of the journal is usually not their main professional responsibility. Reviewing is also a time-consuming process, but it too is not a reviewer's

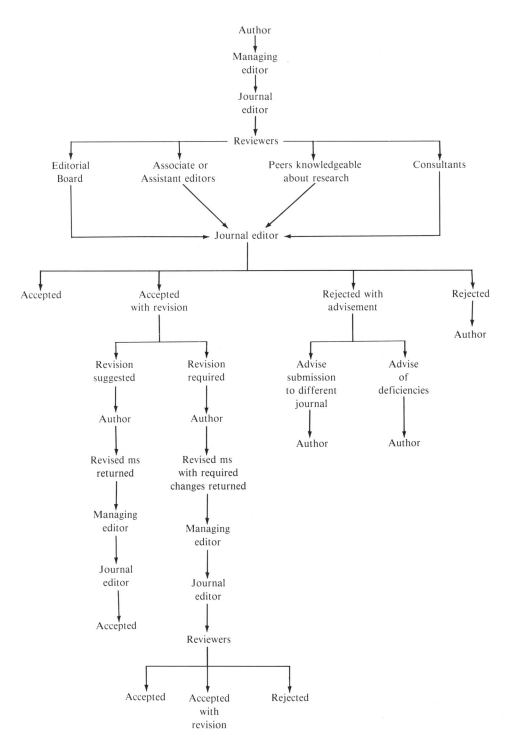

Figure 10.1. Path of manuscript in reviewing process.

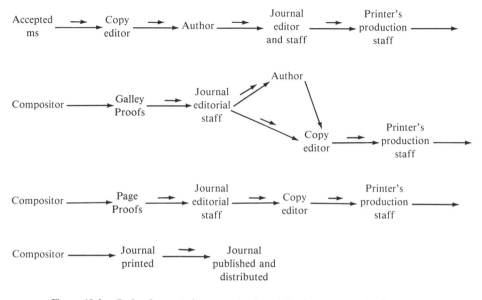

Figure 10.2. Path of accepted manuscript in publication process. Small arrows indicate administrative tasks.

TABLE 10.1   Dates establishing priority of publication

a. *Between Byline and Abstract*

1. (Received 3 November 1986, and in revised form 3 December 1986)
2. (Manuscript received 9 December 1985, in final form 19 May 1986)
3. Received September 2, 1986
4. (Received December 15, 1986; Refereed)
5. Received February 15, 1985; revised August 2, 1986
6. Manuscript received 10 July 1985; accepted 3 September 1985 [*published Jan 1987*]
7. (Received 1986 September 2; in original form 1986 May 29)
8. (Received in final form 6 October 1986)
9. (Received August 15, 1983; revised version accepted June 5, 1984)
10. Original version received 3 November 1986 and accepted version received 12 January 1987

b. *In Other Positions*

1. (Accepted for publication August 1986) [*margin, below byline*]
2. Accepted November 11, 1985 [*below abstract*]
3. Accepted: January 9, 1987 [*below abstract, after key words*]
4. Received for publication, 1 May 1986
   Revised for publication 6 October 1986 [*below reprint notice, on title page*]
5. Accepted by David G. Dannenbring; received November 28, 1983. This paper has been with the author 5 months for 2 revisions. [*footnote to title*]

official responsibility. The author's revisions necessarily cause a lag. Moreover, reviewing, editing, and proofreading take much more time than is expected or allowed for, and so cause delay. Finally, procrastination adds unnecessary delay at each step. It is not surprising therefore that a paper may take a year to publish. To reduce publication time, editors may eliminate the review of some papers, or reduce the number of editing and proofreading stages, but this affects quality adversely.

Editorial and printing deadlines are real deadlines. A delay of one week in one step may be converted to a delay of a month or more because subsequent exchanges cannot be made to correspond. Authors and reviewers should therefore consider a deadline as fixed a part of their schedule as meeting a class or a research deadline. They should begin to review the manuscript or proofs immediately on receipt of them and allow twice the time expected for the task.

10.2 REVIEW OF MANUSCRIPT

10.2.1 Objectives of Review

The formal review process begins when the journal editor reads the manuscript. Most journal editors receive many more papers than they can publish and accept 15–85 percent of those received.

Journal editors usually share the task of selection with peers in the discipline. The process is called "peer review" and the peer scientists "reviewers," sometimes "referees." The latter term is inaccurate, because reviewers are advisors, not decision makers; if anyone is to be called a "referee," it would be the journal editor. Besides, the function of reviewers is not merely to advise about publication; they have as important a function in suggesting changes to improve papers. The term "referee," therefore, misrepresents the review function, shifting the focus from empiricism to authority and moving it out of the domain of scholarship into the arena of competitive sports and conflict. Journal editors share the review of papers because (1) they cannot be specialists in all branches of the discipline, (2) the number of manuscripts received is too burdensome to review alone, and (3) a joint decision to accept or reject a paper is likely to be more impartial.

Journal editors have various scientific criteria for selecting papers. Some criteria are related directly to the research: (1) originality, (2) validity, and (3) importance; and some are related to the discipline: (1) the suitability of the paper for the audience and scope of the journal and (2) the representativeness of the paper relative to research in the discipline. Journal editors also have practical constraints, which may affect their scientific determinations. They must have an adequate number of accepted papers to fill regular scheduled issues, to allow for special issues or for grouping related papers in the same issue, and for delays, omissions, and substitutions. The journal editor must also take the schedule of others into account. Whether a paper is accepted for publication, therefore, depends not only on scientific criteria but also on practical constraints.

10.2.2 Review Process

Journal editor and reviewers.   Journal editors give papers a first reading and decide whether they should be considered for publication or rejected. They may make this decision alone (1) if the paper clearly does not fall within the scope of the journal, (2) if the research is patently not original or not rigorous, or (3) if the paper is so badly written that it is not worth salvaging. For papers to be considered for publication, journal editors select appropriate reviewers, usually two. Reviewers may be associate or assistant editors, members of the editorial board, or peers in the discipline. When a paper presents research in a highly specialized or developing field, journal editors may ask a research scientist competent in the author's field of research to review the paper.

The review process is a confidential process, and reviewers are expected to keep everything about the paper confidential until it is published (*Appendix 10A*). This confidentiality is an important protection for the journal as well as the author. The editor sends the reviewers a copy of the manuscript, often a blind copy, so that the author remains anonymous. Reviewers are similarly not made known to the author. The dual anonymity is designed to obtain an unbiased evaluation of the paper, but the anonymity is not always absolute. In a narrow field of research, the reviewer can often guess the author of the paper from the research, and the author may guess the reviewer from the type of comments and suggestions made.

Journal editors provide reviewers with guidelines, criteria, or suggestions for reviewing papers, and may give reviewers a form to complete (see *Appendix 10B*). However, these criteria are intended only as points of departure for any other comments, so that reviewers have a great deal of latitude. They may address editorial as well as scientific problems, and highly specific details as well as general problems; they may make comments or corrections, offer suggestions, and so on.

Function of reviewers.   The reviewers' role is primarily advisory, but they may have a strong influence on the acceptance of papers and on their final form, especially where the journal editor is not as knowledgeable about the subject. Actually, reviewers have a dual advisory role: to editors in evaluating a paper for publication and making suggestions about its publication, to authors in making suggestions for improving the paper. Reviewers therefore have a general editorial function as well as a scientific function.

The first responsibility of reviewers is to decide whether they are qualified to review a paper and whether they can review it promptly. If they lack competence or time, they should return the manuscript to the journal editor by return mail. If they cannot review it promptly but are interested in reviewing it, they should communicate with the editor upon the receipt of the paper to discuss a later deadline. Reviewers have a responsibility to read the paper *carefully* and *objectively*. Because they are acting as scientific and editorial advisors to the journal editor and author, they should put themselves in the positions of the editor and author, so that they can address their concerns. Assuming an advisory role also promotes a facilitative or participatory role, rather than the adversarial role of "referee" versus author. They can then respond to what the author has actually written, rather than to what they believe the author thinks or intends.

   Review of manuscript.   The scientific review of the paper is both general and specific. Reviewers evaluate papers for their contribution to the discipline—the originality of the research, its significance, and how rigorously the research was performed, whether the research may have been published elsewhere in another form or whether similar research has been published by others. More specifically, they examine the various sections of the paper critically. Throughout their review of the various sections, they are reading the paper in the light of earlier research. They also consider the limitations or deficiencies of the research and the paper, and make suggestions for remedying them.

   The editorial function of reviewers is more elastic and more general. Beyond general suggestions, reviewers may evaluate the writing to whatever extent they have the time and competence. After the reviewers have completed their review of the manuscript, they return it, signed, together with their comments and suggestions. Those that are intended only for the editor and are confidential are kept separate from those that are intended primarily for the author. The confidential comments evaluate the scientific merits of the paper and make recommendations about the advisability of accepting the paper for publication.

   The comments and suggestions addressed primarily to the author are designed to improve the paper and are much more detailed than those to the editor. They address specific substantive scientific or editorial problems. This commentary is one of the most useful contributions that reviewers make to the discipline. Not only do their comments and suggestions help the author and editor to publish a rigorous paper, but they lead to better papers in the future. It is important that reviewers be as specific as possible in their comments and suggestions and that they explain them, especially when they disagree with the author. This gives the editor and the author a reasonable basis for evaluating them. For example, if reviewers consider a paper too long, recommending that the paper be shortened by five pages or ten percent is not as useful as is specifying what parts can be condensed or deleted. If reviewers' comments are indefinite or global, the editor may interpret them differently from the reviewer and the author differently from both.

   After journal editors receive the reviews from the reviewers, they can consider the paper in the light of their own review. If the reviewers submit strongly contradictory recommendations, journal editors may call on a third reviewer and then decide about the disposition of the paper and the revisions to suggest or require. In making such decisions they take into account both the reviewers' and their own evaluation. Journal editors are therefore the first and last reviewers. Occasionally manuscripts are not sent out for review, for example, invited papers or a paper of great current interest; then the review process may take only a few weeks.

10.2.3 Position of the Review Process in Scientific Inquiry

Because the review process is a practical method for selecting papers for publication, it is usually taken as an essential part of scientific inquiry. However, there are criticisms of the review process. It has been criticized (1) for not guaranteeing anonymity and confidentiality—and so not guaranteeing unbiased evaluations of papers, (2) for promoting orthodoxy, (3) for being elitist, and so on. It seems worthwhile to examine the review process at least in the light of its function in scientific research.

It can be argued that there is no scientific basis for the review of scientific papers, because it is the research scientist who is held ultimately responsible for the research, not the editor or reviewers. This perspective places the review process in a more nearly accurate position relative to the scientific method and gives authors the collegial status central to scientific inquiry. From this point of view, reviewers, being peer scientists, have no authority over their peers. On the other hand, most journals do not have enough funds to publish all the papers they receive; therefore some process of selection is mandatory. Even if all papers could be published and the review process eliminated, the evaluation of papers that is done by reviewers would have to be done by readers—an overwhelming task.

Therefore, from a *scientific* standpoint, the review of scientific papers is not an inherent part of the scientific method and is not intrinsic to scientific inquiry or to science. Science is antithetical to authority. Nor does reviewing have an inherent function in the dissemination of scientific knowledge, except as it improves the presentation, and so the communication of research. From a scientific standpoint, anyone who applies the scientific method to study a natural phenomenon can report the study. Those who read the report can then evaluate it from a scientific standpoint. This is the paradigm for scientific research, and there is no expectation of evaluation as a prerequisite to publication. Reviewing papers actually encroaches on the free dissemination of scientific knowledge implicit in the paradigm.

Reviewing is therefore an adventitious, incidental function outside of science. It is a method of exercising *professional* responsibility, that is, establishing and maintaining standards for the profession by screening the papers of members of the profession. However, the *professional* role of reviewers must not be mistaken for a *scientific* role and function. The professional objectives have been allowed to predominate because of the practical convenience of the professional means. As "nonscientific" as the review process may be, it is not likely to be abandoned, as long as the practical problems of costs of publication are pressing and as long as the literature continues to burgeon. Recognizing that it is a function outside of science, however, may serve as a corrective to conferring scientific authority on the process.

The function of reviewers should be seen not as approving papers, but as selecting and improving some that should be made available to readers for *their* evaluation. Reviewers are convenient surrogates for readers, performing a preliminary screening for them. The acceptance of a paper that does not make a definite contribution to the discipline is a serious responsibility because of the waste of resources.

The rejection of a paper that does make a contribution is a graver responsibility, however, because of the *scientific* consequences: lack of dissemination and loss of scientific research. Reviewers who are alert to this position of the review process will be able to view their role as facilitative rather than exclusionary and see their function as *improving* papers for publication.

10.2.4 Disposition of Paper After Review

Editor's response. After a paper is reviewed, journal editors decide whether to accept the paper, then notify the author. When a manuscript is accepted, they may make general comments and suggestions and summarize and integrate those of the reviewers, or they may enclose a copy of the reviewers' comments for the author. They also indicate the revisions to be made, explaining the reasons for them and specifying the revisions that are mandatory and those that are optional. When a paper is rejected, the editor may suggest a more appropriate journal. If the research is not considered publishable, the editor may indicate its deficiencies.

Author's reaction. Since even optional revisions require taking old research in hand to respond to the suggestions, the author's immediate response to the editor's communication is not likely to be wholly positive. When major revisions are required, authors may not be able to undertake effective revision without some adjustment of their perspective. The feedback provided by editor and reviewers can be viewed as part of one's continuing education. Also, the editor and reviewers may not be as knowledgeable about one's branch of research, and they are neither infallible or omniscient. However, although authors may dismiss the editor's and reviewers' evaluations, they should not dismiss their comments and suggestions.

Editors and reviewers do have responsibilities to peers in the discipline and those outside the discipline for the advancement of scientific knowledge. Moreover, they read the papers with more care and attention than most readers will. Finally, their objective—a clear presentation of rigorous research—is one that the author can hardly quarrel with. Therefore, before making a case against a revision or rejection, authors should be certain of the caliber of their research and its presentation in the paper, and the validity of their objections to comments. They may find it helpful to consult a respected colleague for his or her reaction to the comments and their suggestions.

Author's revision and reply. When a paper is accepted conditional upon revision, the revisions suggested are likely to be optional if they are minor, but mandatory if they affect important issues. Authors may find them so fundamental or excessive that they may consider withdrawing the paper and submitting it to another journal. This, however, entails at least accepting a further delay in an additional review. If the author decides to withdraw the paper, he or she should notify the editor.

After an author has decided to revise the manuscript, he or she should review it carefully and revise it as much in accordance with the reviewers' and editor's suggestions as is consistent with his or her own judgment. It is well to remember that readers may have a response similar to that of the reviewers, if the revision is not made. First, one can identify optional revisions that are apposite or acceptable. Then, the changes that are less acceptable can be considered. Authors can now consider the required revisions in the same way. These are more serious revisions and if the author does not wish to make some of them, he or she should consult the editor. After authors revise the manuscript, they return it to the editor with a letter of transmittal, indicating the extent to which the suggestions were accepted. When an author has not accepted important suggestions, he or she should give reasons or explanations for not making the changes, referring to the specific comments or suggestions.

If the paper is rejected unconditionally, the author must determine what further action to take on the basis of reasons given for rejection, if any, or on assessments by supportive but discerning colleagues. The author may wish to ask the editor to suggest changes or additions that might make the paper acceptable for publication. However, even if the journal editor agrees to this or to reconsider the paper, he or she is not guaranteeing acceptance. If the author decides to submit the paper to another journal, it should be revised at least to adapt it to the journal selected, and preferably more substantively, so that it has the best chance of acceptance. The research may be suitable for a short communication or note in a journal. This may be preferable to expending further effort trying to reclaim the paper.

If the deficiency is in the writing, then the writing is likely to need serious attention, because editors usually try to salvage poorly written papers if the research makes an important contribution. Authors may of course employ an editor, but in the long run, they gain more by rewriting and revising papers themselves.

10.3 EDITORIAL PROCESS

When the journal editor finally accepts the manuscript, it is considered to be in its final form, *scientifically,* but it must then be put in its final form *editorially.* The editorial process actually begins with the journal editor and reviewers, but formal editing begins with the copy editor.

10.3.1 Types of Editing

The immediate objective in editing a manuscript is to prepare it for the printer, that is, to provide an edited manuscript with complete directions for setting the manuscript into type. The ultimate objective is to prepare it for readers, that is, to make the writing clear, consistent, and concise. The overriding objective is to prepare it so that no changes will have to be made after it is set in type that could have been made before, in order to keep printing costs low.

The editing of a scientific paper comprises three kinds of editing. *Substantive editing* is part of the review process, the domain of journal editors and the reviewers. They are concerned with the author's scientific message, and their focus is on whether the writing is scientifically accurate and whether the structure best transmits the scientific message effectively.

Structural editing deals more directly with form as it affects the meaning or the message. It includes the structure and reorganization of sentences, paragraphs, sections, and the paper as a whole, as well as the effectiveness of the form of its various sections. Structural editing is the responsibility of the manuscript editor, if the journal has one, or of the managing editor if he or she functions as manuscript editor. When a journal editor has only a copy editor, the reviewers and journal editor attend to the more substantive aspects of structural editing and the copy editor to the aspects more related to form and less dependent on scientific expertise.

Copy editing, the final editing of the manuscript, is the responsibility of copy editors, who concentrate on form and consistency rather than major structure or content. They are responsible for correcting all errors and inconsistencies in the manuscript. They may also mark the copy for the printer, to show how it must be set into type. Copy editors are therefore the most directly responsible for attending to the printer's concerns. The copy editor indicates changes in the manuscript using copyreaders' symbols. Some of these, like the caret, are familiar to most writers; some resemble proofreader's marks.

10.3.2 Copy Editing the Manuscript

Reviewers, editors, and authors may make the kind of editorial changes or corrections that copy editors make, but copy editors have a more comprehensive responsibility; they are responsible for any previous editorial changes, as well as for any further changes or corrections required. When the copy editor has completed the editing of the manuscript, it is ready for the printer.

Copy editors are responsible for ensuring that the manuscript conforms to standard usage and to the style of the journal. This includes details of typography and layout, but also citation of references, and the conventions adopted for scientific and mathematical elements, symbols, and terminology. Editorial decisions have also been established for usage in spelling, abbreviations, acronyms, hyphenation, and so on. All these together determine the style format of the journal. They are written down and become the "style sheet" of the journal, which may be expanded at the disciplinary level into a booklet, for example, the *Style Manual* of the American Institute of Physics, or a book, for example, the *Publication Manual* of the American Psychological Association.

Form and consistency. In ensuring that the manuscript conforms to standard usage and the style of the journal, the copy editor must ensure consistency, but consistency is a special and separate responsibility. Copy editors have primary responsibility for form, where form is important in itself, as in the reference section, in the style of tables or graphs, or for symbols or abbreviated forms. They further ensure that all headings conform to the style of the journal and that coordinate headings are coordinate in form and are consistently ranked.

References, footnotes, scientific and mathematical material.   The references are foremost among a copy editor's responsibilities because of the many details of form, and so the increased likelihood of errors. It is not his or her responsibility to convert references that are in a different style or in a variety of styles to the style of the journal. That is the author's responsibility. The copy editor's function is to ensure that the references conform to the style of the journal in every detail. The actual verification of references is also entirely the author's responsibility, since this is a bibliographic, not an editorial, task. Authors pressed for time to verify the accuracy of the reference citations should employ a bibliographic assistant. Copy editors compare the references in the text with those in the list of references and note any discrepancies. Depending on the editorial policy of the journal, copy editors also eliminate or condense substantive footnotes or incorporate them into the text.

Scientific nomenclature is primarily the responsibility of authors. When nomenclature or abbreviated terms are standardized and widely used, copy editors can edit for consistency. When nomenclature is not standard or temporary abbreviations are used, the copy editor can edit them for consistency if provided with a list of the various terms. Mathematical material is unquestionably the responsibility of the author, who must make equations and mathematical terms clear *before* the compositor sets type rather than *after,* when making the corrections is very costly. The copy editor can be more alert than the author to confusing pairs of letters, and symbols and terms (*Table 10.2*), and can mark them for the compositor.

When the manuscript is copy edited, it is returned to the author for approval. The author then (1) examines the copy editor's changes for scientific accuracy and for form, (2) answers any queries, and (3) makes any other changes he or she wishes to make in the manuscript. All major changes should be made *before* the manuscript is set into print; after the manuscript is typeset, changes are costly. The author then returns the manuscript to the editorial office.

The copy editor examines the author's responses and consults with the manuscript or journal editor about scientific or major changes. When all questions are resolved, the copy editor marks the manuscript for the printer, unless this is done by the copy editor at the print shop. In this "printer's markup," the copy editor marks the different sizes of type, different fonts and so on, for headings, the spacing of headings, the various indentations, and so on, for the various parts of the text. This marking includes various other typographical instructions.

10.3.3 Author-Editor Relationship

Some tension is inherent in the author-editor relationship, the degree of tension depending on the extent to which authors consider their writing "finished" or the

TABLE 10.2 Symbols and Greek and Latin letters (printed) requiring identification because of similarity in handwritten form

Hand written	Identification	Printed Symbol	Hand written	Identification	Printed Symbol
a, α	cap, lc ay	A, a	T	cap tee	T
α	lc alpha	α	T, τ	cap, lc tau	T, τ
$\propto$	proportional to	$\propto$	t	lc tee	t
σ	lc sigma	σ	$\dagger$	dagger	$\dagger$
γ	lc gamma	γ	$+$	plus sign	$+$
B, β	cap bee, lc beta	B, β	u, μ	lc you, mu	u, μ
C, c	cap see, lc see	C, c	V, v	cap, lc vee	V, v
d, ∂, δ	lc dee, delta	d, δ	ν	lc nu	ν
ϵ	lc epsilon	ϵ	υ	lc upsilon	υ
ξ	cap xi	Ξ	x, χ	cap ex, chi	X, X
i, ι	lc i, iota	i, ι	χ	lc ex	x
K, κ	cap kay, kappa	K, K	$\times$	multiplication sign	$\times$
κ, κ	lc kay, kappa	k, κ	Z, z	cap, lc zee	Z, z
l, ℓ	lc el, lc ee	l, e	2	numeral 2	2
$1, ^1$	numeral one, superscript one	$1, ^1$	w	cap sigma	Σ
$/$	vertical bar	$\mid$	w	summation sign	Σ
$', ', '$	apostrophe, single quote, prime	$', ', '$	3	lc double-you	w
			3	lc omega	ω

418

style personal, characteristic, and inviolate and the extent to which editors consider usage fixed or themselves certain of usage.

How much influence an author has over the manuscript and the copy editor's changes depends on the nature of the changes—whether they are substantive—and on the journal editor's policy about editorial changes. Journal editors are the final arbiters, but they leave most editorial decisions to the author and the manuscript or copy editor. In general, authors are not under as great constraint to accept the copy editor's changes that affect scientific meaning. If the editor has changed the meaning of a passage in editing, however, the author must consider whether perhaps the writing was not clear enough to begin with. Where scientific meaning is not in question, authors should try to understand the reasons for changes and not hesitate to ask questions. Querying an expert copy editor is instructive and contributes to the improvement of subsequent manuscripts. However, they need not accept all editorial changes.

10.4 PROOFREADING PROCESS

After the editing is completed, the manuscript is sent to the printer. A copyreader marks the "copy" (manuscript) for the printer, if this has not already been done by the copy editor of the journal. He or she also does any copy editing that has been agreed upon by printer and journal editor. The manuscript is then distributed, the artwork (i.e., illustrations) to illustrators and photoengravers, the manuscript copy to the typesetter or compositor.

10.4.1 Preparing Galley Proofs

In setting type, the typesetter's responsibility is to set the copy into type *exactly* as it is typed, edited, and marked. Typesetters have no editorial responsibilities, and so make no changes or corrections, no matter how obvious, though they may query the author if they notice an error. This limitation of responsibility is based on the expectation that the edited manuscript is in the final and correct form, ready for printing.

The type may be set in the traditional manner in letter press in long galley trays. The proofs drawn from the galley, the galley proofs, are therefore long sheets in which the type is set in one column, which is the width of the column or the printed page of the journal and about two times the length of the printed page. When authors receive galley proofs, therefore, they do not see the type as it will appear in page form. The first proofs are increasingly being set up in page form, so that authors do see the text in pages close to the final pages.

The typesetter pulls several sets of proofs from the galleys for the printer and editor to correct. The proofreader at the print shop proofreads one set of proofs against the manuscript to mark the printer's typographical errors, that is, "typos," and corrects the proof accordingly. This set of proofs becomes the master set and forms the basis for assigning charges for changes in excess of those allowed. The master copy is sent to the editorial office, together with a second set of proofs and the manuscript.

10.4.2 Correcting Versus Revising Proofs

At the editorial office, the master set is retained for the copy editor to read against the manuscript. The second set is sent to the author, accompanied by the manuscript and a covering letter. The letter gives the author a deadline for the proofs, instructions on reading and correcting proofs, and a form for ordering reprints. The ordering of the reprints is not linked to the resetting of type from the corrected galleys and so should not be allowed to delay the return of the proofs. The author should read the proof promptly and meet the deadline.

The instructions to the author emphasize two objectives in reading and correcting proofs: (1) to *correct* the proofs, not to revise them, and (2) to return the proofs *promptly*. Authors tend to mistake the proofreading process for part of the editorial process, that is, as an opportunity to improve the paper. This should have been done *before* the manuscript was sent to the printer. The purpose of proofreading is to *correct* proofs, not to *improve* the paper. The cost of changes is too high to allow the luxury of rephrasing or restructuring.

A change entails not only the time required for the actual change by the typesetter, which is more time consuming than the original typesetting, but also the time and attention of editors, copy editors, proofreaders, and so on before and after the correction is made. A small percentage of the cost of such corrections, called "author's alterations," is allowed without charge; the excess is charged to the journal. There is a further, practical reason for not making substantial changes. The first or galley proofs are the only proofs that authors see. If they make extensive changes, they will have no opportunity to proofread the final proofs; yet the chances of printer's errors are greater in resetting type than in the original typesetting.

10.4.3 Marking Proofs

Authors should use the conventional proofreaders' marks in correcting proofs. They should be familiar at least with those that are especially useful to authors, that is, those having to do with changes in the text, and with the typographical conventions in the discipline. Since there is little space between words and lines, proofreaders' marks are written in the margin together with the necessary correction, and conventions are followed for positioning them in the text. The conventional proofreaders' marks are listed in *Fig. 10.3,* and their use in a proof is shown in *Fig. 10.4.*

Two proofreaders' marks are made for each change: one in the line of type and one at the closest margin to the change. The changes themselves should not be made in the line of type or above it, nor should lines be drawn from marginal notes into the line of type. Marks in the line of type are necessarily minimal. Their primary function is to mark the location of the alteration, though they may also indicate the

Instruction for correction	Mark at margin and in line		Corrected type
Insertions			
Insert material at margin	*annual*	the cycle	the annual cycle
Insert superscript	32	P	^{32}P
Insert subscript	2	HO$_2$	H$_2$O$_2$
Insert space	#	wave front	wave front
Deletions			
Delete word		the ~~annual~~ cycle	the cycle
Restore deletion	*stet*	the ~~annual~~ cycle	the annual cycle
Delete letter and close up		histaddine	histadine
Delete space and close up		two fold study	twofold study
Substitutions			
Reset as superscript	59, 4,5	59Fe; gyroscope 4,5	^{59}Fe; gyroscope^{4,5}
Reset as subscript	2	H2O	H$_2$O
Spell out numeral	*sp*	in ②parasites	in two parasites
Substitute numeral	20	twenty control rats	20 control rats
Substitute material in margin	*protozoan*	the ~~parasite~~ enters	the protozoan enters
Changes in location			
Transpose letters	*tr*	hybrid colonels	hybrid colonies
Transpose material circled	*tr*	was, to some extent (at least)	was at least to some extent
New paragraph	¶	⌐In particular, the Raman laser can	In particular, the Raman laser can
No paragraph	*no ¶*	was attempted.⌐ In particular the Raman laser can	was attempted. In particular, the Raman laser can
Center		] *Discussion* [	*Discussion*
Move to left	⊏	⊏ to minimize	Special care was taken to minimize
		Special care was taken	
Move to right	⊐	Special care was taken to minimize	Special care was taken to minimize
Changes in type			
Set in capitals	*cap*	10 l	10 L
Set in capitals and small capitals	*cap + sm cap*	Synthesis of Enzyme	SYNTHESIS OF ENZYME
Set in Roman type	*rom*	*Salmonella* (sp.)	*Salmonella* sp.
Set in italic type	*ital*	Nicotiana Tabacum	*Nicotiana Tabacum*
Set in boldface type	*b.f.*	l·s	**l·s**
Wrong font	*wf*	(Materials)	Materials
Lower case letter	*lc/lc/lc*	Biological Oxygen Demand ⊏ (BOD)	biological oxygen demand (BOD)
Punctuation			
Period	⊙	Green et al (1986) tried	Green et al. (1986) tried
Comma		However the sources	However, the sources
Semicolon		Many sources were noted however, none	Many sources were noted; however, none

Figure 10.3. Proofreader's marks illustrating instructions for corrections, marks used in proofs, and corrected type.

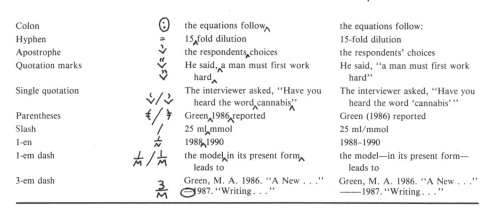

Colon	(:)	the equations follow∧	the equations follow:
Hyphen	=	15∧fold dilution	15-fold dilution
Apostrophe	˅	the respondents∧choices	the respondents' choices
Quotation marks	" "	He said,∧a man must first work hard∧	He said, "a man must first work hard"
Single quotation	˅/˅	The interviewer asked, "Have you heard the word∧cannabis∧"	The interviewer asked, "Have you heard the word 'cannabis'"
Parentheses	(/)	Green∧1986∧reported	Green (1986) reported
Slash	/	25 ml∧mmol	25 ml/mmol
1-en	⊥̃	1988∧1990	1988–1990
1-em dash	⊥/⊥	the model∧in its present form∧ leads to	the model—in its present form— leads to
3-em dash	³⁄ₘ	Green, M. A. 1986. "A New . . ." ⊝1987. "Writing . . ."	Green, M. A. 1986. "A New . . ." ——1987. "Writing . . ."

Figure 10.3   (Continued)

(Presented in Atlantic City, NJ, on 5 May 1976 at the Annual Meeting of the American Society for Microbiology (J. Fiore and R. Babineau, Abstr. Annu. Meet. Soc. Microbiol. 1976, Q 39, p. 197).

Materials and Methods

Collection and treatment of water samples. A volume of 0.1 ml of 10% (wt/vol) sodium thiosulfate solution was added to each 125-ml sampling bottle before sterilization at 121°C for 15 min to provide an approximate concentration of 10 mg/100 ml of sample for neutralization of residual chlorine. Water samples in 100 to 200 ml quantities were obtained from each filtering unit. Just before making dilutions the samples were shaken vigorously 25 times. Ten fold dilutions were then prepared by transferring 10-ml quantities to 90 ml of phosphate-buffered distilled water (pH 7.2). Pour plates (two replicates per dilution) were prepared with a 1-ml quantity of each three appropriate dilutions and plate count agar (Difco Laboratories, Detroit, Michigan). Plates were incubated at 30°C for 48 h. We chose 30°C instead of 35°C for our plate count experiments because it has been our experience that this incubation temperature is more suitable than 35°C for supplying the

Figure 10.4.   Corrected proof marked with proofreader's symbols. [Appl. Environ. Microbiol. 34, 1977, 541]

type of change. For example, a vertical line in the line of type may indicate the deletion of a letter, its replacement with a correct letter, a lowercase letter, or a space between two letters. The mark at the margin explains exactly which change is intended. Most of the symbols used beside the line of type are more specific and less abbreviated. They may confirm or explain the mark in the line of text, "ital" for a single underlining, "cap" to reset the circled letters in capital letters. Where there is more than one correction to be made at a margin, each proofreaders' mark is separated from the text by a long vertical line (*Fig. 10.4*).

The changes made in the proofs may be typographical or substantive. The latter consist largely of insertions, deletions, and transpositions and should be minimal—a word, or phrase, if required for correction. Authors are responsible for *all* errors, although they may not be as acute as the proofreader in detecting typographical errors. However, they must try to focus their eye on characters and spaces rather than read in their usual mode, unconsciously interpreting or correcting errors as they read.

10.4.4 Reading and Correcting Galley Proofs

The manuscript is the basis for the proof, and the proof is the source for printed page. Therefore, no change should be made on the manuscript after proofs are supplied, and any changes, marks, or queries on the proofs made at the printer's or by editors must not be altered.

Reading proofs.   Proofreading is following a text word by word in the manuscript or proof and examining the corresponding printed text character by character. It requires focusing on minute, often meaningless, characters with concentration, patience, and a keen eye. Galley proofs are best read by two proofreaders; one who holds the edited manuscript, that is, the copyholder, and reads it aloud; and another, who holds the galley proof and compares the printed text with the text being read by the copyholder, and then makes corrections accordingly. It is preferable for authors to hold the galley proof, since this keeps them from repeating the original misreadings or misinterpretations.

In reading, copyholders read aloud slowly and carefully, and describe as well as read; that is, they announce headings—their position and typographical form—paragraphs, punctuation marks, capitalization, and so on. Wherever confusion is possible, they spell out words or names. This slow reading allows proofreaders to follow the printed text closely with a pen or pencil, character by character. Otherwise, they will not see typographical errors, short but important words, missing terms, and the like. The time spent in reading proofs can be shortened by marking the proofs in pencil or pencilling a note in the margin and later marking it in final form. When authors read galley proofs alone, they can read more accurately by reading the edited manuscript line by line, keeping a ruler or strip of colored paper under the line of print and the line of typescript being read. Again as they read, they follow the printed text character by character with pen or pencil. If they read for meaning, they are bound to miss errors. When it is difficult to proofread attentively, they should stop and mark the manuscript at a point before the stopping place to allow for inaccurate proofreading because of developing weariness or boredom.

The galley proofs should be read at least twice, but preferably three times. The first time the proofs are read to make certain that they match the text of the edited manuscript. This is the reading for which two proofreaders are particularly desirable. A second reading of the proofs is then desirable to look for errors. The proofreader should be alert to omission resulting from the typesetter returning to the second appearance of the word or phrase thus omitting the intervening text. The third proofreading is a reading without reference to the edited manuscript to determine whether the text makes sense. The objective of this reading is not to *revise,* but to detect errors, such as a missing verb or text, or misplacement of text.

Correcting proofs.   In correcting proofs, authors use a different colored ink from that used by others; they must not erase or change anything written on the proofs by the printer, proofreaders, or editors (*Table 10.3*). They should answer any queries to the author briefly, but definitely and unambiguously, and then circle the answer. They can add directions or notes and circle them. In making corrections, the author should make a special effort to make the number of characters (including

TABLE 10.3   Steps in reading and correcting proof

a. *Examining proofs*

1. Examine proofs for marks, notes, and queries; do not erase or alter these.
2. *Correct* the *proofs* against the manuscript or earlier proofs; do not *revise* the *text,* except to correct errors.
3. Use different colored ink from that used on the proofs.
4. Answer any queries, briefly and specifically in the margin; then draw line through question or incorrect matter.

b. *Proofreading*

1. Proofread, preferably with another person to hold the copy (i.e., manuscript). Have the copy holder read slowly, describing punctuation, capitalization, underlining, and so on.
2. As the copy holder reads, follow in the proof, *character* by *character* with point of pen or pencil; *or* read a line of text (or proof), then compare with proof.
3. When the proof differs from the manuscript, correct it, using proofreader's symbols.
4. Make the corrections next to the line of type, at the margin nearest the correction.
5. For deletion, draw a line through the type to be deleted and place the deletion and close up symbol at the margin.
6. For insertion, place a caret at the point of insertion, and write the material to be inserted at the nearer margin.
7. In making corrections, add, subtract, or change words in the nearby text so that correction fits within space of original text.
8. Type long insertion on separate sheet, paste it at the edge of the proof, and mark point of insertion with caret.

spaces) in the correction equal the number of characters in the proofs. A difference of a few characters may necessitate resetting one or more lines, increasing the chances of error. To make the insertion match the deletion, it may be necessary to modify the text.

Any additions are written in the margin of the galley horizontally, next to the line in which they are to be inserted, despite the inviting length of the galley proof. If the insertion is long, and absolutely necessary, it should be typed and pasted to the galley near the place of insertion, if possible, and with a note to the printer both on the note and at the place of insertion, for example, "Insert A on galley 5," and in the margin of galley 5 "Insert A here."

After the final proofreading, authors should return the proofs promptly to the editorial office. There, the copy editor examines the author's corrections and submits questionable additions or changes to the appropriate editor. The copy editor then transfers the corrections that have been accepted to the master set of galleys. A copy is made, and the master set is returned to the printer, together with the manuscript.

10.4.5 Proofreading Special Parts

Some parts of the manuscript should be given special attention in proofreading, either because it is particularly needful that they be accurate, as for data, or because they are particularly subject to error in printing, as in references and equations.

1. *Structural parts.* The parts of the paper that help to identify its subject and structure, such as the title, byline, abstract, headings, and running heads should be examined separately for accuracy, for conformance to the style of the journal, and for consistency.

2. *References, footnotes, and quotations.* These should have been meticulously verified against the manuscript, which should have been meticulously prepared.

3. *Tables and figures.* There is greater chance of error in printing tables than ordinary running text because setting type is more complicated, and they are insertions in the text that are processed separately. Title headings, captions, legends, and so on should follow the style requirements of the journal. The numerical data in tables should be aligned with the decimal point, and should be meticulously proofread against the original. Footnotes should be verified for numerical order at the foot and from left to right and top to bottom in the table.

Authors must examine the proofs of the illustrations to verify layout, labels, shape, proportions, dimensions, units, magnification, values on graphs, and so on. If figures are drawn from sketches submitted with the paper, or if the editor has had them redrawn, the proofs must be scrutinized especially critically. Figure titles should correspond with the subject of the illustrations, and the legends with the labeling of the parts of the figure and with the description in the text. Line reproductions should be examined for lines broken in reproduction or the appearance of black imprintings not in the original drawing. Photoengravings should be examined for cropping, sharpness, even tone, and blemishes. Defects in figures should be circled, and comments relative to them made in the margin. If the first proof is page proof, the positions of the tables and figures relative to the text can be verified.

4. *Numerals, names, and symbols.* Wherever numbers appear, for example, in data, units of measurement, dates, and sections of the manuscript, they should be proofread for accuracy, and then for consistency. Whenever a number appears in more than one section, for example, a datum in the text of the results, a table, and the abstract, it should be identical to that in the other sections. Names of persons and places and scientific names of organisms should be accurate, correctly spelled, and consistent. Cross-references to tables, figures, appendixes, and sections should correspond.

5. *Equations.* These are particularly liable to error, which are more easily detected by authors than copy editors or proofreaders, who are almost wholly dependent on comparing the proof with the manuscript. Therefore authors should proofread the equations themselves, *symbol by symbol.*

The equations should be set in italics, except for the few standard roman or boldface symbols or abbreviations. The main operational signs for the equation ultimately determine the position of every symbol in the equation (see *Chapter 6*); therefore, they should be used as the point of reference for verifying the alignment of all other operational signs and for terms, fractions, subscripts, superscripts. Symbols that include several levels of terms, for example, parentheses and integration signs, $\int$, should not be disproportionately large or small, bold or light, or too distant or too close to related matter, though the choice may be limited by the availability of type. In a long displayed equation, the second and subsequent lines of the equation should be uniformly treated, each line beginning with a sign.

10.4.6 Reading and Correcting Page Proofs

When the printer receives the corrected master set of galley proofs, the typesetter makes the corrections and divides the type in the galley trays into pages the length of the printed page and prints several sets of page proofs. The sets of page proofs are sent to the appropriate editor together with the corrected galley proofs for proofreading. The copy editor proofreads the page proofs against the corrected galleys to make sure that every correction in the galley proofs has been made in the page proofs. Special attention is paid to every correction, to ensure that errors have not been introduced into lines that have been reset or added, or that lines have not been lost. Although authors rarely see page proofs, it is important to remember that page proofs are read for necessary *corrections* to the *type,* not improvement of the text. It is important to find errors in page proofs, because it is difficult to recall an error in print. Errata sheets are lost or not read, and errors take on the aura of authority that surrounds printed matter.

Page proofs are most easily read by aligning the galley proofs and page proofs side by side and using a long ruler to underline the line to be proofread, reading first one, then the other. Reading backward avoids any mistakes resulting from familiarity or inattention. As in galley proofs, special checks are desirable. Also, wherever the printer must divide the text, there is a chance of error; therefore, the proofreader should examine the continuity from one page to the next, the beginning and end of each line, and the lines before and after figures and tables, and corrections. If references have been made to pages of the text, the page number is now verified and the zeros inserted in the galley proof are now converted to the correct page number. The position of tables and figures should be examined relative to their first mention, and the titles should correspond to the subject of the table or figure. Illustrations should be examined for orientation, especially when this is not obvious.

After the page proofs are corrected, they are returned to the printer. The typesetter makes the corrections and the actual printing follows. The pages are then arranged in sets (signatures), printed, and the sets folded, assembled, and bound to the cover. The issue is then ready for distribution.

Notes added in proof.   When it is essential to make an addition to the text in page proofs, which would require extensive changes in the proofs, a note can be appended after the acknowledgments:

Ex. 10.1

   a. *Note added in proof.*—The static limit of equation (6.1) has been derived by: HUNT, K.L.C., ZILLES, B.A., and BOHR, J. E., 1981, *J. Chem. Phys.,* 75, 3079. [Mol. Phys. 58, 1986, 51]
   b. NOTE ADDED IN PROOF

   We are indebted to Prof. B. Alder for informing us of a similar research being done for a dilute hard-sphere gas at DFVLR in Gottingen (Meiburg, 1985). In that study, Meiburg tows a 45° inclined plate through a very dilute MD gas. He observes *transient* vortex shedding behind the plate at Reynolds number 70 and Mach number 0.7. In contrast, we have dense, interacting fluid flowing past a fixed plate. We observe a steady-state vortex (120 Å in diameter) at a Reynolds number of about 15 and a Mach number of about 0.2. (Again, given the large variation of density and velocity in our system, these numbers should be taken as order of magnitude.) [J. Sci. Comput. 1, 1986, 150]

10.5 PUBLICATION

10.5.1 Reprints and Charges

Authors are allotted a certain number of free copies of their paper. They are asked to specify the number of extra copies they wish, and are charged for them. The copies may be "offprints," copies made by printing extra pages when the issue is being printed and having associated blank pages, or "reprints," copies made by reprinting the required pages from the printed pages by photo-offset. Reprints can be ordered at any time, whereas offprints must be ordered before the journal is published.

10.5.2 Copyright

An author's writing is his or her property. It should not be used by others without the author's permission. If the writing is copyrighted, it is the author's *legal* property, and it is unlawful to use it without his or her permission. Authors of scientific papers are concerned with copyright primarily as owners of the copyright to their paper. Most authors own the copyright to their paper; that is, they have the exclusive and legal right to publish, reproduce, or sell their paper, or any part of it. If the author has prepared the paper as a specific responsibility of his or her position, then the paper is considered "work for hire," and the employing organization owns the copyright. When the author is an employee of the U.S. government, neither the authors nor the government own the copyright. The paper is said to be in the public domain; that is, anyone may publish, copy, reprint, translate, or otherwise use the paper.

Although authors own the copyright to their paper, many journals require that their authors sign a statement transferring their copyright to the journal. This allows the journal to make reprints and offprints, authorize copying or reproduction of material in the journal, compile collections of articles, and so on. With the transfer of the copyright, authors no longer own the copyright to their paper. If they wish to use their paper, for example, as a chapter in a book, or reproduce the paper and offer it for general sale, they must obtain permission from the journal.

As writers wishing to use published material, research scientists can make "fair use" of copyrighted material without requesting permission of the holder (owner) of the copyright. However, the definition of "fair use" is not very clear or easily interpreted. The U.S. Copyright Law allows consideration of the work, its purpose (e.g., nonprofit or educational use), the amount of material used or how substantive it is, and the effect of use on the possible sales of the copyright work. When in doubt, one can consult the U.S. Copyright office, which issues brochures on various aspects of copyright and has a public information telephone service.

10.5.3 Permissions

Permission must be obtained for copyrighted material beyond "fair use," although such extensive use of copyrighted material is very uncommon in writing research papers. One rarely needs to quote a passage of more than 200 words, but if one does, failure to obtain permission constitutes plagiarism and makes one subject to legal action for violation of copyright. Permission to cite such copyrighted material is obtained from the copyright holder, usually the publisher of the journal, and by courtesy from the author also. Most scientific journals give permission freely, but some journals and most publishers operating for profit charge a fee.

In writing for permission, writers should clearly identify the material requested, indicate the use to be made of it, and of course, thank the grantor. They should provide the publisher or author with the statement of permission appended or in a separate form which the author or publisher can sign. Permission should also be obtained for publishing unpublished material.

Formal written permission giving informed consent must be obtained from persons who serve as subjects of research. The permission must be specific to the research, not a general permission. Institutions usually have committees for monitoring research with human subjects, and they should be consulted for guidelines and standard forms for permissions. Permission should also be obtained from persons who appear in photographs that are used in the paper, whether they appear intentionally or accidentally. The signed permission form should be kept on file and a statement provided to the editor that written permission has been obtained.

10.5.4 Ethical Responsibility

Scientists have an ethical responsibility to other scientists that derives from their commitment to the values of science and its advancement; they have a responsibility to the general public that derives from their commitment to the general welfare. These responsibilities are embodied in a code of behavior that forms the standard of ethical behavior that both scientists and the public expect scientists to adhere to. It encompasses the scientist's activities within science, in research and publication, and in communicating the results of research to the general public.

The code is not a written code, though the ethical treatment of human beings and animals used as subjects in research is usually incorporated into written regulations or principles in the appropriate disciplines, for example, "Ethical Principles for Psychologists" (American Psychological Association, 1981). Such codes require that human participants be fully informed about the research and give their "informed consent" before participating, and that they be debriefed after the research and every effort made to nullify all possible adverse effects of the research. They are also designed to ensure the humane treatment of experimental animals, and a statement that such requirements have been fulfilled is required by the editor and included in the paper:

Ex. 10.2

a. Participants were fully informed of the purpose of the study and gave their informed consent. The study protocol was approved by the Cornell University Human Subjects Committee.

b. Written informed consent was obtained from each volunteer. Experimental procedures were approved by the Baylor College of Medicine Institutional Review Board for Human Research and the Texas Children's Hospital Committee on Investigations and Publications.

The responsibility to the general public includes the *accurate* reporting of research, in language that is generally understood by an audience outside of scientific circles. This responsibility leads to expectations about the research itself. As any member of society, research scientists must assume responsibility for the consequences of applying their particular knowledge. Some scientists take this to include the responsibility for *undertaking* research that they consider unethical.

Ethical behavior as it applies to publication of scientific research may touch on the authorship of a research paper, the publication of *original* research, multiple publication of the same research, and so on. The publication of fabricated or manipulated data is considered so infamous a fraud that it is given wide publicity when it is discovered.

Theft from colleagues or peers may be more easily detected, because the person whose research is stolen is likely to be alert to the publication of similar work; however, it may be more difficult to establish the original source without controversy. A more subtle theft of a colleague's or peer's research may be hidden in multiple authorship. The other common problem arises in multiple publication of research. The same data with or without minor additions, subtractions, amendments, or modifications may be presented from a slightly different perspective and offered for publication to a different journal. This wastes the time of editors and reviewers, and if the duplication escapes their notice, of numerous peer scientists. The editors of the journals published by the American Chemical Society have presented a set of ethical guidelines for persons publishing chemical research. Most of these apply to other scientific disciplines, and an abridgment of them is presented in *Appendix 10A*.

Appendix 10A

*Ethical Guidelines for Scientists**

AUTHORS OF SCIENTIFIC PAPERS

1. To present an accurate account of research performed and an objective discussion of its significance.
2. To use the space in a journal wisely and economically; journal space is a very valuable and costly resource and readers have many calls on their time.
3. To provide sufficient detail and refer to public sources of information to allow peers to evaluate and repeat the research.
4. To cite only those publications that have been pertinent to the research reported and that will guide the readers easily to the earlier work necessary for understanding the research.
5. To report unusual hazards inherent in the materials, equipment, and procedures used in the investigation.
6. To avoid using excessive journal space and complicating literature searches by fragmenting the research papers derived from an extensive study.

---

*Adapted from *The ACS Style Guide;* Dodd, J. S., Ed. 1986 Amer. Chem. Soc: Wash. DC, 218–22; © 1986, with permission of Amer. Chem. Soc.

  a. To organize publication so that each report is a well-rounded account of a particular aspect of the general study.

  b. To publish related studies in one journal or in only a few journals.

7. To inform the editor about related manuscripts that are being considered for publication or are in press, and to indicate the relationship of such manuscripts to the one submitted.

8. Not to submit manuscripts describing the same research to more than one journal *except* if the manuscript is a

  a. Resubmission of an article rejected or withdrawn from publication in a journal.

  b. Report of overlapping work to a journal in another discipline, if the research scientists in that discipline are not likely to see the article published in the first, and provided that both editors are notified.

  c. Full paper expanding on a previously published brief or preliminary account.

9. To identify accurately and completely, the source of all information offered or quoted, except that which is common knowledge.

  a. Not to use information obtained privately in conversation, correspondence, or discussion, or in reviewing manuscripts or applications for grants, unless explicit permission is obtained from the originating investigator.

10. To restrict criticism of a study to the research; any personal criticism is irrelevant and invidious.

11. As submitting author, to name as coauthors of the paper all those, and only those, who have made significant scientific contributions to the research and who share responsibility and accountability for it.

  a. Not to name as coauthor anyone who has had only an administrative relationship to the research.

  b. To include deceased persons who meet the criteria for authorship.

  c. Not to list a fictitious name as author or coauthor.

  d. To obtain coauthors' approval of the final draft of the manuscript before submitting the paper for publication.

EDITORS OF SCIENTIFIC JOURNALS

1. To give unbiased consideration to all manuscripts offered for publication, judging each on its merits without regard to race, religion, sex, nationality, seniority, or institutional affiliation of the authors.

2. To consider manuscripts submitted for publication with all reasonable speed.

3. To seek the advice of reviewers, chosen for their expertise and good judgment, as to the quality and reliability of manuscripts submitted and to consider the charge or mission of the journal set by the scientific society in reaching a final decision.

4. To disclose no information about a manuscript submitted to anyone other than those from whom advice is sought, except to disclose the title and authors' names after the manuscript has been accepted for publication.

5. To respect the authors' ownership of the paper and intellectual independence.

6. To delegate editorial responsibility and authority for a self-authored paper or for a paper closely related to the editor's present or past research to another editor of the journal or a member of the editorial board.

  a. To inform the author about this delegation and possibly about the editor's research and research plans.

7. To obtain the consent of the author if the editor wishes to use in his or her own research information, arguments, or interpretations disclosed in a submitted manuscript, except that the editor may discontinue research that a submitted manuscript indicates may be unfruitful.

8. To facilitate the publication of a report that points out or corrects erroneous material or conclu-

sions in a report published in the journal; the correction may be reported by the person who discovered the error or by the original author.

REVIEWERS OF MANUSCRIPTS

1. To assume one's professional responsibility to contribute to science by reviewing manuscripts as one's competence and time allow.
2. To return a manuscript promptly if lacking competence in the subject; the reviewer may suggest a competent reviewer.
3. To review a manuscript objectively and rigorously, but to respect the intellectual independence of the author.
4. To confer with the editor about possible bias or conflict of interest when a manuscript is closely related to the reviewer's research.
   a. To return the manuscript promptly without review or suggest reviewing it jointly with a colleague.
   b. To furnish a (signed) review that the editor may submit to the author.
5. To avoid reviewing a manuscript authored or coauthored by a person with whom the reviewer has a personal or professional relationship.
6. To treat a manuscript as a confidential document.
   a. Not to show a manuscript to others or discuss it with them, except someone from whom specific advice is sought, and to disclose the identity of anyone consulted to the editor.
7. To explain and support judgments adequately, so that the editor and author can understand the basis for them.
   a. To accompany references to previous reports with the full citation for the source.
8. To be alert to failure of authors to cite relevant work by other scientists, and to avoid being self-serving in doing this.
9. To alert the editor of any substantial similarity between the manuscript being considered and any published paper or any manuscript submitted concurrently to another journal.
10. To review the manuscript promptly and submit a report expeditiously. If circumstances preclude a prompt review,
    a. To return the unreviewed manuscript promptly.
    b. To notify the editor of probable delays and propose a revised review deadline.
11. To use or disclose unpublished information, arguments, or interpretation contained in a manuscript being reviewed only with the consent of the author, except to discontinue one's research if the manuscript indicates that the reviewer's research may not be fruitful.
    a. To write the author, with a copy to the editor, about the reviewer's research and research plans, when these are related to the paper being reviewed.

SCIENTISTS ADDRESSING GENERAL PUBLIC

1. To be as accurate in reporting observations and unbiased in interpreting them as in a scientific journal.
2. In using common words of less precision than scientific terminology, to keep writing, remarks, and interviews as accurate as possible.
3. To announce the results of research only if the experimental, statistical, or theoretical support for them warrant publication of the resarch in a scientific journal.
   a. To support such a public announcement by promptly submitting a report of the research for publication in a scientific journal.

Appendix 10B

Example of Form for Reviewers' Comments on Manuscript

[Name of Journal]

Reviewer:

Title of Paper:

Manuscript no.:

Date mailed:

Data returned:

Acceptance of review deadline

Please help us to expedite the publication of papers by responding promptly. Much of the delay in publication is due to the lags between steps in the process rather than the process itself.

1. If you cannot review the paper for any reason, please note this below and return this form promptly in the self-addressed stamped envelope, so that we can select another reviewer.

——————— I cannot meet the deadline.

——————— The paper is not in my field of competence.

——————— I prefer not to review the manuscript.

Other comments:

2. If you cannot review the paper in time to meet the deadline, but are interested in reviewing it, please call us so that we can determine whether the schedule of publication can be revised to accommodate you.

3. If you cannot review the paper, it will expedite the reviewing process if you return the manuscript to us. You may keep it, however, if you wish, but please keep its contents <u>confidential</u> until it is published.

Returning comments

Please return this form promptly in the enclosed, self-addressed envelope. Retain a copy in your files until the paper is published or rejected. Address correspondence to the following address:

[Managing editor]

[Address]

Thank you for generously giving your time to review this paper.
Editor   Associate editor   Managing editor

[name and address]   [name and address]   [name and address]

Comments to editors

    This section is for your comments to the editor(s); it will not be seen by the author(s). The ''Suggestions to Authors,'' which follows, is for your comments to the authors; they will be sent to the author.

1. Suitability for publication in Journal
_____ not suitable for publication   _____ definitely publishable
_____ marginally publishable   _____ sound contribution
_____ publishable, but not excep-   _____ highly suitable for publica-
      tional   tion
2. Contribution to the discipline
_____ none _____ little _____ moderate _____ important _____ significant
3. Innovativeness
__ routine __ variation on __ innovative __ very   __ very
            a type   innovative   original
   4. Interest of article
_____ little _____ some _____ moderate _____ general _____ great
   5. Scientific treatment (explain deficiencies in Further Notes)
_____ major defects _____ minor defects _____ adequate _____ rigorous
   6. Title
_____ does not reflect research reported _____ incomplete
_____ too long _____ informative and concise

    Suggested revision:

   7. Abstract
_____ descriptive, not informative _____ inaccurate
_____ incomplete _____ wordy _____ concise and complete
   8. Review of literature
_____ inadequate _____ partly irrelevant _____ adequate and relevant
   9. Other sections: Methods, procedures, statistical analyses, tables, illustrations, mathematical derivations, results, discussion.

    Please comment as appropriate in ''Further Notes'' and ''Suggestions to Authors(s).''

   10. Recommendation for revision:
_____ only routine editing   _____ major revisions required
_____ minor revisions desirable _____ revision and rewriting too extensive to
                                consider paper for publication

Please comment on clarity, conciseness, wordiness, organization of parts, etc. in ''Suggestions to Author(s).''
11. *Further Notes*

Suggestions to author(s) (These comments can be written on separate sheet.)

- -

11

Order and Structure

A scientific paper must be unified and coherent. The writer achieves this by internal and external structuring. Internal structuring is the logical grouping and ordering of related elements so that they are coherent. External structuring is the introduction of elements from outside the text to reinforce or maintain the internal structuring.

11.1 STRUCTURE OF SECTIONS

11.1.1 Unity and Coherence

The focus of the main sections of the paper is largely predetermined, and that of smaller divisions is established by the author. These then serve as boundaries for including and excluding material. The section must also have unity of form. In a descriptive section like the methods or results, sentences must describe. Unity is readily verified in revising the paper; material not belonging in the section (see *Table 7.1*) can be deleted or transferred to the appropriate section.

Unity is a necessary precursor of coherence. The assigning of the pertinent material to the sections is the first step in organizing the paper. The various topics included are identified and then ordered into subsections. Consider an example, in which the research includes two *experimental procedures,* a *statistical analysis,* and special *apparatus* for the experiments, and the experimental *animals* were subjected to pre-experimental *treatment.* In organizing these into a methods section, the writer may have to decide whether there are two subsections or four:

Ex. 11.1

Apparatus	*Pretreatment*	*Apparatus*	*Materials*
Pretreatment	*Apparatus*	*Procedures*	Apparatus
Experimental	*Experimental*	Pretreatment	Animals: Pretreatment
procedures	*procedures*	Experimental	*Methods*
Analysis	*Analysis*	procedures	Experimental
		Analysis	procedures
			Analysis

and how these are related to each other. Whether the subsections are then labeled with headings depends on their length and importance.

In writing, the writer integrates the sections by linking each sentence and each paragraph to the next. Integration should not be imposed externally:

Ex. 11.2

   a. The next section describes the methods.

   b. The results are discussed in the next section.

   c. The next section will describe the theoretical structure, then the methods and results are described, and finally the discussion.

Such statements are superfluous in papers that have the conventional structure, but may be useful, if they are substantive, in descriptive or theoretical papers, or in long or complicated papers (*Table 4.6*).

11.2 PARAGRAPH STRUCTURE

The paragraph is the basic organizational unit of the paper—internally as well as externally. Internally, it develops one unit of the paper as part of the overall development of the paper. Externally, the visual delimitation of these units of development not only signals the units, but in breaking up the solid block of type, makes the text more accessible.

The writer has much control over the paragraph. The main sections of the paper are largely determined by convention, and the structure of sentences is determined by the syntax of the language. The paragraph however has no such formal constraints; the chief constraint is content.

11.2.1 Unity

As a unit of organization, the paragraph must at least have unity; that is, it must address one subject, and everything included in the paragraph must be related to that subject and contribute to its development. Assorted ideas and information do not become related just by being included in the same paragraph. Unity can be readily recognized on revision by literally reading through the paragraph, sentence by sentence, identifying the subject matter of each sentence, and determining whether it addresses the subject of the paragraph.

11.2.2 Coherence: Paragraph Development

As the organizing unit of the paper, the paragraph must itself be coherent if it is to contribute to the coherent development of the paper. The coherence of a paragraph results from ordering and connecting sentences for clarity of meaning and logical development. A paragraph develops a line of thought from the beginning to the end of the paragraph; that is, there is movement across the paragraph. But even this is not an absolute criterion. A paragraph presenting a set of results may not show much movement. Usually, however, even results have some meaningful order, which can contribute to the coherence of the paragraph.

The line of thought developed in the paragraph depends on the function of the paragraph, which may be to describe, compare, define, analyze, explain, discuss, and so on. The function then determines the function of the sentences in the paragraph, and the function of each sentence relative to neighboring sentences and to the paragraph. Any orderly development, for example, from important to unimportant, from abstract to concrete, from the complex to the simple—or vice versa—will help to organize the paragraph. In scientific papers, for example, the development may be sequential, then temporal or spatial, or analytical, then deductive or inductive (see *Chapter 3*). In the deductive order of development, the paragraph begins with the main point or general statement of the paragraph and is then followed by supporting or explanatory details (*Ex 11.3a*):

Ex. 11.3

   a. The density of nesting males may influence female choice in a number of ways. First, females may actively seek high densities of males because there are more males from which to choose (Bradbury, 1981), and/or because eggs may suffer less predation (Dominey, 1981). Second, higher densities of nesting males may simply be easier to locate, or may expedite female receptivity (De Boer, 1981). Third, this pattern may result from an overflow effect which would be greatest in areas of high male density. For example, a female who is rejected from a nest with yellow eggs may simply go to the nearest acceptable empty nest. On the only occasion in which the events immediately preceding a spawn in an empty nest were observed, the female was rejected by a male with yellow eggs and then spawned with his nearest neighbor. [Copeia, 3, 1988, 717]

   b. When a sample containing N nuclei of a given type is bombarded by a flux of neutrons, and when a particular reaction having a cross-section ø takes place, the rate of formation of product nuclei is given by Nø. Some radioactive product nuclei will have decayed during the reaction, but if the bombardment by neutrons of irradiation is sufficiently long compared to the half-life, a steady state occurs. This is the optimum condition for a nuclear reaction.

An "inductive" development is used for the unfolding and development of the details of the argument, leading to the general statement or main point at the end of the paragraph (*Ex. 11.3b*). A paragraph with this type of development is sometimes called an inverted paragraph. Such a development may be used in a paragraph draw-ing an inference, since the closure of the development is the logical outcome of the evidence. In a complex or extended development, or where the emphasis is on explaining the conclusions, it may be more helpful to readers to follow a deductive development and begin with the conclusion, general statement, or point of the section or paragraph and then provide the details to explain or support it.

Topic sentences.   Topic sentences, commonly recommended as a means of ordering paragraphs, make a general statement, the central statement of the para-graph. This is then developed in the paragraph:

Ex. 11.4

> A number of workers have described techniques that protect enzymes from phenol and quinone damage during their extraction from whole plant material. Staples and Stahmann (1964) have suggested the use of a cysteine hydrochloride and ascorbic acid in small quantities to reduce the quinones and therefore lessen their inhibition. Others have suggested the removal of the phe-nolics using polyvinylpyrrolidone (PVP) (McCown *et al.* 1968) or an anion exchange resin (Loomis and Battaile 1966). Mercaptoethanol is also commonly used as an alternate to ascorbic acid and cysteine hydrochloride. [Bull. Torr. Bot. Cl. 105, 1978, 318]

However, most scientific papers are not written about topics; they are reports about specific procedures, measurements, findings, and their interpretation. The paper does not have a topic or theme that can be successively divided into subtopics, so that finally each paragraph has its "topic." More important, in scientific writing one cannot develop a topic at will; the development is constrained by the content. Consequently, generalizing sentences such as topic sentences (*Ex. 11.4*) are not as common or as appropriate as they are in academic writing, and authors attempting to develop each paragraph from a topic sentence may write sentences with little substance:

Ex. 11.5

> a. In the present study the U.S. Standard Industrial Classification (CSIC) system plays an important role in establishing the environmental unit of analysis.
> b. Green's (1985) discussion of the six dimensions provides a good starting point for research on density measurements.

11.2.3 Coherence: Sentence Relations

The key to the coherence of a paragraph is the ordering and relationship of the sentences. In scientific writing the objective in the development of the paragraph is the logical connection of each sentence to the sentence preceding it and that follow-ing it. To achieve this, every connection must be made explicitly, by transitional devices or by internal structuring, for example, the hook-and-eye connection (see *Chapter 3*). The objectives in writing and revising a paragraph, therefore, are (1) to order the sentences logically and (2) to structure them internally and connect them so that they develop the line of thought intended through and across paragraphs.

Connections between paragraphs.   The connection or transition to the next paragraph is effected by the first or last sentence(s) of a paragraph. If a connecting sentence is more closely related to the sentences of the preceding paragraph, it closes the paragraph. If its thrust to the next paragraph is stronger than its relationship to the preceding sentences, then it opens the following paragraph. Whenever the last sentence serves as the end point of the development as in an inductive development, the connection with the next paragraph must be effected by the opening sentence of the next paragraph. A paragraph that begins with a generalizing sentence and ends with a closing general statement should be avoided. It is a closed unit, which must then be integrated into the line of development by the paragraph preceding it and following it. The closure in such a paragraph is likely to be artificial and redundant. Connecting or transitional sentences should be substantive, not simply rhetorical:

Ex. 11.6

> This is by no means the full extent of the subject. The next section describes the development of the model [heading of section would show this].

The connection of one paragraph to another may be too complex or too extensive to be made in a single sentence; then several sentences may serve as the connecting link.

Connections between sentences.   Transitional words or phrases are a device generally urged on writers as a means to achieving coherence. They may show conditionality, causality, disjunction, concession, exception, or correlation. In a scientific paper, with its highly specific content and its close logical relationships, internal connections are often more effective. Transitional words and phrases are therefore less commonly used and are largely the causal "therefore," "consequently," and the "hence" of logic and mathematics; the intensives "furthermore" and "moreover"; and the ubiquitous "however." The more expansive transitions,

Ex. 11.7

> on the one hand . . . on the other hand   on the contrary   in spite of this
> even granting the position that   on the basis of   from the point of view of

tend to sound vacuous in the midst of the concrete, detailed subject matter of the paper. When used, the transitional word or phrase should make a *necessary* connection and express *exactly* the relation intended.

Connections internal to sentences.   The close logical relation of sentences in scientific writing calls for more intrinsic connections, such as the *hook-and-eye* connection described in *Chapter 3* (*Fig. 3.1*), subordination, and more subtle contextual and referential devices. In the following hook-and-eye type of construction,

Ex. 11.8

> Relaxation takes place when *calcium* ions are pumped back into the sarcoplasmic reticulum. *ATP*, required directly for contractions and pumping *calcium* ions back into the sarcoplasmic reticulum, is derived from ADP by oxidation of *muscle* glycogen. This source of *ATP* may be depleted, however, by *tetanization* [condition of muscle].

Here no transitional words are used to connect sentences, only the referential words, which maintain the line of thought while advancing it.

Pronouns are referential words that function as reminders from sentence to sentence of the subject of discourse. Relative pronouns, for example, *which, that, who,* and demonstrative pronouns (and adjectives), *this, these, that,* and *those* refer back to a noun, noun phrase, or noun clause and so provide continuity. In general writing, such demonstratives can be used for indefinite, general reference, if the implicit reference is absolutely clear. In scientific writing, such indefinite reference falls short of the accuracy, specificity, and clarity required; therefore they should have explicit antecedents. Demonstrative adjectives should modify the noun referred to or an equivalent noun. The noun modified should not be used to shift to another meaning. In revising, writers can readily identify these pronouns and make certain that their reference is explicit and accurate.

Subordination.   The referential connections just discussed are not necessarily hierarchical. To show hierarchical relations, the writer must use subordinating devices. The two sentences in *Ex. 11.9a* are clearly related,

Ex. 11.9

 a. The fibers were long truncated cells. They showed secondary thickening of the wall but no lignification.
 b. The fibers were long, truncated cells, which showed secondary thickening of the wall but no lignification.
 c. The fibers were long truncated cells with secondary thickening of the wall but no lignification.

and should be joined in a subordinating relationship (*Ex. 11.9b,c*) for coherence. A succession of sentences with the same subject also indicates relationship (*Ex. 11.10a*) but connecting them makes for greater coherence (*Ex. 11.10b*):

Ex. 11.10

 a. *Fixed action patterns* are specific sequences of behavior that animals inherit rather than learn. These *sequences of behavior,* which include movements and coordinations, are usually present at birth. *These sequences* are necessary for the survival of the species. *They* are characteristic of the species. In addition, *fixed action patterns,* which are initiated by some external stimuli, run from start to finish automatically.
 b. Fixed action patterns are specific sequences of behavior that animals inherit rather than learn and that are usually present at birth. These sequences of behavior, which include movements and coordinations, are initiated by some external stimuli and run from start to finish automatically. They are characteristic of the species and are necessary for its survival.

A sentence beginning with the same word or phrase that terminates the preceding sentence indicates a relationship which should probably be made explicit:

Ex. 11.11

   a. The flow rate of water was varied and the Wilson method was used to calculate the individual film heat-transfer coefficients. The film coefficients were compared to those predicted by the Nusselt and Donohue equations.

   b. The flow rate of water was varied, and the Wilson method was used to calculate the film heat-transfer coefficients, which were compared to. . . .

A short simple sentence may be embedded in a paragraph in isolation, unconnected to the sentence preceding or following it. This "loner" may not have enough content to warrant a separate sentence, as in sentences stating a measurement or value:

Ex. 11.12

   a. The simplicity of this procedure makes it particularly suitable for analyses in which large numbers of samples must be examined, e.g. in plant breeding programs. *Ninety-six samples can be tested at one time.*

   b. The simplicity of this procedure makes it particularly suitable for . . . plant breeding programs, because 96 samples can be tested at one time.

   c. The kestrel is the smallest falcon. *Its wingspan is 21 in.*

   d. The kestrel, the smallest falcon, has a wingspan of 21 in.

Such a "loner" is often related to one of the adjacent sentences and can be revised to become a subordinate clause or phrase (*Ex. 11.12b,d, 11.13b,d*):

Ex. 11.13

   a. Beneath the A horizon is a layer that is not highly weathered and that has little organic matter. *This is the B horizon.* The B horizon often has minerals that have weathered and leached down from the A horizon. The B horizon is usually lighter and has more color than the other horizons.

   b. Beneath the A horizon is a layer that has little organic matter, the B horizon, which often has. . . .

   c. Liang et al. (1972) studied the effects of Pacific mites on grape yield and quality. Using a vine-by-vine analysis, they found no significant correlation. . . . *The study was performed in two vineyards.* In the first, average mite density on the vines ranged from. . . .

   d. Liang et al. (1972) studied the effects of Pacific mites on grape yield and quality *in two vineyards.* Using a vine-by-vine analysis, they found no. . . .

The "loner," however, may be a general statement like a topic sentence that serves to generalize neither what precedes nor what follows, and so should be deleted.

11.2.4 Length

The visual delimitation of the paragraphs often reflects their internal structuring, but the writer can combine or subdivide paragraphs into more effective units. A paragraph may be too short for coherence, and a series of such short paragraphs may look and read like a series of items; a long paragraph may appear impenetrable and forbidding to the reader, as well as being difficult to read. The writer should reexamine such paragraphs to determine whether to combine or divide them. This reshaping should not be done simply for visual effect. The objective is to achieve a balance among accurate reporting, coherent structuring, and visual accessibility. In scientific writing, one may not have a choice about the length of a short paragraph. In a two- or three-sentence paragraph of results, the writer may have no other data to report and may not be able to expand the paragraph or incorporate it in another paragraph. Nevertheless, two- or three-sentence paragraphs should be closely scrutinized to determine whether they can be justified.

When a paragraph is very long, the writer may consider dividing it at some logical point of division even though it is entirely coherent as one paragraph— simply because such a long paragraph is rather an assault on the eyes, and readers are apt to find it intimidating. Such a paragraph should first be examined, however, to eliminate wordiness, repetition, digressions, and irrelevant or superfluous material.

11.3 SENTENCE STRUCTURE

11.3.1 Unity and Cohesion

The coherence of a paragraph depends on the cohesion (i.e., coherence) of the sentences in the paragraph. To be cohesive, a sentence must at least have unity. It must make a statement about one subject, and all parts of the sentence must contribute directly to that subject. One should not use subordinate or coordinate parts of the sentence to add tangential or unrelated material. In structuring sentences for cohesion, the writer's chief means is the ordering of words. Until words are ordered in a sentence, they cannot make a statement. In an uninflected language such as English, the ordering of words in the sentence determines the syntactic relations in the sentence, and, therefore, its meaning. In the following sentences,

Ex. 11.14

    a. The fungus destroyed the cells.
    b. The cells destroyed the fungus.

merely interchanging the words at the beginning and end of the sentences changes the meaning of the sentences drastically. In more complex sentences, the difference in meaning with different ordering may not be as obvious but may confuse the reader. The writer must therefore position words carefully to meet various constraints: the meaning to be conveyed, the syntactic structures of the language, the function of the sentence in the paragraph, and the relationship of the sentence to neighboring sentences. Such positioning requires attending carefully to syntactic structure, because almost any word, phrase, or clause may be modified by one or more words, phrases, or clauses, and these modifying elements may be coordinate or subordinate and restrictive or nonrestrictive. Modifying elements must be positioned close to the elements they modify, and the positioning must follow the idiomatic ordering for such modifiers. Moreover, the grouping of words into phrases, or clauses must also follow accepted or idiomatic ordering.

Position of elements and emphasis. The position of elements in the sentence determines the emphasis in the sentence and so contributes to the coherence. In the usual subject-verb-object order, the position of the subject makes it dominant. This opening position is useful, therefore, for focusing on the subject or an important word, phrase, or clause:

Ex. 11.15

a. *Consistent with previous research,* the elderly were not abandoned by their family and friends.

b. *Providing that there exists some (as yet unknown) means of retaining resident proteins in each location along the way,* bulk forward transport could conceivably provide all that is needed until the last station in the Golgi is reached.

c. *To explain the observation that increasing the He or NO pressure in the headspace over the PDMS liquid increases the yield of gaseous hydrocarbon products,* it is proposed that vaporization of PDMS accompanies the temperature jump.

The importance of this position is also reflected in the differing functions of the active and passive voice (see *Chapter 3*).

The end of the sentence also functions as a position of emphasis. Frequently however, this position may be taken by a weak verb or unimportant element:

Ex. 11.16

a. In this paper a deterministic model for population growth, some applications both theoretical and empirical, and suggestions for population strategies in hazardous environments *are presented.* [This paper *presents.* . . .]

b. The Z-score of length was the dependent variable, and district, sex, age, year (AGECAT), categories of the mother's education (EDMONCAT), and all two-way interactions were *possible dependent variables.* [The dependent variable was the Z-score of length and the *possible dependent variables* were district, sex,. . . .]

c. Measurements of the proton spin-lattice relaxation time (T_1) in whole blood, plasma, blood cells, *in vitro* liver, live and necrotic tail samples of adult newts and spin-lattice relaxation time (T^2) in human whole blood plasma and blood cell samples *have been carried out.* [Measurements of the proton spin-lattice relaxation time (T_1) *have been carried out* for. . . .]

The sentence in *Ex. 11.15a* keeps readers expectant for 23 words before they are rewarded with the weak verb at the end. This periodic structure is more demanding of the reader's attention than having the important material supplied from the start, as in the bracketed revisions. It is particularly ineffective, as here, when the sentence collapses at the end with an inconsequential verb.

Complexity, length, and variety. Scientific writing is criticized for its monotony of long, complex sentences, and the remedy prescribed is to use short, simple sentences. However short simple sentences are not adapted to presenting the extensive data or expressing the complex concepts and subtle interrelationships. A succession of short, simple sentences tends to have a primer style:

Ex. 11.17

> A new version of the product is calcined clay with modified hydration content. It is gaining widespread use. Components of any particular kaolin, in addition to alumina and silica, are viewed as impurities. The more valued deposits and, hence, higher priced kaolin products are those of the highest purity.
>
> Kaolin is the most widely used mineral in paper manufacturing throughout the world. Key production centers are located in England and the US. Most of the latter's production is located in Georgia. [Min. Engr. 39, 1987, 247]

Such short simple sentences do not express the relationships among the elements and so are not effective in expressing complex concepts and research data. In *Ex. 11.17,* the subject matter is not complex; still, the excerpt would gain in coherence by combining and subordinating some of the sentences.

The writer may not have great latitude in avoiding a series of short simple or long complex sentences because their length and complexity is directly related to their meaning and function. The explanation of a concept may require a succession of complex sentences. In a series of simple results, the sentences may all be simple. Variety is a direct consequence of the complexity and length, because it depends on the structure of successive sentences in the paragraph; therefore the writer often has little choice about varying the structure.

11.3.2 Revising Sentences for Structure and Meaning

Sentences must state the meaning intended, relate to neighboring sentences in the paragraph, and function appropriately in the paragraph. Meeting these objectives is best attempted on revision, and then out of some understanding of the architecture of the sentence. Sentences with a single bit of information (*Ex. 11.18a*) rarely warrant separate status,

Ex. 11.18

   a. The rows were spaced .5 m apart.

   b. The problem of finding an answer to the question is of great importance.

and trivial sentences like *Ex. 11.18b* do not merit space.

Clauses.   Clauses should be substantive; one should not use a clause when a phrase or word is as effective:

Ex. 11.19

   a. They were able to induce *Bougainvillaea glabra, [which is] a short-day plant,* to flower under noninductive long days by changing the contents of the media.

   b. The diode structure *[that was] used* in a computer simulation had a drift zone *[that was]* 3.8 mm wide and operated at 10 GHz.

Adjective or relative clauses introduced by relative pronouns (*who, whom, whose, which, that*) may be used restrictively or nonrestrictively (*Table 11.1*). For a sharp and rigorous designation of these two types of clauses, nonrestrictive clauses are connected by *which* and preceded by a comma; restrictive clauses are connected by *that* and are not separated by punctuation. In informal writing, *which* may be used for both, but in scientific writing, such usage places too great a dependence on the comma alone for expressing the writer's meaning. In *Ex. 11.20a,*

Ex. 11.20

    a. *Nonrestrictive:* The installations, which were placed below ground, consisted of concrete blocks stacked to form drains.

    b. *Restrictive:* The installations that were placed below ground consisted of concrete blocks stacked to form drains.

TABLE 11.1   Sentences with (a) restrictive, (b) nonrestrictive and (c) adverbial clauses

a. *Restrictive Clauses*

1. *Colonies that* ran into each other were counted separately.
2. Gastric mucus glycoprotein is the chief component of mucus *gel that* protects surface mucosal cells.
3. For those vehicle *trips that* were not classified as work related, several additional questions were asked.

b. *Nonrestrictive Clauses*

1. Also, compositing reduces the influence of observational *errors, which* in individual case studies, can bias the results.
2. The strip *tests, which* depended on pH shifts due to the action of the organisms on the substrate, gave variable results.
3. The spectrum in Figure 2D may be regarded as a *triple-septet, which* indicates that the protons are grouped into two and six equivalent protons.

c. *Adverbial Clauses*

1. *If* the average time the binding site is unoccupied is long compared with the electric and diffusive relaxation times . . . the local concentrations can be related to the bulk concentrations by the Boltzmann equation:. . . .
2. Factors unconnected with mineral nutrition are probably also at work and more evidence is required *before* the disjunct and erratic distribution of this species can be explained.
3. In addition, olefins seem to be produced only on one of the two chain propagation sites on the iron catalysts, *while* [whereas] the cobalt catalyst produces little or no olefins.
4. *Although* other factors within the retinal-thalamic-cortical circuit must also limit visual acuity at these ages, our developmental sequence of foveal cone packing density is in good agreement with many psychophysical measures.
5. It is reminiscent of the differential effect of hormonal treatment on immunoreactive kallikrein, *so that* it may be a primary or secondary response.
6. *Because* the Cl remains in solution, rinsing is rapid and efficient, allowing high solids retention.

the sentence refers to *all* the installations and the nonrestrictive (which) clause is a supplementary explanatory remark; that is, that they were placed underground. In *Ex. 11.20b,* the sentence reports on only some of the installations, and the restrictive (that) clause defines and identifies them: those placed below ground. The meaning of the two sentences is therefore very different.

The pronoun *that* may be mistakenly duplicated because of the complexity of the sentence or, as in *Ex. 11.21,* it may be omitted, in an attempt at brevity or at imitating speech:

Ex. 11.21

a. During the period [*that*] the meadow was observed, no significant differences were observed in the species composition.

b. Fig. 9 shows fluorescence spectra of both peak fractions, indicating [*that*] there was the difference of 4–7 nm at emission maxima.

but the complete form is more appropriate in formal writing.

In adverbial clauses with *where* and *when,* the subordinating conjunctions are sometimes incorrectly used as relative adjectives, as in *Ex. 11.22a,b:*

Ex. 11.22

a. The periods *when* [during which] animals were released coincided with light and dark periods.

b. We had to exchange the water for a solvent *where* [in which] the monomer was soluble.

c. Where [*whereas*] optical microscopes are used for transparent specimens, reflection microscopes are required for opaque specimens.

and *whereas* is missued for *when* (*Ex. 11.22c*). Clauses of concession may also be imperfectly structured. The concession should be placed in the subordinate clause (*Ex. 11.23b*) with the conjunctive adverb for concession, instead of in the main clause, as in *Ex. 11.23a:*

Ex. 11.23

a. Even though his measurements were not consistent with his theory, he was the originator of the method.

b. Even though he was the originator of the method, his measurements were not consistent with his theory.

Misplaced and dangling modifiers.   Because the position of words is important for meaning (*Table 11.2*) modifiers must be placed close to the words that they modify. Misplaced modifiers are likely to cause confusion (*Ex. 11.24a,b*), which is most apparent when the modifiers are in unexpected juxtaposition with the words modified:

Ex. 11.24

a. A segment of the fur industry depends on a good supply of these animals coming in to remain in business.

b. Like all computer languages, the major purpose of Fortran is to aid in communication between. . . . [The major purpose of Fortran, like that of all computer languages, is. . . .]

TABLE 11.2   Change in position of *only* and associated change in meaning

1. The rat had to give the lever two tugs to obtain the pellet.
2. *Only* the rat had to give the lever two tugs to obtain the pellet.
3. The *only* rat had to give the lever two tugs to obtain the pellet.
4. The rat *only* had to give the lever two tugs to obtain the pellet.
5. The rat had *only* to give the lever two tugs to obtain the pellet.
6. The rat had to give *only* the lever two tugs to obtain the pellet.
7. The rat had to give the lever *only* two tugs to obtain the pellet.
8. The rat had to give the lever two tugs *only* to obtain the pellet.
9. The rat had to give the lever two tugs to obtain the *only* pellet.

Modifiers that are unattached, that is those for which there is no word in the sentence that they can modify, are said to be dangling. Dangling participles, which have a familiar notoriety, are verbals in search of a noun. They are most easily recognized when they are at the beginning of the sentence:

Ex. 11.25

    a. *Making* appropriate changes in the load and displacement vectors, *equation* (5) can be rewritten as. . . .

    b. *Fastened* in this manner, the intake *charge* was free to move through the chamber.

    c. *Examining* the distribution of moisture and moist static stability, composites were formed from data on the day of Colorado cyclogenesis and on the preceding and following days.

Such participles modify the subject of the sentence; therefore, one can easily determine whether they are dangling by making the subject of the sentence the agent carrying out the action of the participle. For example, in *Ex. 11.25a* it is obvious that the *equation* cannot carry out the action of the participle. Similarly in *Ex. 11.25b* and *c*, it was not the *intake charge* that was fastened, nor the *composites* that examined.

Dangling participles are common in scientific writing, perhaps because the passive voice often eliminates the agent of the verb in the sentence. For example, in *Ex. 11.26a,* because of the passive voice, the agent who knew the temperature is not named; there is no agent and the participle *knowing* dangles. There are three main ways of correcting dangling participles (*Ex. 11.26b–d*):

Ex. 11.26

    a. *Dangling participle: Knowing* the initial temperature, the speed of the drum, and the rate of heating, the *temperature* could be calculated at any point on the graph.

    b. *Provide agent for action of participle: Knowing* the initial temperature, the speed of the drum, and the rate of heating, *one* could calculate the temperature at any point on the graph.

    c. *Convert participial phrase to clause: If* the initial temperature, speed of the drum, and rate of heating *are known,* the temperature can be calculated at any point on the graph.

    d. *Convert participle to gerund acting as subject: Knowing* the initial point temperature, the speed of the drum, and the rate of heating makes it possible to calculate the temperature at any point on the graph.

Providing the agent for *knowing* converts the verb to the active voice (*Ex. 11.26b*). However, since the passive voice was used in order to focus on the mathematical operation rather than the scientist, revisions centering on the calculations rather than the calculator are preferable (*Ex. 11.26b,c*). Converting the participle to a gerund is one of the best ways of revising dangling participles commonly associated with equations:

Ex. 11.27

a. *Calculating* the electric field in the device,
$$E_i\,(x,\ y)\ =\ -\operatorname{grad}\,[\phi_i\,(x,\ y)].$$

b. *Calculating* the electric field in the device *gives*
$$E_i\,(x,\ y)\ =\ -\operatorname{grad}\,[\phi_i\,(x,\ y)].$$

In *Ex. 11.27a, calculating* has no agent; in fact, it is not a sentence, because it lacks an explicit verb (see *Chapter 6*). In *Ex. 11.27b, calculating,* is the gerund, the subject of the verb *gives,* with the equation as its object.

Prepositional phrases.   Prepositions (e.g., *in, on, from, with, by*) are the commonest devices for connecting a word, phrase, or clause that is the object of the preposition to another part of the sentence. Since much of scientific writing has to do with relationships, prepositions are used extensively for their versatility and economy. They can modify almost any word in the sentence and can frequently replace a clause:

Ex. 11.28

a. A color film sensitive *in* the 700–900 mm near-infrared *region* has been used *as* a *tool in* disease *detection by* many *investigators, after* Colwell's (1956) *detection of* cereal crop *diseases.*

b. While there appears to be no earlier studies *of* this *problem in* magnetic *systems,* a problem similar *to it* has been studied *in* recent *times in* multiphase classical fluid *systems near* critical end *points* and *in* the equivalent ^4He *near* the λ *point.*

When two prepositions have the same object, writers, in an effort to avoid repetition, may use an elliptical construction:

Ex. 11.29

a. The temperature was recorded *during* or at the end *of* the *extraction.* . . . [*during* the extraction or at the end of it]

b. These particles moved to the region of lower temperature, where they mixed *with* and transferred part of their energy *to* other *particles.* . . . [they mixed *with* other *particles* and transferred part of their energy to them]

One can avoid the cumbersome, indirect construction by completing one prepositional phrase before completing the other, repeating the object or substituting a pronoun for it, as in *Ex. 11.29.*

Weak constructions.   Weak constructions misplace or distort the emphasis of a sentence. Weak verbs, which provide only indefinite or unspecific action, are common in scientific writing:

Ex. 11.30

| make | carry out | accomplish | indicate | occur | proceed |
| do | conduct | implement | involve | perform | give |

Because the verb carries the thrust of the action from the subject to the object, a sentence with a weak verb lacks strong action or direction. However, the use of weak verbs is often unavoidable in scientific writing because of the frequency of abstract concepts, the passive voice, description, and nominalization. Weak verbs should, therefore, be avoided when stronger verbs can be used.

Frequently, the weak verb is controlled by a noun, often an abstract noun, which contains embedded in itself, the verb action. The sentence is, therefore, doubly weakened by the indefinite or static verb and by the unspecific, vague, indefinite character of the abstract noun, as in *Ex. 11.31*:

Ex. 11.31

   a. The recessive gene codes for a defective enzyme, and *accumulation* of the cerebroside *occurs* in the nerve cells of the brain and liver.
   b. The recessive gene codes for a defective enzyme, and the cerebroside *accumulates* in the nerve cells of the brain and liver.

By extracting the verb *accumulate* from the abstract noun *accumulation* and substituting it for the weak verb *occurs* (*Ex. 11.31b*), the writer can eliminate two weak elements at the same time. However, in *Ex. 11.32a*, the writer may not want to replace *takes place* with the stronger verb *fertilize*:

Ex. 11.32

   a. *Fertilization* of the eggs in most mammals *takes place* in the oviduct [with the exception of the Madagascar hedgehog (*Setifer setosus*), in which *fertilization occurs* in the ovarian follicle].
   b. The eggs of most mammals are *fertilized* in the oviduct [with the exception . . . in which the eggs *are fertilized* in the ovarian follicle].
   c. In most mammals, sperm *fertilize* the eggs in the oviduct, except . . . in which the sperm *fertilize* the eggs. . . .

Fertilization is a scientific term for an important biological process, and the conceptual structure for the term is substantive enough that it is not a mere abstraction. Consequently, "fertilized" (*Ex. 11.32b,c*) may not accurately substitute for it. Also, in a passage or paragraph on the process of *fertilization,* to shift the subject to *eggs* is to shift the subject of discourse and so disturb the line of thought. If one wishes to go further to avoid a weak construction, one can avoid the passive voice and use in the active voice (*Ex. 11.32c*), but this shifts the subject of discourse even farther.

Indefinite pronouns, for example, *there, it,* are semantically empty pronouns that may precede or anticipate the real subject. They thus preempt the strong opening position and thus deemphasize the specific subject and weaken the sentence. Sentences beginning with the anticipatory "there" (*there is, was, there are, were*),

Ex. 11.33

    a. There are spines along the mouth edge, which prevent hairs from interfering with feeding.

    b. Spines along the mouth edge prevent hairs from interfering with feeding.

    c. There were two major events that shaped the geology of Precambrian New York State.

    d. Two major events shaped the geology of Precambrian New York State.

    e. There are differences in the incidence of spontaneous Leydig cell tumors.

can be easily corrected by replacing the pronoun with the specific subject (*Ex. 11.33b,d*) and eliminating the relative pronoun, which the anticipatory subject made necessary. The construction is useful for generalizations, however, or annunciatory constructions as in *Ex. 11.33c,e,* if used to prepare the reader for the elaboration on the subject.

    Adjectives and adverbs of indefinite reference (*Table 11.3*) are too weak and too imprecise for use in scientific writing.

    Agreement of inflected forms.   English has retained some inflected forms; for example, number is indicated by inflection in pronouns (*he, she, they*), nouns (animal, *animals;* ox, *oxen;* goose, *geese*) and verbs (*indicates, indicate*), and gender by inflection in some pronouns (*her, his, him, it*). Where related words in a sentence are inflected, the inflected forms must agree.

    Subject-verb agreement.   Verbs must agree with the subject in number (*Ex. 11.34a*):

Ex. 11.34

    a. This glass *window acts* as a light conductor.

    b. The lizard has a net gain of water at low body temperatures and the increasing *rate* of evaporative water loss at higher body temperatures *reduce* [reduces] this gain.

    c. Their *implication* for the radiation effects *were* [was] also discussed.

TABLE 11.3   Modifiers indicating indefinite degree*

almost	distinct	large	possible
apparent	dramatic	low	pronounced
appreciable	elevated	marked	quite
approximate	extreme	massive	rather
certain	fair	moderate	reduced
circa	good	much	relative
comparative	great	multiple	slight
considerable	gross	nearly	small
decreased	high	on the order of	somewhat
decreasing	increasing	over (quantity)	under (quantity)
diminished	intense	poor	very

*Also the corresponding -ly adverbs, e.g., *apparently, appreciably, approximately,* and so on.

d. The *identification* of PCDFs and PCSSa *were* [*was*] based on the fact that the signals were observed at characteristic mass numbers. . . .

When the verb does not agree with the subject, it is often due to other words intervening between the two (*Ex. 11.34b–d*). The agreement of subject and verb requires special attention to words having Latin or Greek plurals (see *Appendix 11A*):

Ex. 11.35

| a *datum is* | an *alga* accumulates | the *criterion* is used | the *phenomenon* was observed |
| *data* are | algae accumulate | *criteria* are used | *the phenomena* were observed |

Collective nouns may be used as either singular or plural depending on the meaning intended (*Ex. 11.36a,b*):

Ex. 11.36

a. The flock *arrives* at the head of the lake in March.
b. The flock *eat* the gleanings from the harvested corn.
c. Then 10 ml of the solution was [were] pipetted out for the assay.
d. Then a 10-ml aliquot [sample] was pipetted out for the assay.

Quantities may be mistakenly treated like collective nouns, as in *Ex. 11.36c;* the verb should be plural to agree with 10 ml. Finally, particular words may have a (1) plural form but are used with a singular verb, for example, *economics, physics;* (2) a singular form, but used in the singular or plural, for example, *deer, sheep;* (3) or singular form but used in the plural, for example, *cattle*. However even nouns that are not collective nouns may require making the verb agree with their meaning, rather than their form:

Ex. 11.37

a. Root *rots* [*plural subject*] *are* considered to be important in the lack of persistence of alfalfa stands and the low productivity of older stands.
b. Root *rot* and crown *rot* [*compound subject*] *are* considered to be important. . . .
c. Root [rot] and crown *rot* [*elliptical compound subject*] are considered to be important. . . .
d. Root and crown rot [*singular subject*] *is* considered to be important. . . .

Pronoun-Antecedent Agreement.   Pronouns must agree with their antecedent noun in gender as well as number (*Ex. 11.38a*):

Ex. 11.38

a. But the hairs are also sensitive, so that as the *insect* moves, the hairs and leaf close over *him* [*it*].
b. Dissolved solids release many ions necessary for fish to survive, but *they* [solids or ions?] may raise concentrations in tailwaters high enough to preserve fish. [but *they* may raise the concentration of *ions* in tailwaters high enough to preserve fish.]

but when the reference for the pronoun is distant, intervening nouns make it difficult for the reader to determine the antecedent for the pronoun (*Ex. 11.38b*); then one must make the antecedent noun clear.

A common pitfall in making pronouns agree with their antecedents is not providing *definite* antecedents. Often a demonstrative pronoun is used to refer to a general notion or concept that is not explicitly stated, but must be deduced by the reader from the text. Such implicit reference is not accurate enough or clear enough for scientific writing. In *Ex. 11.39a, this* does not modify any noun or noun phrase that precedes it; it actually refers to the whole clause preceding it, that is, to the *finding* stated in the clause:

Ex. 11.39

    a. Green's (1985) first-degree group showed the highest risk of death, but *this* [finding] was not consistent with DeCanet's results.

    b. Most digital systems were designed to be binary. The reason for *this* [design? characteristic?] was that information could be represented as high and low voltages.

    c. Hence, radial design is not a random or diffusional process as has been indicated by previous workers. [2, 12, 15]
    This is further substantiated from a careful observation of the segregation process on the high-speed films.

The demonstrative pronoun at the beginning of a paragraph is particularly likely to lead to ambiguity, as in *Ex. 11.39c,* and it should be supplied with a noun that clarifies it.

The relative pronoun, *which,* may also be used for indefinite reference:

Ex. 11.40

    a. Genetic material in the chromosome can be added to a chromosomal strand that is being formed, *which* is called duplication.

    b. Genetic material . . . being formed, *a process that* [*this process, this addition*] is called duplication.

Here *which* does not refer to any word or phrase preceding it, but to the process of addition described by the whole main clause.

11.4 PARALLELISM AND SHIFTS

Parallelism and seriation, a special form of parallelism, are means of making order in a paper; shifts are breaks in its orderly development. Both can act at any level, from the sentence to the section.

11.4.1 Parallel Structure

''Parallel structure'' refers to the similar structure of two or more coordinate elements in a sentence, paragraph, or even section. Parallel structure is the writer's way of making the structure match the coordinate character or meaning of two or more elements. In *Ex. 11.41a* ''oil pollution'' might be parallel with the various prepositional phrases in *Ex. 11.41b–d,*

Ex. 11.41

   a. Marine pollution is also caused by precipitation of trace elements from the atmosphere and oil pollution.

   b. *from* the atmosphere and [from] oil pollution

   c. *of* trace elements from the atmosphere and [of] oil pollution

   d. *by* precipitation of trace elements . . . and [by] oil pollution.

   e. caused *by* oil pollution and precipitation. . . .

but it is actually parallel with the farthest (*Ex. 11.41d*); therefore inserting *by* before *oil,* identifies the two parallel elements. If *oil pollution* were placed first, before *precipitation* (*Ex. 11.41e*), the *by* would not have to be repeated.

In *Ex. 11.42a*, the three coordinate phrases are expressed as conditional and declarative statements:

Ex. 11.42

   a. If two strains are *quite different at their H-2 loci,* they will undoubtedly undergo acute rejection after grafts are exchanged. Two strains *similar at their H-2 constitution* will have rejection patterns most often toward the chronic end of the time scale; and if they are *H-2 identical,* of course, they will not be rejected at all.

   b. *If two strains are quite different* at their H-2 loci, they will undoubtedly . . . are exchanged. *If they are similar,* they will have rejection . . . time scale; *if they are H-2 identical,* they will of course not be rejected at all.

Making the structure of the sentences reflect the parallel relation of the three highlights the comparison (*Ex. 11.42b*). Parallel structure is particularly useful within sentences:

Ex. 11.43

   a. *Before cultivation* of the prothoracic glands, little or no ecdysone activity was found in the culture medium; *after cultivation,* a great deal of ecdysone activity was found in the medium.

   b. With experience, planktivorous fish *improve* their capture success, *learn* to ignore prey that are difficult to capture, and *increase* their foraging efficiency by learning to specialize on certain type of prey.

   Series.   Members of a series share one or more attributes and so are coordinate, and the series can therefore signal their coordinate character (*Ex. 11.44–11.46*). A series may summarize the material or discussion preceding it, thus keeping the line of development clear. Often it introduces elements that are to be discussed further, and thus the subsequent development:

Ex. 11.44

   Three criteria are used in measuring trees by the point cruise method: (1) height, (2) form class, and (3) total volume cruised. *Height* is more important than any other measure in determining volume. . . . Diameter measures do not have to be as accurate as height measures since the volume of a tree will approximately equal hs^2, where . . . *Form class,* the taper of a tree, can be estimated from the species and region. . . . The *total volume* can be estimated from data compiled. . . .

Members of a series are rarely assorted, and so should not be listed in random order. Any series has an inherent order of position, with the first and last members more dominant. In a long series, therefore, some of the intervening members may be glossed over. The author may order the members beginning with the most important, general, longest, shortest, or whatever order accords with his or her intent explicitly, but often implicitly. In *Ex. 11.45a,* the type of ordering is explicit.

Ex. 11.45

   a. In Algonquin Park, the prey of the wolf in order of decreasing frequency are: deer, moose, beaver, muskrat, hare, red-backed vole, and black bear.

   b. Temperature data were collected in degrees and in the form of growing degree days during the critical periods of shoot elongation, bud initiation, and flowering.

   c. Habitat is sometimes confused with related concepts, such as territory, environment, ecological niche, and range.

In *Ex. 11.45b,* the order is implicit—developmental; *Ex. 11.45c* has no particular order, although a more meaningful order seems possible. Any ordering will highlight the elements, but a meaningful ordering will help to communicate relations as well.

Enumeration.   A long series or one with long, complex, or confusing members should be enumerated to help the reader identify the members. In *Ex. 11.46* the structure is parallel in both examples:

Ex. 11.46

   a. Linear regression was used to evaluate the relationships between nutrient and sediment concentration and streamflow, nutrient and sediment load and streamflow, and nutrient and sediment load and land use.

   b. Linear regression was used to evaluate the relationships between (1) nutrient and sediment concentration and streamflow, (2) nutrient and sediment load and stream flow, and (3) nutrient and sediment load and land use.

but the enumeration in *Ex. 11.46b* spares the reader the task of assigning the *and's* to the appropriate members.

Enumerated elements can be designated numerically or alphabetically. Numerals are readily differentiated from letters and so are more effective in ordering words; letters are more effective with numbers. When the elements of a series are displayed, arabic numerals are more common than letters. Bullets (large centered dots) or dashes are common in promotional material, but do not order the members and do not permit reference to particular members of the series.

In the text, the numerals or letters used for enumerating the elements are enclosed in both opening and closing parentheses, to set the numeral or letter off from the word preceding and following it (*Ex. 11.47a*):

Ex. 11.47

   a. The choice of this arbitrary definition was due to (1) lack of data on other definitions of PEM, (2) general acceptance of Wt/A and Wt/ht measures, (3) their ease of measurement, and (4) the availability of mortality risk data associated with Wt/A.

b. These biased errors do not cancel each other out and are often caused by
  1) Environmental factors that vary greatly between individuals and are either unknown or inaccurately estimated
  2) Variation in selection intensity from sire to sire
  3) Samples or daughters not randomized
  4) Selection of daughters from a special group.

In a displayed list (*Ex. 11.47b*), only the closing parenthesis is needed; the blank space before the numerals separates them from the preceding text, although a period after enumerating elements is more common in print.

11.4.2 Shifts

A shift is a break in the line of development due to omissions, either of substantive material or of internal or external connections, or to mismatches in substance or structure. These may be shifts in point of view, terminology, tense, or subject.

Point of view and topic.   The writer may shift his or her point of view from the third person to the first person, for example, "is helping *us* to understand." This shift may include readers in a point of view, but in some contexts may give the statement a platform speech quality, which is not appropriate for a scientific paper. The shift to the imperative, from the third person to the second, "note that . . ." or "recall that . . ." is a more definite shift to a lecture format. This may be useful to emphasize a point but should be used sparingly in a scientific paper. Rhetorical questions, by addressing the reader directly, also shift the point of view, but in a scientific paper, they often seem contrived.

The author may shift the subject of discourse. For example, the subject of a paper may be the effectiveness of institutions of higher education as *organizations,* but it may shift to the effectiveness of *undergraduate education,* which is only one aspect of the institution as an organization.

Terminology.   Shifts in terminology, that is, using a different term for the same entity, are not common for scientific terms that are distinct from words in the general vocabulary; they are more likely in words from the general vocabulary. In scientific writing, a term must be used consistently for a referent as long as the writer is referring to that referent. In the following excerpt, about insecticides, three terms are used for them:

Ex. 11.48

> Plots were treated with *insecticides* for the first-generation attack or for both first- and second-generation attacks. In 1985 both the plots treated once or twice with Furadan showed the least amount of damage, and this was the only *material* that differed from the control in level of damage. In 1986, the *formulations* that controlled the insect population best were Sevin and Pounce. None of the *materials,* however, showed a significant level of damage in any of the treatments.

Here *insecticides* might be distinguished from *formulations,* the compositions of the solutions of the insecticides, but if the two terms are being used synonymously, then the most accurate of the two should be selected and used consistently. *Materials* is too broad to be accurate here. The use of closely related terms to make a distinction can be effective, provided that they are differentiated consistently; if not, the meaning will shift throughout the paper (see Chapter 2).

Tense.   Writers may mistakenly try to make tense conform closely to an actual time. As a result they may present findings that have not been contradicted or describe an instrument that still exists in the present tense, even though they are being reported as past research. The focus should be on the time relationship in the context of the research and the report of the research.

Subject.   The subject of the sentence is a focal point in the line of development. However, in a long or complex sentence, the writer may shift to a different noun structure, which then functions as subject (*Ex. 11.49a*):

Ex. 11.49

a. The *death rate* for smokers who had smoked more than one pack of cigarettes per day but who had stopped for less than 10 years, still on the average had a higher death rate than men who had continually smoked less than one pack of cigarettes per day.

b. A recent *report* by Hayes et al. (10) further *improved* the *depositions* onto the belt *by spraying* the HPLC effluent on the transfer surface at a 60° angle.

Simply eliminating "the death rate for" corrects this sentence. In *Ex. 11.49b* the subject shifts implicitly from the *report* to *Hayes et al.*

11.5 REDUNDANCY AND TAUTOLOGY

Wordiness, whether through redundancy, tautology, or circumlocution, obstructs the clear presentation of the research findings and concepts, and wastes the reader's time. In a scientific paper, wordiness is an indication of ineffective structure. Such verbiage must be identified and eliminated to make the paper concise and coherent.

Repetition can be used for emphasis; in a book or long report, it is used as a reminder to readers of material dealt with earlier; however, in scientific papers, such repetition is not appropriate and is to be avoided:

Ex. 11.50

Green and DeBrion (1986) reported that solutions applied to nonsterile river mud lost 20% *of their effectiveness* (as tested by the snail bioassay) after 5 hr, 35% *of their effectiveness* after 20 hr and 70% *of their effectiveness* after 5 days.

Words may also be redundant, as in "inexperienced primiparous ewes" and in certain types of phrases, such as, *smaller in size, moment in time* (see *Table 11.4*).

TABLE 11.4　Wordy phrases with redundant words (in parentheses)

at this (moment in) time	equally (as well)	(percentage of) 3.4%
at this (point in) time	(female) ovary	the (pH of the) soil was pH 8
(both) alike	fewer (in number)	repeat (again)
bright green (in color)	filled (to capacity)	(repeated) daily injections
(completely) filled	hydroxylation (reaction)	throughout the (entire) region
contemporaneous (in age)	if and when [either but not both]	two (equal) halves
(continued) hourly observation	large (in size)	(voltage of) 3 V
(definitely) proved	oval, round, or square (in shape)	years (of age)

Sometimes the same idea or information is presented in different parts of the paper, in different words:

Ex. 11.51

    a. The final section of the paper briefly considers the research implications of the present typology and the underlying assumption about choice and organizational behavior. [in introduction.]

    b. Figure 2 presents some interesting research implications of the typology presented in this paper. The figure also suggests the issues or problems. . . . [in results]

Similarly writers may state results in the results section and then restate them in the discussion section—usually within a page or two.

    A sentence may essentially repeat the idea or point presented in the preceding sentence in different words, perhaps with a minor addition or omission, or modification. One sentence may express an idea in general terms and the following sentence repeat it in specific terms, or vice versa (*Ex. 11.52a,b*):

Ex. 11.52

    a. In most studies of innovation, different categories of innovations are not distinguished. All possible innovations are usually included in one category.

    b. If society owes members freedom from a particular harm, it follows that each organization has an obligation to avoid inflicting the harm on any participant. The wrongfulness of harms in society seems logically to imply their wrongfulness in every substructure of society.

    c. These findings suggest that the results obtained here are in fact authentic; they are not simply artifacts.

In another type of tautology one statement is essentially the same statement as the other, except that one is positive and the other is negative (*Ex. 11.52c*). Sometimes, the first and last sentences of a paragraph may make essentially the same general statement. Such tautology may be due to the writer's not recognizing that although the words and structure are different, the essential meaning is not. Also, the sentences may both be a felicitous expression of one's thought, so that it is difficult to give up the felicitous structure of one for the equally felicitous structure of the other. Tautology is a greater pitfall in papers in the social sciences than in the natural sciences, because the former include much discussion, and the terminology is derived mostly from the general language. Wordiness within sentences may result in circumlocution; that is, in the use of an unnecessarily large number of words to express an idea or in stating it in an indirect or roundabout way:

Ex. 11.53

> *This is because of the fact that* particle dynamics effects may inhibit the current flow before the fundamental power balance limit is reached.

In reviewing a sentence to eliminate circumlocution, the writer should make certain that modifying words and phrases are close to the words modified, that there are no long, confusing series of prepositional phrases, that clauses clearly modify definite words, and that repetition is avoided.

11.6 EXTERNAL STRUCTURING

The structuring of sentences, paragraphs, and sections for coherence is a fundamental *internal* structuring of the paper. This can be supported or reinforced by forms of *external* structuring, such as headings and referential elements.

11.6.1 Headings

Function and content.   Headings are external ordering elements at the beginning of a unit, which designate the subject and show the rank of the unit that they head. They serve as titles for them, and by their position and typographical differences, they divide the text, signaling the subject and the ordering of the various units. Because of their visual prominence in print, they also break up the solid block of type and keep the content and structure of the paper before readers—and writers—and so help them to follow the text in the context of the overall structure. They also help readers to find *easily* parts that are of particular interest to them. Headings are particularly important when the relationship of the different parts of the paper diverge from the conventional arrangement and when the development of a paper or section is complex.

Headings stand outside the text, just as signposts stand beside the road. Like signposts, they are unidirectional and refer only to the text that follows; the text is a continuation of the text *before* the heading, not of the heading. Also, as far as the development of the text is concerned, the headings do not exist (*Ex. 11.54a,b*); therefore they cannot be used for integration. The sentence following the heading may, therefore, begin with the same words as the heading (*Ex. 11.54a*),

Ex. 11.54

> a. *Olfactometer.* An olfactometer was used to study stimuli that might give the parasite information on the host habitat.
> b. *Olfactometer.* The instrument used was an airflow olfactometer, which provided stimuli that. . . .
> c. *Olfactometer.* This was used to study stimuli that. . . .

or, to avoid such repetition, a referential term (*Ex. 11.54b*). Making a demonstrative pronoun refer to the heading (*Ex. 11.54c*) breaks the continuity of the text. In writing, headings should be used whenever they help to identify or segregate a topic. Because they are not integrated into the text, they can be removed, if the writer or editors wish, without requiring any revision of the text.

Rank and form.   Headings should consist of a word or phrase; they should not be sentences or questions. They should be as short as possible, even telescopic, and the initial article, for example, "*The* Model," "*A* Discourse Analysis" should be omitted (see also *Ex. 6.6*). Coordinate headings should be parallel in structure. In *Ex. 11.55a,* the subheadings begin with adverb, verbal, and article:

Ex. 11.55

    a. *Analysis*   b. *Analysis*
       How Validity Checked on Dyadic States   Validity Check on Dyadic States
       Calculating Percentage of Time in   Calculation of Percentage of Time
         Dyadic States   in Dyadic States
       The Transactions in Dyadic States   Transactions in Dyadic States

in *Ex. 11.55b,* all begin with a noun. This structural parallelism helps to identify corresponding headings and reinforces the coordinate position in the hierarchy.

    Headings should be hierarchical. The ranks of the hierarchy constitute a series or set, in which the elements are coordinate. Headings of the same hierarchical level should therefore have the same form—position, typeface, capitalization, and indentation, and the form should reflect their rank. A heading should not be more prominent than superordinate headings. Usually large type is dominant over small type; a displayed heading is dominant over a heading that is not displayed; boldface is dominant over italics, and italics over roman; a heading in capitals and small capitals is more prominent than one with initial capitals; and headings that have only the first word capitalized are the least prominent.

    Journals have a characteristic style for designating the different levels of headings. The typography and positioning of a heading show its rank and its relation to other headings. In preparing a paper for a particular journal, the author should examine the headings for the various elements of style. Each rank may be distinguished typographically by

Ex. 11.56

    a. *Position: Vertical* spacing relative to preceding and following text; *horizontal* spacing: flush with left margin, indented, or centered; *sideheading displayed* and without punctuation, or not displayed and followed by punctuation (colon, period, or dash) and text.

    b. *Type:* Roman, italic, boldface; different sizes of type.

    c. *Capitalization:* Full capitals, capitals and small capitals, important words capitalized, first word capitalized.

    d. Enumeration or alphabetical ordering.

For a journal that does not use as many ranks of headings as are required for the paper being submitted, the writer can establish a form for the additional ranks and follow it consistently.

11.6.2 Referential Structuring

Some parts of the paper are independent units extrinsic to the text, although they cannot be dissociated from it, for example, figures, tables, references, and appendices. These must be brought into the text by reference to them. Quotations (see Chapter 4) and equations (see Chapter 6), being verbal, are incorporated directly into the syntactic structure of the text.

Tables and figures, because of their graphic character cannot be made an integral part of the text, but they must be included in the development of the argument. Material elsewhere in the paper or in other papers that may be needed for the development of the argument is referenced to avoid repetition or digression. References in the text to the references listed in the reference section are made by number or by author and date (see Chapter 4), but sometimes are an integral part of the sentence, for example, "Green (1986) found. . . ." Appendices, placed at the end of the paper, must be referred to in the text. A definite reference is preferable to indefinite phrases:

Ex. 11.57

as described earlier	as stated above	as will be shown later
as already indicated	as noted	as will be demonstrated later

These should be replaced by more specific references, such as "as is shown in the discussion of current-flow theory." Such references are more easily understood if they refer back to something readers have already read rather than forward to something they have not read.

Directional words like "above" and "below" are best avoided. Not only are they indefinite, but the reference may not be literally "above" or "below." Similarly, "former" and "latter" and "respectively" require that the reader go back mentally and pair corresponding elements. "Respectively" should not be used, unless the pairs are few and clearly associated. When measurements or values are being paired with variables or with quantities, it is preferable to place the paired values in apposition or parentheses:

Ex. 11.58

The 8.3% of those surveyed who were currently members of study groups tended to be between the ages of 36 and 45 (47.4%), to have at least a junior-college education (63.1%), and to have family incomes of 4.5 million yen or more (73.7%).

rather than make the reader pair two sets of values by using "respectively."

Appendix 11A

*Words of Latin or Greek Origin with Latin, Greek, and English Plurals**

1. Latin

alveolus	addendum	minimum	alga	apparatus
alveoli	addenda	minima (math)	algae	apparatus
		minimums	algas	apparatuses
bacillus	agendum			
bacilli	agenda	momentum	amoeba	hiatus
		momenta	ameba	hiatus
bronchus	bacterium		amoebae	hiatuses
bronchi	bacteria	optimum	amoebas	
		optima (math)		species
cactus	cilium	optimums	antenna	species
cacti	cilia		antennae	
cactuses		ovum	antennas	
	colloquium	ova		
focus	colloquia		formula	
foci	colloquiums	phylum	formulae	genus
focuses		phyla	formulas	genera
	datum			
fungus	data	plasmodium	larva	opus
fungi		plasmodia	larvae	opera
	erratum		larvas	
gladiolus	errata	quantum		
gladioli		quanta	media	
gladioluses	flagellum		mediae	
	flagella	septum		caput
locus		septa	nebula	capita
loci	maximum	septums	nebulae	
	maxima (math)		nebulas	nomen
nucleus	maximums	symposium		nomina
nuclei		symposia	placenta	
nucleuses	medium	symposiums	placentae	lumen
	media		placentas	lumina (cavity)
radius	mediums	stratum		lumens (unit)
radii		strata	seta	
radiuses		stratums	setae	
stimulus		vacuum	septa	
stimuli		vacua	septae	
		vacuums		
streptococcus				
streptococci				

*The order of plurals does not reflect order of preference. The Latin or Greek plural is given first and the English plural follows.

			vertebra
thrombus			vertebrae
thrombi			vertebras

2. Greek

criterion	appendix	analysis	neurosis
criteria	appendices	analyses	neuroses
	appendixes		
ganglion		basis	parenthesis
ganglia	cortex	bases	parentheses
	cortices		
mitochondrion		hydrolysis	psychosis
mitochondria	index	hydrolyses	psychoses
	indices		
phenomenon	indexes	mitosis	synopsis
phenomena		mitoses	synopses
	matrix		
protozoon	matrices	thesis	synthesis
protozoa	matrixes	theses	syntheses
protozoan			
protozoans	vortex	axis	
	vortices	axes	
taxon	vortexes		stoma
taxa		crisis	stomata
	calyx	crises	stomas
automaton	calyces		
automata	calyxes	diagnosis	triponema
automatons		diagnoses	triponemata
	apex		triponemas
	apices	necrosis	
	apexes	necroses	lemma
			lemmata
			lemmas

Appendix A

Verbiage: Wordiness and Circumlocution

Scientific writing must be concise to be clear; readers should not have to interpret away words and phrases that do not contribute to meaning. The wordiness is often due to empty phrases, which *sound* meaningful, but are usually familiar fillers or cliches. Some form the opening of sentences, weakening them by their vacuousness. Some of these wordy phrases are listed together with concise alternatives that might be equivalents.*

Phrases

accounted for the fact that: *because*
a majority of: *most*
a number of: *several, some*
a very limited number of: *few*
an innumerable number of spines: *many spines*
are of the same opinion: *agree*
as it were: [omit]
as shown in Table 3: *see Table 3, (Table 3)*
as yet: [omit]
ascertain the location of: *find*
conducted inoculation experiments on: *inoculated*
created the possibility of: *made possible*
decreased number of: *fewer*
definitely showed: *showed*
despite the fact that: *although*
due to the fact that: *because*
exhibited good growth: *grew well*
give rise to: *result in, cause*
goes under the name of: *is called, is termed*
greater number of: *more*
has been the subject of study: *has been studied*
has the capability of: *can*
lacked the ability to: *could not*
large number(s) of: *many*

large proportion of: *much, most*
lesser extent, degree: *less, smaller*
made a count: *counted*
not as yet: *not yet*
owing to the fact that: *because*
referred to as: *called, termed*
pertaining to: *about*
prior to: *before*
relative to: *about*
serves the function of transferring: *transfers*
similar in every detail: *the same, identical*
small(er) number of: *few, fewer*
species in which spines are lacking: *spineless species, species without spines*
sufficient number of: *enough*
subsequent to: *after*
take into consideration: *consider*
tenacious in character: *tenacious*
the fact that: *that*
throw more light on: *make clearer*
the analysis in question: *this analysis, the analysis*
the fact that: *because*
the treatment having been performed: *after the treatment*
was of the opinion that: *believed, thought*

*In part derived from O'Connor and Woodford (1977).

Prepositional Phrases

along the lines of: *like, similar to*

at about: *about*

at some future time: *later*

at the present moment (time), at this time (moment), at this point in time: *now, at present, currently*

by means of: *by, with*

due to the fact that: *because*

during the course of: *during, while*

during the time that: *while*

for the purpose of studying: *for* studying, to study

for the reason that: *because*

from the standpoint of: *according to*

in a considerable number of cases: *often, frequently*

in a few cases: *infrequently*

in a position to: *can, may*

in a satisfactory manner: *satisfactorily*

in all cases: *always, invariably, (usually)*

in an adequate manner: *adequately*

in case: *if*

in case that: *if, when*

in this case: *here*

in close proximity to: *near, close to*

in connection with: *about, for*

in a very real sense: *actually*

in excess of: *more than*

in most cases: *mostly, usually*

in nature: [omit]

in no case: *never* [state what was not found]

in order to: *to*

in question: [omit, can use *studied*]

in regard to: *about*

in relation to: *about*

in respect to: *about*

in terms of: *for, with*

in the case of: *for, in*

in the context of: *about*

in the course of: *during*   .

in the event that: *if*

in the near future: *soon*

in the present communication: *here, in this paper*

in the vicinity of: *near*

in view of the fact that: *because*

of great importance: *important, very important*

of large size: *large*

of such strength that: *so strong that*

on account of: *because*

on the basis of: *because, from, by*

on the part of: *by, with*

on the grounds that: *because*

on the order of: *about*

on the other hand: *however, but*

through the use of: *by, with*

to bear in mind: [omit], *to recall*

with a view to: *to*

with regard to: *about, in, to, for*

with reference to: *about* [omit]

with the exception of: *except*

with the result that: *so that*

Introductory Phrases and Clauses

As a consequence of: *Because*

As already stated: [omit]

As a matter of fact: *In fact, Indeed*

As can be seen from Figure 5, the tissue: *Figure 5 shows that the tissue; The Tissue . . . (Figure 5); The tissue . . . (see Figure 5).*

As far as these observations are concerned, they: *These observations show*

As far as this particular species is concerned, it: *This species*

As for these experiments, they are: *These experiments are*

As of now: *Now*

Be that as it may: *Regardless, Nevertheless*

As regards these measurements, they: *These measurements*

As to whether: *Whether*

At the present writing: *Now, Currently*

Concerning these results, it must be borne in mind (remembered) that: *These results, Recall that these results*

For the purpose of this research, it was found necessary to make three replications: *Three replications were necessary*

From the standpoint of: [omit]

Considering all the evidence: *The evidence*

Despite the fact that: *Although*

If conditions are such that: *If*

In a considerable number of cases: *Often, Frequently*

In connection with this procedure: *This procedure*

In spite of the: *Despite the*

In the case that, In the event that: *If, When*

In this connection, it may be stated that: [omit and make statement]

In view of the fact that: *Because,*

It appears that: *apparently,* [make statement and qualify verb]

It has been reported by Green: *Green has reported*

It has been found convenient: *For convenience*

It has long been known that: [omit, include in statement if time known is important]

It is abundantly clear that, it is clear that: *Clearly*

It is apparent therefore that: *Therefore*

It is at this point that: *Here*

It is also of importance that: *Also*

It is believed that: [omit; incorporate as verb in statement of belief]

It is important to note that: [omit], *Note that*

It is noted that if: *If*

It is noted that the main effect: *The main effect*

It is noteworthy: [omit]

It is observed that the situation: [specify observation]

It is obvious that: *Clearly, Obviously*

It is of interest to note that; it is interesting to note that: [omit]

It is often the case that: *Often*

It is possible that the cause is: *The cause may be,* [incorporate cause in verb]

It is seen ultimately: *Eventually, Ultimately*

It is the intention of this writer: *The purpose is, The intent here is*

It is the purpose of this research to address: *This research addresses* [state purpose]

It is this that: *This*

It is thought by various investigators: *Various investigators think, Some think*

It is worth pointing out that: [omit, make statement with intensive], *Note that*

It may, however, be noted that: *However*

It may be said that: *Possibly, Perhaps*

It seems likely that: *Perhaps*

It seems to the present writer that: [omit and make statement; if personal opinion is important, use *I (we) think*]

It should be noted (mentioned) that: [omit and make statement]

It should be emphasized that: [omit, make statement in emphatic form]

It was demonstrated that: [omit, state what was demonstrated]

It was found that: [omit, state what was found]

It was observed that the tissue: *The tissue*

It will be seen upon examination of Figure 5 that: *Figure 5 shows, . . . (Figure 5)*

It would appear that: *Apparently*

It would thus (further) appear: *Therefore, Furthermore*

It is reported in the literature that: [omit, give report and cite source(s)]

Owing to the fact that: *Because*

The question as to whether (if): *Whether, If*

There can be little doubt that this: *This is probably, Undoubtedly this*

This is the value that: *This value*

With regard to the question of fusion: *Fusion*

With this in mind, it is clear that: *Clearly*

Appendix B

Some Common Abbreviations*

Recommendations for abbreviations of scientific terms in the broad disciplines are published by several organizations, for example, the American Institute of Physics (AIP), the American Society for Testing and Materials (ASTM), Bio Sciences Information services (BIOSIS), Chemical Abstracts Service (CAS) and in discipline-wide manuals like the *CBE Style Manual* and *The ACS Style Guide*. The exact form recommended for an abbreviation may differ slightly among these various sources. The following abbreviations do not follow one source; they represent an attempt to select a common and shared form.

a: atto (10^{-18})
a: absorbtivity
A: ampere, absorbance, anticlockwise
Å: angstrom
abs: absolute
ac: alternating current
Ac: acetyl
AcCh: acetocholine
ACTH: adrenocorticotropin
A.D.: anno Domini
Ade: adenine
Ado: adenosine
ADP: adenosine 5′-diphosphate
af: audio-frequency
Ah: ampere hour
Ala: alanyl, alanine
alt.: altitude
a.m.: ante meridiem
AM: amplitude modulation
AMP: adenosine 5′-monophosphate
anal.: analysis
anhyd: anhydrous
antilog: antilogarithm
approx.: approximately
aq: aqueous
Ara: arabinose
Arg: arginyl
AS: absorption spectroscopy
Asn: asparginyl

Asp: aspartyl
Atm: atmosphere, standard atmosphere
ATP: adenosine 5′-triphosphate
at. wt: atomic weight
au: atomic units
av: average

b: barn
bar: unit of pressure (not abbreviated)
bbl: barrel
bcc: body-centered-cubic
BeV: billion electron volts, GeV preferred
Bi: biot
b.i.d.: twice each day
biol: biological (-ly)
BMR: basal metabolic rate
BOD: biological oxygen demand
bp: boiling point
Bq: bequerel
br: broad
BSA: bovine serum albumin
Btu: British thermal unit
Bu: butyl
BWO: backward-wave oscillator
Bz: benzoyl

c: centi (10^{-2})
C: coulomb, Celsius
°C: degrees Celsius (centigrade)

*Other abbreviations, e.g., abbreviations of Latin terms, can be found in Appendix D.

ca.: circa
cal: calorie, gram calorie
calc (calcd): calculated
cAMP: adenosine cyclic 3', 5'-monophosphate
Cbz: carbobenzyloxy, benzyloxycarbonyl
cc: cubic centimeter (cm^3 preferred)
c.c.: complex conjugate
cd: candela
CD: circular dichroism
CDP: cytidine 5'-diphosphate
cf.: compare
cfm: cubic feet per minute
cgs: centimeter-gram-second (system)
CHE: cholinesterase
Ci: Curie
c.m.: center of mass
cm: centimeter
cm^3: cubic centimeter
CM: carboxymethyl
CMP: cytidine 5'-monophosphate
CNS: central nervous system
CoA: coenzyme A
COD: chemical oxygen demand
coef: coefficient
colog: cologarithm
compd: compound
concd: concentrated
concn: concentration
const: constant
cor: corrected
cos: cosine
cosh: hyperbolic cosine
cot: cotangent
coth: hyperbolic cotangent
cp: candlepower
CP: chemically pure
cpd: contact potential difference
cpm: counts per minute
cps: cycles per second (H_z preferred, c/s)
counts/s: counts per second
crit: critical
cRNA: complementary ribonucleic acid
CRT: cathode ray tube
csc: cosecant
csch: hyperbolic cosecant
CTP: cytidine 5'-triphosphate
cryst: crystalline
cu: cubic
CW: continuous wave
cwt: hundredweight
Cyd: cytidine
Cys: cysteinyl

Cyt: cytosine

d: day (or spell out), deci ($\times 10^{-1}$)
3-D: three-dimensional
da: deka ($\times 10^1$)
dAMP: deoxyadenosine monophosphate
dB: decibel
dc: direct current
deg: degree (plane angle)
dec: decomposition
det: determinant
dev: deviation
df: degrees of freedom
diam: diameter
dil: dilute
dis/s, dis/sec: disintegrations per second
distd: distilled
div: divergence
DNA: deoxyribonucleic acid
DNP: deoxynucleoprotein
Dns: dansyl
dopa: dihydroxyphenylalanine
dpm: disintegrations per minute
DPN: diphosphopyridine nucleotide
    (NAD)
dyn: dyne

e: exponential
ECG: electrocardiogram
ED: effective dose
ED_{50}: effective dose in 50% of test subjects
EDTA: ethylenediaminetetracetate
EEG: electroencephalogram
EM: electron microscopy
emf: electromotive force
eq., Eq.: equation
equiv: equivalent
erf: error function
e.s.d.: estimated standard deviation
ESR: electron spin resonance
Et: ethyl
e.u.: electron unit
eV: electron volt
exp: exponential
expt: experiment, (-al)

f: femto ($\times 10^{-15}$), function
F: farad, fermi
°F: degrees Fahrenheit
F_1: first filial generation
F_2: second filial generation
FAD: flavin adenine dinucleotide

fc: foot-candle

Fig.,: figure

fm: fermi

FM: frequency modulation

FMN: flavin mononucleotide

fp: freezing point

Fru: fructose

FSH: follicle-stimulating hormone

ft lb: foot-pound

g: gram

G: gauss, giga ($\times 10^9$)

gal: gallon

GC: gas chromatography

GDC: gas displacement chromatography

Gdn: guanidine

GDP: guanosine 5′-diphosphate

GFC: gas frontal chromatography

GeV: giga-electron-volt

Glc: glucose

GLC: gas-liquid chromatography

Gln: glutaminyl

Glu: glutamyl

Gly: glycyl

GMP: guanosine 5′-monophosphate

GPC: gel permeation chromatography

grad: gradient

GSC: gas solid chromatography

GTP: guanasine 5′-triphosphate

Gua: guanine

Guo: guanosine

Gy: gray

h: hour, hecto ($\times 10^2$)

H: henry

ha: hectare

Hb: hemoglobin

hcp: hexagonal-close-packed

hf: high frequency

His: histidyl

hp: horsepower

HPLC: high-pressure liquid chromatography

Hyl: hydroxylysyl

Hyp: hydroxyprolyl, hypoxanthine

Hz: hertz

I: electric current

IAA: indolacetic acid

i.d.: inside diameter

ICSH: interstitial cell-stimulating hormone

ID: infective dose

IDP: inosine 5′-diphosphate

IEC: ion-exchange chromatography

IF: intermediate frequency

Ig: immunoglobulin

Ile: isoleucyl

i.m.: intramuscular

IMP: inosine 5′-monophosphate

in.: inch

Ino: inosine

insol: insoluble

i.p.: intraperitoneal

IR: infrared

ITP: inosine 5′-triphosphate

IU: international unit

i.v.: intravenous

J: joule

k: kilo (10^3)

K: kelvin, equilibrium constant

kat: katal

kcal: kilocalorie

KE: kinetic energy

kg: kilogram

km: kilometer

Km: Michaelis constant

kn: knot (wind velocity)

kX: crystallography unit

l: liter

L: liter, lambert, ligand

lat: latitude

lb: pound

lc: lowercase

LC: liquid chromatography

LC_{50}: lethal concentration, 50%

LCAO: linear combination of atomic orbitals

LD_{50}: dose lethal to 50% of test subjects

Leu: leucyl

lim: limit

LLC: liquid-liquid chromatography

lm: lumen (unit of measure)

ln: natural logarithm (base e)

log: logarithm

long.: longitude

LSD: lysergic acid diethylamide

lx: lux

Lys: lysinyl

m: meter, milli (10^{-3})

M: molar, moles per liter (mol L^{-1}), mega (10^6)

Mal: maleyl, maltose

Man: mannose
MAO: monamine oxidase
max: maximum
mb: myoglobulin
Me: methyl
meq: millequivalent
Met: methionyl
mg: milligram
mi: mile
MICR: magnetic ink character recognition
min: minimum, minute (time)
mks: meter-kilogram-second system
ml: milliliter
MLD: minimum lethal dose
mm: millimeter
mm Hg: millimeters of mercury
mmp: mixture melting point
mo: month
MO: molecular orbital
mol: mole
mol wt: molecular weight
mp: melting point
mph: miles per hour
mRNA: messenger ribonucleic acid
MSH: melanocyte-stimulating hormone
MW: molecular weight

n: nano (10^{-9})
N: newton, nucleoside (unspecified), normal
    concentration
NAD: nicotinamide adenine dinucleotide
    (also DPN)
NADP: NAD phosphate (also TPN)
neg: negative
neut equiv: neutralization equivalent
Nle: norleucyl
NMN: nicotinamide mononucleotide
NMR: nuclear magnetic resonance
no.: number
NS: not significant
Nuc: nucleoside

obs: observed
Oc: octyl
o.d.: outer diameter
Oe: oersted
Orn: ornithyl
oxidn: oxidation
oz: ounce

p: pico- (10^{-12})
P: probability, pita- (10^{15})
Pa: pascal
PAGE: polyacrylamide gel electrophoresis

PBS: phosphate buffered saline
pc: parsec
PC: paper chromatography
PCB: polychlorabiphenyl
pd: potential difference
pe: probable error
PE: potential energy
PG: prostaglandin
pH: (negative logarithim of) hydrogen ion
    concentration
Ph: phenyl
Phe: phenylalanyl
pK: negative logarithm of equilibrium (dis-
    sociation) constant
p.m.: post meridiem
p.o.: per os
poly: polymer of
ppb: parts per billion
ppm: parts per million
ppt: precipitate
pos: positive
Pr: propyl
Pro: prolyl
psi: pounds per square inch
pt: pint
Pur: purine
pyr: pyridine
Pyr: pyrimidine

Q_{10}: change in rate of process with increase
    of 10°C
qt: quart

R: roentgen
R_f: retardation factor (chromatography)
rad: radian, radiation (see rd)
Rbu: ribulose
rd: radiation dose
ref.: reference
rem: roentgen equivalent man
REM: rapid eye movements
rf: radio frequency
Rib: ribose
rms: root-mean-square
RNA: ribonucleic acid
RPLC: reversed-phase liquid chromato-
    graphy
rpm: revolutions per minute
RQ: respiratory quotient
RT: room temperature

s: second (time)
S: siemens

Sar: sarcosyl
s.c.: subcutaneous
SD: standard deviation
SE: standard error
sec: secant
sect.: section
Ser: seryl
SI: Système Internationale des Unités
sin: sine
sinh: hyperbolic sine
soln: solution
sp gr: specific gravity
sp ht: specific heat
sp vol: specific volume
sq: square
sr: steradian
std: standard
STP: standard temperature and pressure
Suc: succinyl
Sv: sievert
sym: symmetrical

t: metric ton
T: tesla
tan: tangent
TCA: tricarboxylic acid cycle (citric acid
    cycle, Krebs cycle)
TEAE: triethylaminoethyl
temp: temperature
tert: tertiary
theor: theoretical
TFA: trifluoroacetyl
thr: threonyl
Thy: thymine
TLC: thin-layer chromatography
Tryp: tryptophyl, tryptophan
TMV: tobacco mosaic virus
TOD: total oxygen demand
TPN: triphosphopyridine nucleotide
    (NADP)
tr, TR: trace

Trt: trityl
TSH: thyroid-stimulating hormone
Tyr: tyrosyl

u: unified atomic mass unit
UDP: uridine 5'-diphosphate
uhf: ultrahigh frequency
UMP: uridine 5'-monophosphate
Ura: uracil
Urd: uridine
USP: United States Pharmacopeia
UTP: uridine 5'-triphosphate
UV: ultraviolet

V: volt
Val: valyl, valine
VB: valence bond
vol: volume
Vol.: volume (bibliography)
vol %: volume percent
v/v, vol/vol: volume ratio, volume per
    volume
vs: versus

w: watt
Wb: weber
wk: week
wt: weight
wt/vol: weight per volume
w/w, wt/wt: weight ratio, weight per weight

Xan: xanthine
Xao: xanthosine
XMP: xanthosine 5'-monophosphate
xu: x-ray unit
Xyl: xylose

yr: year

Z: benzydoxycarbonyl

APPENDIX C

Notes on Usage

This book follows *Webster's Third New International Dictionary* as the authority for spelling and meaning, except when scientific usage appears to be more specific or specialized or when the dictionary lacks an entry for a scientific term. When Webster's Third does not list scientific terms, manuals in the major scientific disciplines and scientific dictionaries have been consulted. The definitions are condensed, selective in meaning, and modified for the particular purposes of the entry. The objective has been to try to follow standard American usage. Usage may differ in disciplines having conventions and particular usages specific to the discipline and in other countries in which English is the spoken language or the language for scholarly studies.

able, -ible: see *Table D.17.*

above, below: avoid for cross-reference to material preceding or following reference; may not be in position specified on printed page; make specific reference or use *previously mentioned, just mentioned, discussed in the next section.*—In a scientific paper, such internal cross-reference rarely needed.

absorbance, absorbancy: ability of layer of a substance to absorb radiation, expressed mathematically as the negative common logarithm of transmittance. **absorptance:** the proportion of radiant energy absorbed before it reaches the further boundary of a layer of absorbing matter, being equal to 1 minus the transmittance.

absorbency: quality or state of being absorbent. **absorbent:** having capacity or tendency to absorb.

absorption: assimilation, incorporation, or imbibition of one substance into another. **adsorption:** taking up of substances on the surfaces of solids or liquids, by physical or chemical forces.—*Adsorption* is a surface phenomenon that has to do with the relation of particles at interfaces; *absorption* refers to the movement of particles into a substance.

accordance: agreement—used chiefly in the phrase "in accordance with." **accord (n):** agreement, as in opinion, will, action; balanced interrelationship. **accord (v):** to reconcile, harmonize, bring into agreement; allow, concede; allot—Note that *to accord with* is not synonymous with "to agree."

accurate: correct, free from error, conforming to some standard. **precise:** having distinct and often close limits, exactly delimited.—A measurement may be very *precise* but not be an accurate measure of the variable of interest. **accurate estimate:** an estimate is an approximation; use *close* or *reliable estimate,* or explain how estimate was obtained.

ad hoc: for a particular purpose without reference to wider application—use sparingly, be specific about purpose or limitations.

affect (v): to act upon, influence, e.g., jogging *affects* the heart rate. **affect (n):** the conscious subjective aspect of an emotion. **effect (n):** the result, outcome, or influence, e.g., the *effect* of jogging is to increase the heart rate. **effect (v):** to bring about, e.g., "passage was *effected* by dissecting out the calcified nodules in the duct.''—*Affect (n)* is the least commonly used because of its specialized meaning in psychology; *effect* (v) is not commonly used, synonyms or more specific verbs being used for it; *affect (v)* and *effect (n)* are therefore the most commonly used, frequently incorrectly.

agreement: see *subject-verb agreement.*

albumen: white of egg. **albumin:** any of large class of simple proteins.

aliquot: contained an exact number of times in something else, e.g., 10 ml is an aliquot of 20, 30, 40, . . . ,100 ml but not of 25, 35, . . ., 75 ml. **sample:** a representative portion of a whole. **subsample:** a portion of a sample without regard to size.—*Sample* and *subsample* are not synonymous with *aliquot;* see *portion.*

all: do not use to generalize; used accurately only to designate a definite, known number or quantity, e.g., ''*all* the samples were incubated at 28°C.''

all of, both of: all or both.

allow: see *enable.*

also see: used in reference to source, e.g., ''(Greene, 1985; DeBrion, 1986; DeLacor, 1987; also see DeBretor, 1985)''—use for a secondary or related source, not simply for additional source.

alternate (adj): in succession, by turns; e.g., ''the *alternate* rows were covered with plastic.'' **alternative:** offering a choice between two or more objects, courses of action, statements, situations, e.g., ''the *alternatives* were to interview respondents in their home, to interview them by telephone, or to have them respond by written questionnaires.''

ambience, ambiance: a surrounding or pervading atmosphere—*ambience* is preferred to *ambiance,* which verges on the pretentious in scientific usage.

amine: one of a class of basic compounds derived from ammonia by replacement of hydrogen by hydrocarbon radicals. **ammine:** molecule of ammonia in an amine regarded as a coordination complex.

among: related to several elements. **between:** related to two; not be used for singular events, e.g. ''between doses,'' *not* ''between each dose.'' **amongst:** not differentiated enough from *among* to be preferred in scientific writing.

analogue, also **analog (n):** something similar to something else; organ similar in function to organ of another animal. **analog (adj):** operating with numbers representing measurable quantities, as in *analog* computer.

analysis, analyses (pl).

-ance, -ence: see *Table D.18.*

and/or: means ''and *or* or''; use the more accurate or appropriate of the two.

and then: use *and* or *then.*

ante-: prior, earlier, preceding, anterior, e.g., *antemeridian.* **anti-:** opposing, opposite, against, e.g., *antibiotic.*

anthropomorphic: ascribing human characteristics to nonhuman things, e.g., ''current long-term persistence models *can have trouble* reproducing the Hurst effect''; see *Table 3.2.*

anticipate: to consider in advance, foretell, prevent—avoid using for *expect.*

a posteriori: see *a priori.*

apparent(ly): seeming, ostensible, misleading, plausible—avoid using for observable, obvious, clear, plainly evident.

appear(s): being or coming in sight. **seem(s):** giving the impression of being.—''They *appear* in the epidermal layer but do not *seem* to disrupt the cells.''

appendix, appendixes or **appendices (pl).**

approximately, comparatively, relatively: modifiers of indefinite degree; avoid using when more definite quantity available; see *Table 11.3*. **approximately:** avoid using as elevated or overprecise term for *about.*

a priori: has numerous definitions, therefore preferable to use English term that expresses exactly the meaning intended. **a posteriori:** reasoning deriving propositions from observation or from experience; more restricted in meaning than *a priori.*—Both are part of the special terminology in logic and should be used with restraint in scientific papers.

arbitrary: arising from unrestrained exercise of will, caprice, or personal preference; selected at random or as a typical example; based on random or convenient selection of choice rather than on reason. **random:** lacking regular plan, purpose, or pattern; having same probability of occurring as every other member of a set.—*Random* is not a synonym of *arbitrary* but the more specialized statistical term.

area: physical, spatial, geographical, geological surface or structure; overused for general, indefinite, intangible entities or concepts—find specific, definite, concrete terms for conceptual uses.

as: not a synonym for *because* or *inasmuch as;* colloquial when used for *that* or *whether.*

as already stated: usually not needed. **as a matter of fact:** wordy for *in fact* or *indeed;* may be mistakenly taken to imply that other statements are less factual. **due to the fact that, in view of the fact that:** because.—Wordy introductory phrases; see *Appendix A.*

as opposed to: avoid using for contrast; use *in contrast to, by contrast.*

assay: see *test.*

assure: see *ensure.*

as well (as):   use as intensive; avoid using *as well* for *and,* especially at beginning of sentence.

author(s):   a convention followed to avoid use of the first person in referring to the authors of a paper, usually in reference to an earlier paper, e.g., "in a previous paper by the *authors,* . . .''; can be avoided by citing paper, e.g., "in a previous paper (Green and DeBretor 1985) . . . .''

below: see *above.*

beside (prep.):   at or by the side of; along or by one side of, e.g., "the foal stood *beside* the mare.'' **besides (prep):** in addition to; other than, e.g., "*besides* the data on length and weight, data were collected on arm circumference.'' **besides (conj. adv.):** in addition, over and above furthermore, moreover; intensive, e.g., "*besides,* the data collected in the two studies were not comparable.''

better: do not use for *more,* as in "*better* [more] than half the sample.''

between: see *among.*

brackets: pair of marks, [ ], used to enclose an author's comments or to enclose parentheses in mathematical terms; also called *square brackets;* the pair of angular marks, ⟨ ⟩, are called *angle brackets, pointed brackets.* **braces:** the pair of marks, { }, used to enclosed mathematical terms that include brackets and parentheses.

Buchner: press. **Büchner:** funnel.

build up (v), buildup (n).

by means of: *by, with,* usually all that is needed; use only when *means* are difficult, important, or merit emphasis.

carried out: avoid for *performed, conducted.*

case: a special set of circumstances or conditions considered as an entity; a set of circumstances constituting a problem; appropriately used in "the history (or record) of the *case,*'' "*case* study''—should not be used indiscriminately for vague or indefinite reference; can often be replaced by single word or by more specific designation in a prepositional phrase, e.g., "in this *case* (*treatment, method, example, patient*)."

cause: a person, thing, condition, agent that brings forth an effect. **result:** consequence, effect, conclusion.—A *result* is observable or measurable, and one can refer a *result* to a process or event preceding the result; the *cause* may not be observable or measurable and may not be definitely attached to the *result*. Therefore *cause* should be used only when a result is known to be a direct consequence of a cause. **etiology:** all the factors that contribute to the occurrence of a disease—Note that the *cause,* i.e., the agent, of the disease, is included in the meaning of *etiology.*

cell line: culture of cells derived directly from primary culture. **cell strain:** culture of cells derived by selection or cloning for specific characteristics.—Often referred to as simply *line* or *strain.*

certain: fixed, settled, stated, determined—should not be used to indicate indefinite degree; e.g., a *"certain* amount" should not be used for an indefinite or unknown amount. **certainly:** see *surely.*

check: to stop, hinder, block, restrain—not be confused with *examine, verify, compare.*

chromatography: not chromotography, but in many other combinations the prefix has a linking *o,* e.g., *chromomere, chromoplast, chromosome.*

circa: too pretentious except for dates in bibliography; use *about.*

circle: do not use for *sphere.*

cite: to quote or refer to for evidence, authority—used for referring to earlier research. **site:** position, place, location; in scientific writing akin to *locus.*

cliché: a trite, stereotyped, hackneyed phrase; too imprecise or inaccurate for use in scientific writing, e.g., "in nature."

commence: avoid as elegant variant for *begin.*

comparatively: reserve for real comparison; do not use as modifier of indefinite degree, e.g., "*comparatively* small"; see *Table 11.3.*

compare to: to address similarities, e.g., "arms in humans can be *compared* to the forelegs in other mammals." **compare with:** to address both similarities and differences, e.g., "birds are often compared with reptiles in evolutionary studies."

comprise: to include, contain, hold, enclose, encompass, e.g., "the encyclopedia *comprises* 24 volumes." **compose:** to form by putting together two or more elements or parts, e.g., "opium is *composed* of several alkaloids." **constitute:** to make up, form, compose, e.g., "forty-eight chromosomes *constitute* the chromosome complement in humans."

concerning: avoid as elevated form for *about.*

congenital: see *genetic.*

connotation: see *denotation.*

considerable, considerably: used for indefinite high degree; **considerable amount:** much. **considerable number:** many. **considerable proportion:** high percentage.—Mostly too indefinite for scientific writing; replace with more definite terms or with quantities if possible; see *Table 11.3.*

constant: see *continual.*

constitute: see *comprise.*

contain: hold, have within—too specific for *have.*

continual: close or increasing occurrence or recurrence over long period of time, e.g., "the *continual* droughts exterminated some species and led to the outmigration of others." **continued:** stretching out in time or space without interruption, e.g., "the *continued* grazing eventually exposed the land to erosion." **continuing:** needing no renewing, lasting, enduring, e.g., "the *continuing* rains kept the farmers from plowing their fields." **continuous:**

having an uninterrupted continuity of objects, events, or parts, e.g., "the *continuous* flow of the water over the fall wore a pothole in the stone." **constant:** uniform, steady persistent recurrence, connoting lack of change or variation, e.g., "the *constant* beating of the heart." **repeated:** renewing or recurring again and again, e.g., "the *repeated* overflow of the river deposited a layer of fertile silt along its banks."

correlate: to relate as a necessary or invariable accompaniment with or without the implications of causality—because of statistical associations the term can no longer be used in its general loose sense in scientific writing; indicate the degree and direction of the correlation (negative or positive) and the level of significance; also indicate the correlation explicitly, e.g., "the number of insect species decreased with depth."

criterion, criteria (pl).

crosshatch (n):   pattern made up of one series of parallel lines crossing another series of parallel lines with the spaces between the lines being identical. **crosshatch (v):** to mark with crosshatch. **cross-hatching (n):** the process or effect of marking with crosshatch. **hatching:** pattern made by drawing fine lines in close proximity.

cross section (n), cross-section (v), cross-sectional (adj.).

dalton: a unit of mass; one-twelfth of the mass an atom of carbon-12. **molecular weight:** a ratio; between one molecule of a substance and one-twelfth of the mass of an atom of carbon-12.—The two terms are not synonymous; a writer may refer to the *molecular weight* of a molecule, which is simply a number, but to the molecular mass, which is in *daltons;* the *dalton* is not an accepted international unit of measure, though it is convenient for use in molecular biology for the mass of molecular structures.

data: plural of *datum*—a single observation, measurement or value constitutes a *datum;* several or all the observations, measurements, or values constitute the *data,* e.g., "the data *were* indicative of . . ."; also since a *datum* or the *data* refer to discrete observations, measurements, or values, there may be *too few data,* not *too little data.*

deduce: to derive by logical process, infer, to draw a necessary conclusion from given premises, to infer something about a particular case from a general principle for all cases. **induce:** to conclude or infer from particulars or by induction.—As a verb, *deduce* is the more commonly used; *deduction* and *induction* are more commonly used together with reference to the scientific method.

definite: having distinct or certain limits. **definitive:** most authoritative, final, and complete, with implication of final and perfected completeness or precision—used for review of literature, research, and scholarship.

delineate, depict: reserve for artists' or illustrators' activities. **exhibit:** reserve for displaying objects to public view. **display:** exhibit—also used for equations and headings that are not incorporated in the line of printed text.—For all these terms, when used in the general sense of make evident or apparent, *show* is simpler and plainer.

demonstrate: to make evident by reasoning, concrete facts or evidence, experimentation, operation, or repeated examples—avoid as elegant variant for *show.*

denotation: direct, specific meaning; name. **connotation:** the suggesting of meaning apart from the thing the word explicitly names or describes, e.g., "the *denotation* of pig is a farm animal; the *connotations* of pig may be a dirty person, a policeman, a greedy person, and so on.—Connotation is too inaccurate and imprecise for scientific writing.

determinant: serving to determine, e.g., "crowding is a *determinant condition* in the migration of the colony." **determinate:** having defined limits, definite, definitive, e.g., "the inflorescence is *determinate,* because flowers do not continue to develop apically."

determine: to settle, decide; to set bounds or limits; to obtain definite and firsthand knowledge of, as to character, location, magnitude, quantity, category. **establish:** to make firm or stable, to bring into existence, create, make, start, originate, set up, found; to confirm, validate.—One may *determine* (by analysis) the toxicity of an alkaloid (one cannot *determine* the alkaloids for toxicity); one can *establish* procedures for disposing of toxic substances after the nature of their toxicity has been *determined.*

differ from: to be unlike or distinct in one or more characters. **differ with:** to disagree with. **different from:** unlike or distinct in one or more characters—preferred to *different than.*

different, diverse, divergent, disparate: unlike in kind. **different:** partly or totally unlike in form or quality; not the same, distinct or separate—implies little more than separateness. **diverse:** differing from one another, having or capable of having various forms or qualities—implies marked difference and decided contrast. **divergent:** moving or extending in different directions from a common point; differing from each other or from a standard—implies movement away from sameness or similarity, usually implying impossibility of coming together again in close association. **disparate:** distinct in quality or character, unequal, dissimilar; comprising markedly dissimilar and unequal elements, not homogeneous—implies unequivocal difference as between incongruous and incompatible things.—The species in a genus are usually slightly *different;* the species in a large family or order are usually quite *diverse.* Their *disparate* character makes it difficult to recognize as members of the same family. As species evolve from primitive forms to more advanced forms, they become more *divergent.*—See *various.*

digit: see *number.*

dilation: act of dilating, stretching, or enlarging an organ. **dilatation:** the condition of being stretched beyond normal dimensions.—Though the distinction is useful, they are often used interchangeably because of their similarity.

disclose: carries connotation of hidden or secret; avoid using as elegant form for *show.*

discreet: not to be misused for *discrete;* rarely pertinent in scientific writing.

discrete: see *distinct.*

display: see *delineate.*

distinct: characterized by individualizing or distinguishing qualities, as apart from, unlike, or not identical with others. **distinctive:** individualizing, characteristic, special. **discrete:** possessed of definite entity; constituting a separate entity; consisting of distinct, unrelated parts, noncontinuous.—*Distinct* is likely to stress characteristics that distinguish or that indicate that the thing modified is apart from or different from others; *discrete* forcefully indicates lack of connection despite apparent similarities.

dosage: the regular administration of medicine in some definite amount. **dose:** the amount of medicine to be administered at a given time.

double negative: as in *not inconsistent with, not incorrect; not unimportant,* and so on; confusing and so to be avoided.

doubtless: without doubt, unquestionably, with certainty; in all probability, presumably—since *doubtless* can clearly be interpreted as free from doubt, it is best to retain it for that meaning in scientific writing; however since little is unquestionable in science, it is not often required; therefore for the second more commonly needed meaning, *in all probability, probably, presumably* is preferable.

drug: substance used for medicinal purposes; a narcotic substance or preparation. **narcotic:** a drug that allays sensibility, relieves pain, and produces profound sleep; in poisonous doses produces stupor, coma, or convulsions.—In scientific writing *drug* should be reserved for the general meaning; *narcotic* should be used when the more restricted meaning is intended.

effect: see *affect.*

effective: capable of bringing about an effect; influential or exerting positive influence; having power to produce effect. **efficient:** operating with a maximum of work or output accomplished with a minimum of effort.—"The instrument was cumbersome, slow, and not well designed, but it was *effective* in measuring the flow; the present commercial counterparts are more *efficient* and can record many more measurements per minute."

effluent: something that flows out. **effluence:** the act or process of flowing out; something that flows out.—For the shared meaning, *effluent* is preferred in scientific writing.

either . . . or; neither . . . nor: used in pairs, apply to no more than two elements; structure of the two elements should be parallel.

elevated: avoid using for *increased* in scientific writing; in general language used for body temperature.

elucidate: elegant variant used to avoid the cliché *throw light on;* use *explain, make clear, clarify.*

elute: to wash out, to remove from absorbent by means of solvent. **eluate:** the washings obtained by *eluting,* such as a solution containing a formerly absorbed substance. **eluent** or **eluant:** the solvent used in *eluting.*

eminent: standing out to be readily noted, conspicuous, evident, noteworthy. **imminent:** ready to take place, near at hand, impending.

employ: avoid as elegant variant for *use.*

enable: to render able, to give power, strength, or competency. **permit:** to give leave, to allow, to make possible, to give an opportunity. **allow:** to permit.—One *permits* (allows) a reaction to come to equilibrium; *enable* is used with persons, usually with making it possible for them to develop or achieve; it has a more positive permissive sense; *permit* and *allow* are used with things as well as persons and are more closely associated with facilitation, authorization, toleration, or granting privilege.

encounter: to meet as adversary, confront, come face to face; to come upon unexpectedly—in scientific writing may be used for reporting difficulties or unexpected results; inaccurate for simply reporting results.

endeavor: try to work with a set purpose; make an effort—avoid as an elegant variant for *try* or *attempt.*

enhance: advance, augment, elevate, heighten, increase—has become a fashionable term for both *increase* and *improve;* should be replaced by the more precise term; *enhance* is part of the special terminology in enzyme studies, in which it is used in the sense of *reinforce* or *heighten.*

ensure: to make sure by pledging, guaranteeing, convincing, or declaring; assure; to make sure, certain, safe. **insure:** to assure against loss by stipulated conditions, enter into contract of insurance on; underwrite. **assure:** to reassure, encourage, strengthen; to put beyond doubt; to make certain the attainment of.—*Insure* has become too restricted in meaning for scientific writing outside of economics; *assure* expresses the notion of removal of doubt, uncertainty, or worry from the person's mind; *ensure* is more commonly pertinent in scientific writing.

equally as good: *equally good* preferred.

establish: see *determine.*

estrous (adj.), estrus (n).

etc.:   reserve for use where subsequent elements are serial or obvious, e.g., at 12:15, 1:00, 1:45, 2:30, 3:15, etc.; use a full form when members of the series are not in a regular relation to one another.

etiology: see *cause.*

exhibit: see *delineate.*

exist: have actual or real being, or life—pretentious and often inaccurate for verb *to be.*

facilitate: tending to make easier or less difficult, or to lessen the labor of. **facultative:** involving permission rather than compulsion; having characteristics that permit alternate responses, e.g., "*facultative* parasites can live as parasites or as free-living saprophytes."

fact: not rigorous or accurate enough for the universe of discourse in scientific writing; use specific and precise term, e.g., *measurement, datum, value, observation, result, finding.*

false: not corresponding to truth or reality, not true, erroneous, incorrect—since science has little to do with truth, and reality is subject to varied interpretation, *false* is rarely appropriate in scientific writing; if used, it should be used circumspectly.

farther: to a greater distance in space; at a more remote place. **further:** going or extending beyond what exists, additional.—*Farther* tends to refer to physical space, time, or quantity; *further,* originally the comparative of *fore* or *forth,* refers to advance, addition, or progression in nonphysical as well as physical degree or quantity.

fast: to abstain from food—used as an intransitive verb; a person may *fast,* but one cannot *fast* animals, and animals cannot be *fasted;* they can only be starved, deprived of food, or put on a restricted diet.

feed: to give food to—one can *feed* animals or *feed* them on a diet, but one cannot *feed* a diet.

feel, felt: avoid for *believe* or *think.*

few(er): used with plural nouns indicating number, e.g., *fewer* chromosomes, *fewer* ions. **less(er):** used with singular nouns indicating quantity, e.g., *less* agar, *less* solvent. **more:** used to compare either size or number, e.g., *more* chromosomes, *more* solvent.

figures of speech: avoid in scientific writing.

firstly, secondly, thirdly: use *first, second, third,* and so on—when *first* is used, may be followed by *then, next,* and so on; when *second* and *third* are used, the preceding enumerator *first* must be used.

flammable: capable of being easily ignited and burning with extreme rapidity; preferred to *inflammable* for scientific use.

following (adj): next after, succeeding, ensuing—use as an adjective of reference, e.g., "the animals were assigned to the *following* groups"; avoid using as a noun, e.g., "the animals were assigned to the *following:* controls, high performers, low performers."

formal, ponderous, words: aforementioned, hereafter, herein, therein, thereof, whence, wherefor, wherefrom, wherein, whereof: substitute plain simple term; see *Ex. 3.3.*

former, latter: can refer only to two elements—avoid using because reader must turn back to match two elements. **latter:** the second of two elements, cannot be used for the last element of a series comprising more than two elements, when *last* must be used.

forward: toward the front, near or belonging to front—not to be misused for *foreword.* **forwards:** forward—used to indicate actual direction in movement.

foreword: front matter in a book likely to be of interest but not essential for understanding the book; commonly written by someone other than the author.

fungous (adj), fungus (n).

generally: mostly, chiefly, mainly, principally, largely. **usually:** customarily, commonly, habitually, regularly, prevalently.—*Generally* has to do with degree or frequency; *usually* has in addition the notion of regularity or a norm.

genetics: study that deals with heredity and variations of organisms and the mechanisms by which these are effected. **genetic:** of or relating to genetics. **hereditary:** genetically transmitted or capable of being transmitted from parent to offspring. **heredity:** the transmission of qualities from ancestor to descendent through chromosomes of germ cells in sexually reproducing organisms. **innate:** existing or belonging to an organism from birth. **congeni-**

tal: existing or dating from birth.—*Congenital* and *innate* need not be *hereditary* or *genetic* and are not part of the terminology of the science of *genetics*.

group: a relatively small number of individuals, objects, organisms sharing some similar character. **series:** a group of three or more objects or events, succeeding in order and having a like relationship to each other; a spatial or temporal succession.—*Series* has the narrower meaning and is included in *group*.

hereditary, heredity: see *genetic*.

higher, lower: do not use indiscriminately for *larger* and *smaller*, *greater* and *lesser*, *more* and *fewer* or *less*.

homogeneous: of uniform structure or composition throughout; of a single type, showing no variation. **homogenous:** relating to correspondence between parts or organs due to descent from the same ancestral type, homologous; homogeneous—*Homogeneous* is the more general and commonly used term; *homogenous* is the more specialized and infrequently used.

hypothesis: a proposition tentatively assumed in order to draw out its logical or empirical consequences and so test its accord with facts that are known or may be determined; the antecedent clause is a conditional, and so predictive, statement. **proposition:** a formal statement of mathematical truth to be demonstrated; theorem, a mathematical statement of an operation to be performed; a declarative sentence, a statement—note that neither the mathematical meanings nor the linguistic meanings encompass the predictive, tentative, or empirical meanings of *hypothesis*. **theory:** the coherent set of hypothetical, conceptual, and pragmatic principles forming the general frame of reference for a field of inquiry, as for deducing principles, formulating hypothesis for testing, undertaking action. **postulate:** a proposition advanced with the claim that it can be taken as axiomatic.—*Hypothesis* and *theory* are in the domain of empirical science; *proposition* and *postulate* are in the domain of logic and mathematics; *theory* should not be used for *hypothesis*, as in "her *theory* is that . . .".

hypothesize: make or adopt a *hypothesis*. **hypothecate:** to pledge without delivery of title or possession—sometimes humorously misused for *hypothesize*.

identical with: preferable to *identical to*—identicalness allows of no degree of difference in the comparison.

if it was: use *if it were* for statement contrary to fact.

imminent: see *eminent*.

impact (n): forceful contact or collision; concentrated force producing change. **impact (v):** to fix firmly as by packing or wedging; to press together; to transmit with force; to crowd. **influence (n):** act, process, or power of producing an effect; capacity of causing an effect. **influence (v):** to affect or alter by indirect or intangible means; to have an effect on condition or development.—*Impact* frequenctly misused; use *influence, affect,* or *effect,* unless a driving impetus or momentum, as in a collision, is intended.

imply: to convey or communicate by illusion or reference; to suggest, hint, e.g., "the paper seems to *imply* that the causative agent might be a mycoplasm rather than a virus." **infer:** to derive by reasoning or implication, to conclude from evidence or premises, e.g., "because of the precipitate, one could *infer* that a sugar was present." **implicate:** to involve deeply or unfavorably, involve as a consequence, corollary or natural inference, e.g., "the research *implicates* cats as carriers of the disease." **indicate:** to point at or point to with more or less exactness, to show probable presence or existence of, e.g., "the evidence seems to indicate parallel evolution in the two groups."—*Indicate* makes explicit reference whereas *imply* does not; *infer* gives the consequence of reasoning, and *implicate* makes *unfavorable* reference implicitly.

in: preposition commonly used in wordy phrases; see *Appendix A*.

inaugurate: see *initiate.*

in nature: cliché; usually a superfluous empty phrase; see *Appendix A.*

incidence: rate of occurrence per unit of time. **prevalence:** extent; percentage of population affected at a given time.

indeterminant: mistakenly used for indeterminate.

indexes, indices: both plurals used but Latin plural conventional for particular scientific usages, e.g., crystallography and mathematics; see *Appendix 11A.*

individual (n): use *person* for human being except when contrasting the single human being with social group or institution.

induce, induction: see *deduce.*

infection: invasion by bacteria, protozoa, viruses. **infestation:** invasion by metazoan parasites such as mites, ticks, insects. **inflammation:** swelling, often red and hot, resulting from irritation.

infer: see *imply.*

inflammable: see *flammable.*

inflammation: see *infection.*

influence: see *impact.*

inherent: structural or involved in the constitution or essential structure of something, so deeply infixed that it is apparently part of its nature or essence, e.g., "roots have an inherent tendency to respond to gravity." **intrinsic:** that which is a property of a thing itself, apart from external relations, conditions, or connections that affect its value, usefulness; originating or due to causes within a body, organ, or part, e.g., "it is an *intrinsic* asthmatic condition, not arising from external irritants."

initiate: facilitate first steps or stages. **inaugurate:** to start or begin formally or ceremoniously—Avoid using as elegant variants for *begin.*

inject: to introduce a substance into a vessel, cavity, muscle, and so on—do not use for the organism receiving the substance; i.e., one injects a drug into the peritoneal cavity or an antitoxin into a muscle; one does not inject rabbits, rats, muscles, cells, and so on.

innate: see *genetic.*

in order to: *to* usually adequate, unless means or objective being emphasized; see *Appendix A.*

in question: do not use unless there is a question; use the *issue,* the *problem,* the *difficulty,* as appropriate, or specify the question.

insanitary: preferred to *unsanitary* in some medical usage.

insanity: use *mental disease.*

insure: see *ensure.*

intensive: highly concentrated, relating to concentration of efforts or materials, e.g., "*intensive* cultivation of land." **intense:** existing in an extreme degree; showing characteristic trait in extreme degree; extremely marked or pronounced, e.g., "*intense* color."

interesting, interesting to note: omit; a judgment to be left to the reader—if it is interesting, the reader will probably find it so; if it is not, saying so will not make it so.

interface (n): a common boundary. **interface (v,t):** to make a garment with interfacing.—Avoid using as intransitive verb, as in "the oil layer *interfaces* with the water layer"; use instead "at the *interface* of the oil and water . . .".

intrinsic: see *inherent.*

invariably: *always*—do not use if not always.

invariant singulars and plurals: singular and **plural:** bison, deer, elk, moose, sheep, trout, fowl, fish. **plural:** bellows, measles, mumps, scissors. **singular:** smallpox; economics, physics, linguistics; beef, lamb, veal, pork.

irregardless, irrespective: see *regardless.*

it: pronoun commonly used in wordy introductory clauses; see *Appendix A.*

it should be emphasized, mentioned, noted, pointed out: omit and make the statement that is to be emphasized or noted. **it is a commonly accepted fact that:** wordy; avoid, not a rigorously scientific statement; state what is generally accepted.—Wordy introductory clauses; see *Appendix A.*

juvenile: young, immature in development. **juvenal:** used to refer to the young, often different, forms in higher-order animals.

kind: see *type.*

Latin words and phrases: see *Table D.4, D.13, D.21.*

latter: see *former.*

less: see *few.*

level: too imprecise for *concentration, percentage, proportion.*

localize: to fix in, assign or consign to a definite place or locality. **locate:** to determine or indicate place of; to set or establish a particular position; to seek out or discover the position. **locus:** place, locality, especially the place connected with a particular event or structure.

majority: means *more than half of;* usually applied to people; too specific for *most;* avoid hyperbole, *vast majority,* for large number.

man, mankind: sex-biased; use *humans, human beings, humankind.*

many: consisting of a large but definite number; e.g., *many* participants, variables. **much:** existing in large quantity or amount, e.g., "*much* sediment, *much* activity about the nest." **most:** greatest in quantity, extent, or degree—can refer both to number, e.g., "*most* mammals," or amounts, e.g., "*most* milk, *most* limestone."

medium, mediums or media: nutrient system for artificial cultivation of microorganisms or cells; something through which something is accomplished, conveyed, carried on, transmitted.—*Media* is often used loosely, but erroneously, in singular construction for *medium; mediums* is not commonly used.

microphotograph: see *photomicrograph.*

misspelled words: see *Tables D.15–D.19.*

mitigate: to soften, alleviate, mollify, lessen. **militate:** to have weight or effect against or in favor of something.—"The medication may *mitigate* the symptoms, but the continued smoking and stress *militate* against halting the progress of the disease."

molal, molar: both refer to molecular concentration; *molal, molality* by weight (1,000 g) of solvent and *molar, molarity,* by volume (1,000 l) of solution. **normal, normality:** concentration expressed in gram equivalents of solute per liter—its use as unit of measure discouraged in preference to *mole.*

mucous (adj), mucus (n):

multiple: use for several parts, not several or numerous elements; avoid as elevated form for *many.*

narcotic: see *drug.*

negative: use for minus; avoid using for undesirable, unexpected, unwanted or none; avoid "*negative* results"; state or describe specific results.

non-:   prefix, usually unhyphenated, for *not, none;* can be used to form temporary words, e.g., *nonriparian.*

normal: according to, constituting, or not deviating from an established norm, rule, or principle—avoid for *usually;* see *molal.* **standard:** something established by authority, custom, or general consent as a model to be followed. **usual:** such as accords with usage, custom, or habit, in ordinary practice; ordinary, common, prevalent.

number, numeral, digit, figure: a character by which an arithmetical value is designated. **number:** character or word, e.g., the *number 27* (twenty-seven); character with an affix, e.g., the ordinal numbers 2d, 3d, 4th. **numeral:** applies to characters as numbers as distinguished from words standing for numbers; stresses characters rather than numbers, e.g., roman *numerals,* "a postal code with letters and *numerals.*" **digit:** one of the characters in arabic notation, i.e., 1–9, though zero often included as a digit. **figure:** numeral, digit; value as expressed in numbers; stresses characters as characters, e.g., "in law, numbers written out are accepted over those expressed in *figures.*"—*Number* is the general term interchangeable with the others; *numeral* refers only to cardinal numbers; *digit* is the most restrictive; *figure* is the most general and descriptive, not dealing with numbers in an arithmetical sense; in scientific writing, should be reserved for illustrations in a paper.

once: avoid using for *when* or *after.*

only: should be placed next to word modified; see *Table 11.2.* **both:** as for *only.*

over, under: use for position in space, not for quantity; i.e., not for *more than* (over) or *less than* (under).

paradigm: example, archetype, or prototype—avoid using for model.

parameter: an arbitrary constant, characterizing by each of its values some member of a system, e.g., expression, curves, surfaces, functions; an independent variable through functions of which other functions may be expressed—avoid using term inaccurately for *measure, criterion, index, factor, value, variable, characteristic.*

part: see *portion.*

percent: literally *per hundred;* used with numbers, e.g., "ten *percent* of the trees in the lot were sampled." **percentage:** proportion expressed in hundredths.—"The *percentage* of oxygen saturation," *not* "the *percent* oxygen saturation"; "The *percentage* [not percent] of plants with chlorotic leaves was much higher than the 10 *percent* expected."

permit: see *enable.*

phenomenon, phenomena (pl):   a fact or event of scientific interest susceptible of scientific description and explanation; common usage retaining the implication of change of mode or being, especially illustrating the operation of some general law—overused; often used loosely for minor observations, changes, events and so tending to erode the consequence of the word for fundamental scientific events; use sparingly; substitute more specific and less grandiose term.

phosphorous (adj), phosphorus (n).

photomicrograph: photograph of magnified image of small object, e.g., of microscopic organisms. **microphotograph:** a small photograph usually magnified for reading or viewing.

play a role: cliché, indefinite; avoid, may imply more than it can or should in scientific writing.

plurals: see *invariant singulars and plurals.* **Latin and Greek plurals:** see *Appendix 11A.*

plus: do not use for *and* or *besides.*

pompous, pretentious writing: writing too elevated and formal for subject matter, purpose, or audience; drawing attention to itself or to writer; see *Ex. 3.3.*

portion: any definite part of limited amount of the whole, a share, e.g., "a *portion* of the sample was analyzed for heavy metals." **part:** something less than the whole, a constituent element, e.g., "*part* of the plot had only coniferous trees on it."—Part is the more general term; in scientific writing, the more specific *aliquot* (see) or *sample* is used for *portion*.

possess: own or have as attribute—too specific to use for *have*.

postulate: see *hypothesis*.

practical: actually engaged in course of action; relating to, or manifested in practice; given or disposed to action. **practicable:** capable of being put into practice, done, or accomplished; feasible, usable.—*Practicable* applies chiefly to something immaterial, e.g., plan, project, scheme, which has not been tested in practice or has not proved successful in operation. *Practical* applies to things, concrete and immaterial, and also to persons; the term stresses opposition to theoretical, speculative ideal, unrealistic, imaginative.

precise: see *accurate*.

preclude: prevent, make impossible—not exclude, or rule out.

predominate (v): to exert controlling power or influence, exercise superiority. **predominant (adj):** holding an ascendancy; having superior strength, influence, authority, position; controlling, dominating, prevailing.—"The *predominant* tree species at high elevations were conifers; at lower elevations hardwoods *predominated*.

preposition: see *Table D.14* for prepositions associated with particular words.

present: to introduce, to bring before or into the presence of, to offer—avoid using in writing for onsight diagnosis when a patient is first seen.

prevalence: see *incidence*.

previous to: use *before*.

principal (adj): most important, consequential or influential, e.g., "the *principal* reason for using the old procedure was its reliability." **principal (n):** chief officer of a school; capital sum placed at interest. **principle (n):** comprehensive and fundamental law, doctrine, or assumption, e.g., "the *principle* underlying the diffusion of substances through a semipermeable membrane."—Only the first and last meanings are pertinent in scientific writing; *principle* used only as a noun.

prior to: avoid as elegant variant for *before*.

problem: a question for inquiry, investigation, consideration, discussion, or decision—avoid using as unspecific, undefinite noun, e.g., "the lipid problem in bivalves" for "the relation of lipid turnover to spawning in bivalves."

proof (n), prove (v): not accurate for empirical research; not possible to *prove* or obtain *proof* in empirical studies; one may demonstrate or provide evidence to support a hypothesis, theory, explanation, or interpretation; one cannot provide *proof*.

proper nouns: avoid using as temporary modifiers, as in "two types of *Macroleon* larval pits"; instead emphasize head nouns, i.e., "larval pits of *Macroleon*."

proposition: see *hypothesis*.

quantity: a determinate or measured amount. **value:** a particular quantitative determination in mathematics.

quasi-: almost, seemingly, as if it were, approximately. **semi-:** precisely half of, half in quantity or value; to some extent, partly, incompletely, partial, incomplete.

quite: completely, totally, wholly, altogether, positively—not a synonym for *very*.

radical: a root part, basic principle, basis, foundation, support; a fundamental constituent of a chemical compound. **radicle:** the lower portion of the axis of a plant embryo or seedling; the hypocotyl and the root.

random: see *arbitrary*.

rather: too indefinite for use in scientific writing.

reality: the quality or state of being real; the actual nature or constitution of something; what has objective existence—avoid in scientific writing; although research scientists deal with objects and the actual, real world, they cannot make statements about *reality*.

reason why: replace *why* with *for,* or omit it, e.g., "the reason for the gradual increase is . . . ." or "the reason is that the sample was too heterogeneous."

redundancy: repetition of words, phrase, or same idea, e.g., **equally (as well); repeat (again); throughout the (entire) region**—makes for wordiness; see *Table 11.4.*

re-form: to form again; take form again. **reform:** restore, renew, amend, return to good state, often used for behavioral change—not pertinent in most scientific writing.

regardless of: without regard to. **irrespective of:** without respect or regard to, independent. **irregardless:** nonstandard for *regardless* or *irrespective*.

regime: a regular pattern of occurrence, e.g., seasonal rainfall; a method of management. **regimen:** a systematic plan (diet, therapeutic procedures, sanitary measures, medication) designed to improve and maintain health of patient.

relatively: adverb of indefinite degree; avoid using in scientific writing; see also *Table 11.3.*

reliability: The extent to which an experiment, test, or measuring procedure yields the same result on repeated trials. **validity:** the capability of measuring, predicting, or representing according to the prediction or design.

repeated: see *continual*.

respective, respectively: avoid using especially for long or complicated elements or many simple elements.

results: outcome of experiment or procedure—not an attribute of a person, e.g., *not* "Green's results indicate . . .".

reveal: has connotations of hidden, secretive; avoid as elegant form for *show*.

review: to examine again critically or deliberately—in scientific writing, to be distinguished from *study* or *research*.

sacrifice: euphemism; use *kill* or specify method of killing, e.g., *decapitate, etherize*.

said: relates to speech; use *stated, wrote, noted, indicated, suggested,* and so on.

sample, subsample: see *aliquot*.

see: avoid for *determine, discover, perceive, ascertain*.

seem: see *appear*.

semi: see *quasi-*.

sensitivity: the rate of displacement of the indicating element relative to the change in measured quantity. **specificity:** the state or quantity of being specific, peculiar to, or limited to a particular condition, structure, individual or group of organisms.—*Sensitivity* is related to precision within a test, to ability to measure precisely; *specificity* is related to accuracy relative to other tests, to the ability to distinguish, differentiate, or identify.

series: see *group*.

significant(ly): important(ly)—though correctly used for importance, should be avoided in scientific writing because of possible confusion with the common scientific use of the term for *statistical* significance; use synonyms, e.g., *important, suggestive, notable, marked, indicative*.

since: has time connotation, though used for *because* in scientific writing to avoid the commitment to cause; avoid for *because* when possible, especially at beginning of sentence, when time or cause meaning not immediately evident.

singular: see *invariant singulars and plurals.*

site: see *cite.*

sophisticated: use *advanced, highly developed,* or other more specific and accurate scientific term.

specificity: see *sensitivity.*

spelling: see *misspelled words.*

square in shape: redundant; square is enough; similarly for *red in color, large in size, fewer in number;* see *Table 11.4.*

standard: see *normal.*

stationary: standing still, immobile, fixed in a station or course. **stationery:** materials for writing or typing—not usually used in scientific writing.

subject-verb agreement: verb must agree with subject in number not with a nearby preceding noun (see Chapter 11).—In scientific names of plants and animals, species and lower taxa take a singular verb, e.g., "*Salmoni gairdneri was* used to study mitochondrial oxidation; genera and higher taxa take a plural verb; e.g., "the Orchidaceae *have* mycorhizae on their roots."

such that: do not use for *so that.*

supernatant (adj): floating on the surface. **(n) supernate:** substance floating on the surface.

surely: without doubt, certainly, undoubtedly. **certainly:** as in *surely.—Surely* used for subjective judgments and *certainly* for judgments that rest on indubitable evidence.—See *doubtless.*

temperature: avoid using for fever.

test: procedure used to identify or characterize a substance or constituent; diagnostic procedure. **assay:** a chemical test to determine the absence, presence, or more often quantity of one or more components of a material (e.g., ore, alloy, drug, antibiotic).—*Assay* has the narrower meaning; both are standardized known procedures to be distinguished from experimental methods or procedures and the "*testing*" of a hypothesis.

that, which: see *Ex. 11.20, Table 11.1* for use.

theme: not usually pertinent to scientific research; use *subject* or *topic;* where the reference is recurrent or regular events or relationship, express the regularity specifically.

theory: see *hypothesis.*

this, these: demonstrative adjectives, e.g., "*these* viruses, *this* structure," and pronouns, e.g., "*these* [cultures] were incubated for 48h"; "*this* [solution] was decanted slowly."—Pronoun should refer clearly to specific noun, noun phrase, or noun clause preceding it, *not* to an implicit idea or notion inferred from text preceding pronoun; see *Ex. 11.39.*

thusly: use *thus.*

total: use *whole, entire, all,* except when addition or sum is the point of interest; superfluous in phrases such as "a *total* of 10 samples"; where used takes a singular verb, e.g., "the total *is* slightly less than 100 because . . . ."

-trophic: relating to or functioning in nutrition. **-tropic:** turning, changing, or tending to turn or change in response to stimulus, e.g., "the sunflower is *heliotropic.*"

true (adj), truth (n): not rigorous enough for general use in scientific writing; use *actual* for *true;* avoid *truth*—having special uses, e.g., "seed *true* to type" in breeding programs; "*true* value" for values after calculations for error, etc.; see *false.*

type: a group having strong, clearly marked similarities throughout the group, so that distinctiveness of group cannot be ignored. **kind:** a group in which shared characteristics are likely to be indefinite and involve any criterion of classification whatever.—The *type* specimen is the basis of comparison for identifying a plant, but when one refers to the *kinds* of plants one may be referring to their habit, habitat, growth requirements, morphology, anatomical structure, and so on.

ultraviolet light: use ultraviolet radiation.

under: see *over.*

under circumstances: use *in* circumstances.

unique: the only *one* of its kind—avoid using for very unusual, uncommon, different.

unsanitary: see *insanitary.*

upon: avoid using as elegant variant for *on.*

usual: see *normal.*

vague, woolly, empty words and phrases: words and phrases too general or indefinite in meaning to be accurate or specific enough for scientific writing; may conceal lack of meaning; see *Appendix A.*

validity: see *reliability.*

verbiage: see wordiness.

very (adv):   to a high degree, extremely, exceedingly—too indefinite for scientific writing.

viable alternative: redundant; use *alternative.*

where: adverbial conjunction; indicates direction, location, position—avoid using for relative pronoun, e.g., *in which, to which.*

which: see *that.*

while: adverbial conjunction of time; during the time that, as long as, at the same time as— avoid using for *whereas* or *although.*

wordiness: use of excess words that do not carry their weight in meaning; see *Appendix A.*

work up (v), workup (n).

worth (prep):   having the value of, equal in value to. **worthy (adj):** having worth, value, or importance, estimable, marked by qualities warranting honor, respect, or esteem.—*Worth* is used in balancing value, e.g., "It is not worth the lower cost if it is not compatible with the other components; *worthy* is the affirmative judgment, e.g., "a worthy successor," "a worthy objective."

Appendix D

Tables of Symbols, Abbreviations, Units of Measure, and Usage

SYMBOLS

TABLE D.1 Letters of Greek alphabet and their names

Lowercase	Capital	Name	Lowercase	Capital	Name
α	A	alpha	ν	N	nu
β	B	beta	ξ	Ξ	xi
γ	Γ	gamma	o	O	omicron
δ, ∂	Δ	delta	π	Π	pi
ϵ	E	epsilon	ρ	P	rho
ζ	Z	zeta	σ, ς	Σ	sigma
η	H	eta	τ	T	tau
θ, ϑ	Θ	theta	υ	Υ	upsilon
ι	I	iota	ϕ, φ	Φ	phi
κ	K	kappa	χ	X	chi
λ	Λ	lambda	ψ	Ψ	psi
μ	M	mu	ω	Ω	omega

TABLE D.2 Common statistical symbols

Symbol	Definitions	Symbol	Definitions
Sample			
n, N	Total number of individuals or variates	t	Student's t test
$\overline{x}$	arithmetic mean	χ^2	chi-square test
s, SD	standard deviation	F	variance ratio
s^2	sample variance		
SE	standard error	*Population*	
p, P	probability (level of significance)		
r	correlation coefficient	μ	mean
R	regression coefficient	s	standard deviation
C.V.	coefficient of variation	s^2	sample variance
		β	regression coefficient

488

ABBREVIATIONS

TABLE D.3 Abbreviations of bibliographic terms

anon.	anonymous	par.	paragraph
app.	appendix	pl.	plate
ch., chap.	chapter	pt.	part
comp.	compiler	pub.	published by, publisher
ed.	edition, editor	repr.	reprinted, reprint
fig.	figure	rev.	revised, revision, review
intro., introd.	introduction	sect.	section
n.d.	no date	supp., suppl.	supplement
no.	number	vol.	volume
p., pp.	page, pages		

TABLE D.4 Abbreviations of some common Latin terms

ca.	*circa,* about, approximately	m.m.	*mutatis mutandis,* the necessary changes being made
cf.	*confer,* compare (not see)		
e.g.	*exempli gratia,* for example	N.B.	*nota bene,* note well, take careful note
et al.	*et alii,* and others		
etc.	*et caetera,* and so forth	op. cit.	*opere citato,* in the work cited
ibid.	*ibidem,* in the same place	Q.E.D.	*quod erat demonstrandum,* which was to be demonstrated
i.e.	*id est,* that is		
inf.	*infra,* below		
loc. cit.	*loco citato,* in the place cited	q.v.	*quod vide,* which see
		viz.	*videlicet,* namely
		vs., v.	*versus,* against

TABLE D.5 Abbreviations of the months and days

Jan.	Feb.	Mar.	Apr.	May	June	July	Aug.	Sept.	Oct.	Nov.	Dec.
Ja	F	Mr	Ap	My	Je	Jl	Ag	S	O	N	D
J	F	M	A	M	Je	Jl	A	S	O	N	D

Sun.	Mon.	Tue.	Wed.	Thurs.	Fri.	Sat.
Su	M	Tu	W	Th	F	Sa
S	M	T	W	Th,R	F	Sa

TABLE D.6 Common abbreviated forms from different sciences

Abbreviated form	Scientific term	Abbreviated form	Scientific term
Geology		*Physiology*	
C_z	Cenozoic period	BMR	basal metabolic rate
M_z	Mesozoic period	MMR	maximum metabolic rate
P_z	Paleozoic period	Q	radiant energy
		T_{dp}	dew point temperature
Genetics		T_b	mean body temperature
AA	genotype, dominant	T_a	ambient temperature
aa	genotype, recessive		
AA	phenotype dominant	*Immunology*	
aa	phenotype recessive	Ig	immunoglobulin
arabic numeral	autosome	IgG	immunoglobulin G
		IgA	immunoglobulin A
roman numeral	sex chromosome	Rh	antigen from rhesus monkey

UNITS OF MEASURE

TABLE D.7 List of SI units

Name	Abbreviation	Quantity	Name	Abbreviation	Quantity
		Base Units			
ampere	A	electric current	kilogram	kg	mass
candela	cd	luminous intensity	meter	m	length
kelvin	K	thermodynamic temperature	mole	mol	amount of substance
			second	s	time
		Supplementary Units			
radian	rad	plane angle	steradian	sr	solid angle

TABLE D.8 Some named SI-derived units

Name of unit	Abbreviation	In terms of other units	In terms of base units	Quantity
becquerel	Bq		s^{-1}	activity (of radionuclide)
gray	G	J/kg	$m^2\,s^{-2}$	absorbed dose
hertz	H		s^{-1}	inductance
newton	N		$m\cdot kg\cdot s^{-2}$	force
joule	J	N·M	$m^2\cdot kg\cdot s^{-2}$	energy, quantity of heat, work
lumen	lm	lm	$cd\cdot sr$	luminous flux
pascal	Pa	N/M²	$m^{-1}\cdot kg\cdot s^{-2}$	pressure, stress
sievert	Sv	J/kg	$m^2\cdot kg\cdot s^{-2}$	dose equivalent
watt	W	J/s	$m^2\cdot kg\cdot s^{-3}$	power, radiant flux

TABLE D.9 Some SI-derived units in terms of base units

Derived unit	Expressed in base units	Quantity	Derived unit	Expressed in base units	Quantity
square meter	m^2	area	meter per second	m/s	velocity, speed
cubic meter	m^3	volume	meter per second squared	m/s^2	acceleration
kilogram per cubic meter	kg/m^3	density	mole per cubic meter	mol/m^3	concentration (of amount of substance)

TABLE D.10 Existing units acceptable to use with SI units

Unit	Abbreviation	Value in SI units	Quantity
hour	h	60 min, 3600s	time
minute	min	60s	time
liter	l,L	dm^3, $10^{-3}m^3$	volume
curie	Ci	3.7×10^{10} Bq	activity
rad	rd	0.01 Gy, 1cGy	absorbed dose
roentgen	R	2.58×10^{-4} C/kg	exposure

TABLE D.11   Metric multiplying prefixes and their abbreviations*

Factor	Prefix	Abbrev.	Factor	Prefix	Abbrev.
10^1	deka	da	10^{-1}	deci	d
10^2	hecto	h	10^{-2}	centi	c
10^3	kilo	k	10^{-3}	milli	m
10^6	mega	M	10^{-6}	micro	μ
10^9	giga	G	10^{-9}	nano	n
10^{12}	tera	T	10^{-12}	pico	p
10^{15}	peta	P	10^{-15}	femto	f
$10^{18\dagger}$	exa	E	$10^{-18\dagger}$	atto	a

*Prefixes can combine with SI symbols, e.g., kilogram (kg), picofarad (pF), preferably in multiple of 10^3 or 10^{-3}.

†10^{18} = 1 000 000 000 000 000 000

10^{-18} = 0. 000 000 000 000 000 001

TABLE D.12   Some physical quantities with their mathematical symbols and abbreviation of unit of measure

Quantity	Symbol	Unit of measure	Quantity	Symbol	Unit of measure
mass	m	kg	absorptance	α	l
ampere	I	A	capacitance	C	F
energy	E	J	molar mass	M	kg mol^{-1}
volume	V	m^3	wavelength	λ	m

USAGE

TABLE D.13   Latin words and phrases used in scientific and scholarly writing with their English equivalents

ab initio: from the beginning

ab ovo: from the egg (beginning)

addenda: things to be added

ad hoc: to or with respect to this (object)

ad infinitum: to infinity

ad libitum: at will, pleasure; as much as one pleases

a fortiori: with stronger reason

a posteriori: from effect to cause

a priori: from cause to effect

ceteris paribus: other things being equal

de facto: in fact, actual or actually

de jure: from the law

de novo: anew

errata: errors, list of errors

et caetera (cetera): and the rest

ex cathedra: from the chair, with high authority

in re: in the matter of

in situ: in its original position

inter alia: among other things

in toto: the whole, entirely

in vacuo: in empty space, in a vacuum

in vitro: in glass, i.e., in a test tube or artificial environment

ipso facto: by that very fact

non sequitur: it does not follow

passim: everywhere, throughout, in all parts of the book, chapter

pro et contra: for and against

(continued)

TABLE D.13 (Continued)

post hoc ergo propter hoc: after this, therefore on account of this; subsequent to, therefore due to this—illogical reasoning and a particular pitfall in scientific writing

quod vide: which see

reductio ad absurdum: a reduction to the absurd; a method of proof for a proposition by demonstrating its contradictions

seriatim: in a series, one by one

sine qua non: without which nothing; something indispensable

status quo: the state in which; the existing condition

stet: let it stand, do not delete

terminus ad quem: term or limit to which

verbatim: word for word

vice versa: the terms of the case being interchanged or reversed, conversely

TABLE D.14 Words regularly associated with particular prepositions

accommodate *to* or *with*

accompanied *with* (things); *by* (persons)

adapted *to* (a use); *for* (a purpose); *from*

adjacent *to*

adjusted *to*

advantage *of, over*

affinity *between, with, to*

agree *with* (persons); *to* (suggestions); *in* (thinking); *upon* (a course of action)

alternate *with*

amalgam *of*

amalgamate *into, with*

analogous *to*

analogy *between, with*

antecedent *to*

anterior *to*

append *to*

appendage *of*

apportion *to, among, between*

approximate *to*

associate *with*

assure *of*

attempt *at*

augmented *by, with*

based *on, upon, in*

basis *of, for*

blend *with*

border *on, upon*

capable *of*

capacity *for* (ability); *of,* (volume)

careful, careless *with, of, about*

caused *by*

caution *against*

characteristic *of*

coalesce *with*

coincide *with*

compatible *with*

concur *with* (persons); *in* (a measure)

conducive *to*

consequent *to, on, upon*

consistent *with*

consonant *with*

contrast (n) *to, with*

correspond *to, with*

culminate *in*

decide *on, upon;* (legal) *for, against*

deficient *in*

depend *on, upon*

deprive *of*

derive *from*

deviate *from*

differentiate *from, between, among*

distinguish *between, from*

divide *between, among*

drench *with*

emerge *from*

end *with, in*

equal *to*

equivalent (adj) *to, with;* (n) *of*

essential (adj) *to;* (n) *of*

estimated *at*

exception *to, from, against*

exclude *from*

exclusive *of*

(continued)

TABLE D.14 (Continued)

expect *of, from*	migrate *to, into*
expel *from*	mix *with, into*
extract *from*	necessary *to, for*
exude *from*	necessity *of, for*
fail *in, at, of*	occupied *by, with*
favorable *for, to, toward*	opposite (adj) *to;* (n) *of*
feed *on, off*	originate *in, with*
founded *on, upon, in*	parallel *to, with*
full *of*	part *from, with*
identical *with, to*	participate *in*
identify *with*	permeate *into, through*
impenetrable *to, by*	permeated *by*
impervious *to*	pertinent *to*
implicit *in*	possibility *of*
improve *on, upon*	precedent (adj) *to;* (n) *for, of*
incidental *to, upon*	precluded *from*
incongruous *with*	preface (n) *of, to*
inconsistent *with*	preference *to, over, before, above*
incorporate *with, into*	preparatory *to*
independent *of*	prerequisite (adj) *to;* (n) *for, of*
infer *from*	productive *of*
inferior *to*	proficient *in, at*
infested *with*	profit *by, from*
infiltrate *into*	prohibit *from*
infiltration *of*	provide *with, for, against*
influence (n) *over, upon, with*	question (n) *on, about, as to, concerning, of*
infuse *with*	range *through, with, along, between*
innate *in*	reason *for*
inquire *for, about, after, into*	receptive *to, of*
inseparable *from*	resemblance *to, between, among*
insert *in, between*	similar *to*
insight *into*	solution *of, to*
intervene *in* (dispute); *between* (disputants)	subject *to, of*
introduce *to, into*	suitable *to, for, with*
invaded *by*	superior *to*
invest *with, in*	tendency *to, toward*
isolate *from*	theorize *about*
join *with, to*	unfavorable *for, to, toward*
justified *in*	useful *in, for, to*
lacking *in*	variance *with*
made *from, out of, of*	vary *from*

TABLE D.15 Commonly misspelled words

absence	controlled	fundamentally	occurrence	propagate
acceptable	controversial	gauge	opportunity	pursue
accessible	council	grammar	particular	receive
accommodate	definitely	guaranteed	performance	recommend
accuracy	definition	height	permanent	reexamine
achievement	dependent	incidentally	persistent	referring
acquire	description	ingredient	persuade	relative
allotted	discussion	interest	phenomenon	relieve
already	dominant	interference	possession	repetition
apparent	effect	interpretation	possible	separate
argument	eighth	interrupt	precede	separation
basically	environment	irrelevant	predominant	significance
believe	equipped	led	preferred	similar
beneficial	exaggerate	lose	presence	succeed
cannot	exercise	losing	prevalent	succession
category	existence	maintenance	primitive	susceptible
commission	explanation	noticeable	principal	transferred
comparative	field	occasion	principle	unnecessary
conceivable	forward	occurred	proceed	vacuum
consistent	foreword	occurring	prominent	yield

TABLE D.16 Scientific terms and other words often misspelled*

abscissa	azimuthal	cross-link	eyepiece	infrared
adiabatic	backscatter	crystallography	feedback	input
aerobic	bandwidth	cuvette	ferrous	least squares
aging	base line	data base	gassed	lead-in
air-dry	blackbody	desiccator	guideline	lemma
alkylize	capacitance	detectable	half-life	lifetime
analogous	chromatography	determinant	halfway	maintenance
anion	clear-cut	discrete	heat-treat	manageable
anomalous	coauthor	dissymmetry	heuristically	mean life
antioxidant	coaxial	divergence	histogram	measurable
aperture	collimate	echoes	horsepower	midpoint
aqua regia	collinear	eigenfunction	hybridization	noticeable
aquatic	concomitant	eigenvalue	hydrodynamics	ordinate
aqueous	controlled	eluate	hysteresis	outflow
asymmetry	cooperate	end point	inadvertent	overall
asymptote	coordinate	environment	inasmuch as	photocell
audio frequency	corollary	exercise	infinitesimal	photo-ionization
auxiliary	co-worker	explicit	inflection	piezoelectric

(continued)

TABLE D.16   (Continued)

procedure	reevaluate	solenoid	threshold	wavelength
radioactive	reflectance	space-time	transit time	wave packet
radiocarbon	reinforced	steam-distilled	transmittance	wave shape
radiocobalt	resistivity	step-up	un-ionized	wave surface
radio frequency	rock salt	stepwise	viscous	wave system
radioiodine	side band	stochastic	wave front	wave theory
recurrence	sinusoidal	test tube	wave function	X ray

*For some journals, the preferred spelling for some of these words may differ.

TABLE D.17   Words ending in *-able* and *-ible**

-able

acceptable	detectable	incurable	operable
advisable	dissolvable	inevitable	perceivable
alterable	distinguishable	inseparable	perishable
applicable	durable	justifiable	predictable
available	excusable	manageable	receivable
conceivable	filterable	measurable	solvable
considerable	hatchable	notable	suitable
controllable	imaginable	noticeable	unaccountable
desirable	improbable	objectionable	unmistakable

-ible

accessible	defensible	indestructible	reducible
audible	digestible	inexhaustible	reproducible
combustible	divisible	intelligible	responsible
compatible	edible	invisible	reversible
comprehensible	eligible	perceptible	suggestible
convertible	feasible	permissible	susceptible
corruptible	flexible	plausible	tangible
credible	impossible	possible	visible

variants

admissible *also* admissable   irresistible *or* irresistable
condensable *also* condensible   negligible *also* negligeable
discernible *or* discernable   preventable *also* preventible
indispensable *also* indispensible

*Corresponding adverbs omitted.

TABLE D.18   Words ending in *-ance (-ant)* and *-ence (-ent)**

-ance -ant

abundance	dominance	medicant	redundant
appearance	dormant	mordant	reflectance
attendance	incessant	operant	reluctance
capacitance	irrelevance	perseverance	resemblance
conductance	locant	predominance	resistance
determinant	luminance	protuberance	significance

-ence -ent

adolescence	convenience	inadvertence	negligence
apparent	emergence	incidence	permanence
competence	environment	independence	persistence
complement	excrescence	luminescence	precedence
confidence	existence	lutescent	prevalence
consequence	flavescence	medicament	respondent
consistent	implement	nascent	turbulence

variants

ascendance *also* ascendence	resistant *also* resistent
dependence *or* dependance	pendant (n) *also* pendent
descendant *also* descendent	pendent (adj) *also* pendant
insistence *also* insistance	

**Corresponding adjectives or nouns omitted.

TABLE D.19   Words with variant spellings*

ambience *or* ambiance	disk *or* disc
ampul (also ampule, ampoule)	distill (also distil)
analog (also analogue)	electron volt [electronvolt]
artifact *or* artefact	eluent *or* eluant
autoxidation (also auto-oxidation)	embedded *or* imbedded
base line [baseline]	enclose *or* inclose
burette *or* buret	enzymatic *or* enzymic
by-product [byproduct]	filterable *or* filtrable
catabolism *or* katabolism	fitted *or* fit
catalog *or* catalogue	flavin *or* flavine
deoxy *or* desoxy	fulfill *or* fulfil
dialogue (also dialog)	gases (also gasses)

(continued)

TABLE D.19 (Continued)

gauge *or* gage	sizable *or* sizeable
gray *or* grey	spatial *or* spacial
homologue *or* homolog	stoichiometric (also stoichiometrical)
leukocyte (also leucocyte)	sulfur *or* sulphur
[line width, linewidth]	supersede *or* supercede
liquefy (also liquify)	syrup *or* sirup
liter *or* litre	thiamine (also thiamin)
meter [metre]	tropine [tropin]
neuron (also neurone)	urethane *or* urethan
pharmacopoeia (also pharmacopeia)	wave form [waveform]
pipette (also pipet)	wave guide [waveguide]
plasmapheresis (also plasmaphoresis)	wave number [wavenumber]
quadrupole *or* quadripole	wholly (also wholely)
riboflavin (also riboflavine)	zeros (also zeroes)
side band [sideband]	

*This list and others follow *Webster's Third New International Dictionary*. In variant spellings, when spellings are separated by ''or,'' the two spellings are equal variants; when spellings are separated by ''also,'' the parenthetical forms are secondary variants. When words are in brackets, they are not listed in Webster's, but are included because they appear in style manuals for scientific disciplines and seem to be in frequent use. This is not a list of recommended or preferred forms. It is intended to alert authors to the variants so that they can verify the form preferred in their discipline or the journal of interest.

TABLE D.20 Plant taxa and rank, with Latin endings for ranks above genus and abbreviations for ranks below genus

Taxon	Latin ending	Taxon	Abbreviation
Division	-ophyta	Genus	
Subdivision	-icae	Subgenus	subg.
Class	-opsida	Section	sect.
Subclass	-idae	Subsection	subsect.
Order	-ales	Series	ser.
Suborder	-ineae	Subseries	subser.
Family	-aceae	Species	sp., spp. (pl)
Subfamily	-oideae	Subspecies	subsp. sspp. (pl)
Tribe	-eae	Variety	var.
Subtribe	-inae	Subvariety	subvar.
		Form	f.
		Subform	subf.

TABLE D.21 Latin terms used in designating taxon of new scientific name in Latin with their abbreviation and translation

Taxon	Abbreviation	Translation
auctorum	auct.	of authors
combinatio nova	comb. nov.	new combination
familia nova	fam. nov.	new family
forma	f.	form
genus novum	gen. nov.	new genus
morpha	m.	form
nomen conservandum	nom. cons.	conserved name
nomen dubium	nom. dub.	doubtful name
nomen novum	nom. nov.	new name
nomen nudum	nom. nud.	invalid name
sensu lato	s.l.	in the broad sense
sensu stricto	s.s.	in the narrow sense
species nova	sp. nov.	new species
varietas nova	var. nov.	new variety

TABLE D.22 Words generally capitalized

a. *Names of groups of human beings*

Creationist	Hispanic	Oriental	Senecas
Essenes	Latino, -a	Scandinavian	Zulu

b. *Titles preceding a person's name*

Professor (Prof.) Einstein   Doctor (Dr.) Schweitzer

c. *Names (or pseudonym) of persons used in noun or adjective form in compound terms*

Gram stain	St. Elmo's fire	Müllerian mimicry
(but gram positive)	tetrad of Fallot	quadratic Stark effect

d. *Official names of organizations and groups*

Cornell University   American National Standards Institute
Hudson Bay Company   Botanical Society of America

e. *Names of the months and days of the week; not seasons*

Monday	Friday	July	November	spring	winter

f. *Trade names*

Plexiglass	Tween 80	Kimwipes	Mel-Temp II
Neodene	Vitride	Tetronic	Chem-Dry

TABLE D.23   Hyphenation of terms with words having various functions

Noun-verb + noun
   water-drive reservoir
   doll-play procedure

Noun verb-preposition
   ionization arc-over
   table hook-up

Noun-and-noun + noun
   ball-and-socket joint
   vegetation-and-ecosystem mapping

Noun-preposition-noun + noun
   head-to-head condensation
   rate-of-change method

Verb-noun + noun
   pour-plate culture
   yield-pillar system

Verb-verb + noun
   transmit-receive tube
   read-write channel

Verb-adverb-verb
   track-while-scanning
   read-while-writing

Verb-preposition + noun(s)
   boil-off assistant
   built-in antenna
   building-out network
   set-forward point
   breach-in transportation hypothesis

Number-adjective + noun
   three-dimensional illustration

Number-noun + nouns(s)
   three-phase model
   two-cell compartment
   first-order Markhov process

Adjective-preposition-adjective + noun
   white-to-black amplitude range

Adverb-participle + noun
   sharp-crested weir
   well-known study
   open-ended question

Adverb-verb + noun
   back-haul rate

Adverb-adjective-noun + noun
   most-probable-number technique

Preposition-noun + noun(s)
   on-line operation
   on-line disk file

Preposition-article-noun + noun
   beyond-the-horizon communication
   on-the-job training

Preposition-preposition + noun
   before-after experiment

TABLE D.24 Punctuation of introductory phrases and clauses set off from rest of sentence

a. *Dependent clauses*

1. *Because the conditions of this experiment made across-sessions weight losses probable,* a licensed veterinarian periodically checked the subjects' health.
2. *When the sample was removed from the pressure vessel,* the fracture was found to be healed so that the sample pieces were loosely bonded together.

b. *Participial phrase*

1. *Using rats with jejunal Thiry-Vella loops,* we have shown that direct infusion of gut contents into the loop activates ODC 10-fold in loop mucosa without increasing enzyme activity.
2. *Taken together,* these observations suggest that frequent A* of T4 DNA contains a gene which codes for the 10-kDa RNA P-binding protein.

c. *Adverbial phrase*

1. *Prior to use,* the tissue was again cleaned and washed in sterile seawater.
2. *At each site,* identical instrumentation at two levels was maintained by staff meteorologists.
3. *Only in exceptional cases,* e.g. trichomes, is there even the simplest form of cell differentiation.

d. *Word or short phrase that is sharply set off from rest of sentence or that might cause confusion*

1. *In particular,* granular "shims" of silica may support the sample pieces at different contact points. *Second,* the silica distribution may be less uniform.
2. *Rather,* discrete random secretory bursts operated upon by exponential clearance mechanisms are sufficient.

e. *Connectives sharply set off from the clause that follows*

1. *Consequently,* knowledge is bound to be selective and abstract.
2. *For example,* the labor market is split into higher-paid male labor and lower-paid female labor.

TABLE D.25   Punctuation of internal parenthetical elements

a. *Adjectival or participial phrases and clauses*

1. Indeed, ΔT, *averaged for the 3 h preceding sunrise,* was 5.20°C at SHNP and 2.99°C at RNP.
2. The four panels, *from top to bottom,* present the high- and low-income conditions in the order in which they occurred. ITI duration, *which determines the income level in each condition,* is presented within each panel.

b. *Adverbial phrases and clauses*

1. Jackson concluded, *based on late 1950 records,* that only vigorous frontal cyclones would provide the necessary upward vertical motion.
2. Discrimination effects can occur, *although quantitative information is extremely rare,* for the exit slit of the ion source and mass spectrometer.

c. *Explanatory connectives: that is (i.e.), for example (e.g.), viz., when not followed by a clause*

1. In the last one, an energy-resolved one-photon counting technique similar to that used in light spectroscopy, *viz., time-resolved fluorescence,* was utilized.

d. *Word, phrase, or clause in apposition to a noun*

1. Young trees, *clones,* were chosen because selection during self-thinning had not begun to affect genotypic structure.
2. The reason for this difference, *use of tubular specimens instead of solid specimens,* is known with certainty as a result of more recent tests conducted on a high strength, low-alloy steel.

e. *Word, phrase, or clause out of its normal word order*

1. The interesting point about the laser is that, *because of the GVD,* wavelength-tunable pulses can be obtained with a Raman bandwidth by changing the cavity length.
2. A necessary first step, *therefore,* is to correct for the effects of the shallow-level fractionation.

f. *Complementary or antithetical phrase or clause*

1. Food, *but not water,* was removed 48 h before surgery.

Bibliography

REFERENCES

"Abbreviations and Symbols," *Eur. J. Biochem.* 74 (1977), 1–6.

Abbreviations for Use on Drawings and in Text. New York: The American Society of Mechanical Engineers, 1972 (reaffirmed 1984).

Ad Hoc Working Group for Critical Appraisal of the Medical Literature, "Academia and Clinic," Annals of Internal Medicine, 106 (1987), 598–604.

American Psychological Association, "Ethical Principles of Psychologists," *American Psychologist,* 36 (1981), 633–38.

ARNOLD, CHRISTIAN K., "The Construction of Statistical Tables," *IRE Transactions on Engineering Writing and Speech,* EWS-5 (1962), 9–14.

BARZUN, J. and HENRY G. GRAFF, *The Modern Researcher* (3rd ed.). New York: Harcourt Brace Jovanovich, 1977.

BECKER, A. L., "A Tagmemic Approach to Paragraph Analysis," *College Composition and Communication,* 16 (1965), 237–42.

BORKO, HAROLD and CHARLES L. BERNIER, *Abstracting Concepts and Methods.* New York: Academic Press, 1975.

BROSS, IRWIN D. J., "Prisoners of Jargon," *American Journal of Public Health,* 54 (1964), 918–27.

BROSS, IRWIN D. J., P. A. SHAPIRO, and B. B. ANDERSON, "How Information Is Carried in Scientific Sub-Languages," *Science,* 176 (1972), 1303–07.

BROAD, WILLIAM J., "Fraud and the Structure of Science," *Science,* 212 (1981), 137–41.

BURMAN, KENNETH D., "Hanging from the Masthead: Reflections on Authorship," *Annals of Internal Medicine,* 97 (1982), 602–05.

CHASE, J. M., "Normative Criteria for Scientific Publication," *The American Sociologist,* 5 (1970), 262–65.

CHRISTMAN, RUTH C., "Illustrations for Scientific Publications," *Science,* 119 (1954), 534–36.

CLARK, EVE V. and HERBERT H. CLARK, "When Nouns Surface as Verbs," *Language,* 55 (1979), 767–811.

CLEVELAND, WILLIAM S., "Graphs in Scientific Publications," *The American Statistician,* 38 (1984), 261–69.

————"Graphical Methods for Data Presentation: Full Scale Breaks, Dot Charts, and Multibased Logging," *The American Statistician,* 38 (1984), 270–80.

————*The Elements of Graphing Data,* Monterey, Cal.: Wadsworth, 1985.

CREMMINS, E. T., *The Art of Abstracting.* Phila., Pa.: ISI Press, 1982.

CULLITON, BARBARA J., "Coping with Fraud: The Darsee Case," *Science,* 220 (1983), 31–35.

CHASE, JANET M. "Normative Criteria for Scientific Publication," *The American Sociologist,* 5 (1970), 262–65.

DEBAKEY, LOIS, *The Scientific Journal. Editorial Policies and Practices.* St. Louis, Mo.: C. V. Mosby, 1976.

DIRCKX, JOHN H. "Hybrid Words in Medical Terminology," *Journal of the American Medical Association,* 238 (1977), 2043–45.

DOWNING, PAMELA, "On the Creation and Use of English Compound Nouns, *Language,* 53 (1977), 810–42.

EISENHART, CHURCHILL, "Expression of the Uncertainties of Final Results," *Science,* 160 (1968), 1201–04.

ELDER, JOSEPH D., "Jargon—Good and Bad," *Science,* 119 (1954), 536–38.

ENKE, C. G., "Scientific Writing: One Scientist's Perspective," *The English Journal,* 67 (1978), 40–43.

FLOOD, W. E., *Scientific Words. Their Structure and Meaning.* New York: Duell, Sloan and Pearce, 1960.

FOWLER, H. W., *Modern English Usage* (2nd ed.), rev. and ed. by Sir Ernest Gowers. New York: Oxford University Press, 1965.

GEORGE, JAMES E. and ADERS VINBERG, "The Display of Engineering and Scientific Data," *IEEE Transactions on Professional Communication,* PC-25 (1982), 95–97.

HATCHER, ANNA GRANVILLE, *Modern English Word Formation and Neo-Latin.* A Study of the Origins of English (French, Italian, German) Copulative Compounds. Baltimore, Md.: Johns Hopkins University Press, 1951.

HARRIS, ELIZABETH, "Applications of Kinneavy's *Theory of Discourse to Technical Writing,"* *College English,* 40 (1979), 625–32.

HARRIS, ZELLIG S., *Mathematical Structures of Language.* New York: Interscience, 1983.

HESS, EUGENE L., "Effects of the Review Process," *IEEE Transactions on Professional Communication,* PC-18 (1975), 196–99.

HOWARTH, RICHARD J. and MARK ST. J. TURNER, "Statistical Graphics in Geochemical Journals," *Mathematical Geology,* 19 (1987), 1–24.

HUTH, E. J., "Authorship from the Reader's Side," *Annals of Internal Medicine,* 97 (1982), 613–14 (editorial).

IEEE Standard Letter Symbols for Units of Measurement. New York: The Institute of Electrical and Electronics Engineers, 1978.

Illustrations for Publication and Projections. New York: American National Standards Institute, 1959.

INTERRANTE, C. G. and F. J. HEYMANN, eds., *Standardization of Technical Terminology: Principles and Practices,* ASTM Special Technical Publication 806. Phila., Pa.: ASTM, 1983.

ISO, *List of Serial Word Title Abbreviations.* Paris: CIEPS, ISDS, International Center, Geneva, 1985.

JORDAN, MICHAEL P., *Fundamentals of Technical Description,* Malabar, Fl.: Robert E. Krieger, 1984.

JUHASZ, STEPHEN, EARL CALVERT, TOM JACKSON, DAVID KRONICK, and JOSEPH SHIPMAN, "Acceptance and Rejection of Manuscripts: *IEEE Transactions on Professional Communications,* PC-18 (1975), 177–85.

KEETON, W. T., *Biological Science* (2nd ed.), p. 236, New York: Norton, 1972.

KINNEAVY, JAMES L., *A Theory of Discourse: The Aims of Discourse.* Englewood Cliffs, N. J.: Prentice-Hall, 1971.

KLARE, GEORGE R., "A Second Look at Readability Formulas," *Journal of Reading Behavior,* 8 (1976), 129–52.

LARKIN, JILL H. and HERBERT A. SIMON, "Why a Diagram is (Sometimes) Worth Ten Thousand Words," *Cognitive Science,* 11 (1987), 65–99.

LEES, R. B., *The Grammar of English Nominalizations.* Bloomington, Ind.: Research Center in Anthropology, Folklore, and Linguistics, Publication 12, 1960.

MCALLISTER, D. T., "Is There Accepted Scientific Jargon?" *Science,* 121 (1955), 530–32.

MACGREGOR, A. JEAN, "Selecting the Appropriate Chart," *IEEE Transactions on Professional Communication,* PC-25 (1982), 102–04.

MICHAELSON, HERBERT B., "How to Write and Publish a Dissertation," *Journal of Technical Writing and Communication,* 17 (1987) 265–74.

———"The Incremental Method of Writing Engineering Papers," *Transaction Professional Communications,* PC-17 (1974), 21–22.

O'CONNOR, MAEVE, *The Scientist as Editor.* Guidelines for Editors of Books and Journals. New York: John Wiley, 1979.

PETERSON, MARTIN S., *Scientific Thinking and Scientific Writing.* New York: Reinhold, 1961.

POPKEN, RANDALL L., "A Study of Topic Sentence Use in Scientific Writing," *Journal of Technical Writing and Communication,* 18 (1988), 75–86.

RICKETT, H. W., "The English Names of Plants," *Torrey Botanical Club,* 92, (1965), 137–39.

RIDGWAY, R., *The Birds of North and Middle America, Pt. I.* Washington, D.C.: GPO, 1901.

SELZER, JACK, "Another Look at Paragraphs in Technical Writing," *Journal of Technical Writing and Communication,* 10 (1980), 283–301.

———"Certain Cohesion Elements and the Readability of Technical Paragraphs," *Journal of Technical Writing and Communication,* 12 (1982), 285–300.

SIMON, JULIAN, "A Plan to Improve the Attribution of Scholarly Articles," *The American Sociologist,* 5 (1970), 265–67.

SMITH, MARY LEE, "Publishing Qualitative Research," *American Education Research Journal,* 24 (1987), 173–83.

STEVENSON, DWIGHT W., "Toward a Rhetoric of Scientific and Technical Discourse," *The Technical Writing Teacher,* 5 (1977), 4–10.

SWANSON, ELLEN, *Mathematics into Type.* Providence, R.I.: American Mathematical Society, 1971.

SZOKA, KATHRYN, "A Guide to Choosing the Right Chart Type," *IEEE Transactions on Professional Communication,* PC-25 (1982), 98–101.

TUFTE, E. R., *The Visual Display of Quantitative Information,* Cheshire, Conn.: Graphics Press, 1983.

WAINER, HOWARD, "How to Display Data Badly," *The American Statistician,* 38 (1984), 137–47.

WALSTER, WILLIAM G. and T. ANNE CLEARY, "A Proposal for a New Editorial Policy in the Social Sciences," *The American Statistician,* 24 (1970), 16–19.

WEBER, MAX, *Economy and Society,* ed. Guenther Roth and Claus Wittich, vol. 1, p. 362. Berkeley, Cal.: University of California Press, 1978.

WEIL, B. H., ed., *Technical Editing.* New York: Reinhold, 1958.

WHITTAKER, R. H., S. A. LEVIN, and R. B. ROOT, "Niche, Habitat, and Ecotope," *The American Naturalist,* 107 (1973), 321–38.

DICTIONARIES

ATKINSON, BERNARD and FERDA MAVITUNA, *Biochemical Engineering and Biotechnology Handbook*. New York: Nature Press, 1983.

CIRCLOT, JUAN E., trans. by J. Sage, *A Dictionary of Symbols*. New York: Philosophical Library, 1962.

COLLOCOTT, T. C. and A. B. DOBSON, eds. Rev. ed., *Chambers Dictionary of Science and Technology*. New York: Barnes and Noble, 1984.

DURRENBERGER, R. W., ed., *Dictionary of the Environmental Sciences*. Palo Alto, Cal.: National Press Books, 1973.

ETTER, LEWIS E., *Glossary of Words and Phrases Used in Radiology and Nuclear Medicine*. Springfield, Ill.: Charles C. Thomas, 1960.

FAIRCHILD, H. P., ed., *Dictionary of Sociology*. Westport, Conn.: Greenwood Press, 1970.

GOULD, JULIUS and WILLIAM L. KOLB, eds., *A Dictionary of the Social Sciences*. New York: The Free Press of Glencoe, 1964.

GREENWALD, D., ed., *The McGraw-Hill Dictionary of Modern Economics*. New York: McGraw-Hill, 1983.

HUSCHKE, RALPH E., ed., *Glossary of Meteorology*. Boston, Mass.: American Meteorological Society, 1959.

ISAACS, A. and others, eds., *Concise Science Dictionary*. New York: Oxford University Press, 1984.

JABLONSKI, STANLEY, *Illustrated Dictionary of Eponymic Syndromes and Diseases, and Their Synonyms*. Phila., Pa.: W. B. Saunders, 1969.

JAY, F., ed.-in-chief, *IEEE Standard Dictionary of Electrical and Electronics Terms* (3rd ed.). New York: Institute of Electrical and Electronics Engineers, 1984.

KENDALL, M. G. and W. R. BUCKLAND, *A Dictionary of Statistical Terms*. London: Longman, 1982.

KNIGHT, ROBERT L., *Dictionary of Genetics: Including Terms Used in Cytology, Animal Breeding and Evolution* (1st ed.). Waltham, Mass.: Chronica Botanica, 1948.

LARKIN, R. P. and G. L. PETERS, *Dictionary of Concepts in Human Geography*. Westport, Conn.: Greenwood Press, 1983.

MALISOFF, WILLIAM M., ed.-in-chief, *Dictionary of Biochemistry and Related Subjects*. New York: Philosophical Library, 1943.

MILLER, P. McC. and M. J. WILSON, *A Dictionary of Social Science Methods*. Chichester, Eng.: John Wiley, 1983.

MITCHELL, G. D., ed., *A New Dictionary of the Social Sciences*. New York: Aldine, 1979.

PARKER, SYBIL B., ed.-in-chief, *McGraw-Hill Dictionary of Scientific and Technical Terms* (3rd ed.). New York: McGraw-Hill, 1984.

PAXTON, JOHN, *Dictionary of Abbreviations*. Totowa, N.J.: Rowman & Littlefield, 1974.

READING, HUGO F., *A Dictionary of the Social Sciences*. Boston, Mass.: Routledge & Kegan Paul: 1977.

ROE, K. E. and R. G. FREDERICK, *Dictionary of Theoretical Concepts in Biology*. Metuchen, N.J.: Scarecrow Press, 1981.

ROSENBERG, J. M., *Dictionary of Computers, Data Processing, and Telecommunications*. New York: John Wiley, 1984.

STENESH, J., *Dictionary of Biochemistry*. New York: John Wiley, 1975.

THEODORSON, G. A. and A. G. THEODORSON, *A Modern Dictionary of Sociology*. New York: Crowell, 1969.

TOWELL, JULIE E. and HELEN E. SHEPPARD, eds., *Acronyms, Initialisms & Abbreviations Dictionary. 1986–87*. Detroit, Mich.: Gale Research, 1985.

WEAST, ROBERT C., ed., *Handbook of Chemistry and Physics.* Cleveland, Oh.: CRC Press, 1976.

Webster's Third New International Dictionary. Springfield, Mass.: G. & C. Merriam, 1976.

WINDHOLZ, MARTHA, et al., eds., *Merck Index.* Rahway, N.J.: Merck, 1983.

ZADROZNY, J. T., *Dictionary of Social Science.* Washington, D.C.: Public Affairs Press, 1959.

WRITING MANUALS

ANDERSON, C. et al., *American Medical Association Manual of Style* (8th ed.). Baltimore, Md. Williams & Wilkins, 1989.

BECKER, HOWARD S., *Writing for Social Scientists,* with a chapter by Pamela Richards. Chicago: University of Chicago Press, 1986.

BOOTH, VERNON, *Writing a Scientific Paper* (4th ed.). London: The Biochemical Society, 1978.

CBE Style Manual Committee, *CBE Style Manual* (5th ed.). Bethesda, Md.: Council of Biology Editors, 1983.

The Chicago Manual of Style (13th ed.). Chicago, University of Chicago Press, 1982.

DAY, ROBERT A., *How to Write and Publish a Scientific Paper* (3rd ed.). Phila., Pa.: Oryx Press, 1979.

DIRCKX, JOHN H., *Dx + Rx. A Physician's Guide to Medical Writing.* Boston, Mass.: G. K. Hall, 1977.

DODD, JANET S., ed. *The ACS Style Guide:* A Manual for Authors and Editors. Washington, D.C.: American Chemical Society, 1986.

EBEL, H. F., C. BLIEFERT, and W. E. RUSSEY, *The Art of Scientific Writing:* From Student Reports to Professional Publications in Chemistry and Related Fields. New York: VCH, 1987.

FISHBEIN, M., *Medical Writing:* The Technic and the Art (4th ed.). Springfield, Ill.: C. R. Thomas, 1978.

HATHWELL, DAVID and A. W. KENNETH METZNER, eds., *Style Manual* (3rd ed.). New York: American Institute of Physics, 1978.

HUTH, EDWARD J., *How to Write and Publish Papers in the Medical Sciences.* Phila., Pa.: ISI Press, 1982.

MCMILLAN, VICTORIA E. *Writing Papers in the Biological Sciences.* New York: St. Martin's Press, 1988.

MENZEL, DONALD H., HOWARD MUMFORD JONES, and LYLE G. BOYD, *Writing a Technical Paper.* New York: McGraw-Hill, 1961.

MICHAELSON, HERBERT B., *How to Write and Publish Engineering Papers and Reports.* Phila., Pa.: ISI Press, 1982.

MONROE, JUDSON, CAROLE MEREDITH, and KATHLEEN FISHER, *The Science of Scientific Writing.* Dubuque, Ia.: Kendall/Hunt, 1977.

O'CONNOR, MAEVE and F. PETER WOODFORD, *Writing Scientific Papers in English.* New York: Elsevier North-Holland, 1977.

Publication Manual of the American Psychological Association (3rd ed.). Washington, D.C.: American Psychological Association, 1985.

TRELEASE, SAM F., *How to Write Scientific and Technical Papers.* Cambridge, Mass.: MIT Press, 1969.

WOODFORD, F. PETER, ed., *Scientific Writing for Graduate Students:* A Manual on the Teaching of Scientific Writing. Bethesda, Md.: Council of Biology Editors, 1988.

Index

E

F

G

Graphics (*cont.*)
  uses of, 161–62
  visual effectiveness of, 162–63
Graphic symbols, in equations, 295, 296, 310–11
Graphs
  advantages of, 245
  bar type, 245–52
  circular, 252
  labeling in, 245
  line type, 252–94
  use and preparation of, 241–45
Greek, scientific terms from, 46
Greek alphabet letters
  clarification, in manuscript, 402
  in manuscript, identification of, 418
  table of, 488
  use in equations, 295, 307
Greek words, plurals of, 452, 462–63

H

Halftone reproductions, of photographs, 230–33
  combined with line engravings, 231
  description of process used in, 230
Hatching, in bar graphs, 248
Headings, 83
  for abstract section, 351
  copy editing and ranking of, 416–17
  in discussion section, 330, 331, 333–34
  function and content of, 459–60
  in manuscript, review of, 397
  in materials and methods section, 136, 137, 142, 143, 156, 322
  rank and form of, 460–61
  in results section, 316
  in social science papers, 156
  in tables, 173–80
    repetition avoidance in, 198, 199
    spacing in, 200–206
    table titles based on, 188–90
Headnotes, in tables, 180–83
Hierarchical relations, subordination role in, 441
Hook-and-eye connections, role in paragraph structure, 67, 439, 440
Human beings, permission for studies on, 379, 429
Humor
  avoidance in scientific writing, 64
  in titles, avoidance, 372
Hyphenation
  in compound terms, 38–42
  of terms (table), 58, 59, 500
Hypotheses
  discussion in discussion section, 328, 329
  discussion in results section, 321
  experimentation and, 5–6

in introduction section, 103, 104
role in scientific research, 4–6
in scientific papers, 97
in social science papers, 124–25
in table footnotes, 209, 210
vs. propositions, 124–25
"working" and "predictive" compared, 7–8

I

Illustrations, 216–94. *See also* Graphics.
  in abstracts, 356, 358
  criteria for, 218–24
  design and execution of, 218–26
  labeling of, 220–21
  mailing of, 402
  planning and preparation of, 217–18
  preparation for printing and publication, 400, 419
  professional quality necessary for, 217, 218, 220, 221
  proofs of, proofreading of, 425–26
  purpose and advantages of, 216–17
  reduction of, 226
  size of, 224–26
  space economy in, 223–24
  titles and legends for, 226–30
  writer's responsibility for quality of, 217
Indefinite pronouns, in weak constructions, 450–51
Independent-related studies, reporting of, 87
Independent studies, reporting of, 85–86
Indexing
  abstracts prepared for, 358
  key words for, 351, 358, 364
Indicative titles, 371, 372
Inductive method, use in scientific research and writing, 8, 69, 438–39
Inflected forms, agreement of, 451–53
Informative abstracts, 348
  description and examples of, 350–52, 383–88
Informative titles, 371–72
Informed consent, by persons participating in studies, 403, 429
Initialisms
  in appendices, 382
  definition and formation of, 43–44, 45
Instruments, description in social science papers, 157
Interdependent studies, reporting of, 86–87
Interpretation
  avoidance, in results section, 324–25
  in discussion section, 337, 338
Introduction (section), 96–110. *See also* Literature review.
  abstract representation of, 352–53

Q

R